Sustainable Use of Genetic Diversity in Forage and Turf Breeding

Christian Huyghe
Editor

Sustainable Use of Genetic Diversity in Forage and Turf Breeding

Springer

Editor
Christian Huyghe
INRA - Poitou Charentes
86600 Lusignan
France
christian.huyghe@lusignan.inra.fr

ISBN 978-90-481-8705-8 e-ISBN 978-90-481-8706-5
DOI 10.1007/978-90-481-8706-5
Springer Dordrecht Heidelberg London New York

Library of Congress Control Number: 2010925360

Printed on acid-free paper

Springer is part of Springer Science+Business Media (www.springer.com)

Preface

Grassland produces feed for livestock, improves soil fertility and structure, protects water resources and may contribute to climate change mitigation through carbon storage and to biodiversity preservation. It simultaneously maintains sustainable economic outputs for farmers and provides ecosystem services. Turf similarly considerably contributes to our environment by adding beauty to our surroundings, providing a safe playing surface for sports and recreation.

The species diversity present in most grasslands and turfs is a functional diversity contributing to the previously mentioned agronomic and environmental benefits. The species belong to different functional groups and the adequate species composition may maximise the agronomic performance through a higher production and a better quality and the environmental benefits through symbiotic nitrogen fixation or sources of pollen and nectar to pollinators. In a given grassland or turf, the genetic diversity available in each variety contributes to this economic and environmental performance, but also to the stability of these performances including the stability of the resistance against pathogens and pests.

Natural grasslands share many species with the sown swards. They may be regarded as favourable sites for in situ preservation of genetic diversity as well as valuable sources of diversity for breeding.

Breeding programs in forages have resulted in large genetic improvements in forage yield, quality and disease resistance. Similarly, in turf, large improvements were achieved in aesthetic value and resistance to diseases. Registration criteria and systems play a key role to validate these improvements and to release them to the market for the benefit of the end-users.

Resources available for breeders become increasingly large, with more access to better characterised materials, rapid and accurate methods for phenotyping and genotyping, expanding molecular resources, bioinformatics and computational resources. This huge amount of resources requires to clearly defining breeding objectives and optimum variety structure and to integrating phenotyping and genotyping. These are prerequisite conditions for a sustainable use of genetic diversity in forage and turf breeding.

Previous conferences of the Eucarpia Fodder and Amenity species sections were held in Perugia, Italy, in 2006 and Copenhagen, Denmark, in 2007. On this occasion, the 27th meeting was held in La Rochelle, France in 2009. It was organised by the French Association of Forage and Turf breeders (ACVF) in partnership with the National Institute of Agronomic Research (Inra), Poitou-Charentes Research Centre. Attendees included genebank curators, breeders, geneticists and molecular biologists of both public institutes and private breeding and seed companies from 40 countries. The program featured plenary addresses from leading international speakers, selected oral presentations, volunteered poster presentations, as well as tours of research centres, private breeding companies and cattle or dairy goat farms.

This book includes papers from the plenary lectures and offered papers presented either orally or as posters during the Conference. A wide variety of themes is included and offers a valuable overview of the present knowledge in forage and turf breeding. Five main sections will be found in the present book. After an introduction on the stakes for forage breeding, a section is dedicated to genetic resources with characterisation of the germplasm through various techniques and various innovative approaches such as biogeography. The second section investigates the changes which may occur in grassland swards. Population genetics and genetic ecology may be beneficial to question the paradigm of variety and sward stability.

The third section reviews the genetic improvements and the role of registration systems to stimulate and validate genetic progresses. The numerous achievements and findings in molecular biology are presented in the fourth section, while the fifth one will focus on new variety structures and prospects offered by interspecific hybridizations.

The 27th Conference of the Eucarpia Fodder and Turf Section and the publication of this book were supported by the Poitou-Charentes Regional Council, the Conseil Général de Charente-Maritime, the Xavier Bernard Foundation, GNIS, RAGT and Jouffray-Drillaud breeding companies, Agri-Obtentions, La Rochelle municipality.

The scientific program of the conference, the selection of the oral presentation among the numerous offered papers, and well as the critical reviewing of papers, sometimes leading to rejection, were made by the scientific committee which I had the pleasure to chair: Michael Abberton, Beat Boller, E. Charles Brummer, Jean-Louis Durand, Ulf Feuerstein, Marc Ghesquière, Marie-Christine Gras, Jean-Marie Prosperi, Isabel Roldan-Ruiz, Daniele Rosellini. They are sincerely thanked for their outstanding contribution to the success of the conference and to the quality of the present book.

The local organising committee made a great job to set a fruitful conference in a friendly atmosphere: Claude Tabel, head of R2n (RAGT breeding company), who chaired the committee, Nathalie Bonnet (Inra), Marc Lécrivain (Sicasov), Antoine de la Soujeole (Sicasov), Jean-Paul Sampoux (Inra), Fabien Thierry (Sicasov). They are sincerely thanked for their valuable work.

We thank Jacco Flipsen and Noeline Gibson of Springer for their assistance and cooperation in the publication of this book. Finally, we express our gratitude to the participants for their contribution to discussion and to authors whose dedication, work and responsiveness made this book possible.

Lusignan, France Christian Huyghe

Contents

Contributors

Michael T. Abberton Institute of Biological, Environmental and Rural Sciences, Aberystwyth University, Penglais, Aberystwyth Ceridigion, Wales SY23 3DA, UK

Chaouki Al Faiz Institut National de la Recherche Agronomique (INRA), Rabat B.P. 415, Morocco

Snežana Anđelković Institute for forage crops, Trg Kosturnice 50, Kruševac 37 000, Serbia

Ulrike C.M. Anhalt Teagasc Crop Research Centre, Oak Park, Carlow, Ireland; Department of Biology University of Leicester, Leicester LE1 7RH, UK

Paolo Annicchiarico CRA – Centro di Ricerca per le Produzioni Foraggere e Lattiero-Casearie, viale Piacenza 29, Lodi 26900, Italy

Akira Arakawa National Institute of Livestock and Grassland Science, 768 Senbonmatsu, Nasushiobara Tochigi 329-2793, Japan

Rita Armonienė Lithuanian Institute of Agriculture, Instituto a.1, Akademija, Kėdainiai r. LT-58344, Lithuania

Torben Asp Department of Genetics and Biotechnology University of Aarhus, Århus, Denmark

Geoff Auricht South Australian Research and Development Institute, Urrbrae, SA, Australia; CRC for Future Farm Industries, Crawley, WA, Australia

Snežana Babić Institute for forage crops, Trg kosturnice, Kruševac, Serbia

Vincent Badeau INRA, Centre de Nancy, UMR Ecologie et Ecophysiologie Forestières, Champenoux F-54280, France

Joost Baert ILVO Plant, Caritasstraat 21, Melle 9090, Belgium

James H. Baird Department of Botany and Plant Sciences University of California, Riverside, CA 92521, USA

Ross Ballard Cooperative Research Centre for Future Farm Industries, Crawley, WA 6009, Australia; South Australian Research and Development Institute, Adelaide, SA 5001, Australia

Thierry Bariac CNRS, Unité BIOEMCO Bâtiment EGER, Grignon 78 850, France

Philippe Barre UR4, Unité de Recherche Pluridisciplinaire sur les Prairies et les Plantes Fourragères, Lusignan BP6, 86600, France

Ed Barrett-Lennard Cooperative Research Centre for Future Farm Industries, Crawley, WA 6009, Australia; Department of Agriculture and Food Western Australia, South Perth, WA 6151, Australia; School of Plant Biology The University of Western Australia, Crawley, WA 6009, Australia

Susanne Barth Teagasc Crop Research Centre, Oak Park, Carlow, Ireland

Jan Bartoš Laboratory of Molecular Cytogenetics and Cytometry Institute of Experimental Botany, Olomouc CZ-77200, Czech Republic

Pierre Baudouin ACVF (Association des Créateurs de Variétés Fourragères), 7, rue du Coq Héron, Paris Cedex 01 F-75030, France

Bernard Bayle ACVF (Association des Créateurs de Variétés Fourragères), 7, rue du Coq Héron, Paris Cedex 01 F-75030, France

Paul R. Berg Department of Animal and Aquacultural Sciences Centre for Integrative Genetics (CIGENE), Norwegian University of Life Sciences, Ås N-1432, Norway

Karine Bernard INRA, UR 4, BP 6, Lusignan 86600, France

Annick Bertrand Soils and Crops Research and Development Centre, Agriculture and Agri-Food Canada, Québec, QC G1V 2J3, Canada

Hem S. Bhandari The Samuel Roberts Noble Foundation, Inc., 2510 Sam Noble Parkway, Ardmore, OK 73401, USA

Andy Binley Lancaster University, Lancaster LA1 4YW UK

Philippe Biron CNRS, Unité BIOEMCO Bâtiment EGER, Grignon 78 850, France

Helene Blois Diversity Arrays Technology, Yarralumla, ACT 2600, Australia

Beat Boller Research Station Agroscope Reckenholz-Tänikon ART, Reckenholzstrasse 191, Zürich CH-8046, Switzerland

Amanda Bonython Cooperative Research Centre for Future Farm Industries, Crawley, WA 6009, Australia; South Australian Research and Development Institute, Naracoorte, SA 5271, Australia

Abdelaziz Bouizgaren Institut National de la Recherche Agronomique, INRA, Marrakech BP: 533. Av Mohamed VI., Marrakesh, Morracco

Pierre Bourdon ACVF (Association des Créateurs de Variétés Fourragères), 7, rue du Coq Héron, Paris Cedex 01 F-75030, France

Thierry Bourgoin Agri-Obtentions, Technological Platform of Variety Innovation (PTIV), Lusignan BP 6, 86600, France

Gilles Boutet Genotyping platform of the department of genetics and plant breeding National Institute of Agronomical Research (INRA), 234 Avenue du Brézet, Clermont-Ferrand 63000, France

Joseph H. Bouton Forage Improvement Division The Samuel Roberts Noble Foundation, Inc., 2510 Sam Noble Parkway, Ardmore, OK 73401, USA

Gintaras Brazauskas Lithuanian Institute of Agriculture, Instituto a.1, Akademija, Kėdainiai r. LT-58344, Lithuania

Koos de Bruijn ACVF (Association des Créateurs de Variétés Fourragères), 7, rue du Coq Héron, Paris Cedex 01 F-75030, France

E. Charles Brummer Institute for Plant Breeding, Genetics, and Genomics, University of Georgia, 111 Riverbend Rd., Athens, GA 30602, USA

Vincent Béguier Jouffray-Drillaud, Station expérimentale La Litière, 86 600 Saint Sauvant, France; ACVF (Association des Créateurs de Variétés Fourragères), 7, rue du Coq Héron, Paris Cedex 01 F-75030, France

Bohumír Cagaš OSEVA PRO Ltd., Grassland Research Station Rožnov-Zubří, Hamerská 698, 756 64 Zubří, Czech Republic

Vanessa Caig Diversity Arrays Technology, Yarralumla, ACT 2600, Australia

Jean-Christophe Caillet GIE GRASS, Station expérimentale La Litière, 86 600 Saint Sauvant, France

Andrew Cairns Crop Genetics, Genomics and Breeding, Institute of Biological, Environmental and Rural Sciences (IBERS), Aberystwyth University, Gogerddan, Aberystwyth SY23 3EB, UK

Isabelle Cameleyre Breeder Association of Forage Varieties (ACVF), 7, rue du Coq-Héron, Paris Cedex 01 75030, France

Stefano Capomaccio Department of Applied Biology University of Perugia, Borgo XX giugno 74, Perugia 06121, Italy

Lucien Carlier Plant, Crop Husbandry and Environment Institute for Agricultural and Fisheries Research (ILVO), Burg. Van Gansberghelaan, Merelbeke 109, 9820, Belgium

Serge Carré INRA – UR4 Unité de Recherche Pluridisciplinaire Prairies et Plantes Fourragères (UR P3F), 86 600 Lusignan, France

Antonio M. Carroni Unità di ricerca per i sistemi agropastorali in ambiente mediterraneo, Consiglio per la Ricerca e sperimentazione in Agricoltura (CRA), Podere Ortigara Loc. Sanluri Stato, Sanluri (VS) 09025, Italy

Yves Castonguay Soils and Crops Research and Development Centre, Agriculture and Agri-Food Canada, Québec, QC G1V 2J3, Canada

Vladimír Černoch Šlechtitelská stanice Hladké Životice, s.r.o., Hladké Životice CZ-74247, Czech Republic

Nigel Charman Cooperative Research Centre for Future Farm Industries, Crawley, WA 6009, Australia; South Australian Research and Development Institute, Adelaide, SA 5001, Australia

Fabienne Chataigner INRA, UR 4, Unité de Recherche Pluridisciplinaire Prairies et Plantes Fourragères, Lusignan BP6, 86600, France

Barbara Chaves Plant, Crop Husbandry and Environment Institute for Agricultural and Fisheries Research (ILVO), Burg. Van Gansberghelaan, Merelbeke 109, 9820, Belgium

Jean François Chosson R2n (filiale R.A.G.T.), Centre de Recherche de Montours, F-35460 Montours (Bretagne), France; ACVF (Association des Créateurs de Variétés Fourragères), 7, rue du Coq Héron, Paris Cedex 01 F-75030, France

Jean Cloutier Soils and Crops Research and Development Centre, Agriculture and Agri-Food Canada, Québec, QC G1V 2J3, Canada

Gerda Cnops Plant Sciences Unit – Growth and Development, Institute for Agricultural and Fisheries Research (ILVO), Caritasstraat 21, Melle 9090, Belgium

Philip Cocks School of Plant Biology, The University of Western Australia, Crawley, WA 6009, Australia; Centre for Legumes in Mediterranean Agriculture, The University of Western Australia, Crawley, WA 6009, Australia

Timothy Colmer Cooperative Research Centre for Future Farm Industries, Crawley, WA 6009, Australia; School of Plant Biology The University of Western Australia, Crawley, WA 6009, Australia

Andrew Craig Cooperative Research Centre for Future Farm Industries, Crawley, WA 6009, Australia; South Australian Research and Development Institute, Naracoorte, SA 5271, Australia

Branko Ćupina Department of Field and Vegetable Crops University of Novi Sad, Trg Dositeja Obradovića 8, Novi Sad 21000, Serbia

Elzbieta Czembor IHAR Radzikow, Blonie PL-05-870, Poland

John De Boever Institute for Agricultural and Fisheries Research (ILVO), Plant Sciences Unit, 21 Caritasstraat, 9090 Melle, Belgium

Jan De Riek Institute for Agricultural and Fisheries Research, Applied Genetics and Breeding, Plant Sciences Unit, Caritasstraat, Melle 9090, Belgium

Alex De Vliegher Plant, Crop Husbandry and Environment, Institute for Agricultural and Fisheries Research (ILVO), Burg. Van Gansberghelaan, Merelbeke 109, 9820, Belgium

Klaus J. Dehmer Leibniz Institute of Plant Genetics and Crop Plant Research (IPK), Genebank, Satellite Collections North, Inselstrasse 9, Malchow/Poel 23999, Germany

Dušica Delić Institute of Soil Science, Teodora Drajzera 7, Belgrade 11000, Serbia

François Deneufbourg FNAMS (Fédération Nationale des Agriculteurs Multiplicateurs de Semences), impasse du Verger, Brain-sur-l'Authion F-49800, France

Kerstin Diekmann Teagasc Crops Research Centre, Oak Park, Carlow, Ireland; University of Dublin, Botany Building, School of Natural Sciences, Trinity College Dublin, Dublin, Ireland

Phillip J. Dix Department of Biology National University of Ireland, Maynooth, Ireland

Dragoslav Djokic Institut for forage crops, Trg Kosturnica, Krusevac 37000, Serbia

Dejan Dodig Maize Research Institute, Slobodana Bajica 1, Zemun Polje 11185, Serbia

Jaroslav Doležel Laboratory of Molecular Cytogenetics and Cytometry Institute of Experimental Botany, Olomouc CZ-77200, Czech Republic

Guro Dørum Department of Chemistry, Biotechnology and Food Science Agricultural University of Norway, Ås N-1432, Norway

Jeffrey J. Doyle Institute for Plant Breeding, Genetics, and Genomics, University of Georgia, 111 Riverbend Rd., Athens, GA 30602 USA

Sheena Duller Institute of Biological, Environmental and Rural Science Aberystwyth University, Aberystwyth SY23 3 EB, UK

Jean-Louis Durand INRA – URP3F Unité de Recherche Pluridisciplinaire Prairies et Plantes Fourragères, BP6, Lusignan 86600, France

Fred Eickmeyer Saatzucht Steinach, Steinach D-94377, Germany

Nadia Elboutahiri Institut National de la Recherche Agronomique (INRA), Rabat, Morocco; Mohammed V University, Rabat, Morocco

Markku Farrell Crop Genetics, Genomics and Breeding, Institute of Biological, Environmental and Rural Sciences (IBERS), Aberystwyth University, Gogerddan, Aberystwyth SY23 3 EB, UK

Melanie Febrer Teagasc, Crops Research Centre, Oak Park, Carlow, Ireland; The Genome Analysis Centre, Norwich Research Park, Norwich NR4 7 UH, UK

Ulf Feuerstein Euro Grass Breeding GmbH, Asendorf D-27330, Germany

Veerle Fievez Laboratory for Animal Nutrition and Animal Product Quality (LANUPRO), Ghent University, Melle 9090, Belgium; Ghent University, LANUPRO, Proefhoevestraat 10, Melle 9090, Belgium

Sandrine Flajoulot Jouffray-Drillaud, Station expérimentale La Litière, 86 600, Saint Sauvant, France; Breeder Association of Forage Varieties (ACVF), 7, rue du Coq-Héron, Paris Cedex 01 75030, France

Michael Fothergill Institute of Biological, Environmental and Rural Sciences (IBERS), Aberystwyth University, Aberystwyth SY233EB, UK

Clive Francis Centre for Legumes in Mediterranean Agriculture, The University of Western Australia, Crawley, WA 6009, Australia

Mi Fugui College of Ecology and Environmental Science, Inner Mongolia Agricultural University, Hohhot 010018, China

Christophe Galbrun ACVF (Association des Créateurs de Variétés Fourragères), 7, rue du Coq Héron, Paris Cedex 01 F-75030, France

Jérôme Garon GIE GRASS, Station expérimentale La Litière, Saint Sauvant 86 600, France

François Gastal INRA – UR4 Unité de Recherche Pluridisciplinaire Prairies et Plantes Fourragères (UR P3F), F86600 Lusignan, France

Vincent Gensollen Groupe d′Etude et de contrôle des Variétés et des Semences (GEVES), La Valette 711, Rue Jean-François Breton, Montpellier 34090, France

Marc Ghesquière UR4 (Unité de Recherche Pluridisciplinaire Prairies et Plantes Fourragères), INRA, Centre Poitou-Charentes, Lusignan, F-86600, France

An Ghesquiere Institute for Agricultural and Fisheries Research, Applied Genetics and Breeding, Plant Science Unit, Caritasstraat 21, Melle B 9090, Belgium

Chrystel Gibelin INRA, UR 4, BP 6, Lusignan 86600, France

Trevor J. Gilliland Agri-Food Biosciences Institute (AFBI), Plant Testing Station, Belfast BT6 9SH, UK

Jevtić Goran Institute for forage crops, Trg Kosturnice 50, Kruševac 37000, Serbia

Pavlína Gottwaldová Agricultural research, Ltd., Zahradní 1, Troubsko 664 41, Czech Republic

Stéphane Grenier INRA – Unité de Recherche Pluridisciplinaire Prairies et Plantes Fourragères, Route de Saintes, Lusignan BP 6, 86600, France

Badara Gueye UR4, Unité de Recherche Pluridisciplinaire sur les Prairies et les Plantes Fourragères, Lusignan, France

Wang Guihua College of Ecology and Environmental Science, Inner Mongolia Agricultural University, Hohhot 010018, China

Eric Hall TIAR: Tasmanian Institute of Agricultural Research, Launceston, TAS, Australia

Yuanhong Han Forage Improvement Division, The Samuel Roberts Noble Foundation, Sam Noble Parkway, Ardmore, OK 73401, USA

Stefan Hartmann Institut für Pflanzenbau u. Pflanzenzüchtung der Bayerischen Landesanstalt für Landwirtschaft, Freising D-85354, Germany

Richard Hayes Institute of Biological, Environmental and Rural Science Aberystwyth University, Aberystwyth SY23 3 EB, UK

Phil Haygarth Lancaster University, Lancaster, LA1 4YW UK

Doris Herrmann INRA, UR 4, BP 6, Lusignan 86600, France

J.S. (Pat) Heslop-Harrison Department of Biology University of Leicester, Leicester LE1 7RH, UK

Kurt Heungens ILVO (Institute for Agricultural and Fisheries Research), Unit Plant, Burg., Van Gansbergenlaan 96, Merelbeke 9820, Belgium

Trevor R. Hodkinson University of Dublin, Botany Building, School of Natural Sciences, Trinity College Dublin, Dublin, Ireland

Stephen Hughes Cooperative Research Centre for Future Farm Industries, Crawley, WA 6009, Australia; South Australian Research and Development Institute, Adelaide, SA 5001, Australia

Thierry Huguet ENSA, Toulouse, France

Mike Humphreys Crop Genetics, Genomics and Breeding, Institute of Biological, Environmental and Rural Sciences (IBERS), Aberystwyth University, Gogerddan, Aberystwyth SY23 3 EB, UK

Mervyn Humphreys Crop Genetics, Genomics and Breeding, Institute of Biological, Environmental and Rural Sciences (IBERS), Aberystwyth University, Gogerddan, Aberystwyth SY23 3 EB, UK

Alan Humphries South Australian Research and Development Institute, Urrbrae, SA, Australia; CRC for Future Farm Industries, Crawley, WA, Australia

Susanne Hünmörder Leibniz Institute of Plant Genetics and Crop Plant Research (IPK), Genebank, Satellite Collections North, Inselstrasse 9, Malchow/Poel 23999, Germany

Christian Huyghe Inra, de recherche Poitou-Charentes, Route de Saintes, Lusignan 86600, France

Hana Jakesova Plant breeding, Hladke Zivotice 100, Hladke Zivotice CZ-74247, Czech Republic

Milenkovic Jasmina Institute for Forage Crops Krusevac, Trg Kosturnice 50, Krusevac 37000, Serbia

Glyn Jenkins Institute of Biological, Environmental and Rural Sciences, Aberystwyth University, Penglais, Aberystwyth, Ceridigion, Wales SY23 3DA, UK

Marković Jordan Institute for forage crops, Trg Kosturnice 50, Kruševac 37000, Serbia

Bernadette Julier INRA, UR 4, BP 6, 86600 Lusignan, France

Rajae Kallida Institut National de la Recherche Agronomique (INRA), Rabat, Morocco

Đura Karagić Institute of Field and Vegetable Crops, M. Gorkog 30, Novi Sad 21000, Serbia

Slobodan Katić Institute of Field and Vegetable Crops, M. Gorkog 30, Novi Sad 21000, Serbia

Vilma Kemešytė The Lithuanian Institute of Agriculture, Instituto al. 1, Akademija, Kaunas, Lithuania

Dong-Man Khu The Samuel Roberts Noble Foundation, 2510 Sam Noble Parkway, Ardmore, OK 73401, USA

Andrzej Kilian Diversity Arrays Technology, Yarralumla, ACT 2600, Australia

Julie King Crop Genetics, Genomics and Breeding, Institute of Biological, Environmental and Rural Sciences (IBERS), Aberystwyth University, Gogerddan, Aberystwyth SY23 3 EB, UK

Takako Kiyoshi National Institute of Livestock and Grassland Science, 768 Senbonmatsu, Nasushiobara Tochigi 329-2793, Japan

Roland Kölliker Agroscope Reckenholz-Tänikon Research Station ART, Reckenholzstrasse 191, Zürich CH-8046, Switzerland

David Kopecký Laboratory of Molecular Cytogenetics and Cytometry, Institute of Experimental Botany, Olomouc CZ-77200, Czech Republic

Bernhard Krautzer Bundesanstalt für alpenländische Landwirtschaft Gumpenstein, Irdning A-8952 Austria

Đorđe Krstić Department of Field and Vegetable Crops University of Novi Sad, Trg Dositeja Obradovića 8, Novi Sad 21000, Serbia

Đorđe Kuzmanović Institute of Soil Science, Teodora Drajzera 7, Belgrade 11000, Serbia

Serge Laberge Soils and Crops Research and Development Centre, Agriculture and Agri-Food Canada, Québec, QC G1V 2J3, Canada

Meriem Laouar INRAA, El-Harrach Alger, Algeria

Arild Larsen Graminor AS, c/o Bioforsk Nord Bodø, Torggården, Bodø N-8049, Norway

Hennietha Leenheer DLF Trifolium Netherlands B.V., Hellegatsestraat 19, Moerstraten NL-4727 ST, The Netherlands

François Lelièvre INRA, UMR System, Montpellier cedex 1, Montpellier 34060, France

Hans Lellbach BAZ Institut für Landwirtschaft und Kulturen, Gross Luesewitz D-18190 Germany

Nijolė Lemežienė The Lithuanian Institute of Agriculture, Instituto al. 1, Akademija, Kaunas, Lithuania

Isabelle Litrico Institut National de la Recherche Agronomique – Unité de Recherche Pluridisciplinaire Prairies et Plantes Fourragères, Route de Saintes, Lusignan 86600, France

Angelo Loi Department of Agriculture and Food Western Australia, Locked Bag 4, Bentley Delivery Centre, Bentley, WA 6983, Australia; Centre for Legumes in Mediterranean Agriculture, The University of Western Australia, Crawley, WA 6009, Australia

Alessandro Longu Unità di ricerca per i sistemi agropastorali in ambiente mediterraneo, Consiglio per la Ricerca e sperimentazione in Agricoltura (CRA), Podere Ortigara Loc. Sanluri Stato, Sanluri (VS) 09025, Italy

J. Alan Lovatt Institute of Biological, Environmental and Rural Science, Aberystwyth University, Gogerddan, Aberystwyth SY23 3EB, Wales, UK

Matthew Lowe Institute of Biological, Environmental and Rural Sciences (IBERS), Aberystwyth University, Aberystwyth SY233EB, UK

Adam J. Lukaszewski Department of Botany and Plant Sciences University of California, Riverside, CA 92521, USA

Radek Macháč OSEVA PRO Ltd. Grassland Research Station Rožnov-Zubří, Hamerská, Zubří, Czech Republic

Kit Macleod Cross Institute Programme for Sustainable Soil Function (SoilCIP), Rothamsted Research (RRES), Okehampton, Devon EX20 2SB, UK

Marianne Malengier Institute for Agricultural and Fisheries Research, Applied Genetics and Breeding, Plant Sciences Unit, Caritasstraat 21, Melle 9090, Belgium

Patrizia Manunza Unità di ricerca per i sistemi agropastorali in ambiente mediterraneo, Consiglio per la Ricerca e sperimentazione in Agricoltura (CRA), Podere Ortigara Loc. Sanluri Stato, Sanluri (VS) 09025, Italy

Athole Marshall Institute of Biological, Environmental and Rural Sciences (IBERS), Aberystwyth University, SY233EB, Aberystwyth, Ceredigion, UK

Bram Marynissen Plant, Crop Husbandry and Environment, Institute for Agricultural and Fisheries Research (ILVO), Burg. Van Gansberghelaan, Merelbeke 109, 9820, Belgium

Darryl McClements Cooperative Research Centre for Future Farm Industries, Crawley, WA 6009, Australia; Department of Agriculture and Food Western Australia, South Perth, WA 6151, Australia

Rémi Métral UR4 (Unité de Recherche Pluridisciplinaire Prairies et Plantes Fourragères), INRA, Centre Poitou-Charentes, Lusignan F-86600, France

Réal Michaud Soils and Crops Research and Development Centre, Agriculture and Agri-Food Canada, Québec, QC G1V 2J3, Canada

Vojislav Mihailović Forage Crops Department, Institute of Field and Vegetable Crops, Maksima Gorkog 30, Novi Sad, Serbia

Aleksandar Mikić Forage Crops Department, Institute of Field and Vegetable Crops, Maksima Gorkog 30, Novi Sad, Serbia

Dan Milbourne Teagasc, Crops Research Centre, Oak Park, Carlow, Ireland

Dragan Milić Institute of Field and Vegetable Crops, M. Gorkog 30, Novi Sad 21000, Serbia

Bogić Miličić Institute of Soil Science, Teodora Drajzera 7, Belgrade 11000, Serbia

Nadjette Missi-Bakour Centre Poitou-Charentes, Institut National de Recherche Agronomique, Route de Saintes, Lusignan 86600, France

Maria J. Monteros Forage Improvement Division, The Samuel Roberts Noble Foundation, 2510 Sam Noble Parkway, Ardmore, OK 73401, USA

Christy M. Motes Forage Improvement Division, The Samuel Roberts Noble Foundation, 2510 Sam Noble Parkway, Ardmore, OK 73401, USA

Hilde Muylle Plant Sciences Unit – Growth and Development, Institute for Agricultural and Fisheries Research (ILVO), Caritasstraat 21, Melle 9090, Belgium

Jan Nedělník Agricultural research, Ltd., Zahradní 1, Troubsko 664 41, Czech Republic

Phillip Nichols Cooperative Research Centre for Future Farm Industries, Crawley, WA 6009, Australia

Bradley Nutt Department of Agriculture and Food Western Australia, Locked Bag 4, Bentley Delivery Centre, Bentley, WA 6983, Australia

Sally O'Donovan Crop Genetics, Genomics and Breeding, Institute of Biological, Environmental and Rural Sciences (IBERS), Aberystwyth University, Gogerddan, Aberystwyth SY23 3 EB, UK

Liv Østrem Norwegian Institute for Agricultural and Environmental Research, N-6997, Hellevik i Fjaler, Norway

Tolis Papadopoulos Lancaster University, Lancaster LA1 4YW, UK

Vanda Paplauskienė Lithuanian Institute of Agriculture, Instituto a.1, Akademija, Kėdainiai r. LT-58344, Lithuania

Laurence Pauly Jouffray-Drillaud, Station expérimentale La Litière, Saint Sauvant 86 600, France

Luciano Pecetti CRA – Centro di Ricerca per le Produzioni Foraggere e Lattiero-Casearie, viale Piacenza 29, Lodi 26900, Italy

Jan Pelikán Agricultural research, Ltd., Zahradní 1, Troubsko 664 41, Czech Republic

Jean-Baptiste Pierre Breeder Association of Forage Varieties (ACVF), 7, rue du Coq-Héron, Paris Cedex 01 75030, France; INRA, UR 4, Unité de Recherche Pluridisciplinaire Prairies et Plantes Fourragères, Lusignan BP6, 86600, France

Wieslaw Pietraszek ACVF (Association des Créateurs de Variétés Fourragères), 7, rue du Coq Héron, Paris Cedex 01 F-75030, France

Maurizio Pitzalis Unità di ricerca per i sistemi agropastorali in ambiente mediterraneo, Consiglio per la Ricerca e sperimentazione in Agricoltura (CRA), Podere Ortigara Loc. Sanluri Stato, Sanluri (VS) 09025, Italy

Laurence Poinsard S.A. Carneau, Orchies F-59310, France

Charles Poncet Genotyping platform of the department of genetics and plant breeding National Institute of Agronomical Research (INRA), 234 Avenue du Brézet, Clermont-Ferrand 63000, France

Ulrich K. Posselt State Plant Breeding Institute (720), University of Hohenheim, Stuttgart 70593, Germany

Sandro Proietti CRA – Centro di Ricerca per le Produzioni Foraggere e Lattiero-Casearie, viale Piacenza 29, Lodi 26900, Italy

Jasmina Radovic Institute for forage crops, Trg Kosturnice 50, Kruševac 37 000, Serbia

Nataša Rasulić Institute of Soil Science, Teodora Drajzera 7, Belgrade 11000, Serbia

Elaine Rees Institute of Biological, Environmental and Rural Sciences (IBERS) Aberystwyth University, Aberystwyth SY233EB, UK

Wilfried Rémus-Borel Soils and Crops Research and Development Centre, Agriculture and Agri-Food Canada, Québec, QC G1V 2J3, Canada

Clinton Revell Department of Agriculture and Food Western Australia, Locked Bag 4, Bentley Delivery Centre, Bentley, WA 6983, Australia

Rafael Reyno Center for Applied Genetic Technologies, University of Georgia, 111 Riverbend Rd, Athens, GA 30602, Greece

Patricia Richard CNRS, Unité BIOEMCO Bâtiment EGER, Grignon 78 850, France

Mary-Jane Rogers Cooperative Research Centre for Future Farm Industries, Crawley, WA 6009, Australia; Department of Primary Industries Victoria, Tatura, VIC 3616, Australia

Odd Arne Rognli Department of Plant and Environmental Sciences, Norwegian University of Life Sciences, N-1432 Ås, Norway

Antje Rohde Plant Sciences Unit – Growth and Development, Institute for Agricultural and Fisheries Research (ILVO), Caritasstraat 21, Melle 9090, Belgium

Isabel Roldán-Ruiz Plant Sciences Unit – Growth and Development, Institute for Agricultural and Fisheries Research (ILVO), Caritasstraat 21, Melle B-9090, Belgium

Massimo Romani CRA-FLC, Lodi, LO I-26900, Italy

Michel Romestant Breeder Association of Forage Varieties (ACVF), 7, rue du Coq-Héron, Paris Cedex 01 75030, France

Daniele Rosellini Department of Applied Biology, University of Perugia, Borgo XX giugno 74, Perugia 06121, Italy

Youri Rothfuss CNRS, Unité BIOEMCO Bâtiment EGER, Grignon 78 850, France

Niels Roulund DLF-Trifolium, Højerupvej 31, Store Heddinge DK-4660, Denmark

Heidi Rudi Department of Plant and Environmental Sciences Norwegian University of Life Sciences, Ås N-1432, Norway

Luigi Russi Universita' degli Studi, Perugia I-06121, Italy

Malay C. Saha The Samuel Roberts Noble Foundation, Inc., 2510 Sam Noble Parkway, Ardmore, OK 73401, USA

Muhammet Sakiroglu Institute for Plant Breeding, Genetics, and Genomics, University of Georgia, 111 Riverbend Rd., Athens, GA 30602, USA

Mauro Salis Unità di ricerca per i sistemi agropastorali in ambiente mediterraneo, Consiglio per la Ricerca e sperimentazione in Agricoltura (CRA), Podere Ortigara Loc. Sanluri Stato, Sanluri (VS) 09025, Italy

Jean-Paul Sampoux INRA, Centre Poitou-Charentes, UR4 (Unité de Recherche Pluridisciplinaire Prairies et Plantes Fourragères), Lusignan F-86600, France

Ruth Sanderson Institute of Biological, Environmental and Rural Science, Aberystwyth University, Aberystwyth SY23 3 EB, UK

Simen R. Sandve Department of Plant and Environmental Sciences Norwegian University of Life Sciences, Ås N-1432, Norway

Gajić Sanja University of Belgrade, Belgrade-Zemun 11080, Serbia

Oana Saracutu Plant Sciences Unit – Growth and Development, Institute for Agricultural and Fisheries Research (ILVO), Caritasstraat 21, Melle 9090, Belgium

Franz Xaver Schubiger Agroscope Reckenholz-Tänikon, Research Station ART, Zurich CH-8046, Switzerland

Sabine Schulze Saatzucht Steinach Bornhof, Bocksee D-17219, Germany

Carla Scotti CRA-FLC, Viale Piacenza 29, Lodi 26900, Italy

Yasmina Semiani INRAA, El-Harrach Alger, Algeria

Magdalena Ševčíková OSEVA PRO Ltd., Grassland Research Station Rožnov-Zubří, Hamerská, Zubří, Czech Republic

Naima Shaimi Institut National de la Recherche Agronomique (INRA), Rabat B.P. 415, Morocco; Université Ibn Tofail, Kenitra, Morocco

Dong Shujun College of Bioengineering, Inner Mongolia Agricultural University, Hohhot 010018, China

Aleksandar Simić University of Belgrade, Belgrade, Serbia

Ellen Sizer-Coverdale Institute of Biological, Environmental and Rural Sciences (IBERS) Aberystwyth University, Aberystwyth SY233EB, UK

Leif Skøt Aberystwyth University, Institute of Biological, Environmental and Rural Sciences, Plas Gogerddan, Aberystwyth, Ceredigion SY23 3 EB, UK

Mary K. Sledge Lipscomb University, One University Park Dr, Nashville, TN 37204, USA

Richard Snowball Department of Agriculture and Food Western Australia, Locked Bag 4, Bentley Delivery Centre, Bentley, WA 6983, Australia

Dejan Sokolović Institute for forage crops, Trg Kosturnice, Kruševac, Serbia

Olivera Stajković Institute of Soil Science, Teodora Drajzera 7, Belgrade 11000, Serbia

Rade Stanisavljevic Institut for forage crops, Trg Kosturnica, Krusevac 37000, Serbia

Aleksandra Stanojković Institute of Soil Science, Teodora Drajzera 7, Belgrade 11000, Serbia

Ratibor Štrbanović Institute for forage crops, Trg Kosturnice 50, Kruševac 37 000, Serbia

Vaclovas Stukonis Lithuanian Institute of Agriculture, Instituto a.1, LT-58344, Akademija, Kėdainiai r., Lithuania

Fabien Surault INRA, UR 4, URP3F, Lusignan 86600, France

Claude Tabel R2n, Rue Emile Singla BP 3336 F-12033, Rodez Cedex 9, France

Tadashi Takamizo National Institute of Livestock and Grassland Science, 768 Senbonmatsu, Nasushiobara Tochigi 329-2793, Japan

Vasić Tanja Institute for forage crops, Trg Kosturnice 50, Kruševac 37000, Serbia

Marie Claire Tardin R2n, Druelle F-12510, France

Dragan Terzic Institut for forage crops, Trg Kosturnica, Krusevac 37000, Serbia

Imane Thami-Alami Institut National de la Recherche Agronomique (INRA), B.P. 415, Rabat, Morocco

Bernard Tharel ACVF (Association des Créateurs de Variétés Fourragères), 7, rue du Coq Héron, Paris Cedex 01 F-75030, France

Céline Tomaszewski Teagasc Crop Research Centre, Oak Park, Carlow, Ireland

Ivone Torres-Jerez The Samuel Roberts Noble Foundation, 2510 Sam Noble Parkway, Ardmore, OK 73401, USA

Lesley Turner Crop Genetics, Genomics and Breeding, Institute of Biological, Environmental and Rural Sciences (IBERS), Aberystwyth University, Gogerddan, Aberystwyth SY23 3EB, UK

Kazuhiro Uchiyama National Institute of Livestock and Grassland Science, 768 Senbonmatsu, Nasushiobara Tochigi 329-2793, Japan

Sripada M. Udupa ICARDA-INRA Cooperative Research Project, International Center for Agricultural Research in the Dry Areas (ICARDA), Rabat B.P. 6299, Morocco

Michael Udvardi The Samuel Roberts Noble Foundation, 2510 Sam Noble Parkway, Ardmore, OK 73401, USA

Erik Van Bockstaele Unit Plant, Genetics and Breeding Institute for Agriculture and Fisheries Research, Caritas straat, Melle 9090, Belgium

Rob van den Bekerom Department of Biology National University of Ireland, Maynooth, Ireland

Stephan A.G. van der Heijden Barenbrug Holding, Nijmegen NL-6515 AB, The Netherlands

Steven Van Hulle Plant Sciences Unit – Growth and Development, Institute for Agricultural and Fisheries Research (ILVO), Caritasstraat 21, Melle B-9090, Belgium

Gijs Van Ranst Institute for Agriculture and Fisheries Research (ILVO), Plant Science Unit, 21 Caritasstraat, 9090 Melle, Belgium

Johan Van Waes Plant, Crop Husbandry and Environment, Institute for Agricultural and Fisheries Research (ILVO), Burg. Van Gansberghelaan, Merelbeke 109, 9820, Belgium

Chris Van Waes Institute for Agricultural and Fisheries Research (ILVO), Plant Sciences Unit, 21 Caritasstraat, 9090 Melle, Belgium

Muriel Vandewalle Institute for Agricultural and Fisheries Research (ILVO), Plant Sciences Unit, 21 Caritasstraat, 9090 Melle Belgium

Kirsten Vangsgaard Breeder Association of Forage Varieties (ACVF), 7, rue du Coq-Héron, Paris Cedex 01 75030, France

François VanHee Innoseeds, Dijkwelsestraat 70, Kapelle, AJ NL-4421, The Netherlands

Sanja Vasiljević Forage Crops Department, Institute of Field and Vegetable Crops, Maksima Gorkog 30, Novi Sad, Serbia

Fabio Veronesi Department of Applied Biology University of Perugia, Borgo XX giugno 74, Perugia 06121, Italy

Merja Veteläinen MTT AgriFood Research Finland, Biotechnology and Food Research/Genetic Diversity, Jokioinen FI-31600, Finland

Magnus D. Vigeland Uelands gate, Oslo 59 J, 0460, Norway

Anne Viguié ACVF (Association des Créateurs de Variétés Fourragères), 7, rue du Coq Héron, Paris Cedex 01 F-75030, France

Tim Vleugels ILVO (Institute for Agricultural and Fisheries Research), Unit Plant, Caritasstraat 21, Melle 9090, Belgium

Tomáš Vymyslický Agricultural research, Ltd., Zahradní 1, Troubsko 664 41, Czech Republic

Chris Watts RRES, Harpenden, Hertfordshire AL52JQ, UK

Richard Whalley RRES, Harpenden, Hertfordshire AL52JQ, UK

Fabienne Wichmann Agroscope Reckenholz-Tänikon Research Station ART, Zürich 8046, Switzerland

Franco Widmer Agroscope Reckenholz-Tänikon Research Station ART, Zürich 8046, Switzerland

Evelin Willner Leibniz Institute of Plant Genetics and Crop Plant Research (IPK), Genebank, Satellite Collections North, Inselstrasse 9, Malchow/Poel 23999, Germany

Kenneth H. Wolfe University of Dublin, Botany Building, School of Natural Sciences, Trinity College Dublin, Dublin, Ireland

Lukas Wolters Euro Grass Breeding GmbH, Gennep, AA NL-6590, The Netherlands

Toshihiko Yamada Field Center for Northern Biosphere, Hokkaido University, Kita 11, Nisha 10, Kita-ku, Sapporo 060-0811, Japan

El-Houssine Zaid Mohammed V University, Rabat, Morocco

Yan Zhang Forage Improvement Division The Samuel Roberts Noble Foundation, 2510 Sam Noble Parkway, Ardmore, OK 73401, USA; Rutgers University, New Brunswick, NJ 08902, USA

Bogovid Živković Institute for forage crops, Trg kosturnice, Kruševac, Serbia

Lugic Zoran Institute for Forage Crops Krusevac, Trg Kosturnice 50, Krusevac 37000, Serbia

Zbigniew Zwierzykowski Institute of Plant Genetics of the Polish Academy of Sciences (IPG/PAS), Strzeszynska 34, Poznan 60-479, Poland

Part I
Introduction

Chapter 1
Grasslands and Forage Crops in France: Context and Stakes. Consequences for Breeding

Christian Huyghe and Claude Tabel

Abstract In a first part, the paper will present the socio-economic context of grasslands and forage crops in France over the last decades. The grasslands and forage crops contribute 45% of the total French arable land. Variation in acreage of the various types of soil covers in the various regions will be presented with a special attention to temporary grasslands. The variation in the size of herbivore herds will be described, and especially the number of dairy cows and suckling cows. Data related to seed production of forage species will be presented.

The second part will present the main stakes for the future of grasslands. They are related to (i) balance between income and workload for the farmers, taking into account the common agricultural policy, (ii) quality of animal products in response to end-users' expectations and (iii) combination of economic performance and environment preservation, with a special interest to reduction of soil and nutrient losses, reduction of fossil energy consumption and greenhouse gas emission, preservation of hosted biodiversity.

Meeting these stakes assigns new goals to breeding. It first requires an increased persistence of grassland stands which may be achieved through either more persistent plants or through exploitation of the population dynamics of swards. It also offers good prospects for mixtures of species as sources of overyielding and more yield persistency. Finally, these stakes offer new prospects for forage legumes, source of nitrogen fertility in cropping systems and protein in animal diets with a positive effect on the grassland biodiversity.

Keywords Acreage · Herd size · Economy · Forms · Biodiversity · Energy · Efficiency · Mixtures

---

C. Huyghe (✉)
Inra, de recherche Poitou-Charentes, Route de Saintes, Lusignan 86600, France
e-mail: christian.huyghe@lusignan.inra.fr

C. Huyghe (ed.), *Sustainable Use of Genetic Diversity in Forage and Turf Breeding*,
DOI 10.1007/978-90-481-8706-5_1,

Introduction

Breeding perennial species, such as forage grasses and legumes is a long term enterprise. It is thus necessary to analyse the present and future stakes of the supply chain where the future varieties have to be used. These stakes must include the economic aspects of the supply chain. In the case study of the perennial forage species, the economy must include the animal production as well as the expectations of the farmers. Environment preservation has to be taken into account by all production systems, and it is especially important for the grasslands and herbivore production systems as they valorise a very large part of the arable land in France and Europe. It is also necessary to anticipate the possible long term changes. Among them, the consequences of climate change are likely to be very important for grasslands and for plant species.

The present paper will first present the economic importance and associated stakes of grasslands and related animal production in France, with a description of acreage, animal herds and performance as well as an analysis of changes in number of farms and farm size. Challenges for combining economic performance and environment preservation will be presented. It will be proposed that multi-species swards offer a high potential to meet these objectives. The adequate breeding objectives will be discussed. A special attention will be paid to the consequences of climate changes for the species and breeding objectives and for the preservation of the in situ genetic resources which are abundant in France and Europe for many perennial forage and turf species.

Grasslands and Related Animal Production in France: Economic Importance and Stakes

Acreage of Grasslands and Forage Crops

Grasslands and forage crops contribute 45% of the arable land in France, with a total of more than 14 Mha. In the national statistical databases, two thirds of this area is provided by permanent grasslands, part of this being old temporary grasslands (more than 6 years old). Over the last three decades, the acreage in permanent grasslands showed a strong decrease, with an average annual loss of 140,000 ha (Fig. 1.1). Part of this loss led to an increase of the acreage of annual crops while another part turned into forests and bushes. Most of the permanent grasslands are located in mountainous areas (Massif Central, the Alps, the Pyrenees, Jura and Vosges) where nothing else could be produced except forestry.

The sown grasslands and forage crops showed important variation of the last decades. Acreage in forage maize, which is used for silage production, tends to slowly decline after a sharp increase before 1990. However, the acreage in silage maize still increases in the regions specialised in milk production.

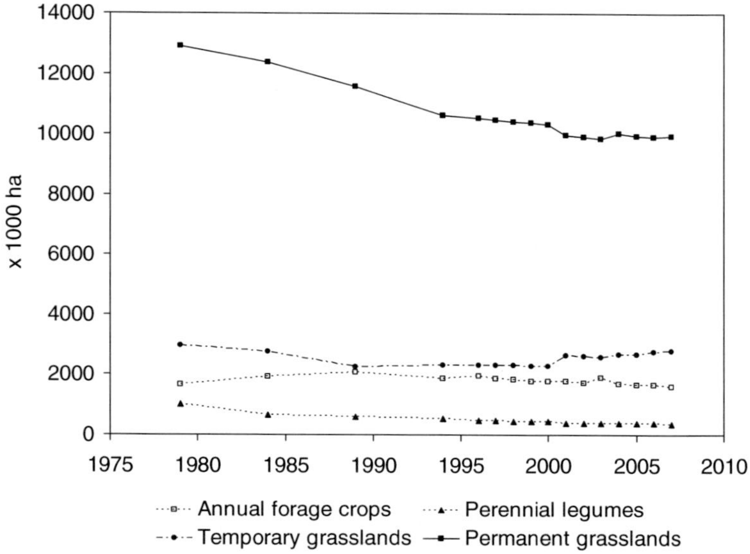

Fig. 1.1 Changes in acreage of annual forage crops, perennial legumes and temporary and permanent grasslands over the last three decades

While the acreage in pure perennial legumes (mainly alfalfa) is still decreasing after a peak at 1.2 Mha in 1961, the acreage of temporary grasslands slowly increases (Huyghe, 2008). Interestingly, a high percentage of the temporary grasslands are sown with mixtures of species, mainly grasses and legumes. This percentage was estimated at 72% in 2007.

Seed Production and Marketing

The volume of seeds sold in France varied accordingly (Fig. 1.2).

As the statistical data before 1990 did not separate turf and forage seeds for the perennial ryegrass, they are presented together on this figure. The sharp increase between 1985 and 1990 corresponds to the creation of the turf market in France.

Seed tonnage of Italian ryegrass varied a lot. This is explained by the use of this short cycle species. It is mainly devoted to the production of stocks over short growth cycle. The variations are due to the needs to quickly build feed stocks when they become small on the farm. This may be due to poor grazing seasons, low production of hay or low production of silage maize. The volume of perennial legumes, mainly alfalfa, red and white clovers is much more stable. However, when considering it in more details, the volumes declined between 1980 and 1990, were then fairly stable and showed a strong increase over the last 5 years. This is due to the increase of the acreage of temporary grasslands sown as mixtures and of the proportion of forage legumes in these mixtures.

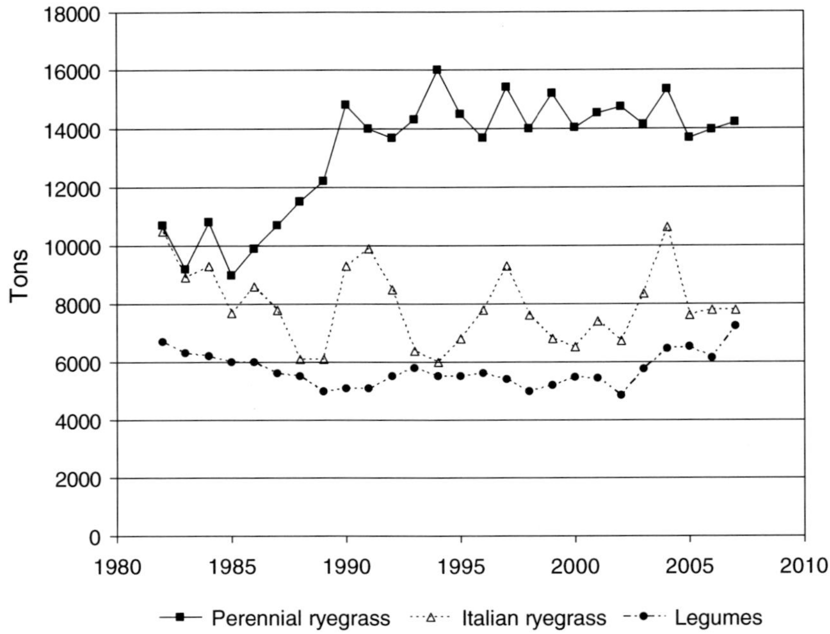

Fig. 1.2 Changes in the volume of seed of perennial and Italian ryegrass and of perennial legumes in France over the last three decades

The French regulation for marketing of seed mixtures in forage species was modified in 2004. The regulation mentions that (1) mixtures must be declared and have a constant species and variety composition, (2) varieties must be registered for a use as forage, (3) seed quality of each component must be the same as when sown as pure crops, and (4) each component must contribute at least 5% of the mixture (in mass). It is then advised not to exceed 6 components. Four years after the implementation of this regulation, the volume of grassland mixtures reached 2700 tons during the 2007–2008 market campaign. The main species used in the mixtures are perennial ryegrass, cocksfoot, tall and meadow fescues and white clover. The proportion of varieties registered on the French catalogue is high (81%) and recently registered varieties are mainly used in the marketed mixtures.

The acreage devoted to forage and turf seed production in France varied a lot over the last 20 years (Fig. 1.3). Production of forage legumes has always been a major commercial activity. It is mainly due to alfalfa as the summer climate is very suitable and the structure of the landscape (numerous small sized seed fields) is adequate to support a large population of wild pollinators which live in the permanent landscape components (hedges, roadsides, lanes,. . .).

The acreage of forage and turf grasses showed large inter-annual variation. In turf, the increasing trend was due to production of tall fescue and perennial ryegrass. In forage, the large waves were due to strong variation in the seed production of cocksfoot, low acreage being due to stocks and to European competitiveness.

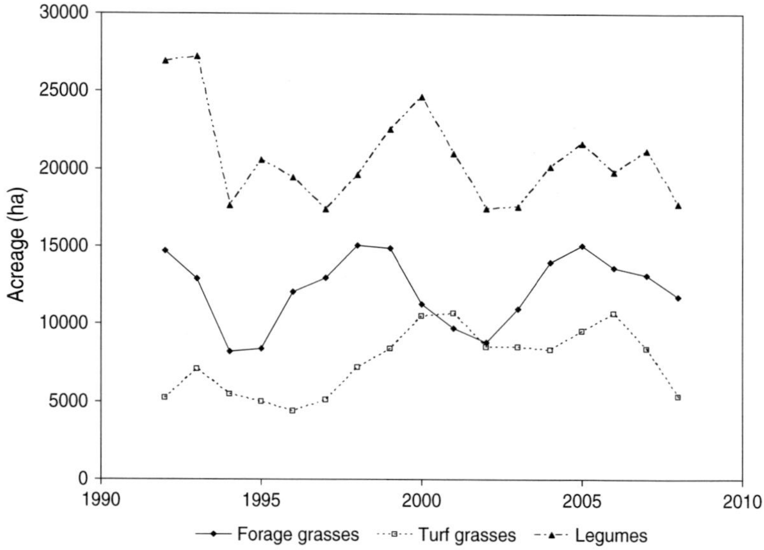

Fig. 1.3 Changes in the acreage dedicated to seed production of forage grasses, turf grasses and legumes

The strong reduction of the last season for all three groups is due to a change in the European support to these crops and large seed stocks all over Europe. Moreover, the 'good' weather conditions, mild temperatures and summer rainfall, induced large forage production and maintained the meadows in good state, with limited need for resowing.

Animal Production

Considering the same time period, the herbivore herds significantly varied. The herd of dairy goats was constant with a sharp increase in milk production per head. The number of sheep steadily declined, this being due to a strong reduction in the number of sheep devoted to meat production. The main variations were due to the number of dairy and suckling cows. As a consequence of the milk quota established in 1984 and of the sharp increase of the milk production per head, the number of dairy cows was dramatically reduced, the animals being predominantly located in west of France, with a small proportion in the mountainous regions where the milk is mainly processed into AOC cheeses. As the bovine meat consumption did not drop much during the same period of time, this opened a possibility for the development of a large herd of suckling cows. Presently, the number of cows is fairly similar with 4 Millions of heads each. These changes have strong consequences for grasslands. Because of the increase in milk production per dairy cow, diets have been modified for a higher concentration in digestible energy and protein. This explains why silage maize is popular for these animals, when soil and climate are suitable. However, high milk production may be achieved with diets based upon grazing, where it requires large biomass availability per grazing animal and adequate species composition. Suckling cows are more located in mountainous and hilly areas and efficiently valorised permanent grasslands through grazing.

Farms and Farmers

The number of dairy farms showed a 80% reduction over the last three decades (Fig. 1.4). Reducing the number of farmers was a main tool in the public policy to reduce milk production and reach the national quota. This mainly targeted the small farms. The reduction has become less pronounced since 1993. Over the last 10 years, the number of farms involved in milk and beef cattle production declined at a similar rate of 5000 farms a year. Due to the variation in number of farms, the production per farm increased, even if it is moderate when compared to the large average herd size observed in Denmark or in the UK. The increased production per farm generated a new stake for the farmers, where reducing work load

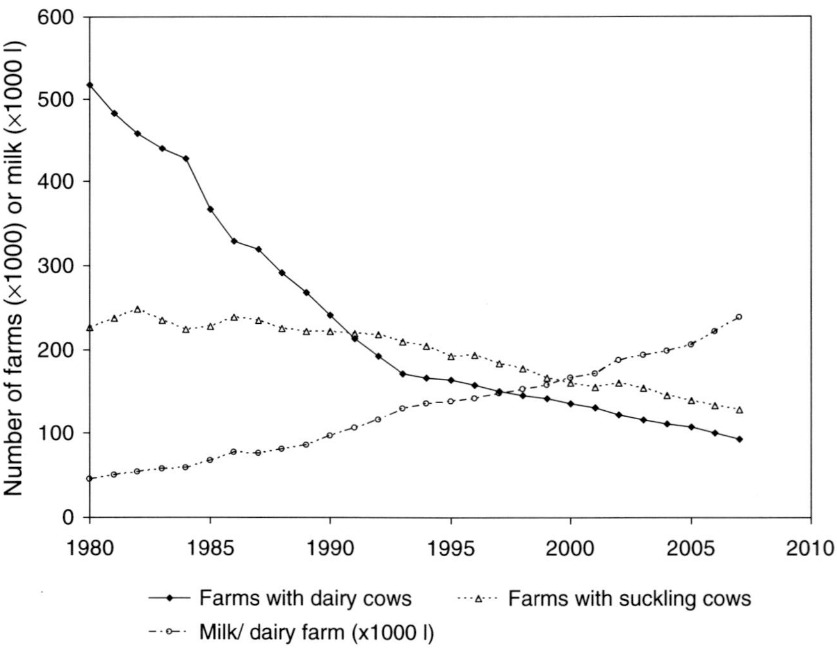

Fig. 1.4 Changes over the last three decades in number of farms with a dairy herd or a beef cattle herd in France and the mean milk production per farm

and above all work uncertainty became a major objective, alongside with economic performance.

Increasing the Environmental Performance

Alongside with the economic stakes which include the main concerns for the farmers, for the industry but also for the end-users, it is necessary to consider the aspects related to preservation of environment, or the environmental performance.

Three aspects are especially relevant to grasslands and herbivore production (Huyghe, 2009).

The first point is related to the losses of nutrients and especially nitrate leaching, with the possible consequences on the quality of water and the subsequent pollution. Mineral nitrogen fertilisation plays an important role, as it must meet requirement of grassland growth. However, the species composition and especially the use of mixtures may provide an adequate answer as it may limit the use of nitrogen fertilisers and nitrate leakages. Grassland management is also a tool to monitor and limit nutrient losses. The exploitation regime (cutting and grazing) and the animal stocking rate are important. Sward persistency and renovation also have to be considered, as a significant part of the losses are likely to occur during ploughing and renovation.

Fossil energy consumption and greenhouse gas emission have to be limited to reduce global warming. Mineral nitrogen fertilisation contributes a large part of the consumption of fossil energy in grassland production and herbivore production. Species composition of grasslands, with presence of legumes thus offers an opportunity to reduce consumption of fossil energy. It also limits the emission of nitrous oxide, as the emission coefficient was estimated to be null for symbiosis while its estimation is 0.7% of the applied nitrogen either through mineral or organic fertilisation. This topics is still very much under debate and requires to be extensively documented.

The drawback of grasslands regarding GHG is the methane produced by ruminants, as a consequence of the rumen functioning. Lassey (2007) estimated that 6% of the carbon in the diet was lost as methane. It was suggested by Martin et al. (2006) that the presence of tanniferous legumes species in the diet was likely to modify the bacterial population in the rumen and may reduce methane emission. Even though this has still to be documented, it would support the idea to promote temporary grasslands with a large number of species.

The last possible contribution of grasslands to environmental performance is the preservation of biodiversity. It was shown on the basis of many studies that the hosted diversity increases with the number of plant species, of functional groups and with the functional diversity in the grasslands (Lavorel and Garnier, 2002). This is true for instance for the number of fungi (Fischer et al., 2008), but we will get back to this item later in the paper. It is also true for the number of pollinating species, and especially wild bees. The abundance of pollinators also increases in grasslands because of the presence of hedges. The management is obviously critical for preserving a rich hosted biodiversity.

Adequate Sward Composition and Varieties to Meet the Stakes

All these elements lead to the definition of a sward ideotype and support the hypothesis that temporary grasslands must be increasingly sown with mixtures of species, and especially with grasses and legumes.

A sward ideotype must have a high persistency as ploughing temporary grassland is a major perturbation generating possible nitrogen leakages and deeply modifying the ecosystem. It must also have a high feed quality, through a high quality of each component of the sward.

Mixtures of species may lead to overyielding (the mixtures on average produce more aerial biomass than the individual species in pure swards) as documented by several authors, a major study being published by Hector et al. (1999), following the seminal work of Sinclar in 1826 reported by Hector and Hooper (2002). The overyielding phenomenon may be explained by either competition where the most competitive species are also the most productive ones, complementarity and/or facilitation.

Management practices must be adapted either to preserve plant species diversity or to facilitate the species replacement over time in the swards. It is also obvious that the characteristics of varieties as well as the variety testing must be adapted to reach this objective.

If mixtures of grasses and legumes offer an adequate answer, it is however necessary to define the optimum species composition. Indeed, overyielding also depends on species identity (Sanderson et al., 2004). It is also necessary to revise the main traits of the varieties to be sown in mixtures. The hierarchy of traits and thus the prevalence in selection index (or registration index) may be different.

This situation may be illustrated by the importance of disease resistance. In variety selection and registration for use in pure swards, a heavy weight was given to disease resistance. This was very relevant because of the losses in biomass production and quality that may be due to fungal symptoms. However, when growing

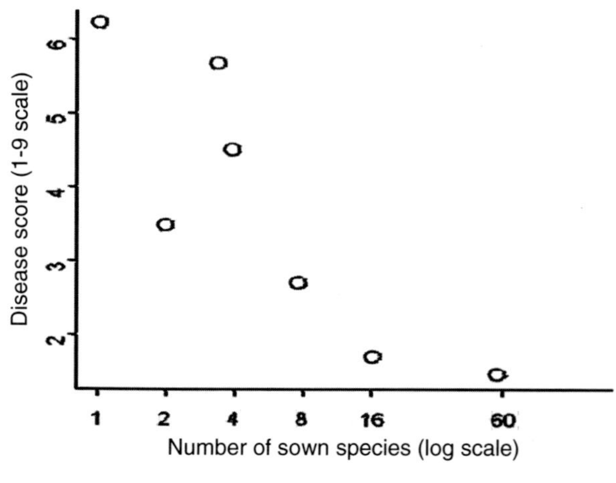

Fig. 1.5 Relationship between the number of plant species sown in a grassland sward and the disease score (adapted from Fischer et al., 2008)

species in mixtures, the pattern of damages due to leaf pathogens may become very different. This was documented on a large ecological experimental design set at Jena (Germany). It was shown by Fischer et al. (2008) that when increasing the number of species or of functional groups in a sward, the number of fungi species increased. But, the damages on each individual plant decreased (Fig. 1.5). This feature may partly explain the overyielding and could be classified as facilitation. These results mean that the requirement in terms of disease resistance may be lower. And, with a constant research investment, this may offer possibilities to have more genetic gains on other traits.

Even though temporary grasslands with a large species diversity offers promising prospects, it opens several questions.

One of them is the difficulty to stabilize, monitor and predict the dynamics of swards over time. It is necessary to understand and model the mechanisms underlying the dynamics of these swards and it is very unlikely that it will be possible to prevent any change. However, it would be very important to be able to predict to some extent the future of a sward sown with many species, and especially the installation period, as it was shown by Korner et al. (2008) that this stage was critical. Reduce sowing density and implement adequate seed technology would offer possibilities for a more predictable species composition. This point is important, as the unpredictable species composition is not relevant with the farmers' expectation of less uncertainty.

Improving the feed value is also a major issue. Improving feeding value requires improving the quality of each component. No positive diversity effect was observed for the main components of feed value such as protein content and dry matter digestibility (Huyghe et al., 2008). So, it is important to improve the quality of each component. For the grass to be used in grass – legume mixtures, it seems important to target high content in water soluble carbohydrates (WSC). Indeed, high WSC content will improve the animal use of the high protein content of such mixtures. It will also improve their ensilability. This would be very important for some grass species with low WSC content, such as cocksfoot or tall fescue.

The context of multispecies swards offers the possibility to exploit the feed interactions, i.e. where a component of one species interacts with a component of another species to improve its valorisation. This could be the case for tannins and polyphenol oxydase (PPO) (Lee et al., 2004), both produced by legumes species which may reduce the protein degradability of the whole diet. These interactions have been mainly documented for tannins (Julier et al., 2002). These possible interactions would suggest that it would make sense to increase the tannins content and the PPO activity to a level which would not make sense for a pure crop.

Climate Change will Modify the Objectives

This analysis of the stakes and of their consequences in terms of breeding is mainly done assuming a constant environment, the possible breakings being mainly social and economic. This is justified due to the considered time interval. However, thinking on a longer term basis, it is necessary to anticipate the effect of climate changes. According to the climate models and scenarios, the mean temperature in the next 50 years is likely to increase in a range of +0.6 to +2.5°C, to be compared to the 0.6°C of increase over the last three centuries. Although a very large uncertainty reigns over the current discussions about the future, meteorologists generally agree that the climate change will also strongly impact the rainfall and its distribution during the year. Changes in rainfall distribution will be very strong for the South and South-West of Europe. with less rain during summer and autumn. As far as grasslands and forage crops are concerned, this means more possibilities for winter growth and much more drought constraints during the summer. In most regions, water availability for irrigation will not increase.

This might lead to large changes in the species to be bred. Cocksfoot and tall fescue are likely to have an expanded area of cultivation, while species such as perennial ryegrass will offer the opportunity to exploit the winter growth.

Selection criteria will be influenced and more concern should be paid to winter growth and drought resistance. A special attention should also be given to pest and disease resistance as the spectrum of disease and pests is likely to be affected.

Climate changes will also affect the genetic resources. Indeed, Europe is the centre of primary or secondary diversification for numerous perennial forage species, such as perennial ryegrass, cocksfoot or fine-leaved fescues. There are a lot of natural populations which are locally adapted and whose characteristics were shaped by the soil and climate of the site. The climate change will likely be quicker than the capability of these perennial plants to adapt. As a consequence, part of this diversity is likely to be wiped out. It would be necessary to model the possible impact of climate changes on *in situ* genetic resources and to promote concerted actions to preserve this diversity.

Conclusions

There are numerous new stakes of the grasslands and herbivores supply chain, especially due to the need to combine economic and environmental performance and due to the on-going climate change. Because it is possible to use a wide range of species and to combine species in multi-species swards, forage species offer a wide range of answers. Breeding new varieties as well as registration of these varieties have to take into account this context, exploiting all available tools and concepts from a wide range of disciplines including ecology and population genetics. This must include new tools from molecular genetics and possibly transgenesis when relevant and when this does not induce risks of dissemination towards environment and local plant populations.

The changing climate may strongly impact in situ genetic resources. This deserves a specific effort to analyse it and possibly to promote concerted actions.

References

Fischer, M., Rottstock, T., Marquard, E., Middelhoff, C., Roscher, C., Temperton, V.M., Oelmann, Y., Weigelt, A. 2008. L'expérience de Iéna démontre les avantages de la diversité végétale pour les prairies. Fourrages 195:275–286.

Hector, A., Schmid, B., Beierkuhnlein, C., Caldeira, M.C., Diemer, M., Dimitrakopoulos, P.G., Finn, J.A., Freitas, H., Giller, P.S., Good, J., Harris, R., Hogberg, P., Huss-Danell, K., Joshi, J., Jumpponen, A., Korner, C., Leadley, P. W., Loreau, M., Minns, A., Mulder, C.P.H., O'Donovan, G., Otway, S.J., Pereira, J.S., Prinz, A., Read, D.J., Scherer-Lorenzen, M., Schulze, E.D., Siamantziouras, A.S.D., Spehn, E.M., Terry, A.C., Troumbis, A.Y., Woodward, F.I., Yachi, S., Lawton, J.H. 1999. Plant diversity and productivity experiments in European grasslands. Science (Washington) 286:1123–1127.

Hector, A., Hooper, R. 2002. Darwin and the first ecological experiment. Science 295:639–640.

Huyghe, C. 2008. La multifonctionnalité des prairies. I. Les fonctions de production. Cahiers Agricultures 17:427–435.

Huyghe, C. 2009. La multifonctionnalité des prairies. II. Conciliation des fonctions de production et de préservation de l'environnement. Cahiers Agricultures 18:7–16.

Huyghe, C., Baumont, R., Isselstein, J., 2008. Plant diversity in grasslands and feed quality. In: Hopkins, A. et al. (eds.), Biodiversity and Animal Feed (Vol. 13, pp. 375–386). Grassland Science in Europe.

Julier, B., Lila, M., Huyghe, C., Morris, P., Allison, G., Robbins, M. 2002. Effect of Condensed Tannin Content on Protein Solubility in Legume Forages (Vol. 7, pp. 134–135).In Grassland Science in Europe.

Körner, C., Stöcklin, J., Reuther-Thiebaud, L., Pelaez-Riedl, S., 2008. Small differences in arrival time influence composition and productivity of plant communities. New Phytol. 177:698–705.

Lassey, K.R. 2007. Livestock methane emission: from the individual grazing animal through national inventories to the global methane cycle. Agr. Forest Meteorol. 142:120–132.

Lavorel, S., Garnier, E. 2002. Predicting the effects of environmental changes on plant community composition and ecosystem functioning: revisiting the holy grail. Funct. Ecol. 16:545–556.

Lee, M.R.F., Winters, A.L., Scollan, N.D., Dewhurst, R.J., Theodorou, M.K., Minchin, F.R. 2004. Plant-mediated lipolysis and proteolysis in red clover with different polyphenol oxidase activities. J. Sci. Food Agric. 84:1639–1645.

Martin, C., Morgavi, D., Doreau, M., Jouany, J.R. 2006. Comment réduire la production de méthane chez les ruminants? Fourrages 187:283–300.

Sanderson, M.A., Skinner, R.H., Barker, D.J., Edwards, G.R., Tracy, B.F., Wedin, D.A.2004. Plant species diversity and management of temperate forage and grazing land ecosystems. Crop Science 44:1132–1144.

Part II
Genetic Resources

Chapter 2
A State of the Art of Germplasm Collections for Forage and Turf Species

Beat Boller and Merja Veteläinen

Abstract Four categories of plant genetic resources (PGR) are important for breeding: Wild relatives, ecotypes, landraces, and cultivars. Forage and turf species differ from field crops in the relative importance of these categories, as well as in the relative importance of in situ vs. ex situ conservation. As they are less domesticated, a continuum of wild and naturalized forms of the cultivated species exists as ecotypes in a great variety of permanent grasslands. An often random fraction of this variety has been either used to aliment active breeding pools or collected in gene banks, while a great range of potentially useful genetic variation remains yet to be exploited.

This paper reviews recent, partly molecular marker based literature pointing to criteria and strategies of collecting PGR in situ in grassland dominated regions. Guidelines are given for establishing and maintaining ex situ germplasm collections and for evaluating them in view of their utilization by breeders. A comprehensive overview of publicly accessible germplasm collections, especially those in the auspices of European Co-operative Programme for Plant Genetic Resources (ECPGR) is given. Their present status and expected future developments are high-lighted. Strategies for breeders to utilize domestic and exotic germplasm in breeding programmes are presented. Also pre-breeding strategies along with a Nordic case study will be presented in order to demonstrate the utilisation of unique alleles outside the breeding pool. The potential of new molecular tools for the management and better utilization of PGR collections is shown.

Keywords Characterization · Conservation · Ex situ · Genetic resources · In situ · Utilization

B. Boller (✉)
Research Station Agroscope Reckenholz-Tänikon ART, Reckenholzstrasse 191, Zürich CH-8046, Switzerland
e-mail: beat.boller@art.admin.ch

C. Huyghe (ed.), *Sustainable Use of Genetic Diversity in Forage and Turf Breeding*, DOI 10.1007/978-90-481-8706-5_2, © Springer Science+Business Media B.V. 2010

Introduction

Forage and turf species differ from most field crops with respect to germplasm collection, conservation and utilization. In important field crops such as maize or wheat, the wild form of the cultivated species no longer exists, since breeding for domesticity has resulted in plant species being unable to reproduce without the helpful hand of humans. Forages are less domesticated. Wild forms of common forage species still exist, as well as feral (naturalized) forms (populations that originated from forage crops, but that escaped to persist in the natural environment). The closeness of wild and cultivated forms of fodder crop species makes a wealth of natural genetic variation readily accessible for use in breeding.

Plant genetic resources (PGR) are indispensable for any breeding effort. They serve to:

- procure initial variation when a breeding programme is started
- create new opportunities for selection in an advanced breeding programme where genetic variance has decreased due to several cycles of recurrent selection

A careful choice of PGR is necessary in both cases. PGR should be adapted to the target growing conditions while maximizing the chances of introducing unique new alleles affecting the traits of interest.

Categories of PGR

Four basic categories are of potential importance for forage and turf breeding programs:

1. *Wild relatives* are representatives of taxa different of, but related to the cultivated species. They are of limited importance for forage and turf breeding. Due to their relatively young history as crops, available genetic variability within the cultivated species themselves is still sufficiently large. Introgression of specific characters from wild ancestors is of some more interest in allopolyploid species with complex systematic like alfalfa or white clover.
2. *Ecotypes* of cultivated species occur in permanent grassland. For forage and turf species, there is no clear borderline between wild, semi-natural and cultivated forms. This is because permanent grassland in most relevant cases exists only as a consequence of human agricultural activity in zones where forests would be the natural vegetation. Adapted native grasses originating from non-agricultural habitats settle in permanent grassland together with naturalized populations of the same species which may have spread from an initial seeding. Rather than trying to assign such populations to an either wild or semi-natural origin, it is more appropriate to address them as *ecotypes*. Using this term implies populations which have adapted to a known environment after many years of natural

selection, usually involving natural re-seeding but without deliberate human interference such as selection, seed harvest, or human-mediated seeding. That is, human interference in the development of an ecotype is limited to actions of usual management practices, such as frequency and type of utilisation, or intensity of fertilization. If sufficient cycles of recombination with local genetic material and natural selection have occurred, an initial sowing of a fodder crop variety may, after 10–20 years, give rise to an ecotype population.

3. *Landraces* are populations which have adapted to a specific region or location, such as a farm (farm varieties, "Hofsorten") by repeated seed harvest and human-mediated re-seeding. The term "landrace" implies that human actions are carried out deliberately to improve local adaptation, e.g. by re-seeding the surfaces with locally produced seed, and by carrying out seed harvest after several years of utilization as forage to improve persistency. Prominent examples of highly valuable, traditional landraces are alfalfa in Italy (Torricelli et al., 2003), timothy in Norway (Schjelderup et al., 1994), and red clover in Switzerland (Kölliker et al., 2003).
4. *Varieties*: any cultivated variety (cultivar), whether freely available on the market, protected by plant breeder's rights, or having become obsolete and stored in gene banks, can be used in breeding without any restriction. The right to freely use even protected varieties as PGR in breeding is called "breeder's exemption" and is an important provision of the international convention for the protection of new varieties of plants (UPOV, 1991). For use in breeding, varieties have the advantage of being precisely described through the registration procedure for distinctness, uniformity and stability (DUS), and usually have been evaluated extensively in official tests for their value for cultivation and use (VCU). Furthermore, commercially successful varieties have proven their ability to give satisfactory seed yields. These properties render cultivars very popular as PGR in fodder crop breeding.

PGR Maintained In Situ

Ecotypes and landraces can be maintained in situ (referred to as "on farm" in the case of landraces). The idea of in situ conservation is to maintain the environment which has allowed the development of the distinctive properties of the PGR. Genetic evolution is explicitly wanted as it allows a further development of PGR to even better match the requirements of their specific environment. PGR growing in situ remain the largest gene pool of interest to forage and turf breeding.

Breeding Importance of Germplasm Collected In Situ

Historically, genetic resources growing in situ in permanent grassland have been by far the most important sources of germplasm used in forage and turf breeding. The bulletin of perennial forage plants listed in the French national catalogue in

Table 2.1 Declared origin of varieties in the official French catalogue of grass varieties accepted 1957–1984 (I.N.R.A., 1984)

	Number of Varieties originating from:			
Species	Ecotypes	Varieties or landraces	Ecotypes and varieties	Breeding material of diffuse origin
Dactylis glomerata	12	0	0	1
Festuca arundinacea	11	0	0	6
Festuca pratensis	4	1	1	1
Phleum pratense	5	0	3	1
Lolium perenne	10	1	8	9
Lolium multiflorum ssp. italicum	5	1	3	10
Total	47	3	15	28

1984 (I.N.R.A., 1984) contains a voluntary declaration of the material of origin of the varieties, and ecotypes clearly predominate where an unequivocal origin was declared (Table 2.1).

Nowadays, most commercial breeders rely more on crosses between varieties as starting materials for their breeding, rather than introducing newly collected material. However, since most of the older varieties have been derived from an original collection in grassland, we can assume that the majority of varieties currently in use trace back at least partly to breeding material originally created from collections of PGR in situ. This is reflected by the small genetic distance between cultivars and ecotypes found in molecular studies, for example in perennial ryegrass (Bolaric et al., 2005; McGrath et al., 2007), meadow fescue and Italian ryegrass (Peter-Schmid et al., 2008a).

Grassland Dominated Regions as Centres of Diversity

Grassland dominated regions provide the most diverse opportunities for collecting PGR of fodder crops and amenity grasses in situ. For most temperate species of interest, non-irrigated permanent grasslands are concentrated in zones with at least 800 mm of annual rainfall and about 9°C annual mean temperature. In Europe, such temperate grassland dominated regions are particularly prominent and diverse in the zone between the northern foothill of the Alps and the coastal zones of the Atlantic and Baltic Sea.

Criteria and Strategies for Collecting Germplasm In Situ

Depending on breeding objectives, two basic criteria need to be considered in making a strategy for collecting fodder crops PGR in situ:

1. When the objective is to enlarge genetic diversity of the breeding population, the relative degree of genetic variation within and among sites will affect decisions about the number of sites to be visited and the number of individuals to sample per site
2. When the objective is to find new genes affecting particular characters, the choice of collecting sites will be influenced by the expected action of environmental or management factors that are selective forces for traits of interest.

Genetic variation within and among ecotype populations has been studied using neutral molecular markers for a number of grassland species (Table 2.2). Variation within populations accounts for at least 60 and up to 98% of the total genetic variation. This suggests that sampling a large number of sites with just a few individuals is less effective for capturing genetic diversity than sampling fewer sites with a higher number of individuals. Theses studies do not point consistently to a specific strategy in the search for sites with a high genetic variability. Genetic variation within ecotype populations subjected to different managment in situ was reported to be negatively affected by a higher managment intensity (Kölliker et al., 1998), but this effect was not observed in a larger collection of ecotype populations of the same species (Peter-Schmid et al., 2008a).

When the variability of morpho-physiological characters of a comparable set of ecotypes was assessed in relation to molecular marker variability, a larger among populations variability was observed for the morphological characters (Fjellheim et al., 2007; Peter-Schmid et al., 2008b) than would have been expected

Table 2.2 Contribution of among and within populations variance in marker based studies of genetic diversity of grassland ecotypes, based on analysis of molecular variance (AMOVA)

					AMOVA: % of variance	
Source	Species	Marker system	No. of ecotype pops.	No. of indiv. per pop.	Among pops.	Within pops.
Bolaric et al. (2005)	*Lolium perenne*	RAPD	22	20	29	71
Rudmann-Maurer et al. (2007)	*Poa alpina*	SSR	54	8	25	75
Reisch et al. (2003)	*Sesleria albicans*	RAPD	25	4	38	62
Peter-Schmid et al. (2008a)	*Lolium multiflorum*	SSR	12	23	2	98
	Festuca pratensis	SSR	12	23	4	96
McGrath et al. (2007)	*Lolium perenne*	cpSSR	61	16	37	63
Fjellheim et al. (2005)	*Festuca pratensis*	AFLP	15	20	31	69

from the variability of molecular markers (Fjellheim et al., 2005; Peter-Schmid et al., 2008a). This implies that sampling ecotype populations from a larger range of sites will increase the chance of including the extremes for the traits of interest.

An obvious way to infer environmental and management factors acting at a potential sampling site is to assess floristic composition of the vegetation present. It is reasonable to assume that factors affecting plant species composition will affect genetic differentiation of ecotypes in a similar way, and therefore, similar ecotypes of a species can be expected in grasslands of similar floristic composition or vegetation classification. A recent study with *Lolium multiflorum* (Boller et al., 2009) suggested that ecotypes from sites the vegetation of which was classified as *Lolietum multiflori* were more productive and showed better resistance against bacterial wilt and snow mould than ecotypes from *Arrhenatherion* sites. This suggests that the chances of finding well adapted germplasm of a species can be increased by choosing collecting sites with a floristic composition pointing to agricultural management similar to that of the target use. Additionally, visiting places with contrasting floristic composition appears to be a good way to increase genetic diversity within a collection of a species.

Protection of PGR Maintained In Situ

In recent years, targeted programmes aiming at a more relaxed grassland management as part of agri-environment measures have been established to increase biodiversity (Marriott et al., 2004). Nösberger et al. (1998) concluded that management for habitat heterogeneity at all scales will conserve most of the biotic diversity at a site. Such a system will allow for intensive management of the most favourable pastures to produce high quality ruminant feed, along with maintenance of infrequently cut hay meadows providing opportunities for environmental services. Whether or not such an overall strategy is sufficient to adequately protect in situ conserved PGR, remains open to question. Programmes for promoting biodiversity with a focus on nature conservation concentrate on extensively managed, species-rich grassland. However, more intensively managed grassland may hold ecotypes which are of greater interest for future breeding. Peter-Schmid et al. (2008a) showed that Swiss ecotype populations of *Festuca pratensis* from extensively managed habitats with a high nature conservation value contained significantly less rare alleles than populations from habitats managed more intensively. Management intensity also had a significant influence on morphological characters, along with geographic location (Peter-Schmid et al., 2008b). A conservation concept taking these considerations into account was presented for Switzerland by Weyermann (2007), suggesting to describe and eventually protect five to nine agriculturally used surfaces for each of 16 vegetation units (associations) within each of 7 bio-geographic regions.

Genetic Resources Maintained Ex Situ

Ex situ conservation of forage crops serves the user community with readily accessible genetic variation for utilisation per se, research and plant breeding. The ex situ conservation programmes involve not only the collection and maintenance of the species and population diversity, but include also characterisation, evaluation and wide range of diversity studies. Proper documentation of the ex situ collections needs to be carried out both for gene bank management purposes and to promote use of the collections. For the latter purpose the development of adequate information and documentation systems is the key issue, as well as quality and quantity of data included in the systems.

Forage germplasm is acquired to the gene bank collections mainly through collecting material found in in situ conditions and requesting cultivars from plant breeders. Sometimes germplasm is also repatriated from other ex situ collections, when not anymore available in the original country or region of origin. The objective in germplasm collection is to sample the largest amount of diversity with a manageable amount of accessions. The strategy proposed by Marshall and Brown (1983) for forage species includes collecting seed from 50–100 individuals at each site, and sampling as many sites as possible to capture the range of environmental diversity. Sackville-Hamilton and Chorlton (1995) outlined strategies for collecting vegetative samples of forage grasses and legumes. Vegetative sampling requires additional effort to produce seed but may reduce effects of sampling time on preferential sampling of either early or late flowering genotypes.

Prior to storage in gene banks seeds are dried to low moisture content and are then stored at subzero temperatures in cold stores or deep freezers in order to achieve good longevity. The long viability of seeds reduces the speed of loosing genetic diversity in storage and reduces the need for to regenerate accessions. Preferred and accepted standards for the different steps of regeneration for forage species are provided by Sackville-Hamilton (1998).

The aim of characterisation and evaluation is to produce information on germplasm collections in order to enhance their use. It can be questioned whether it is a task for gene banks to work with evaluation of agronomic traits with low heritability or rather to concentrate on characterisation that refers to description of traits with relatively high heritability. However, any information on germplasm that supports the use of collections – be it collection data that can be related to adaptive traits or characterisation data for resistance properties – increases the value of the germplasm collections. The present challenge is to provide existing evaluation and characterisation data through gene bank documentation systems in a user-friendly way.

Besides the forage genetic resources conservation programmes implemented by various national and regional gene banks, the European co-operation within the European Co-operative Programme for Plant Genetic Resources (ECPGR) has substantially contributed to the conservation activities for forages. The Working Group on Forages was established in 1984, as one of the original six working groups developed during the first Phase of ECPGR. The regional co-operation has been important, for example, in developing regeneration standards for main perennial forage grasses and legumes of temperate grasslands (Sackville-Hamilton and Chorlton, 1997; Sackville-Hamilton, 1998). This quality oriented work that serves both conservation and use has provided tools to produce seed with good genetic integrity and seed quality. Other Forage Working Group activities that aim to improve the quality and security of forage genetic resources conservation are co-operation in safety-duplication and sharing of conservation responsibilities between the member countries and institutes. The aim in sharing of responsibilities is to "conserve the genetically unique and important accessions for Europe and make them available for breeding and research. Such material will be safely conserved under conditions that ensure genetic integrity and viability in the long term" (ECPGR, 2009).

Table 2.3 Number of forage accessions in ECPGR European central crops databases (ECDDB) and EURISCO catalogue

Database	Available on-line	Number of accessions in ECCDBs	Number of accessions EURISCO catalogue*
Agropyron		56	385
Agrostis	x	1,381	1,720
Arrhenatherum	x	254	389
Bromus		583	918
Dactylis		8,700	10,853
Festuca		7,364	12,211
Forage grasses minor	Under development		
Forage legumes minor		1,090	3,144
Lathyrus		> 4,500	4,049
Lolium	x	9,543	11,276
Lupinus		Info not available	8,872
Medicago, annual		2,321	
Medicago, perennial		2,888	
Medicago, all in EURISCO			8,360
Phalaris	x	344	369
Phleum	x	5,085	5,229
Poa	x	5,311	4,524
Trifolium alexandrinum and Trifolium resupinatum		Info not available	311
Trifolium pratense		2,294	6,080
Trifolium repens	x	1,415	1,902
Trifolium subterraneum		4,776	2,715
Trisetum	x	86	70
Vicia spp.		Info not available	23,684
Vigna		Info not available	4,290

*http://eurisco.ecpgr.org, accessed on 21st April 2009

This means that ex situ conservation of germplasm will be carried out according to common, agreed quality standards, independently of where the germplasm is physically located.

At present the information on major ex situ collections in Europe can be found in local gene bank information systems and especially for forages in so called European Central Crops Databases (ECCDBs). Development of these forage crop specific databases has been a major input of the ECPGR Working Group for Forages. Today, there are 23 different forage crop databases (Table 2.3). They provide passport information on 56,444 accessions. Also EURISCO, a web-based germplasm catalogue based on national germplasm inventories, provides passport data of ex situ plant collections maintained in Europe. In Table 2.3 number of forage germplasm accessions found both in ECCDBs and EURISCO catalogue are given.

Outside Europe important sources for forage germplasm are USDA National Plant Germplasm System in United States with app. 33,142 accessions; International Center for Agricultural Research in the Dry Areas (ICARDA) in Aleppo, Syria with 20,031 accessions and Australian Medicago Genetic Resource Centre in Adelaide, South Australia with 45,640 accessions (Boller and Greene, 2010).

Strategies for Using PGR in Breeding

Although genetic variability is a prerequisite for plant breeding, can plant breeders sometimes find themselves in the situation where enough variability is not readily available in order to achieve a certain breeding goal (Baenziger and Peterson, 1991). In this situation choice of relevant PGR and its incorporation into the breeding programme becomes critical. Ideally new genetic variability is incorporated into a breeding programme continuously, in a long-term manner without disturbing the performance of the existing, high performing, highly adapted elite germplasm (Simmonds, 1993; Kannenberg and Falk, 1995). In addition, an ideal long-term strategy recognises also the need for continued evolution in crop gene pools and genetic conservation (Veteläinen, 1997).

The ease of germplasm choice in germplasm collections is dependent on information at the accession level provided by the collection holder. If environmental data of germplasm origin are known, their adaptation can be matched with the target environment of the breeding programme. Increasingly, germplasm collections are improving on the quality of passport data, especially more precise information on latitude and longitude. Text descriptions of older collection site locations are being converted into geographic map coordinates and new collection sites are being documented with GPS. This allows plant breeders to more easily select germplasm adapted to their target environment. Once GIS-based applications are coupled to gene bank collection databases, selection of adapted germplasm will be further facilitated (Greene et al., 2007). Interesting characteristics are usually less well known than with varieties but may be derived from knowledge of environmental and management factors of the site the ecotypes originate from. Also so called "core collections", that should best represent the range of genetic diversity of the crop and collection in question, have been developed to assist the choice of germplasm in the gene bank collections (Johnson and Hodgkin, 1999).

Pre-breeding (=germplasm enhancement) strategies can cover both short- and long-term actions. Amongst the short-term actions hybridization and backcrossing to the elite germplasm can be mentioned. Test cross evaluation allows breeders to further evaluate unadapted germplasm, especially for quantitative characters such as yield, as well as start the pre-breeding process (Boller and Greene, 2010). Test cross progeny can be bulked to form composite populations, or gene pools. Williams et al. (2007) developed several white clover varieties in New Zealand using this technique. The formation of regional gene pools was proposed to broaden the genetic base of alfalfa in the United States (Barnes et al., 1977). Regional genepools of perennial ryegrass were established in France (Charmet and Balfourier, 1995). In a study based on molecular genetic distance (GD) among German ecotypes of perennial ryegrass distinct genepools (Northern vs. Southern) were identified (Bolaric et al., 2005).

An example of an attempt to widen genetic base of a forage crop in long-term is a Nordic project that connects different Nordic efforts to conserve and utilise timothy (*Phleum pratense* L.) genetic resources in the development of better varieties that benefit farming at the northern borders of the agriculture. The project includes (1) analyses of genetic variation of Nordic timothy germplasm in terms of distribution, dispersion history and important adaptive traits such as vernalization response and frost tolerance, (2) evaluation of Nordic collection of timothy in terms of genetic variability, and improve this by targeted collections and/or proposing in situ conservation, (3) use exotic germplasm to study the bio-geographical history of Nordic timothy, (4) broadening the genetic base of timothy breeding materials by identifying heterotic groups and sources (including exotic germplasm) for important traits, e.g. frost tolerance, (5) definition of in situ conservation methods for old pastures and meadows and (6) improvement of the value and access to the Nordic timothy collection by providing new phenotypic and genotypic data for the users of the collection. This unique project involves all stakeholders in the field of genetic resource management and germplasm utilisation in the Nordic countries i.e. gene bank curators, researchers and plant breeders. The project web-site can be found at http://www.nordictimothy.net/. Several contributions of these conference proceedings highlight the present state of the specific parts of the entire project.

References

Baenziger, P.S., Peterson, C.J. 1991. Genetic variation: Its origin and use for breeding self-pollinated species. In: Stalker, H.T., Murphy, J.P. (eds.), Plant Breeding in the 1990's (pp. 69–92). CAB International, UK.

Barnes, D.K., Bingham, E.T., Murphy, R.P., Hunt, O.J., Beard, D.F., Skrdla, W.H., Teuber, L.R. 1977. Alfalfa Germplasm in the United States: Genetic Vulnerability, Use, Improvement and Maintenance. USDA Technical Bulletin 1571. ARS, Washington, DC.

Bolaric, S., Barth, S., Melchinger, A.E., Posselt, U.K. 2005. Molecular characterization of genetic diversity in European germplasm of perennial ryegrass. Euphytica 146:39–44.

Boller, B., Greene, 2010. Genetic resources. In: Boller, B. et al (eds.), Handbook of Plant Breeding, Vol. 5, Fodder Crops and Amenity Grasses (pp. 13–37), Springer, New York.

Boller, B., Peter-Schmid, M., Tresch, E., Tanner, P., Schubiger, F.X. 2009. Ecotypes of Italian ryegrass from Swiss permanent grassland outperform current recommended cultivars. Euphytica 170:53–65.

Charmet, G., Balfourier, F. 1995. The use of geostatistics for sampling a core collection of perennial ryegrass populations. Gen. Res. Crop Evo. 42:303–309

ECPGR. 2009. A Strategic Framework for the Implementation of a European Genebank Integrated System (AEGIS). A Policy Guide. European Cooperative Programme for Plant Genetic Resources (ECPGR). Biodiversity International, Rome, Italy.

Fjellheim, S., Blomlie, A.B., Marum, P., Rognli, O.A. 2007. Phenotypic variation in local populations and cultivars of meadow fescue – potential for improving cultivars by utilizing wild germplasm. Plant Breeding 126:279–286.

Fjellheim, S., Rognli, O.A. 2005. Molecular diversity of local Norwegian meadow fescue (*Festuca pratensis* Huds.) populations and Nordic cultivars – consequences for management and utilisation. Theor. Appl. Genet. 111:640–650.

Greene, S.L., Minoura, T., Steiner, J.J., Pentecost, C.G. 2007. Webgrms: prototype software for web-based mapping of biological collections. Biodiv. Conserv. J. 16:2611–2625.

I.N.R.A. 1984. Bulletin des variétés 1984 de plantes fourragères. INRA Publications, Versailles, 505 pp.

Johnson, R.C., Hodgkin, T. 1999. Core Collections for Today and Tomorrow. International Plant Genetic Resources Institute, Rome, Italy.

Kannenberg, L.W., Falk, D.E. 1995. Models for activation of plant genetic resources for crop breeding programmes. Can. J. Plant Sci. 75:45–53.

Kölliker, R., Herrmann, D., Boller, B., Widmer, F. 2003. Swiss Mattenklee landraces, a distinct and diverse genetic resource of red clover (*Trifolium pratense* L.). Theor. Appl. Genet. 107: 306–315.

Kölliker, R., Stadelmann, F.J., Reidy, B., Nosberger, J. 1998. Fertilization and defoliation frequency affect genetic diversity of *Festuca pratensis* Huds. In permanent grasslands. Mol. Ecol. 7:1557–1567.

Nösberger, J., Messerli, M., Carler, C. 1998. Biodiversity in grassland. Annales de Zootechnie 47:383–393.

Marriott, C.A., Fothergill, M., Jeangros, B., Scotton, M., Louault, F. 2004. Long-term impacts of extensification of grassland management on biodiversity and productivity in upland areas. A review. Agronomie 24:447–462.

Marshall, D.R., Brown, A.H.D. 1983. Theory of Forage Plant Collection. Genetic Resources of Forage Plants. CSIRO, East Melbourne.

McGrath, S., Hodkinson, T.R., Barth, S. 2007. Extremely high cytoplasmic diversity in natural and breeding populations of *Lolium* (*Poaceae*). Heredity 99:531–544.

Peter-Schmid, M.K.I., Boller, B., Kölliker, R. 2008a. Habitat and management affect genetic structure of *Festuca pratensis* but not *Lolium multiflorum* ecotype populations. Plant Breeding 127:510–517.

Peter-Schmid, M.K.I., Kölliker, R., Boller, B. 2008b. Value of permanent grassland habitats as reservoirs of *Festuca pratensis* Huds. and *Lolium multiflorum* Lam. populations for breeding and conservartion. Euphytica 164:239–253.

Reisch, C., Poschlod, P., Wingender, R. 2003. Genetic differentiation among populations of *Sesleria albicans* Kit. *ex* Schultes (*Poaceae*) from ecologically different habitats in central Europe. Heredity 91:519–527.

Rudmann-Maurer, K., Weyand, A., Fischer, M., Stocklin, J. 2007. Microsatellite diversity of the agriculturally important alpine grass *Poa alpina* in relation to land use and natural environment.Ann. Bot. 100:1249–1258.

Sackville Hamilton, N.R. 1998. The regeneration of accessions in seed collections of the main perennial forage and grasses of temperate grasslands: Background considerations. In:

Maggioni, L. et al (eds.), Report of a Working Group on Forages. Sixth meeting, 6–8 March 1997, Beitostølen, Norway. International Plant Genetic resources Institute, Rome.

Sackville Hamilton, N.R., Chorlton, K.H. 1995. Collecting vegetative material of forage grasses and legumes. In: Guarino, L. et al (eds.), Collecting Plant Genetic Diversity: techical Guidelines. CAB International, UK.

Sackville Hamilton, N.R., Chorlton, K.H. 1997. Regeneration of Accessions in Seed Collections: A Decision Guide. Handbooks for Genebanks No. 5. International Plant Genetic Resources Institute, Rome, Italy.

Schjelderup, I., Aastveit, A.H., Aastveit, K. 1994. Winter hardiness in marginal populations of timothy (pp. 61–68). In: Rognli, O.A. et al (eds.), Breeding Fodder Crops for Marginal Conditions. Proceedings of the 18th Eucarpia Fodder Crops Section Meeting, Loen, Norway, 25–28 August 1993. Kluwer Academic Publishers, Dordrecht.

Simmonds, N.W. 1993. Introgression and incorporation. Strategies for the use of crop genetic resources. Biol. Rev. 68:539–562.

Torricelli, R., Russi, L., Silveri, D.D., Falcinelli, M., Veronesi, F. 2003. Lucerne genetic resources from central Italy. Czech J. Genet. Plant Breed. 39:251–254.

UPOV. 1991. International convention for the protection of new varieties of plants of December 2, 1961, as revised at Geneva on November 10, 1972, on October 23, 1978, and on March 19, 1991. International Union for the Protection of new Varieties of Plants (UPOV). http://www.upov.int/en/publications/conventions/1991/act1991.htm. Accessed 30.01.2009.

Veteläinen, M. 1997. Dynamic Genepools of Barley – Utilisation of Exotic Germplasm in Nordic Plant Breeding. Doctoral thesis, Swedish University of Agricultural Sciences, Alnarp.

Weyermann, I. 2007. Konzept zur in situ Erhaltung von Futterpflanzen. Bundesamt für Landwirtschaft, Bern, http://www.cpc-skek.ch/pdf/ConceptFutterpflanzen_Vprovisoire.pdf

Williams, W.M., Easton, H.S., Jones, C.S. 2007. Future options and targets for pasture plant breeding in New Zealand. New Zealand Journal of Agricultural Research 50:223–248.

Chapter 3
Empirical Niche Modelling of the Spontaneous Diversity of Forage and Turf Species to Improve Collection and Ex Situ Conservation

Jean-Paul Sampoux and Vincent Badeau

Abstract Rational sampling of the spontaneous diversity of forage and turf species requires an *a priori* knowledge of the range of environmental conditions suitable for these species. We introduce some concepts and methods for investigating the environmental range of species by empirical modelling of species ecological niche, and we suggest how such investigations could help to plan collection campaigns and to improve the choice of core-collections. The empirical modelling of the ecological niche of a species consists of building a function of environmental parameters predicting the presence of the species from a calibration dataset including observed presence-absence or abundance records of the species and environmental data at observation sites. We emphasize that data from collection campaigns of plant breeders are valuable information for niche modelling. We introduce two methods for investigating the environmental distribution of species and for niche modelling based on presence-absence data: the canonical correlation analysis and the logistic regression. We give examples combining niche model and GIS software that may contribute to organize collection campaigns. We suggest that models predicting probability of presence of species may be useful for the selection of core-collections. Such models may help to delineate geographically isolated areas of presence of species that should be sampled separately for selecting a core-collection. In each isolated area of presence, we propose to stratify the accessions in clusters according to the predicted probability of presence of the species in collection sites, and to select accessions in each cluster.

Keywords Core-collection · GIS software · Niche modelling · Spontaneous diversity

---

J.-P. Sampoux (✉)
INRA, Centre Poitou-Charentes, UR4 (Unité de Recherche Pluridisciplinaire Prairies et Plantes Fourragères), Lusignan F-86600, France
e-mail: jpsampoux@lusignan.inra.fr

C. Huyghe (ed.), *Sustainable Use of Genetic Diversity in Forage and Turf Breeding*,
DOI 10.1007/978-90-481-8706-5_3,

Introduction

Many forage and turf species grow spontaneously in various environmental conditions. Natural selection in species has promoted the differentiation of a diversity of ecotypes adapted to a variety of local environmental conditions. This ecotypic diversity has been the initial resource of breeding programmes, and it remains the tank in which genetic variability can be found to face new agronomical challenges. Therefore, the spontaneous diversity of forage and turf species has long been extensively collected by breeders and geneticists (Mansat, 1995). Rational sampling in collection campaigns requires an *a priori* knowledge of the range of environmental conditions suitable for species under collection. The plant breeding community has indeed investigated widely and efficiently the spatial and environmental distributions of spontaneous intra-specific diversity for several important forage and turf species (Balfourier and Charmet, 1991; Warren et al., 1998; Prospéri et al., 2005). However, the range of environmental conditions suitable for the presence of a species has not really been a subject of study by this community until now. In this paper, we intend to: (i) introduce some concepts and methods for investigating the environmental range of species by empirical modelling of species ecological niche, and (ii) suggest how such modelling could help to plan collection campaigns and improve the choice of core-collections.

Conceptual Frame and Methods

The Concept of Ecological Niche

Each species requests a particular range of abiotic and biotic conditions to grow and reproduce. The range of environmental conditions suitable for the presence of a species was formalised by Grinnell (1917, 1924) as its ecological niche. Hutchinson (1957) improved the concept as a multidimensional ecological hyper-volume in which each dimension depicts an environmental abiotic or biotic parameter conditional for the presence of a species. He distinguished a *fundamental niche* as including the whole range of environmental parameter values suitable for a species without competition or predation, and a *realized niche* as restricted to the part of the *fundamental niche* where the species actually occurs.

The empirical modelling of the ecological niche of a species consists of building a function of environmental parameters predicting the presence of the species from a calibration dataset including observed presence-absence or abundance records of the species and environmental data at observation sites. As this empirical modelling relies on observed occurrences of species, it has been claimed to model the *realized niche*. However, models better predict combinations of environmental parameters that are expected to be suitable for a species; therefore, it would be more appropriate to consider that they predict *potential habitats* (Araujo and Guisan, 2006). Furthermore, the empirical niche modelling assumes that the modelled species is in equilibrium with its environment, in other words that the species is not in a phase of environmental range extension (Guisan and Thuiller, 2005). However, this cannot always be assumed, for example in the case of invasive species recently introduced.

Data for Niche Modelling

Niche modelling requires a calibration dataset including species data and environmental data. Species data may be presence/absence or abundance records of species over a number of observation sites. In ecological sciences, such records are classically obtained by field surveys over areas of interest. Herbarium information may also be used in addition or in replacement of field survey data. Data from collection campaigns of plant breeders may also be valuable information, although such information has seldom been taken into account for such purpose. Environmental data requested for niche modelling are values at observation sites of environmental parameters expected to be conditional for the presence of the species. Austin (2007) classified physical or bio-physical parameters in different categories. He distinguished indirect parameters (such as altitude or latitude) from direct parameters (such as rainfall, temperature, or solar radiation), and distal parameters (e.g. climatic parameters) from proximal parameters (i.e. involved in intimate ecophysiological processes). He pointed out that niche modelling is as more efficient as more direct and more proximal parameters are used. Other kinds of environmental parameters may be relevant, like soil texture and land-cover (Thuiller et al., 2004), or observation records of other species (competitors or facilitators). Databases of valuable environmental data are more and more easily accessible. Interpolated climatic data at fine resolution are now available for large areas, for example in the Worldclim database at 1 × 1 km resolution (http://www.worldclim.org). Satellite views may also be processed to obtain data for various parameters like global solar radiation or land-cover. Land-cover data for Europe are available in the CORINE land-cover database (http://www.eea.europa.eu/themes/landuse/clc-download). In plant breeder collections, passport data that document accessions may also provide valuable environmental information.

The spatial scale of the calibration dataset is determining for the relevance of species modelling (Guisan and Thuiller, 2005). The resolution of the observational network needs to be fine enough to capture the variation of important environmental parameters conditioning species presence. Furthermore, the extent of the observational area is important. If a species has been sampled in a limited part of its range of presence for an important environmental parameter, the model will only be valid within this domain of variation of the environmental parameter.

Statistical Modelling

Many methods have been proposed for statistical niche modelling. We briefly present two different methods that can be used in the case of presence-absence observations of species. Assuming that the presence and absence of several species have been recorded in a sample of observation sites, the calibration data may be organised in two tables. One table describes the occurrences of species; it has as many lines as observation sites, and each column reports occurrences of a species, with 1 if species is present, 0 otherwise. The other table includes environmental parameter values, each line referring to an observation site and each column to an environmental parameter.

A Multivariate Method: The Canonical Correlation Analysis

With such two tables, it is possible to use the Canonical Correlation Analysis (CANCOR), a multivariable method that handles collinearity between variables (Hotelling, 1936). CANCOR looks for successive pairs of linear combinations of two sets of variables Y and X that are maximally correlated. For each set of variables, successive linear combinations found by the method (canonical variates) are uncorrelated by construction. Regarding our concern, Y and X variables are the columns of the tables of species occurrences and environmental values, respectively. To use the CANCOR analysis, the number of variables in both X and Y sets should be much lower than the number of observations (number of observation sites in this case). Canonical variates associated with X variables could then be regarded as environmental gradients along which species niches are best separated (Gimaret-Carpentier et al., 2003), and niche amplitude of species can be represented by standard deviation for canonical scores along the canonical axes (Carnes and Slade, 1982). The suitability of any site for a species can be predicted by the distance from this site to the barycentre of the species in the space of the major environmental canonical axes. In fact, a true niche modelling is not really possible with CANCOR, unless assuming an *a priori* distribution shape of the species response (classically unimodal and symmetric). Therefore, CANCOR is mainly interesting as a tool providing an overview of species distribution, for example in the case of several related species.

The Logistic Regression Modelling

Logistic regression is a case of Generalized Linear Models (GLM). In GLMs, the expectancy of a response variable is related to a linear combination of predictor variables by a link function. Different distributions of the response variable are possible, as well as different link functions (McCullagh and Nelder, 1997). The logistic regression frame, assuming a binomial distribution of the response variable and a logistic link, is convenient for modelling presence-absence events of a species. The response variable is the binary variable listing the presence and absence of the species, whereas the predictor variables are the environmental parameters. The logistic regression is then implemented to model the probability of presence of the species. As in classical linear regression, collinearity between predictors and over-parametrisation should be avoided. A preliminary multivariate approach, like the CANCOR analysis, may help to select a limited number of no more than weakly correlated environmental predictors. Minimum numbers of observations and of presence events are requested for reliable modelling (Pearce and Ferrier, 2000a; Coudun and Gégout, 2006). Attention should be drawn to the fact that absence observations are requested to perform a logistic regression. Alternatively, techniques generating pseudo-absence observations have been proposed (Zaniewski et al., 2002). Selection of predictors in the final model may be performed by a classical stepwise process. However, more efficient and reliable methods can be used (Wintle et al., 2003; Johnson and Omland, 2004). Finally, the predictive ability of the selected model has to be evaluated through its calibration and its discrimination ability (Pearce and Ferrier, 2000b). The calibration (or goodness of fit) can be assessed for example by the adjusted generalised coefficient of determination of Nagelkerke (1991). The discrimination ability is the ability of the model to discriminate between presence and absence events. It can be assessed by comparing predicted and actual events on the calibration dataset. A better practice is however to use different datasets for calibration and assessment of discrimination ability.

An Example: Modelling the Realized Niche of Fine-Leaved Fescue Taxa in France

We used data from a sample of 382 fine-leaved fescue populations collected over France in 1993 and 1994 by ACVF (French forage and turf plant breeder society) breeders and INRA (Fig. 3.1) to study the ecological niche of fine-leaved fescue taxa. At each collection site, all panicles of fine-leaved fescue plants were picked within a circle of radius 50 m without differentiating between taxa, and seeds from each site were bulked in a unique seed lot. Taxonomic observations were afterwards performed in a common garden for 36 plants grown from each seed lot; all seed lots apart from seven were recognized as including a single taxon, whereas the remaining seven seed lots included two taxa each. Therefore, the lists of collection sites of identified taxa were considered to provide a presence-absence observation dataset, assuming that taxa not collected at a site were absent. Climatic data were extracted from the Aurelhy database of Météo France providing interpolated maps at a 1×1 km resolution for temperature and rainfall parameters. Global solar radiation data were obtained from satellite images. Passport data recorded by collectors provided land-use and soil texture data. Figure 3.2 illustrates the results of a CANCOR analysis of species occurrences and climatic parameters. This analysis pointed to two main climatic gradients that best separate realized niches of taxa (or taxonomic clusters), a summer radiation gradient and a mid-season rainfall gradient, associated with the first and second canonical directions, respectively.

Logistic regression modelling was performed for taxa collected in more than 100 sites with summer radiation, mid-season rainfall, land-use and soil texture as potential predictors. Predictors entering the final model of taxa are given in Sampoux and Huyghe (2009). Figure 3.3 shows the marginal response curve of the probability of presence of two taxa along the summer radiation gradient for each soil texture. The expected response curve of a species along an environmental gradient is expected to be unimodal and to tend towards zero at each end of the species range. It can

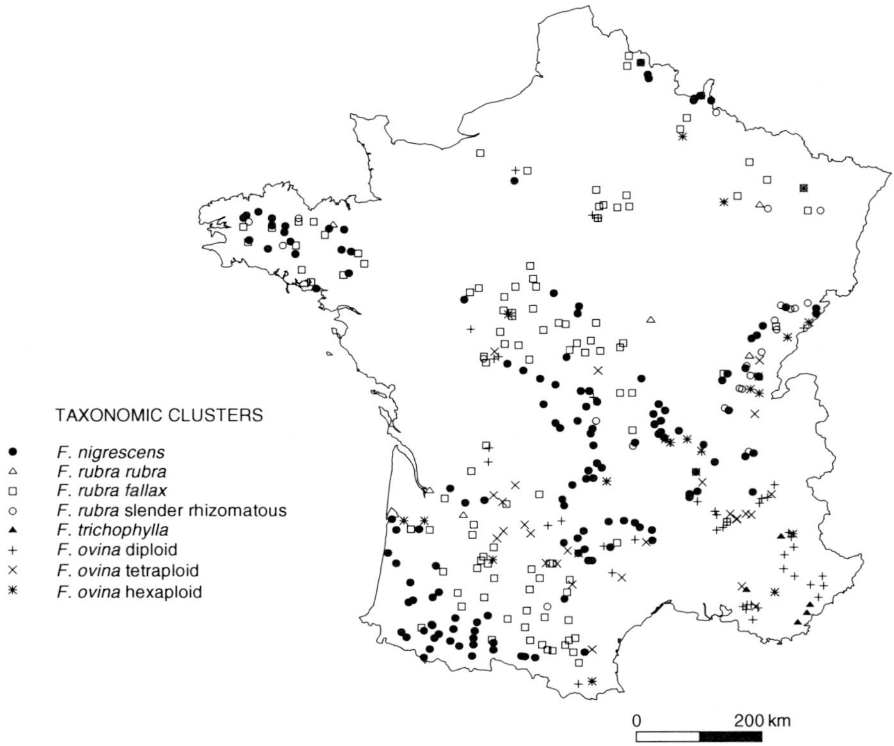

Fig. 3.1 Geographical distribution in France of 382 collection sites of fine-leaved fescue populations and of eight taxonomic clusters set up in the population sample

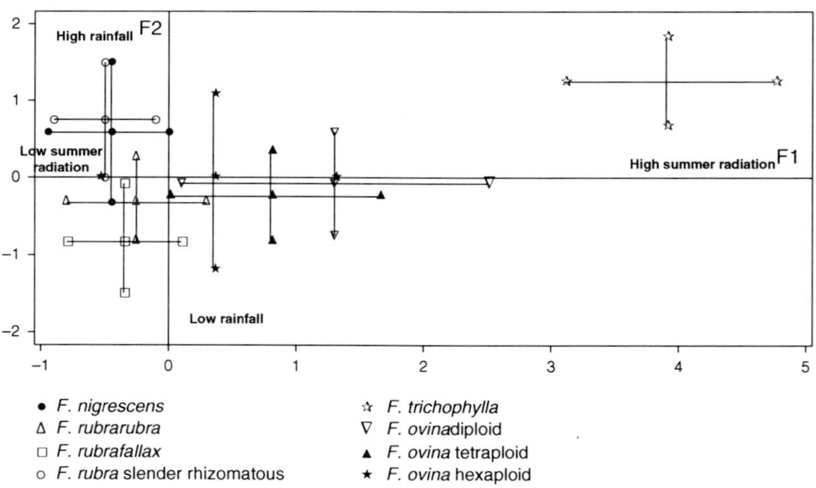

Fig. 3.2 CANCOR analysis between occurrences of taxonomic clusters of fine-leaved fescue populations and climatic parameters. Horizontal and vertical bars represent ± 1 SD for canonical scores

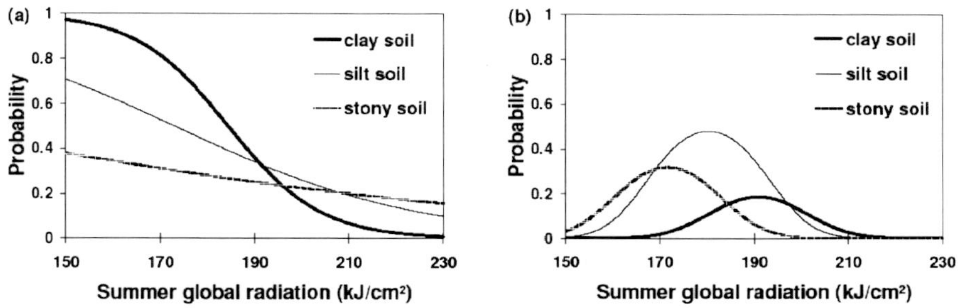

Fig. 3.3 Curves of the probability of presence of taxonomic clusters of fine-leaved fescue populations derived from logistic regression modelling: (**a**) *Festuca nigrescens*, (**b**) *Festuca rubra fallax*

be noted that the full range of summer radiation is present in the sampled areas of France for *Festuca rubra fallax* but not for *F. nigrescens*.

What to do with a Niche Model?

Contributing to Ecological Sciences

Empirical niche models are efficient tools to study the ecological distribution of species. Predicted distributions of species may be used to test hypotheses in ecology and biogeography. For example, the modelling of fine-leaved fescue taxa distribution previously mentioned was used to study the contribution of adaptive trait diversity and ploidy level variation to the diversity of realized niches of fine-leaved fescue taxa (Sampoux and Huyghe, 2009). This study evidenced that adaptive trait diversity contributes more than ploidy level variation in this respect. It also suggested that high ploidy level cytotypes may have been efficient colonisers and competitors, favouring expansion of the lineage and leading to their present prevalence.

Contributing to Genetic Resource Collection

Mapping Species Distribution

Niche models predicting species probability of presence may be built with environmental predictors available on fine resolution grids over large areas. In such cases, it is possible to predict the probability of presence of a species at each grid node. GIS softwares may then be used to map the species probability of presence over an area of interest. For example, we set up a logistic regression model for several fine-leaved fescue taxa using climatic parameters available on a 1×1 km grid as

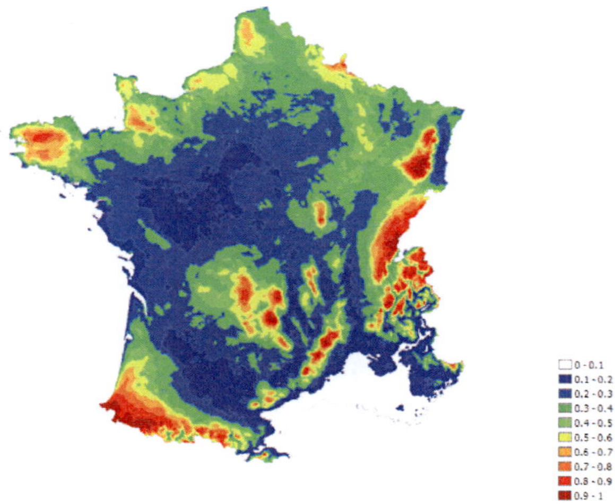

Fig. 3.4 Map of probability of presence of *Festuca nigrescens*

predictors. Maps of probability of presence of these taxa were then generated with GIS software (see for example Fig. 3.4). Such maps are expected to be valuable tools to guide future field campaigns aiming at completing the collection. However, such prediction maps should be used with caution when they are set up for a geographical region different from the region in which the calibration dataset has been sampled. In such cases, it would be worthwhile to cross-validate the model for the new region by comparing predictions with a sample of actual observations from this region.

Mapping Trait Distribution in Species

Because of natural selection within species, correlations do exist between trait variations and environmental parameter variations. As natural selection may lead to similar phenotypes in different areas with similar environmental conditions, these correlations can be used to predict trait value according to environmental parameter values. In the fine-leaved fescue collection, we evidenced a number of linear relationships between traits and climatic parameters. From these linear relationships, we set up linear regression models and derived the probability that a trait exceeds a given threshold at any location as a function of environmental parameter values. As this probability is by construction conditional to the presence of the species, its product with the species probability of presence yields the probability to find at a given location a population that exceeds the trait threshold. Mapping such probabilities with a GIS software may be expected to be a helpful mean to identify collection areas where to find some specific phenotypic features (see for example Fig. 3.5).

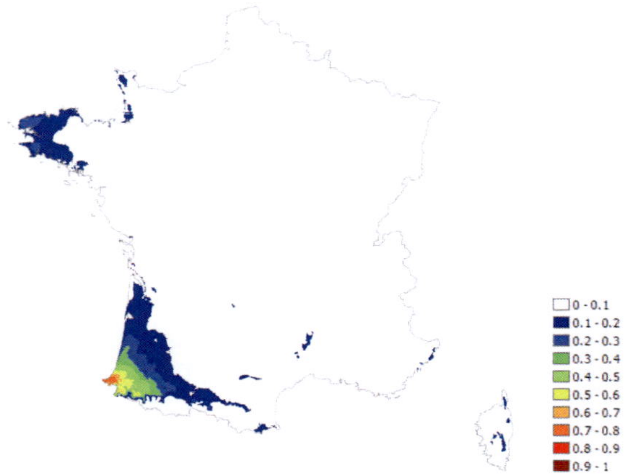

Fig. 3.5 Map of probability of presence of rust resistant *Festuca nigrescens* (predicted rust resistance in or above the best third of the rust resistance range of already collected *F. nigrescens* accessions)

Contributing to Core-Collection Selection in Genetic Resource Conservation Programmes

Ex situ collections with numerous accessions are difficult to handle and maintain in the long term. Frankel and Brown (1984) proposed the concept of core-collection which consists of identifying a sub-collection representing as best as possible the genetic diversity of a larger collection with a limited number of accessions. To set up a core-collection, Brown (1989a) suggested a random sampling of 10% of accessions in the base collection. He also suggested to stratify the base collection into clusters on the basis of political, geographical, or genetic information, and to randomly sample accessions in each cluster (Brown, 1989b). Schoen and Brown (1993) proposed methods aiming at maximising neutral diversity in the core-collection; however, such methods are efficient to sample selected diversity only if linkage disequilibrium exists between neutral and selected diversities and/or if gene flow between populations is low (Bataillon et al., 1996). The choice of a core-collection may also be based on phenotypic data. In this case, the base collection can be stratified into clusters, according to phenotypic distance between accessions; the core-collection is then set up by randomly selecting accessions in each cluster. Noirot et al. (1996) proposed a more direct approach, known as Principal Component Score (PCS) method, which consists of selecting accessions in order to maximise a criterion of phenotypic diversity in the sample of selected accessions. However, some adaptive differences between accessions may not be evidenced when accessions are evaluated in an experimental network including only a few locations. Therefore, maximising observed phenotypic diversity does not necessarily mean that genetic selected diversity is maximised. In order to take into account genetic relatedness between accessions in the selection of a core-collection, Balfourier et al. (1998) proposed several methods based on the use of spatial autocorrelation patterns of phenotypic traits, which are assumed to be caused by variation of environmental selection pressure and isolation by distance.

In the case of a collection of spontaneous populations sampled within a species over a large area, we suggest that species probabilities of presence predicted from a niche model provide additional information that may help to optimise the choice of accessions entering a core-collection.

GIS maps of predicted probability of species presence may indeed evidence spatially disconnected areas of presence. For example, the predicted distribution mapped on Fig. 3.4 shows several areas of potential presence that are geographically isolated by climatic contrasts. In such situations, we propose to stratify the base collection into groups corresponding to the predicted geographically isolated areas from which accessions originate, and to select accessions independently in each group. This practice would be a mean to ensure that genetic diversity is captured from several groups whose genetic differentiation is maintained by isolation by distance. Furthermore, isolation by distance is known to promote reproductive barriers between isolated groups. Noirot et al. (1996) emphasized that the PCS method should be applied separately in groups that do not interbreed; the stratification in geographically isolated groups that we propose could contribute to meet this requirement of the PCS method.

Furthermore, when selecting accessions within each geographically isolated group, it would be worthwhile to take into account species probabilities of presence that are predicted for collection sites. In sites where the species is predicted to have a poor probability of presence, the presence of the species is indeed expected to be rather unstable as small environmental changes are likely to make the site unsuitable for the species. Moreover, in areas where the main environmental factors driving a species presence are sub-optimal, the presence of the species becomes conditioned by the level of several secondary environmental parameters. For example, Fig. 3.3(a), (b) evidence that the presence of fine-leaved fescue taxa is conditioned by soil texture in the unfavourable part of the summer radiation range. As a consequence, a species may be difficult to find in areas where a main environmental parameter is sub-optimal, and collection is likely to be costly in such areas. Therefore, the selection of a core-collection should deliberately retain a part of the diversity collected in sites of poor predicted probability of species presence. With this aim, we suggest that the base collection, or each geographically isolated group of accessions, should be stratified in clusters according to predicted probabilities of species presence at collection sites. As the statistical model involved in predictions has always a reliability limited by the statistical method and the calibration dataset, no more than two or three clusters corresponding to probability intervals of same amplitude should be distinguished.

If trait diversity is distributed along gradients conditioning species presence, then clusters of species probability of presence are expected to partly structure trait diversity, although extreme low and high trait values should both correspond to poor species probabilities of presence. However, trait diversity may also be distributed along other environmental gradients than those conditioning species presence. In such a case, phenotypic variations are not expected to be associated with variations of species probability of presence, and large trait diversity is expected in each cluster of species probability of presence. As a consequence, a wise practice should be to select accessions in each cluster of species probability of presence on the basis of strategies maximising diversity in the core-collection, like for example the PCS method of Noirot et al. (1996) maximising phenotypic diversity.

Conclusion

In the case of collection of spontaneous populations of forage and turf species, the empirical modelling of species niche may be an interesting tool to improve collection strategies and core-collection selection. Niche modelling is facilitated by the development of a number of statistical methods adapted to a variety of situations and by the increasing availability of environmental databases. Furthermore, GIS softwares offer new opportunities to investigate spatial outcomes of species environmental distribution. Data from plant breeder collections are undoubtedly valuable information for niche modelling. Niche modelling is as more efficient as the calibration data cover better the range of environmental situations suitable for a species. In this respect, it could expected that gathering information from different collections at the European scale would be an efficient mean to set up relevant niche models for the improvement of genetic resource collection and conservation.

References

Araujo, M., Guisan, A. 2006. Five (or so) challenges for species distribution modelling. J. Biogeogr. 33:1677–1688.

Austin, M. 2007. Species distribution models and ecological theory: A critical assessment and some possible new approaches. Ecological Modelling 200:1–19.

Balfourier, F., Charmet, G. 1991. Relationships between agronomic characters and ecogeographical factors in a collection of French perennial ryegrass populations. Agronomie 11:645–657.

Balfourier, F., Charmet, G., Prosperi, J.M., Goulard, M., Monestiez, P. 1998. Comparison of different spatial strategies for sampling a core collection of natural populations of fodder crops. Genet. Sel. Evol. 30(Suppl. 1):S215–S235.

Bataillon, T.M., David, J.L., Schoen, D.J. 1996. Neutral genetic markers and conservation genetics: Simulated germplasm collections. Genetics 144:409–417.

Brown, A.D.H. 1989a. Core collections: a practical approach to genetic resources management. Genome 31:818–824.

Brown, A.D.H. 1989b. Size and structure of collection: the case for core collection. In: Hodgkin, T., Brown, A.D.H., Hintum, T.J.L., van Morales, E.A.V. (eds.), The Use of Plant Genetic Resources. John Wiley & Sons, Baffins Lane, Chichester, UK, pp. 136–156.

Carnes, B.A., Slade, N.A. 1982. Some comments on niche analysis in canonical space. Ecology 63:888–893.

Coudun, C., Gégout, J.C. 2006. The derivation of species response curves with Gaussian logistic regression is sensitive to sampling intensity and curve characteristics. Ecological Modelling 199:164–175.

Frankel, O.H., Brown, A.D.H. 1984. Current Plant Genetic Resources – A Critical Appraisal. In: Genetics New Frontiers, Proceedings of the 15th International Congress of Genetics (Vol. 4, pp. 3–13). Oxford and IBH Publishing Co.

Gimaret-Carpentier, C., Dray, S., Pascal, J. 2003. Broad-scale biodiversity pattern of the endemic tree flora of the Western Ghats (India) using canonical correlation analysis of herbarium records. Ecography 26:429–444.

Grinnell, J. 1917. Field tests of theories concerning distributional control. Am. Nat. 51:115–128.

Grinnell, J. 1924. Geography of evolution. Ecology 5:225–229.
Guisan, A., Thuiller, W. 2005. Predicting species ditribution: offering more than simple habitat models. Ecology Letters 8:993–1009.
Hotelling, H. 1936. Relations between two sets of varieties. Biometrika 28:321–377.
Hutchinson, G.E. 1957. Concluding remarks. Cold Spring Harb. Symp. Quant. Biol. 22:145–159.
Johnson, J.B., Omland, K.S. 2004.Model selection in ecology and evolution. Trends Ecol. Evol. 19:101–102.
Mansat, P. 1995. "Préface". In: Prospéri, J.M., Guy, P., Balfourier, F. (eds.), Ressources génétiques des plantes fourragères et à gazon (pp. 9–12). Paris: BRG-INRA.
McCullagh, P., Nelder, J.A. 1997. Generalized Linear Models. Monographs on Statistics and Applied Probability. Chapman & Hall, London.
Nagelkerke, N.J.D. 1991. A note on a general definition of the coefficient of determination. Biometrika 78:691–692.
Noirot, M., Hamon, S., Anthony, F. 1996. The principal component scoring: a new method of constituting a core collection using quantitative data. Genet. Res. Crop Evol. 43:1–6.
Pearce, J., Ferrier, S. 2000a. An evaluation of alternative algorithms for fitting species distribution models using logistic regression. Ecological Modelling 128:127–147.
Pearce, J., Ferrier, S. 2000b. Evaluating the predictive performance of habitat models developed using logistic regression. Ecological Modelling 133:225–245.
Prosperi, J.M., Jenczewski, E., Angevain, M., Ronfort, J. 2005. Morphologic and agronomic diversity of wild genetic resources of *Medicago sativa* L. collected in Spain. Genet. Res. Crop Evol. 53:843–856.
Sampoux, J.P., Huyghe, C. 2009. Contribution of ploidy-level variation and adaptive trait diversity to the environmental distribution of taxa in the 'fine-leaved fescue' lineage (genus *Festuca* subg. *Festuca*). Journal of Biogeography (36):1978–1993.
Schoen, D.J., Brown, A.D.H. 1993. Conservation of allelic richness in wild crop relatives is aided by assessment of genetic markers. Proc. Natl. Acad. Sci. USA 90:10623–10627.
Thuiller, W., Araújo, M.B., Lavorel, S. 2004. Do we need land-cover data to model species distributions in Europe? J. Biogeogr. 31:353–361.
Warren, J.M., Raybould, A.F., Ball, T., Gray, A.J., Hayward, M.D. 1998. Genetic structure in the perennial grasses *Lolium perenne* and *Agrostis curtisii*. Heredity 81:556–562.
Wintle, B.A., McCarthy, M.A., Volinsky, C.T., Kavanagh, R.P. 2003. The use of bayesian model averaging to better represent uncertainty in ecological models. Conserv. Biol. 17:1579–1590.
Zaniewski, A.E., Lehmann, A., Overton, J.M.C. 2002. Predicting species spatial distributions using presence-only data: a case study of native New Zealand ferns. Ecological Modelling 157: 261–280.

Chapter 4
The Genetic Diversity of Fine-Leaved Fescue (*Festuca* L.) Species in Lithuania

Rita Armonienė, Vaclovas Stukonis, Vanda Paplauskienė, and Gintaras Brazauskas

Abstract Fine-leaved fescue (*Festuca* L.) species are valuable breeding material with potential use for marginal land cover. However, the breeding of these species is complicated by morphological similarity among species and high heterogenity within species. Molecular DNA markers are needed to facilitate species identification. Inter simple sequence repeat polymerase chain reaction (ISSR-PCR) was used for the genetic analysis of the five fine-leaved fescue species encountered in Lithuania, namely *Festuca sabulosa*, *Festuca polesica, Festuca ovina*, *Festuca trachyphylla* and *Festuca psammophila*. Fifty six ISSR markers were scored for 21 fescue accessions. Similarity indicies were calculated and UPGMA dendrogram constructed with NTSYSpc 2.2. Three species (*F. ovina*, *F. trachyphylla* and *F. psammophila*) could be distinguished as separate clusters in the dendrogram. Further analysis revealed two ISSR fragments of 600 and 950 bp to be species-specific for *F. psammophila*. These fragments are potential targets for more specific SCAR marker development. The remaining two highly related species (*F. sabulosa* and *F. polesica*) formed one intermixed cluster and could not be distinguished from each other. The results of this work show the suitability of the ISSR method to investigate genetic diversity of fine-leaved fescue species encountered in Lithuania and provide valuable data for the identification of these species.

Keywords ISSR-PCR · Genetic diversity · *Festuca*

Introduction

With over 450 species, *Festuca* L. is one of the largest genera within the *Poaceae* family found in temperate regions throughout the world and fifteen species are encountered in Lithuania (Stukonis and Bednarska, 2007). This large and

G. Brazauskas (✉)
Lithuanian Institute of Agriculture, Instituto a.1, Akademija, Kėdainiai r. LT-58344, Lithuania
e-mail: gintaras@lzi.lt

C. Huyghe (ed.), *Sustainable Use of Genetic Diversity in Forage and Turf Breeding*,
DOI 10.1007/978-90-481-8706-5_4, © Springer Science+Business Media B.V. 2010

variable genus includes nine subgenera and *Festuca* subgenus *Festuca* is the largest, most taxonomically difficult subgenus in *Festuca*. This subgenus also contains the collection of economically important turf species known as the 'fine-leaved fescues' characterized by their fine to very narrow leaves, usually less than 1 mm in width. Fine-leaved fescue species are valuable breeding material with potential use for marginal land cover and prevention of soil erosion. They are adapted to infertile, acid soils, limited moisture and low-maintenance. Watschke (1990) noted their superior advantage for use as low-maintenance roadside cover, while maintaining an attractive aspect.

There is a remarkable morphological similarity among the taxa of *Festuca*, caused to a great extent by the morphological variability of particular characters. Identification of species often depends on the disposition of sclerenchyma strands in the transverse section of a leaf blade. However, based on morphological characters alone, interpretation of many taxa is very problematic and sometimes nearly impossible (Šmarda et al., 2005). Molecular DNA markers are needed to facilitate identification of these polymorphous species.

Inter-simple sequence repeats (ISSR) have become markers of choice for non-model plant species genotyping. Being a simple and quick PCR-based method, ISSRs are informative for numerous loci and are suitable to discriminate closely related genotype variants (Forster et al., 2004). ISSRs have been successfully used to estimate the extent of genetic diversity at the inter- and intra-specific level in a wide range of crop species such as rice, wheat, fingermillet, sweet potato, and others (Reddy et al., 2002). ISSR fingerprinting was useful for developing DNA markers within the *Lolium/Festuca* complex as a tool for marker-assisted selection and for genotype identification (Pašakinskienė et al., 2000).

In this study, ISSR markers were applied for genetic diversity evaluation among 21 accessions from five fine-leaved fescue species encountered in Lithuania.

Material and Methods

Plant Material

Seeds of 21 fine-leaved fescue accessions were collected from various natural growing locations in Lithuania during the period of 1999–2006. Five species were included: *F. sabulosa* (Anderson) H. Lindb., *F. polesica* Zapał, *F. ovina* L., *F. trachyphylla* (Hack.) Krajina and *F. psammophila* (Hack. ex Čelak.) Fritsch. The seeds were sown in experimental field in Dotnuva. DNA samples were isolated from the mixture of 30 plants per accession. DNA was extracted from young leaves following the CTAB-based extraction protocol (Doyle and Doyle, 1990).

ISSR-PCR Analyses

ISSR – PCR were carried out in a volume of 15 μl containing 50 ng of genomic DNA, 1 × PCR reaction buffer, 2 mM $MgCl_2$, 0.2 mM dNTP, 0.2 μM of each primer and 0.3 units of DyNAzymeTM II DNA Polymerase (FINNZYMES, Finland).

PCRs were carried out in the GenAmp PCR System 2700 Thermocycler (Applied Biosystems, USA). The thermal profile for ISSR- PCR was as follows: 95°C initial denaturation for 2 min, then 40 cycles of 95°C for 30 s, 50°C for 1 min and 72°C for 30 s. A final extension step of 6 min at 72°C was followed by 10°C. ISSR amplification products were separated by gel electrophoresis in 1.5% agarose gels, stained with ethidium bromide and vizualized by UV –light. GeneRuler TMDNA Ladder Mix (MBI Fermentas) was used to determine the size of the DNA fragmens.

Data Analysis

Matrix of similarity indicies (Lynch, 1990) for each pair of 21 accessions was generated for ISSR marker presence or absence data and UPGMA dendrogram constructed. It was assumed that similarity of fragment size was an indicator of homology. All calculations were performed with NTSYSpc v.2.2.

Results and Discussion

Twenty three primers with di-, tri- and tetranucleotide repeats were tested for their capacity to differentiate among 21 fine-leaved fescue accessions. Six primers with dinucleotide repeats showed high level of polymorphism among accessions, gave reproducible banding patterns and were subsequently chosen for further analysis. The number of ISSR bands scored per primer varied from 6 (primer UBC 827) to 13 (primer UBC 822) (Table 4.1). An average of 9.5 bands was obtained per primer within size range from 410 to 1150 bp. All primers produced polymorphic fragments with a degree of polymorphism ranging from 83 to 100%.

Fifty seven DNA fragments were obtained with the six chosen primers. The majority of the fragments were polymorphic among fine-fescue accessions under study (Table 4.1). This high level of polymorphism detected shows genetic heterogeneity among species investigated. Being outcrossing species, high level of within-population heterogeneity is expected. The mixture of 30 plants per accession was used for DNA extractions in order to minimise the sampling

Table 4.1 Primers used for ISSR analysis of fine-leaved fescue species, number of identified ISSR bands, their size intervals

Primer	Sequence (5′ to 3′)	Total bands	Polymorphic bands	Polymorphic bands (%)	Size range of DNA fragments (bp)
UBC 822	$(TC)_8A$	13	13	100	500–1120
UBC 823	$(TC)_8C$	11	11	100	550–1150
UBC 824	$(TC)_8G$	9	9	100	410–1100
UBC 825	$(AC)_8T$	8	8	100	550–1100
UBC 827	$(AC)_8G$	6	5	83	580–900
UBC 857	$(AC)_8YG$	10	10	100	510–1020
Total/average:		57	56	98	410–1150

variance. All six primers under study had dinucleotide repeats either 'TC' or 'AC'. Dinucleotide repeats in plant systems are more abundant (70%) than other microsatellite patterns (Varshney et al., 2000). Primers with 'AC' repeats show high polymorphism also in other species, such as wheat (Nagaoka and Ogihara, 1997), potato (McGregor et al., 2000) and clover (Dalla Rizza et al., 2007).

The generated UPGMA dendrogram of all 21 accessions showed four clusters (Fig. 4.1). Two species *F. sabulosa* and *F. polesica* were separated from the other clusters, however, intermixed. These two closely related taxonomic groups sometimes are treated as one species (Markgraf-Dannenberg, 1980). The remaining three species *F. trachyphylla, F. ovina* and *F. psammophila* were in separate clusters. The geographic distances between sites where populations were collected were also reflected in the dendrogram. For example, four populations of *F. trachyphylla* found in the central part of Lithuania (TRA-14, TRA-20, TRA-23, TRA-27), could be easily differentiated from populations coming from distant regions in Lithuania: from Juodkrantė (TRA-26), Biržai (TRA-16) and Marcinkonys (TRA-96), western, northern and southern parts of Lithuania respectively.

The ultimate task of this work was to identify potential diagnostic markers for fine-leaved fescue species identification. Of the 57 ISSR fragments scored two bands were identified as potential targets for SCAR marker development. Species-specific fragments of 600 and 950 bp were obtained from *F. psammophila.* Further analysis is needed to validate the suitability of these markers in fine-leaved fescue species identification.

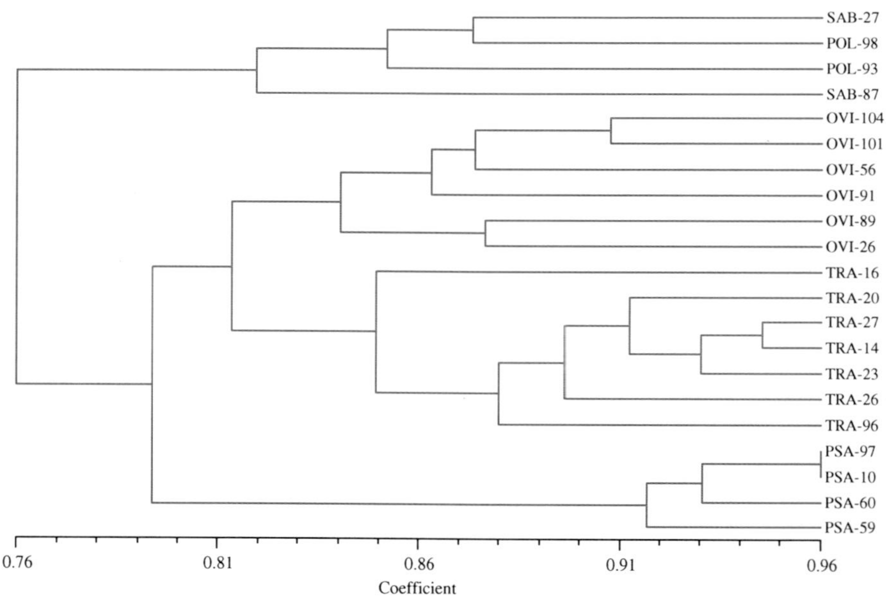

Fig. 4.1 UPGMA dendrogram showing clustering of 2 *F. sabulosa* (SAB) 2 *F. polesica* (POL), 6 *F. ovina* (OVI), 7 *F. trachiphilla* (TRA) and 4 *F. psammophila* (PSA) populations

Acknowledgements This work was partly supported by the Lithuanian Science Council Student Research Fellowship Award (RA) and by the Lithuanian State Science and Studies Foundation (grant N21/2008).

References

Dalla Rizza, M., Real, D., Reyno, R., Porro, V., Buergueno, J., Ericco, E., Quesenberry, K.H. 2007. Genetic diversity and DNA content of three south American and three eurasiatic *Trifolium* species. Genet. Mol. Biol. 30(4):1118–1124.

Doyle, J.J., Doyle, J.L. 1990. Isolation of plant DNA from fresh tissue. Focus 12:13–15.

Forster, J.W., Jones, E.S., Batley, J., Smith, K.F. 2004. Molecular marker based analysis of pasture and turf grasses (pp. 197–238). In: Hopkins, A., Wang, Z.Y., Mian, R., Sledge, M., Barker, R.E. (eds.), Molecular Breeding of Forage and Turf. Kluwer Academic Press, The Netherlands.

Lynch, M. 1990. The similarity index and DNA fingerprinting. Mol. Biol. Evol. 7:478–484.

Markgraf-Dannenberg, I. 1980. *Festuca* L./Flora Europaea (p. 145,147,153). Cambridge, University Press, Cambridge.

McGregor, C.E., Lambert, C.A., Greyling, M.M., Louw, J.H., Warnich, L. 2000. A comparative assessment of DNA fingerprinting techniques (RAPD, ISSR, AFLP and SSR) in tetraploid potato (*Solanum tuberosum* L) germplasm. Euphytica 113:135–144.

Nagaoka, T., Ogihara, Y. 1997. Applicability of inter-simple sequence repeat polymorphisms in wheat for use as DNA markers in comparison to RFLP and RAPD markers. Theor. Appl. Genet. 94:597–602.

Pašakinskienė, I., Griffiths, C.M., Bettany, A.J.E., Paplauskienė, V., Humphreys, M.W. 2000. Theor. Appl. Genet. 100:384–39.

Reddy, M.P., Sarla, N., Siddiq, E.A. 2002. Inter simple sequence repeat (ISSR) polymorphism and its application in plant breeding. Euphytica 128:9–17.

Šmarda, P., Müller, J., Vrna, J., Kočķ, K. 2005. Ploidy level variability of some Central European fescues (*Festuca* subg. *Festuca*, *Poaceae*). Biologia 60(1):25–36.

Stukonis, V., Bednarska, I. 2007. *Festuca pseudovina* in Lithuania. Botanica Lithuanica 13(1): 13–18.

Varshney, R., Kumar, A., Balyan, H., Roy, J., Prasad, M., Gupta, P. 2000. Characterization of microsatellites and development of chromosome specific STMS markers in bread wheat. Plant Mol. Biol. Rep. 18:299–300.

Watschke, T.L. 1990. Low-maintenance grasses for highway roadsides. Grounds Maintenance 25(8):40–41.

Chapter 5
Plant Collection in Kazakhstan and Azerbaijan for Forage Improvement in Australia

Geoff Auricht, Steve Hughes, Alan Humphries, and Eric Hall

Abstract Plant germplasm, totalling 1,344 accessions, was collected during the course of two Australian led seed collection missions in Kazakhstan in 2002 and Azerbaijan in 2004. The collection in Kazakhstan involved members from Australia, China, the Vavilov Institute in Russia and the Aral Sea Experiment Station in Kazakhstan while the mission to Azerbaijan involved scientists from Australia, ICARDA and the Azerbaijan National Academy of Sciences.

Samples collected covered 54 genera and 144 species. A sub-sample of seed of 1,170 of these accessions was brought to Australia and indexed into the South Australian Genetic Resource Centre collection. Perennial pasture legumes were targeted on both missions and 725 samples were collected. Wild ecotypes of lucerne, particularly *Medicago sativa* subsp. *caerulea* and subsp. *falcata*, with associated rhizobia, were collected at 100 sites across the two countries from the mountains to the deserts. Soils ranged from clays to sands with pH from 5.8 to 9.5. The incidence of lucerne decreased at acidic sites as expected. The perennial clover, *Trifolium tumens*, was collected at 36 sites, in Azerbaijan only, quadrupling previous *ex-situ* conservation of this species. The range of adaptation and frequency was far greater than expected and increased drought and grazing tolerance is expected above that in current germplasm holdings. A total of 180 grass accessions, predominantly perennials, were collected, including *Dactylis glomerata* and *Lolium perenne*. Wild cereal relatives collected included ancient and wild wheat (*Aegilops* and *Triticum*) and barley (*Hordeum*) while herb, pulse and oil seed crops included *Cichorium*, *Brassica* and *Vicia*. SARDI has since conducted characterisation, preliminary evaluation and seed multiplication activities on 617 of the accessions including the perennial *Medicago* accessions while TIAR has characterised a further 35 *T. tumens* accessions. Based on the information collected, selected wild lucernes are now in field trials across southern Australia to identify drought tolerant perennials while a *T. tumens* cultivar is being commercialised now in Tasmania based on the accessions collected.

G. Auricht (✉)
South Australian Research and Development Institute, Urrbrae, SA, Australia; CRC for Future Farm Industries, Crawley, WA, Australia
e-mail: auricht.geoff@saugov.sa.gov.au

C. Huyghe (ed.), *Sustainable Use of Genetic Diversity in Forage and Turf Breeding*,
DOI 10.1007/978-90-481-8706-5_5,

In all over 1000 seed samples from 500 of the accessions have been distributed to scientists in Australia and six other countries.

Keywords Forage · Germplasm · Azerbaijan · Kazakhstan · *Medicago sativa* · *Trifolium tumens*

Introduction

Australian agriculture relies heavily on introduced species of forage and crop plants. Pasture legumes, the focus of these collection missions, are used for fodder production, direct grazing and to benefit the crop rotation by providing nitrogen and a break for cereal diseases and weeds. Traditional annual pastures include annual medics, such as: barrel medic, *Medicago truncatula*; strand medic, *M. littoralis* and spineless burr medic, *M. polymorpha*. They also include annual clovers, such as: subterraneum clover, *Trifolium subterraneum*; balansa clover, *T. michelianum* and Persian clover, *T. resupinatum*. Australia also has a long tradition of lucerne, *M. sativa* production. However, current varieties and traditional species have proven inadequate for the full range of environments and farming systems experienced in Australia, particularly in dry seasons, in low rainfall areas and in poor soils. Greater plant diversity is required, both in existing crops and forages and spanning prospective new species (Hughes et al., 2008). Pasture genebanks are looked to, to provide this, either from existing stocks or through new collection missions. Australia has temperate forage germplasm collections in the States of South and Western Australia and a small number of collaborating forage scientists across five Australian States. The SARDI forage collection located in Adelaide, South Australia, comprises over 40,000 accessions gathered from over 100 countries over the past 50 years. This collection has been growing and diversifying to meet the requirements of plant breeders for both new species and new variation of established species, with an increasing emphasis in recent years on Australian native plants with prospects for use in agriculture or natural resource management. Prior to 2004 however, the collection had little representation from the broad genetic diversity of pasture plants available in Kazakhstan and Azerbaijan.

Collection missions were undertaken in these two countries in 2002 and 2004, with the aim of collecting seed of a broad array of forage legume and grass species, with a strong emphasis on perennials, particularly lucerne and on stress tolerance traits such as acid soil and drought tolerance. Soil samples were also collected at many sites for the isolation of legume *Rhizobium*. Following these collection missions SARDI has introduced the new germplasm through quarantine and grown many of the accessions to record characterisation data and produce fresh seed. This seed is now safely conserved in the SARDI cold stores, at –20°C and 2°C and is in active use, with over 1000 small samples being dispatched to date to recipients in Australia and overseas.

Many of the samples collected are of great interest to Australian pasture scientists, including the authors, for a variety of projects, including several supported by the Future Farm Industries Cooperative Research Centre (FFI CRC). These projects have a particular focus on improving perennial forages as these can provide drought hardiness, extended forage production and improved soil cover. Central Asia was targeted as it is an area of great environmental and genetic diversity and a centre of origin for lucerne, including wild relatives, *Medicago sativa* subsp. *falcata* and the diploid ($2n = 16$) *M. sativa* subsp. *caerulea*. This wild relative of traditionally grown tetraploid lucerne is expected to contain genes useful for stresses such as water-logging, aluminium, drought and grazing tolerance.

Materials and Methods

Logistics

Transport for the collections comprised two large, Russian built, four-drive vehicles on each trip. Accommodation was generally in tents with the teams camping out in the field along the collection route. As well as scientific staff, both missions had support staff of two drivers and a cook. Water was taken daily from wells in villages along the route and perishable food supplies were purchased fresh in village markets wherever possible.

Personnel and Routes

Kazakhstan

This mission involved the following participants:

1. Australia: Mr Geoff Auricht, Mr Eric Hall and Mr Steve Hughes.
2. Russia: Vavilov Research Institute. Dr Nickolai Dzyubenko and Dr Sergey Shuvalov.
3. Kazakhstan: Aral Sea Research Station. Dr Auskhan Khusainov and support staff.
4. China: Beijing Forestry University. Professor Lu Xinshi.

The collection in eastern Kazakhstan ran for three weeks from the 5th to the 25th August 2002. The collection route included parts of the Old Silk Route, between Europe and Asia, in a loop north and west of Almaty, near the border with China and running up to near the border with Russia then across to the east before turning back down south to Almaty.

Azerbaijan

Personnel on this mission were:

1. Australia: Mr Steve Hughes, Mr Alan Humphries and Mr Eric Hall.
2. ICARDA, Syria: Dr Ali Ismail.
3. Azerbaijan: Dr Zeynal Akparov, Mr Azer Hasanov and Dr Aliyar Ibragimov and support staff.

This collection was undertaken from 30th June to July 19th 2004. The route covered over 2,700 km and encompassed most of the country with the exception of the Naxcivan province and the Nagorono-Karabakh area of conflict (see Fig. 5.1).

Fig. 5.1 Azerbaijan collection mission route

Collection Sites and Passport Data

A wide diversity of sites was targeted on the collection missions ranging from plains to mountains and from moist, fertile soils to deserts and salt affected areas. Sites sampled spanned many different geographical regions, including mountain passes and meadows, irrigated and non irrigated crops and grasslands, sandy beaches, arid deserts and sodic, wet or saline soils.

Collection sites were chosen as the ecogeography or land use changed at intervals from 10–50 km. Site elevation ranged from 328 to 1 827 m in Kazakhstan and from minus 36 to 1,700 m in Azerbaijan. Detailed site information was recorded including latitude, longitude, altitude and information concerning the soils or terrain, the status of the plant populations and any pressure such as grazing or insects present. Soil surfaces of collection sites were often clays but also included a full range from non-wetting sands, through sandy clays, to loams, clay loams, highly dispersive clay and gravely to very stony soils. Solonetz sites, particularly around or near waterways had extremely high concentrations of alkali and salt as determined with an EC meter. The highest level recorded was > 19 dS/m and *Aeluropus* was collected at this site. Soil pH was measured using CSIRO developed test kits and ranged from 4.5 to over 10 with the majority neutral to alkaline. Grazing pressure of the sites ranged from nil or cut to very heavy. It was generally far heavier in Azerbaijan than Kazakhstan.

Local weather data was determined after each trip, from long-term meteorological records.

Sampling

Collection sites were chosen at intervals that ranged from 20–50 km apart. Wherever possible, seeds from at least 30 plants of the target species were collected at each site where they occurred. In addition to seed, soil samples were obtained from the base of leguminous plants for rhizobial extraction by SARDI rhizobiologists.

Plant Characterisation and Preliminary Evaluation

The South Australian Genetic Resource Centre, SAGRC, has traditionally grown out seed samples from collection missions, in or near Adelaide with up to 1,000 or more accessions grown in a given year. This serves two purposes – generating larger quantities of fresh seed and allowing the recording of a wide range of plant characteristics including a preliminary assessment of performance against standard check varieties according to established protocols (Auricht et al., 1999). Priority species from the collection missions in Kazakhstan and Azerbaijan have been grown in this way and this is summarised in the results.

Results

Samples Collected

Table 5.1 shows the plant types and numbers of accessions collected and indexed into the SA GRC and the number subsequently grown for characterisation.

Kazakhstan

670 samples were collected including 498 forage legumes and 140 forage grass accessions along with small numbers of grain and horticultural crops. The majority

Table 5.1 Number of samples collected in Azerbaijan and Kazakhstan and those grown for characterisation as at June 2009, by plant groups

Plant Groups	Collected	Characterised*
Perennial Legumes	725	524
Annual Legumes	181	44
Perennial Grasses	172	38
Perennial Herbs	48	44
Annual Herbs	20	
Shrubs	13	2
Annual Grasses	9	
Trees	1	
TOTAL	1,170	652

*Samples were characterised in South Australia other than 35 of the perennial herbs, specifically the *T. tumens* accessions, which were characterised in Tasmania

of the legume accessions were wild ecotypes of lucerne, *Medicago sativa* and *Melilotus* species, targeted for salt tolerance. The main forage grass species were *Agropyron cristatum, Bromus inermis, Dactylis altaica* (a progenitor of *D. glomerata*), and *Elymus junceus*. 42 soil samples were obtained from the base of leguminous plants and two Rhizobium bearing nodules from the roots of plants were also obtained.

Forage legumes were not widely cultivated in any of the regions visited. However, around Almaty, we did observe irrigated seed crops of lucerne, most likely US varieties. We also observed crops of safflower, wheat and vegetable crops. We frequently observed the mowing of wild pastures including *Medicago, Trifolium, Onobrychis, Glychyrizza, Lathyrus, Vicia* and other forage grass and legume species with heavily laden hay carts a feature on many rural roads.

Azerbaijan

The mission yielded 674 accessions predominantly perennial pasture legumes. The majority of these were perennial medics, including: diploid wild ecotypes of lucerne, *Medicago sativa* subsp. *caerulea*; tetraploid lucerne, *M. sativa* subsp. *sativa*; glandular lucerne, *M. sativa* subsp. *glomerata* and variable lucerne, *M. sativa* subsp. *varia*. Perennial clovers were also well represented including *Trifolium tumens, T. pratense, T. repens, T. fragiferum, T. ambiguum* and *T. canescens*. Other perennial legume genera collected of interest to Australian breeding programs include *Lotus, Onobrychis, Melilotus, Astragalus, Dorycnium* and *Glycyrrhiza*. Annual forage or fodder legumes collected covered 12 genera and 54 species, predominately *Trifolium, Medicago, Vicia, Lathyrus* and *Trigonella*. The most frequently sighted annual species, *Medicago minima*, was not collected at all sites. Important forage grass species included *Dactylis altaica, Lolium perenne, Agropyron cristatum, Cynodon dactylon, Festuca arundinacea, Phalaris arundinacea* and an extremely salt tolerant grass, *Aeluropus littoralis*.

Grain crops collected included: seven species of ancient wheat relatives, *Aegilops* sp.; two species of wild wheat, *Triticum*; oats, *Avena sativa*; four species of barley, *Hordeum* including the perennial species *H. bulbosum* and a close relative *Taeniatherum caput-medusae*. Pulse and oil seed crops collected included *Brassica*, *Lens*, *Pisum*, *Vicia* and *Lathyrus*. This material is of major interest to ICARDA and Australian breeding programs including the Winter Cereal and Temperate Field Crop GRC's in Tamworth and Horsham. Collection details have been reported to these Centres and seed, currently being held by ICARDA, will be made available upon request.

Legumes, especially tetraploid lucerne, were widely cultivated in Azerbaijan, especially on the central flats under irrigation but also in mountainous meadows. Also observed were crops of wheat, barley, chickpea, lentil, cotton, rice, sunflower, tobacco and numerous vegetable and orchard crops. The cutting of wild pastures comprising *Medicago, Trifolium, Onobrychis, Lathyrus, Vicia* and other forage grass and legume species was frequently observed.

The grazing animals observed in order of magnitude were cattle, sheep, horses and goats. Grazing pressure was generally very high and a large proportion of the country is being overgrazed and colonised by non-palatable species notably *Alhagi pseudoalhagi*. Grasshoppers were in very high numbers, especially in the drier sites, limiting seed available for collection.

Sample Sharing

Seed samples were shared between participating countries and forwarded to mandated Australian Genetic Resource Centres. Due to Australian Quarantine restrictions on the importation of potential weedy species, accessions were also sent to the International Centre for Agricultural Research in Dry Areas (ICARDA) for cleaning, taxonomic confirmation and seed multiplication before possible future importation into the SAGRC. As a result 1,170 of the 1,344 accessions collected were brought directly back to Australia and have now been indexed into the SA GRC. The soil and nodule samples collected have been incorporated into the SARDI Rhizobiology Group collection and will be farmed for matching *Rhizobium*.

Characterisation

From the time of collection to mid 2009, a total of 652 or just over half of the collected lines have been grown in South Australia or Tasmania for characterisation and preliminary evaluation. Table 5.1 shows the breakdown of lines characterised by plant groups and shows the strong emphasis on growing the perennial legumes for fresh seed and a first look at their characteristics. The results, for the lucerne accessions, have been published by Humphries and Hughes (2006). The results for the *T. tumens* lines assessed in Tasmania are given in Table 5.2.

Table 5.2 Site and Characterisation data for *Trifolium tumens*

Region & (#sites)	Altitude m	Rainfall mm	Site pH	Winter vigour 0–9 best	Summer growth 0–9 best	Days to first flowering	Seed yield gms/plant	Field performance 0–10 best
Masali (1)	20	700	6.5	5	4	143	2.6	4.4
Lankaran (2)	10–25	1200	5.8–6.5	4–6	3–4	136–139	3.4–4	4.4–4.5
Astara (1)	140	1200	5.8	4	3	140	1.5	2.8
Lerik (6)	320–1350	500 – 1200	5–8	5–8	6–9	142–161	2.7–7.9	5.7–8.7
Yardimir (1)	650	300	7.5	8	7	150	4.2	4.2
Celibad (1)	540	450–500	5	5	3	153	1.5	6.3
Ganca (2)	820–1300	500	5–7.5	3–6	4–7	172	0.9–2.9	3.4–5
Daskasan (3)	1220–1700	800	5.8–7	4	4–5	174–177	1.4–3.6	4.3–4.9
Tovuz (1)	680	500	7	7	7	149	7.3	2.9
Sheki (2)	1050	1300	7.5	6–7	6–7	137–147	7.3–8.6	3.9–4.6
Oguz (2)	430–540	800	6.5–8	4–6	5	136–143	4.4–5.1	2.6–3.9
Qobalb (1)	800	800	6.5	5	4	127	1.8	4.1
Ismayilli (1)	760	800	7.5	7	6	145	3.6	6.6
Agsu (1)	745	800	6.5	7	7	146	5.1	5.7
Samaxi (2)	550–1090	800	6	7	6–8	137–141	2.4–6.4	5.9–7.8
Davaci (2)	130–530	300–500	7.5–8	6–8	5–7	133–143	2.7–2.8	5.1–5.4
Quba (3)	800–1250	500–800	6–8.5	4–6	6–7	150–165	1–3.3	4.1–6.4
Xacmax (3)	10–370	300–500	6.5–9	4–7	1–4	139–144	0.5–2	2.5–5.7
Cv. 'Permatas'				8	7	151	6.1	5.9

Distribution

Small seed samples, of generally less than 1 g, have been distributed to scientists across five States of Australia and six overseas countries namely; Argentina, Canada, China, France, New Zealand and USA. In total, 586 of the accessions have been distributed on at least one occasion with some accessed up to 9 times. This is particularly in the case of sets of perennials being evaluated at multiple sites across southern Australia. This leads to a total of 1,083 seed samples from accessions collected on the two missions being distributed from the genebank. Table 5.3 summarises this activity.

Table 5.3 Accessions from both countries distributed to June 2009

Plant Groups	Unique Accessions	Total Accessions
Perennial Legumes	418	834
Perennial Grasses	100	142
Annual Legumes	56	60
Perennial Herbs	7	42
Annual Herbs	4	4
Shrubs	1	1
TOTAL	586	1,083

Genera of greatest interest have been the perennial legumes from the genera *Medicago, Trifolium* and *Lotus* along with some of the accessions of *Astragalus, Latyhyrus, Melilotus, Onobrychis, Trigonella*. Also of interest to users have been the grasses *Agropyron, Bromus, Dactylis, Festuca* and *Lolium*. Not yet accessed are the annual *Medicago* accessions, the *Hedysarum* and a number of the *Melilotus* accessions reflecting the emphasis of research programs to seek new perennials rather than annuals or the biennial *Hedysarum*.

Discussion

Notes on Species of Interest

Medicago sativa (Lucerne)

Central Asia is considered the primary centre of origin for lucerne however, prior to these collections relatively few accessions from this region could be found in *ex situ* germplasm collections. A total of 157 accessions of *Medicago sativa* was collected on the two missions. Material collected ranged from cultivated tetraploid lucerne right though to its wild relatives including *M. sativa* subsp. *falcata* and the diploid species, *Medicago sativa* subsp. *caerulea*.

Lucerne is the most important perennial forage legume in Australia where it is grown under irrigation or rain fed conditions. The development of new varieties to extend the boundaries into the cropping country is an important objective of the

SARDI lucerne breeding program. Attributes sought include greater production and persistence and improved tolerance to acidity, salinity, water-logging, cold, drought and grazing. It is anticipated that many of these attributes, especially improved water-logging, drought and grazing tolerance will be found among the lucernes collected.

Trifolium tumens

T. tumens, another target species and was collected at 36 sites in Azerbaijan only quadrupling previous *ex situ* collections of this species. The range of adaptation and frequency was much greater than expected and the collection team are confident that increased drought and grazing tolerance has been collected above that which exists in current germplasm holdings. The species is very drought and grazing tolerant, summer active, with some winter activity, deep rooting, productive and long lived. It has been very promising in Tasmanian Institute of Agricultural Research (TIAR) trials and a cultivar 'Permatas' has been developed by TIAR and is now being commercialised. Table 5.2 shows the range of sites and the range in performance of the *T. tumens* accessions. Collection site mean annual rainfall for example ranged from as low as 300 mm to more than 1300 mm while soil pH ranged from 5.8 to 9. Days to flowering ranged from 127 to 177 days after germination and seed yield ranged from less than 1 gm to more than 8 gms per plant. It has the potential to play a significant role in the pasture/animal system in southern Australia especially in the medium to high (up to 750 mm) rainfall, cool temperate pasture zone, where potential exists to include new perennial legumes.

Trifolium ambiguum (Caucasian Clover)

T. ambiguum is a long lived rhizomatous, summer active perennial. It is well adapted to high rainfall cool environments and tolerates water-logging. It is performing very well in Tasmania and is a very promising species for the NSW tablelands. Three accessions were collected.

Trifolium fragiferum (Strawberry Clover)

T. fragiferum was collected across both countries, at 45 sites and was especially common on inundated and mildly saline sites. This stoloniferous perennial tolerates water-logging, moderate levels of salinity and extended dry periods.

Lotus corniculatus (Birdsfoot Trefoil)

L. corniculatus is the most widely distributed species of perennial lotus and the most genetically variable. Glabrous and pubescent types were collected at 16 sites representing a broad range of elevation and soil types. Despite the significant genetic variation in *L. corniculatus* the current commercial varieties come from a limited genetic background, which may partly explain why domesticated cultivars are restricted in their adaptation compared with wild types. This wild material collected will add significant depth to the genetic diversity present in *L. corniculatus* and will support breeding efforts to develop new cultivars suited to areas receiving 550 mm to 750 mm annual rainfall permanent pasture zones, and 400–650 mm high rainfall cropping zones.

Lotus glaber (Narrow Leaf Trefoil)

Of particular importance to the FFI CRC is *L. glaber*. This species has significant potential for commercial application in the high rainfall permanent pasture zones (600–750 mm) and the cropping systems (400–550 mm) respectively. Very productive types were collected and were persisting well on very wet and heavily grazed sites. As in all the species of *Lotus*, there is a need to select for flowering within a certain period (determinant flowering) and for reduced pod shattering, in order to maximise seed production.

Onobrychis (Sainfoin)

Perennial *Onobrychis* species, particularly *O. viciifolia* are a non-bloating alternative to lucerne. *Onobrychis* has shown good adaptation to well-drained neutral to alkaline soils, good insect resistance, especially aphids and persistence under low rainfall conditions (330 mm). Production has been similar to lucerne and may provide an alternative for dry areas. The germplasm collected, especially *O. radiata* and *O. cyri* will greatly assist breeding efforts.

Melilotus (Sweet Clover)

52 lines of highly water logging and salt tolerant perennial legume *M. officinalis* were collected. *Melilotus* is currently a priority species for Australian pasture plant improvement as it can grow on soils where most other legumes fail in saline, wet or low nutrient soils. These species prefer calcareous or loamy soils with pH of 6.5 or more for optimum root nodulation and growth, although they have been reported to grow on soils with a pH of 4.5. The genus, particularly *M. albus* and *M. sulcatus* has shown excellent performance in salinity trials in South Australia and Victoria. *Melilotus* contains varying amounts of coumarin, which under certain conditions (e.g. spoilt hay) can be converted to dicoumarol, which is an anti-coagulant. Lines have been screened for low-coumarin types. The collected materials will broaden the available gene pool and add value to the breeding programs.

Cichorium intybus (Chicory)

24 accessions of this deep tap rooted herbaceous perennial species were collected. Some lines were very productive and heavily grazed. The species can offer out of season production remaining green over summer. It is best grazed when 30 cm tall down to 10 cm. There is evidence to suggest that it reduces parasites in lambs and

they demonstrate good live weight gains. The species has only 4 g/kg DM of condensed tannins but has another favourable secondary compound called lactone. It is not recommended for dairy cows due to tainting of the milk as a result of the lactone.

Annual forage legumes

While not a target of these expeditions, a very diverse and potentially useful representation of annual legume species was collected. *Trifolium* was the most collected genus and is represented by 13 species. The most promising ecotypes were observed within species of *T. lappaceum, T. diffusum, T. subterraneum and T. resupinataum.* Very productive *T. lappaceum* and *T. resupinatum* were collected at very wet and frequently inundated sites. *T. diffusum* was observed growing extremely well in association with perennial species at higher altitudes. Within the annual *Medicago* complex, 11 species were collected. Most promising ecotypes for the SARDI medic-breeding program were observed within species of *M. polymorpha* and spineless *M. littoralis*, which were growing in highly saline areas (EC 6.0) and very low rainfall sites (150 mm) respectively. The potential as an aerial seeding alternative to traditional medics may exist in *M. orthoceras*. Other annual species of recognised potential include *Trigonella calliceras*. This high yielding, aerial seeder was previously only represented by one accession.

Forage grasses

An extremely useful collection of perennial forage grass species was acquired. The most noteworthy of these were *Agropyron cristatum* collected from areas receiving less than 250 mm of rainfall and *Aeluropus littoralis* collected in highly saline areas. Other important forage grasses collected includes *Dactylis altaica* (a progenitor of *D. glomerata*), *Festuca arundinacea*, and *Lolium perenne*. Many of these grasses were growing in very harsh and sometimes saline environments. This material will be of particular interest to TIAR and the FFI CRC, Grass Improvement Program in Hamilton Victoria.

Pulse and Fodder Species

Over 90 lines representing 29 species of pulse and fodder legumes including *Brassica*, *Lens*, *Pisum*, *Vicia* and *Lathyrus* species were collected. This material is represented by locally sown landraces as well as wild material and will greatly contribute to available diversity. These species are of particular interest to the Australian Temperate Field Crops collection at Horsham. The Centre works closely with Australian breeding programs and has a high priority on the conservation storage, regeneration and distribution of pulse germplasm collections of pea, lentil and chickpea plus their wild relatives. *Vicia*, *Lathyrus* and *Brassica* are of secondary importance.

Cereals and their Wild Relatives

A very representative collection of sown cereal crops and their wild relatives was collected. The collection consisted of 75 lines, representing 5 genera and 15 species of local land races and wild types. In addition to sown landraces of wheat, barley and oats, the collection included seven species of *Aegilops*, a close relative of wild wheat that is important to breeding programs for a whole range of traits. *A. squarrosa* in particular, is important as it contains the ancestral D genome of wheat and is currently being used in Australian wheat breeding programs as a source of rust resistance. Other interesting wild species collected included perennial wild barley, *Hordeum bulbosum* and *H. spontaneum*. Ecotypes of these species were collected in very wet and partially saline soils and have the potential to expand the current range of adaptation for barley. This material is of high priority to Australian breeders and will be supported through ICARDA and the Australian Winter Cereals Collection in Tamworth, NSW.

Acknowledgements The plant collections and subsequent research has been funded by the agencies represented by the authors. The missions were made possible by the support of the host countries particularly the Genetic Resources Institute of the Azerbaijan National Academy of Sciences and the Aral Sea Experiment Station in Kazakhstan.

We also acknowledge the scientists from outside Australia and the host countries, being from China (Beijing Forestry University), ICARDA and Russia (Vavilov Institute).

Funding for the missions was provided by the Australian Centre for International Agricultural Research (ACIAR), the Grains Research and Development Corporation (GRDC), the CRC for Plant Based Management of Dryland Salinity (now the FFI CRC), SARDI and TIAR.

Logistical support from the International Centre for Agricultural Research in Dry Areas (ICARDA), Allepo, Syria is also gratefully acknowledged.

References

Auricht, G.C., Prosperi, J.M., Snowball, R., Hughes, S.J. 1999. The Characterisation and Preliminary Evaluation of *Medicago* and *Trifolium* Germplasm (pp. 141–149). In: Bennett, S.J., Cocks, P.S. (eds.). Kluwer Academic Publishers, The Netherlands.

Hughes, S.J., Snowball, R., Reed, K.F.M., Cohen, B., Gadja, C., Williams, A.R., Groeneweg, S.L. 2008. The systematic collection and characterisation of herbaceous forage species for recharge and discharge environments in southern Australia. Aust. J. Exp. Agr. 47: 397–408.

Humphries, A.W., Hughes, S.J. 2006. Preliminary evaluation of diverse lucerne (*Medicago sativa sspp.*) germplasm to identify new material for livestock and cropping based farming systems in Australia. Aust. J. Agr. Res. 57:1297–1306.

Chapter 6
Effect of Light Quality and Quantity on Leaf Growth in *Lolium perenne* L.

Philippe Barre, Badara Gueye, and François Gastal

Abstract In a sward, the quantity and quality of light are modified by the canopy, with in particular a decrease of red – far red ratio and a decrease of blue light. Plants react to these modifications by altering their shoot morphogenesis. In particular, in grasses, modification of light composition induces leaf growth changes. The objective of this study was to evaluate the effect of light quality and quantity on leaf growth in a set of perennial ryegrass genotypes. Ten clones of ten perennial genotypes highly variable in leaf length were used. Four light treatments were applied: a standard treatment (transparent filter), decreased PAR (neutral filter decreasing all the wavelengths similarly), low red – far red ratio associated with low blue (green filter, simulating the effect of a canopy), and low blue (red filter). After a growing period of three weeks, plants were defoliated and the light treatments were applied. Leaves three and four following defoliation were measured in order to obtain leaf elongation rate (LER), leaf elongation duration (LED) and adult leaf length. Strong genotype and light effects were observed on all traits. The low blue treatment had a particularly large effect, increasing adult leaf length by increasing both LER and LED. The differences in adult leaf length between genotypes were explained by both LER and LED. The genotypic × treatment interaction was significant but low in comparison to the principal effects.

Keywords Light effect · Leaf length · LER · LED · Genetic diversity · Perennial ryegrass

Introduction

Plants are able to detect and morphologically react to the presence of neighbours before actual shading and subsequent reduction in photosynthetic carbon assimilation. In particular, plants perceive and react to the decrease in red/far red ratio and

P. Barre (✉)
UR4, Unité de Recherche Pluridisciplinaire sur les Prairies et les Plantes Fourragères, Lusignan BP6, 86600, France
e-mail: philippe.barre@lusignan.inra.fr

C. Huyghe (ed.), *Sustainable Use of Genetic Diversity in Forage and Turf Breeding*, DOI 10.1007/978-90-481-8706-5_6, © Springer Science+Business Media B.V. 2010

in blue light (Varlet-Granchet and Gautier, 1995), determined respectively by differential transmission and reflection of red and far red by the foliage of other plants, and by the absorption of blue light in parallel to absorption of PAR.

In grasslands a large within and among species diversity is often observed. Leaf length is one of the traits which showed a particularly high degree of genetic variability. In perennial ryegrass swards under a management maximizing biomass production, i.e. infrequent defoliation, long-leaved genotypes were more adapted than short-leaved genotypes and were selected (Hazard and Ghesquière, 1995). However it is unknown whether short leaved or long leaved genotypes are genetically variable in plasticity and response to light.

The objective of the present study was to measure the response of a large range of perennial ryegrass genotypes to light quality and quantity.

Material and Methods

Plant Material and Growing Conditions

Ten genotypes of perennial ryegrass were used:

- three from forage varieties: Brée, Herbie and Ohio
- one from turf variety: Idole
- two from ecotypes (EcoGreece and EcoSweden)
- four experimental genotypes selected for their morphology: LL long leaves, L/S+ large leaf length/sheath length ratio), L/S- small large leaf length/sheath length ratio, T+ many tillers).

Four light treatments were used (Table 6.1):

Table 6.1 Optical properties of filters

	Standard S	Grey filter N	Green filter G	Red filter B-
PAR (400–700 nm)	93%	35%	38%	34%
Blue (350–550 nm)	91%	33%	19%	0.3%
R/FR (660/730)	1.1	1.1	0.3	1.1

- a transparent filter as standard
- a grey filter decreasing each wave length similarly.
- a green filter simulating a canopy
- a red filter without blue.

Each genotype was cloned in 40 plants with 10 plants in two blocks for each light treatment. After a growth period of three weeks, plants were cut to 3 cm and the filters were applied.

Characterization

For one tiller per plant, the length of the third uncut leaf was measured three times a week. These measurements were used to estimate leaf elongation rate (LER) and leaf elongation duration (LED).

Statistical Analysis

A variance analysis was performed with treatment, genotype and replicate as fixed factors, and with the interaction treatment × genotype. For significant factors, means were compared using a Newman and Keuls test.

Results and Discussion

Light treatment and genotype effects and the interaction between light treatment and genotype were significant for all leaf growth parameters (Table 6.2). However, principal effects were much greater than the interaction.

The comparison between the transparent and grey filters showed that shade increased leaf length through an increase of both LER and LED (Table 6.3). The comparison between the grey and green filters showed that, for the same PAR, a decrease in the R/FR ratio increased leaf length through an increase of LER only. The comparison between the grey, green and red filters showed that lack of blue

Table 6.2 Variance analysis with light treatment, genotype and replicate as fixed factor and with treatment × genotype interaction on leaf length, leaf elongation rate (LER) and leaf elongation duration (LED). MS: mean square

Effect	Leaf length (mm) MS/*p* value	LER ($mm.°C^{-1}d^{-1}$) MS/*p* value	LED (°Cd) MS/*p* value
Treatment	347325/0.0001	3.82/0.0001	23211/0.0001
Genotype	230081/0.0001	5.34/0.0001	11532/0.0001
Treatment × Genotype	3834/0.0001	0.09/0.0155	537/0.0013

Table 6.3 Light treatment effect on leaf growth. Mean comparison between transparent (S), grey (N), green (G) and red (-B) filters. Different letters indicate significant difference between means

Filters	Leaf length (mm)	LER (mm/°C.d)	LED (°C.d)
S	197 a	1.14 a	170 a
N	259 b	1.32 b	195 b
G	288 c	1.45 c	196 b
−B	346 d	1.63 d	209 c

Table 6.4 Genotype effect on leaf growth. Comparison of means per genotype based on Newman & Keul's test. Means followed by the same letter are not significantly different ($p = 0.05$)

Genotypes	Leaf length (mm)	LER (mm/°C.d)	LED (°C.d)
Brée	335 f	1.51 d	212 d
Herbie	286 de	1.51 d	190 bc
Ohio	330 f	1.82 f	176 a
Idole	97 a	0.55 a	181 ab
EcoGreece	175 b	0.91 b	194 c
EcoSweden	243 c	1.31 c	181 ab
LL	324 f	1.37 c	229 e
L/S+	305 e	1.63 e	171 ab
L/S–	342 f	1.64 e	203 d
T+	287 d	1.58 de	178 a

led to a strong increase of leaf length through an increase of both LER and LED. Moreover, the effect of absence of blue was stronger than the effect of the green filter.

As expected, leaf length was highly different between genotypes with those sampled in forage varieties having longer leaves than those sampled in the turf variety and the two ecotypes (Table 6.4). These differences were due to LER and/or LED differences. For example, leaf lengths of Brée and Herbie were different of 49 mm due to a difference of LED without any difference of LER. The turf variety Idole had the shortest leaves due to both a small LER and a short LED.

Despite the fact that all genotypes tended to respond to light quality and quantity in the same way with respect to leaf length, some genotypes behaved slightly differently (Fig. 6.1). For example, Ohio had a stronger response to the absence of blue than the other genotypes.

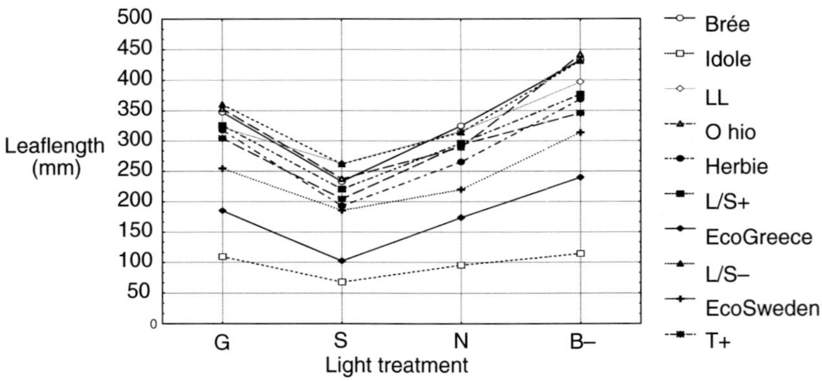

Fig. 6.1 Interaction between light treatment and genotype for leaf length

Conclusion and Perspectives

- Light quality and quantity had a strong effect on leaf length through effects on both LER and LED.
- The absence of blue had a dramatic effect on leaf length. The origin of this effect will be further studied. A similar effect of a decrease of blue light on leaf length on perennial ryegrass was observed by Gautier and Varlet-Grancher (1996) but with a slightly less impact.
- The differences in leaf length among genotypes were due to differences in LER and/or LED.
- The interaction between light treatment and genotype was significant but small, revealing a tendency for all genotypes to react the same way with slight differences. The impact of these differences on the entire plant should be investigated.

Acknowledgements The authors thank all the technical staff and students involved in this experimentation for their patient help.

References

Gautier, H., Varlet-Grancher, C. 1996. Regulation of leaf growth of grass by blue light. Physiologia Plantarum 96:424–430.

Hazard, L., Ghesquière, M. 1995. Evidence from the use of isozyme markers of competition in swards between short-leaved and long-leaved perennial ryegrass. Grass For. Sci. 50:241–248.

Varlet-Grancher, C., Gautier, H. 1995. Plant morphogenesis response to light quality and consequences for intercropping. In: Sinoquet, H., Cruz, P. (eds.), Ecophysiology of Tropical Intercropping (pp. 275–284). INRA, Paris. ISBN 2-7380-0603-5.

Chapter 7
Genetic Variation in Lowland Switchgrass (*Panicum virgatum* L.)

Hem S. Bhandari, Malay C. Saha, and Joseph H. Bouton

Abstract Switchgrass (*Panicum virgatum* L.) is identified as one of the main cellulosic biofuel feedstocks in the USA. Understanding genetic variation for biomass yield in switchgrass would be helpful in determining the appropriate breeding approach for cultivar development. In order to estimate the genetic component of variation in lowland switchgrass, 10 half-sib families and 47 full-sib families produced by making crosses between selected genotypes from one of the Noble Foundation's lowland breeding populations were evaluated in a nested design. The seedlings of these 47 families, and two checks, 'Alamo' and 'Blade™ EG1101', were established in the greenhouse and transplanted in late summer of 2007 at two Oklahoma (USA) locations, Ardmore and Burneyville, using a honeycomb planting design with 1.5 m plant-spacing. Each family was represented by 30 genotypes, including 15 reciprocals, at each location. The biomass from individual plants was harvested separately after the killing frost in 2008. Genetic components of variation were estimated following the mixed model in SAS, and heritability was estimated. Significant effects due to half-sib and full-sib families suggested both additive and non-additive gene actions were important in biomass dry matter yields of lowland switchgrass. The heritability estimates based on family analysis (0.33) and parent-progeny regression (0.18) were low, suggesting that the trait was under the control of many genes with minor effects and influenced by significant environmental effects. Developing high yielding switchgrass cultivars will probably need to exploit both additive and non-additive gene effects and selection and testing will need to be done in the target environments.

Keywords Genetic variation · Heritability · *Panicum virgatum*

H.S. Bhandari (✉)
The Samuel Roberts Noble Foundation, Inc., 2510 Sam Noble Parkway,
Ardmore, OK 73401, USA
e-mail: hsbhandari@noble.org

C. Huyghe (ed.), *Sustainable Use of Genetic Diversity in Forage and Turf Breeding*,
DOI 10.1007/978-90-481-8706-5_7,

Introduction

Switchgrass (*Panicum virgatum* L.), a perennial warm-season native grass of the North American tallgrass prairie, is adapted to a wide range of environments. It is traditionally used for livestock feed, soil conservation and prairie restoration. Switchgrass is now projected to be a main herbaceous biofuel feedstock (McLaughlin et al., 1999).

Switchgrass has two distinct ecotypes; upland and lowland (Hultquist et al., 1996). Switchgrass possesses a significant amount of among and within population genetic diversity for both ecotypes (Narasimhamoorthy et al., 2008). Understanding the nature of genetic variation is important in order to adopt an appropriate approach for cultivar development. The objective of this study was to assess the genetic component of variation and estimate the heritability for dry matter yield in a diverse, lowland switchgrass population.

Materials and Methods

A total of 47 bi-parental crosses were made in spring of 2007 between the genotypes of broad-based lowland switchgrass population whose pedigree included released cultivars, elite breeding lines, and PI accessions. These crosses also included 31 crosses produced following North Carolina design-I that involved 10 common genotypes (half-sib parents), each crossed to two to four other genotypes both selected at random. The reciprocal seeds for each cross were harvested separately.

Seeds were scarified with 50% H_2SO_4 for 2 minutes, which was followed by wet-chilling for 72 hours at 4°C and germination in a growth chamber (26°C) for 2 weeks. Germinated seedlings were planted in a flat containing greenhouse soil and allowed to grow for about 40 days. Seedlings were then transplanted to the fields at two locations Ardmore (34°12′ N, 97°05′W) and Burneyville (33°53′ N, 97°17′ W), in the last week of July 2007. The nursery was planted in a honeycomb design with 1.5 m spacing between the plants (Fasoula and Fasoula, 2000). Each of the families was represented by 30 genotypes that included 15 genotypes from each of the reciprocals. These crosses also included 31 families produced following nested design (North Carolina design-I) (Comstock and Robinson, 1948) by crossing 10 genotypes, each with two to four other genotypes. Two check cultivars, Blade ™ EG-1101 and Alamo, were also included. Parental genotypes were also planted adjacent to the nursery at the Ardmore location. No observations were made during the establishment year. All plants were mowed off in February 2008, and biomass from individual plants was harvested and weighed separately after the killing frost (January 2009). Data were analyzed using GLM procedures to test the effects due to families and reciprocals, and their interactions with the environment. For the 31 families in the nested design, data were also analyzed with mixed procedures to estimate variance components due to half-sib and full-sib families using the SAS (SAS, 2004). Heritability was estimated based on component of variation among half-sib families and mid-parent progeny regression according to Gallais (2003).

Results and Discussion

Full-sib families were different in biomass dry matter yield ($P < 0.01$) (Table 7.1). Three of the 47 families also demonstrated reciprocal effects ($P < 0.05$) (data not shown). These reciprocal effects were likely due to unfavorable effects of non-nuclear genes. Only one family was superior to check variety EG1101, and 14 families were superior to Alamo.

There was significant family × location interaction for dry matter yield as shown by change in ranks of dry matter yields of families between the two locations (Table 7.1; Fig. 7.1). Hopkins et al. (1996) also reported significant g × e interaction for forage yield in switchgrass. However, expected mean square due to family × location interaction was less than half the expected mean square due to family.

Table 7.1 GLM mean square and expected mean squares for dry matter yields among 47 full-sib families of switchgrass

Sources of variation	df	Mean square	Expected mean square
Location (E)	1	32.96**	–
Families (G)	46	2.39**	0.031
Reciprocal/ G (R)	47	0.39**	0.0042
G × E	46	0.67**	0.014
(R/G) × E	47	0.27**	0.003
Residual	2,518	0.23	0.2262

Mean DMY: 1.43 Kg/plant Range: 0.71–1.79 Kg/plant
CV(%) = 32.43 LSD = 0.17 Kg/plant

**Effects were significant at $P < 0.01$

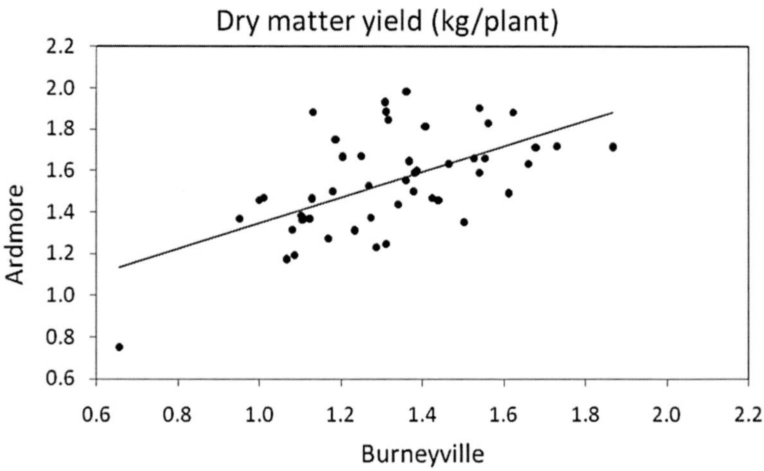

Fig. 7.1 Dry matter yield of 47 full-sib families at Ardmore and Burneyville, Oklahoma. Differences in spread of the plots from the diagonal line in either direction showed the relative magnitude of g × e effects

Table 7.2 Expected mean square for dry matter yield among half-sib and full-sib families for 31 families produced following NC design-I

Sources of variation	df	Mean square	Expected mean square
Location (L)	1	25.1	–
Half-sibs (H)	9	5.6**	0.024
Full-sib (F)/(H)	21	1.6**	0.018
L×H	9	0.8**	0.001
L×F/(H)	21	0.6**	0.015
Residual	1,725	0.24	0.235

** Effects were significant at $P < 0.01$

Expected mean square due to half-sib families was 0.024 (Table 7.2). The expectation of variation due to half-sib families in an autotetraploid is $\frac{1}{4}\sigma_A^2 + \frac{1}{36}\sigma_D^2$. Past research results have shown $\sigma_A^2 > \sigma_D^2$. If we ignore the contribution of dominant genetic variance, then we have $\frac{1}{4}\sigma_A^2 = 0.024$ 'or' $\sigma_A^2 = 0.096$. Then, narrow sense heritability is, $(h_n^2) = \sigma_A^2/\sigma_P^2 = 0.096/0.2930 = 0.33$.

Regression of dry matter yields of 47 families on mid parent value was 0.18. In autotetraploids, the regression of progeny on mid parent value has an expectation of $\sigma_A^2 + \frac{1}{3}\sigma_D^2$ (Gallais, 2003). In the absence of dominance, the regression coefficient is also the estimate of the narrow sense heritability. The results demonstrated a low heritability estimate for biomass yield that suggests that the trait is controlled by several genes with minor effects and significant environmental influences or g × e. The differences between the heritability estimates following two methods could be attributed to the presence of the dominant effects of genes. Dry matter biomass yield heterosis reported for hybrids between upland summer and lowland Kanlow varieties also suggested the presence of dominant gene effects in switchgrass (Vogel and Mitchell, 2008). In the presence of dominance in autotetraploids, variation among half-sib families would provide a better estimate of heritability compared to mid-parent–progeny regression (Gallais, 2003).

Our results suggest genetic gain in switchgrass may be maximized by following procedures that exploit both additive as well as non-additive gene action and that selection and testing in the target environments will be critical.

Acknowledgments This research was supported by funds from CERES, Inc., and The Samuel Roberts Noble Foundation, Inc.

References

Comstock, R.E., Robinson, H.F. 1948. The components of genetic variance in populations of bi-parental progenies and their use in estimating the average degree of dominance. Biometrics 4:254–256.

Fasoula, V.A., Fasoula, D.A. 2000. Honeycomb breeding: principles and applications. Plant Breed. Rev. 18:177–251.

Gallais, A. 2003. Quantitative genetics and breeding methods in autopolyploid plants. Institut National de la Recherche Agronomique (INRA), Paris.
Hopkins, A.A., Vogel, K.P., Moore, K.J., Johnson, K.D., Carlson, I.T. 1996. Genotype effects and genotype by environment interactions for traits of elite switchgrass populations. Crop Sci. 35:125–132.
Hultquist, S.J., Vogel, K.P., Lee, D.J., Arumuganathan, K., Kaeppler, S. 1996. Chloroplast DNA and nuclear DNA content variations among cultivars of switchgrass, *Panicum virgatum L.*Crop Sci. 36:1049–1052.
McLaughlin, S., Bouton, J., Bransby, D., Conger, B., Ocumpaugh, W., Parrish, D., Taliaferro, C., Vogel, K., Wullschleger, S. 1999. In: Janick, J. (ed.), Developing Switchgrass as a Bioenergy Crop (pp. 282–299). ASHS press, Alexandia, VA.
Narasimhamoorthy, B., Saha, M.C., Swaller, T., Bouton, J.H. 2008. Genetic diversity in switchgrass collections assessed by EST-SSR markers. Bioenerg. Res. 1:136–146.
SAS Institute. 2004. Genetic analysis of complex traits using SAS. SAS Institute, Inc., Cary, North Carolina.
Vogel K.P., Mitchell, R.B. 2008. Heterosis in switchgrass: biomass yield in sward. Crop Sci. 48:2159–216.

Chapter 8
Morphological and Molecular Diversity of Branching in Red Clover (*Trifolium pratense*)

Gerda Cnops, Antje Rohde, Oana Saracutu, Marianne Malengier, and Isabel Roldán-Ruiz

Abstract Mixed grass-clover grasslands are an essential element of sustainable farming systems. The presence of clover in the mixture contributes significantly to the reduction of nitrogen fertilizer application needs, and results in improved nutritional value. In red clover, architecture is under genetic and environmental control. Similarly to what has been found in other plant species, we anticipate that architectural changes in red clover will strongly influence traits such as forage yield, re-growth capacity, seed yield and persistence. The genetic aspect of branching has been widely studied in model plants but did not obtain much attention in the past in red clover. Our aim is to translate knowledge from model plants on genes involved in meristem initiation, bud formation, and the activity and determination of the apical meristems to red clover.

Keywords Architecture · Branching · Diversity · Germplasm · Red clover

Introduction

Red clover is a forage crop that offers several possibilities; it can be used as a monoculture or in mixed grasslands as well in conventional as in organic farming. The presence of clover in grassland mixtures contributes significantly to the reduction of nitrogen fertilizer application needs. As grass-clover mixtures are more palatable, they ensure a greater intake and a higher milk production (Bertilsson and Murphy, 2003). The addition of red clover to silage improves the nutritional value of the fodder and, in turn, of the animals' milk and meat (Dewhurst et al., 2003a; Dewhurst et al., 2003b).

Plant architecture is defined by the degree of branching, internode elongation and shoot determinacy and is under genetic and environmental control (photoperiod,

G. Cnops (✉)
Plant Sciences Unit – Growth and Development, Institute for Agricultural and Fisheries Research (ILVO), Caritasstraat 21, Melle 9090, Belgium
e-mail: Gerda.Cnops@ilvo.vlaanderen.be

C. Huyghe (ed.), *Sustainable Use of Genetic Diversity in Forage and Turf Breeding*,
DOI 10.1007/978-90-481-8706-5_8,

density, grazing...). Similarly to what has been found in other plant species (McSteen and Leyser, 2005) we anticipate that architectural changes in red clover will strongly influence traits such as forage yield, re-growth capacity, seed yield and persistence. The genetic aspect of branching has been widely studied in model plants. In red clover, genomic regions containing branching QTLs have been found, but these regions are still very large (containing several genes) due to the low resolution of the maps (Hermann et al., 2006; Abberton, 2007). Furthermore, the markers which define these QTLs have not been validated in other germplasm and cannot be used in breeding applications.

The final aim of this study is to translate knowledge from model plants such as *Arabidopsis* or pea on genes involved in meristem initiation, bud formation, and the activity and determination of the apical meristems to red clover. Analyses of allelic diversity in branching genes will allow us to determine the genetic factors which are responsible for architectural differences in this species. This involves the use of a collection of genotypes that have been thoroughly characterized for branching characteristics. Here we report on the preliminary screening of a set of commercial cultivars and landraces for branching during one growth season. The obtained results provide a good overview of the variability present in red clover germplasm and will be used to select a set of plants and a set of characteristics to follow-up in coming growth seasons using clonal replicates.

Material and Methods

Fifteen commercial cultivars and 6 landraces from the ILVO diploid red clover collection were sown in the greenhouse in January 2008. From each accession 10 plants were chosen at random to be observed during one growing season. Six-week old seedlings were transplanted to 22 cm pots, spaced 15 cm, and kept in a non-conditioned greenhouse with extra illumination during the first two months. The plants were watered once to twice a week and no fertilisation was applied.

The number of leaves and the number and order of branches (primary, secondary, ...) were recorded during the vegetative phase. From bolting onwards only the number of branches was counted monthly.

Results

During the first growing season, the primary shoot remains vegetative in red clover. First order branches derive from the leaf axils of the primary shoot and can either remain vegetative or become reproductive, terminating with an inflorescence, depending on the genotype and the developmental stage of the plant.

During the vegetative phase, the number of leaves was highly correlated with the number of branches. Therefore, only the total number of branches was counted from bolting onwards. A high degree of heterogeneity was found among accessions

Fig. 8.1 Heterogeneity between and within accessions. Top and bottom left are a poor and an abundant branching genotype, representatives for respectively cultivar 1 (cv1) en cultivar 2 (cv2) in the vegetative phase; a representative for cultivar 1 (*top middle*) and cultivar 2 (*bottom middle*) during the flowering phase. The poor branching genotype had an abundant branching phenotype in the generative phase. Cv3 is a poor branching genotype, representative for cultivar 3 compared to another erect representative genotype for cv4 in the post-flowering phase. Architectural heterogeneity between different genotypes within cultivar 5 is shown at the *right*

(Fig. 8.1). The level of heterogeneity found within accessions was also remarkably high. These differences were determined mainly by two factors: (i) whether the axillary meristems grew out, and (ii) the fate of the axilary meristem (shoot, inflorescence or flower).

Figure 8.2 shows the seasonal variation of the average number of branches per accession. Important differences were found among accessions when the branching patterns at three developmental phases (vegetative, flowering and post-flowering) were compared. Taking all these data together, four main 'branching types' could be distinguished among the 21 accessions studied:

- abundant branching during the whole growing season (e.g. cv9, cv12, cv14, lr7)
- poor branching during the whole growing season (e.g. cv3 and cv4)
- abundant branching during the vegetative phase but poor during the generative phase (e.g. cv10 and cv11)
- poor branching during the vegetative phase but abundant during the generative phase (e.g. cv1)

At the end of the growing season branches were cut, counted and classified according to their position in the plant. Branches derived from axillary meristems positioned up to 10 cm above the soil were called 'crown branches'. Branches that originated from meristems located higher than 10 cm above the soil were called 'aerial branches'. A strong heterogeneity within and between accessions was observed. Plants with a similar amount of total branches displayed a very different ratio crown/aerial branches. This illustrates once more the complexity of the

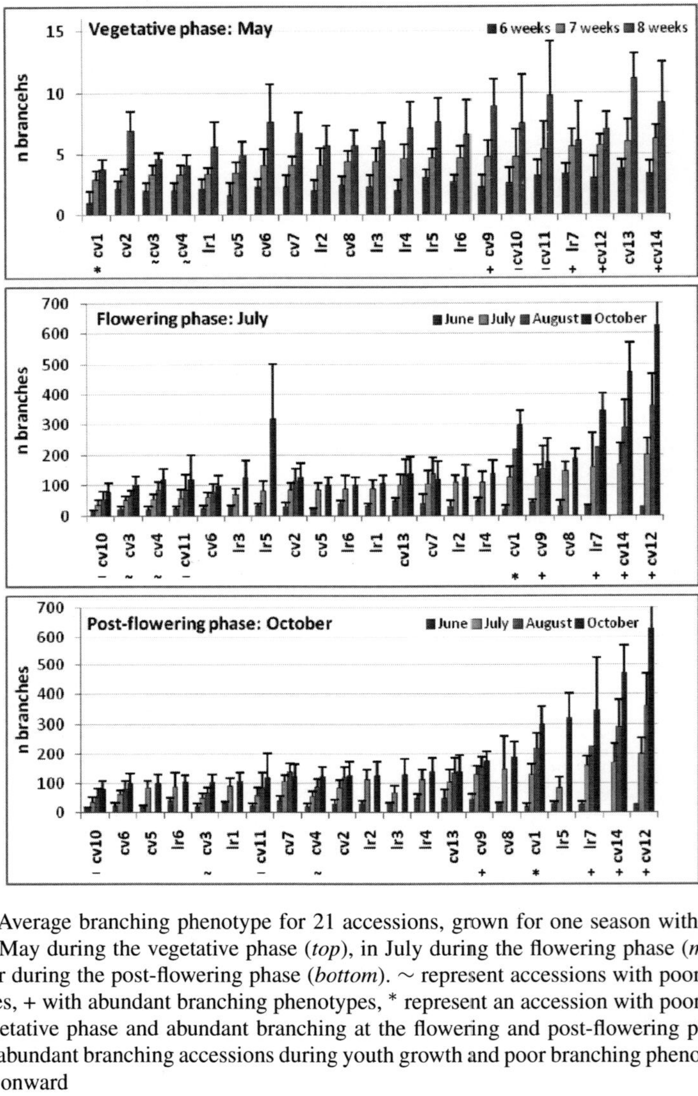

Fig. 8.2 Average branching phenotype for 21 accessions, grown for one season without cutting scored in May during the vegetative phase (*top*), in July during the flowering phase (*middle*) and in October during the post-flowering phase (*bottom*). ~ represent accessions with poor branching phenotypes, + with abundant branching phenotypes, * represent an accession with poor branching at the vegetative phase and abundant branching at the flowering and post-flowering phase and – represent abundant branching accessions during youth growth and poor branching phenotypes from flowering onward

branching pattern in red clover and the necessity to carry out detailed measurements to characterize the branching phenotype of a given plant.

Discussion

The preliminary screening of red clover cultivars and landraces presented here has highlighted a huge diversity for branching between populations. Without the evaluation of clonal replicates of the genotypes within a population it is impossible to separate the true genetic variation from the random variation. However, our data suggest also a huge diversity within a population. We have identified accessions and individual plants that display contrasting branching patterns throughout the season. However, the data available are still rather preliminary and it is essential to compare them with data from field evaluation of replicated genotypes. Therefore, for a number of selected genotypes, we have produced a minimum of 5 clonal replicates. These plants will be characterized in coming growth seasons. The final selection of germplasm to include in this study will also include wild populations.

In a further step we will investigate the relationship between branching patterns and other traits of agronomic relevance.

References

Abberton, M. 2007. Developing Productive and Persistent Red Clover Cultivars for Sustainable Livestock Production. DEFRA, SID 5 Research, final report.

Bertilsson, J., Murphy, M. 2003. Effect of feeding clover silages on feed intake, milk production and digestion in dairy cows. Grass For. Sci. 58:309–322.

Dewhurst, R.J., Fisher, W.J., Tweed, J.K.S., Wilkins, R.J. 2003a. Comparison of grass and legume silages for milk production. 1. Production responses with different levels of concentrate. J. Dairy Sci. 86:2598–2611.

Dewhurst, R.J., Evans, R.T., Scollan, N.D., Moorby, J.M., Merry, R.J, Wilkins, R.J. 2003b. Comparison of grass and legume silages for milk production. 2. In vivo and in sacco evaluations of rumen function. J. Dairy Sci. 86:2612–2621.

Herrmann, D., Boller, B., Studer, B., Widmer, F., Kölliker, R. 2006. Characterisation of Genetic Diversity and Molecular Dissection of Seed Yield and Persistence in Swiss Mattenklee (*Trifolium pratense* L.). TAG Theor. Appl. Genet. 112:536–545.

McSteen, P., Leyser, O., 2005. Shoot branching. Annu. Rev. Plant Biol. 56:353–374.

Chapter 9
Genotypic Differences in Symbiotic N_2 Fixation of Some Alfalfa (*Medicago sativa* L.) Genotypes

Dušica Delić, Olivera Stajković, Jasmina Radović, Aleksandra Stanojković, Đorđe Kuzmanović, Nataša Rasulić, and Bogić Miličić

Abstract In a highly controlled laboratory experiment, four alfalfa (*Medicago sativa* L.) cultivars were grown in all combinations with five strains of *Sinorhizobium* spp. with the aim to assess differences in nitrogen fixation among alfalfa genotypes and compatibility of symbionts. Based on shoot dry weight and total N content, it was noted that each cultivar showed a great compatibility with particular strains. Variety K-28 inoculated with all investigated strains exhibited the highest shoot dry matter, compared to other investigated cultivars. Using symbiotic N fixation, cultivar K-28 inoculated with L3Si strain gained high compatible association due to the 87% of shoot dry weight in relation to uninoculated control with full N content (100%). The significant coefficient of variation (CV%) for shoot dry matter of inoculated treatments was found (19.41–57.83%) depending on bacterial strain and plant genotype. Higher CV% values of inoculated plants compared to uninoculated plants indicated the influence of strain on increasing variability of the plants. As the effectiveness of the N_2 fixation varies widely in different rhizobia-host combinations, it would be possible to identify the highly effective rhizobial strains, which would represent commercial strains of microbiological fertilizers for particular cultivars.

Keywords Alfalfa · Co-selection · Nitrogen fixation · Variability · *Sinorhizobium* spp

Introduction

Alfalfa (*Medicago sativa* L.) is one of the most important forage crops with the ability of symbiotic nitrogen fixation (SNF) with specific nodule bacteria (rhizobia) *Sinorhizobium meliloti* and *Sinorhizobium medicae.* SNF is a significant boost

D. Delić (✉)
Institute of Soil Science, Teodora Drajzera 7, Belgrade 11000, Serbia
e-mail: vukmirdusica@yahoo.com

C. Huyghe (ed.), *Sustainable Use of Genetic Diversity in Forage and Turf Breeding*,
DOI 10.1007/978-90-481-8706-5_9,

to N fertilization and additionally, does not cause any hazard to the environment (Vance, 1998). Legume-Rhizobium association is a genetic complex, so that phenotypic differences can be brought about by genetic variation in both symbionts (Mytton et al., 1984). Various alfalfa genotypes and rhizobial strains differ in SNF capacity measured, inter alia, by particular plant performances like shoot dry matter (SDW). This fact indicates that SNF depends upon both symbionts and their compatibility, resulting in selection of genotypes of host plant with compatible rhizobial inoculant for high effective SNF as an important aim (Mytton and Skot, 1993; Appunu et al., 2008). In order to determine favourable gene combinations, coordinated and coincidental selection of both symbionts is required. In a controlled laboratory experiment, four alfalfa cultivars were grown in all combinations with five strains of *Sinorhizobium* spp with the aim to access differences in N_2 fixation among alfalfa cultivars and compatibility of symbionts.

Material and Methods

One Serbian (K-28), one from Republic of Srpska, entity of Bosnia and Herzegovina (BL-88), one Slovenian (Soča) and one Slovakian (Vanda) alfalfa cultivars were used. Five effective *S. meliloti* strains from the Collection of Institute of Soil Science (224. 218, L5, LR1ks and L3Si) were applied for alfalfa inoculation. Alfalfa seeds were surface-sterilized (0.1% solution $HgCl_2$) and placed into test tubes (2.5 × 25 cm) containing N-free Jansen agar by Vincent, 1970. 48 h later, alfalfa seedlings were inoculated with 0.5 ml suspension of *S. meliloti* pure cultures grown for 48 h on yeast mannitol broth (Vincent, 1970). Plants were cultivated in a growth, light chamber under a 16/8 h light/dark cycle and 28/20°C day/night temperature. The experimental layout was a completely randomised block design with 20 inoculated treatments and 4 uninoculated controlled treatments (4 controls with full N content (0.5% KNO_3-NØ)) with 15 plants per treatment as replication. The experiment was completed 60 days after sowing. The following analyses and measurement were performed: nodule weight and number, SDW and total N content (CNS Elemental Analyzer vario EL III). All data were subjected to ANOVA using the statistical analysis system SPSS 10.0, 1999. Means of all treatments were calculated and the differences tested for significance using the LSD test at the 0.05 probability level. Correlation coefficients were calculated to study the associative relationships among the measured traits while the coefficient of variation (CV) for SDW was determined to assess plant variability within each inoculated cultivar.

Results and Discussion

In our investigation, four *M. sativa* cultivars were evaluated for their symbiotic performances with five *S. meliloti* strains. Based on two factorial variance analysis, it was found that alfalfa cultivars and strains had highly significant ($P < 0.01$) effects

Table 9.1 ANOVA and correlation coefficient for plant characters, number and dry weight of nodules of four alfalfa genotypes inoculated with five *S. meliloti* strains. Correlation coefficients ≥ 0.21 are statistically significant at $P < 0.01$

Source of variance	Shoot dry weight (mg plant^{-1})	DW of nodules (mg plant^{-1})	Nodule number (plant^{-1})	Total N content (mg plant^{-1})
Cultivars	27.51***	12.52***	43.34***	23.09***
Strains	33.90***	0.61^{ns}	2.76*	11.04***
Interaction	2.67***	1.27^{ns}	1.46^{ns}	1.18^{ns}
Correlation coefficients				
Characteristics				
SDW	1	0.114	−0.045	0.946***
DW of nodules		1	−0.085	0.001
Nodule number			1	0.009

*, ***Significant at $P < 0.05$, 0.001, respectively.

on SDW and total N content while, their interaction significantly effected only SDW (Table 9.1), which is relevant with other authors (Mytton et al., 1984; Phillips and Teuber, 1985; Rengel, 2002).

It has been reported that plant SDW is the best parameter to evaluate symbiotic activity of legume-Rhizobium associations. The effects of the symbionts on the variation of SNF were measured by bi-factorial analysis of variance for SDW (Phillips and Teuber, 1985; Hefny et al., 2001). Our previous investigations in alfalfa cloned genotypes underline significant strain influence on variability of SNF measured by SDW (Delić et al., 2006). In our experiment, the nodule number and nodule mass were significantly affected by host plant while the strain had significant influence only on the nodule number. In previous researches, it was found that variation in the nodulation ability depended on both symbionts (Caetano-Anoles and Gresshoff, 1991).

Data presented in Table 9.2 showed that among alfalfa cultivars, K-28 produced the highest SDW and total N content after inoculation with all investigated strains (average SDW value was 53.1 mg plant^{-1}). The other tested *M. sativa* cultivars reached significantly lower average values of these parameters. N mineral fertilizer in NØ control had significantly higher effect on plant performances, than symbiotically fixed N_2 in most inoculated treatments. The effect of mineral N fertilizer on plant quality and quantity is usually higher, than the effect of N_2 fixation. However, there are effective symbiotic interactions resulting in the same or higher plant performance than plants fed with N mineral fertilizer (Phillips and Teuber, 1985), as in our experiment. There was no significant difference between SDW of Serbian K-28 plants inoculated with L3Si and L5 rhizobial strains and K-28 control treatment (NØ).

When investigated rhizobium strains were compared for symbiotic performances on all studied *M. sativa* cultivars the best one appeared to be strain L3Si and L5 as well as 218. The plants inoculated with these strains showed the highest average dry

Table 9.2 Symbiotic performances and coefficient of variation of alfalfa genotypes inoculated with different *S. meliloti* strains

Alfalfa cultivars		Strains						
		224	L5	LR1ks	L3Si	218	NØ	Mean
Shoot dry weight (mg plant^{-1})								
Vanda		36.15	37.78	37.01	32.51	40.89	54.01	39.73
Soca		26.95	32.33	36.55	42.31	31.58	63.03	38.79
BL-88		34.63	41.93	29.09	37.08	39.31	73.42	42.58
K-28		42.24	61.15	43.23	64.71	54.09	74.53	56.66
Mean		34.99	43.30	36.47	44.15	41.47	66.25	
LSD	Cultivars	4.41						
	Strains	5.10						
	Interaction	10.80						
Total N content (mg plant^{-1})								
Vanda		1.57	1.40	1.37	1.15	1.51	1.45	1.41
Soca		0.87	0.98	0.89	1.21	0.98	1.64	1.09
BL-88		1.35	1.51	1.07	1.21	1.42	2.64	1.53
K-28		1.70	2.49	1.55	2.44	2.07	2.75	2.17
Mean		1.37	1.60	1.22	1.50	1.49	2.12	
LSD	Cultivars	0.254						
	Strains	0.312						
	Interaction	0.365						
Coefficient of variation for shoot dry weight (CV%)								
Vanda		34.99	38.04	40.69	38.45	33.06	24.93	
Soca		54.43	35.04	28.72	19.41	57.83	14.74	
BL-88		38.15	19.60	28.93	34.41	34.43	11.18	
K-28		40.84	29.39	36.37	26.69	21.26	21.41	

LSD at 0.05 level of probability.

weight and total N content, compared to plants inoculated with other investigated strains. Similar results have been noted by Mytton and Skot (1993). They found that some rhizobium strains influenced plant yield more than others, due to their nodulating ability (Table 9.1).

Correlation coefficients among symbiosis parameters and nodulation traits for alfalfa-rhizobium associations (Table 9.1) indicated that SDW was not correlated with nodule mass ($r = 0.114$, $n = 18$) and nodule number ($r = -0.045$, $n = 18$), while there was a highly significant positive correlation among SDW and total N content in the plant ($r = 0.946$, $n = 18$). This is in agreement with the literature (Jessen et al., 1988; Vance, 1998; Hefny et al., 2001).

The results in Table 9.2 indicate more plant variability for SDW within each inoculated cultivars (CV% for SDW was 19.4–57.8%) than in fertilized plants-NØ (11.2–24.9%). These values represented variability in control conditions thus mainly influenced by genetic factors. Environmental factors would increase CV%. Alfalfa is highly variable species due to its tetraploidy, cross-pollination and the synthetic status of the varieties. Symbiosis is a supplementary source of variability

(Vitale et al., 1996; Delić et al., 2006). Based on CV% (Table 9.2), there is a condition for the selection of plant genotypes with high SNF potency.

Variability in the N fixating ability among strains and plant genotypes has already been reported (Rengel, 2002). The effectiveness of the N fixation varies in different rhizobia-host combinations. Based on LSD-test for SDW, 20 investigated symbiotic pairs could be grouped in 3 significant different N fixing active groups: the most effective group consists of 3 pairs (K-28 with L3Si and L5 as well as with 218), the significantly lower effective group consists of 6 pairs and the lowest effective group of 11 symbiotic pairs. We identified the most effective rhizobial strains for the various cultivars: L5, LR1ks and 218 strains for Vanda; L3Si strain for Soča; L3Si, 218 and L5 strains for BL-88. Symbiotic pairs of the three groups fixed respectively around 87%, more than 50% and less than 50% N_2 in comparison to NØ. It was noted that intensity of N fixation of one strain was different with different host genotypes due to the compatibility of symbionts (Appunu et al., 2008). Taking into account these results, it can be concluded that the increase of SNF efficiency should be found by selection of highly compatible plant genotypes of host and strain.

Conclusions

Our results indicate that Serbian K-28 cultivar has a high potential plant growth and could be good parental genotype in breeding programme for improved N_2 fixation capability by inoculation with compatible strains L3Si and L5, as well as with 218 strains which would represent commercial strains of microbiological fertilizers for K-28 cultivar.

Acknowledgements This research was supported by the Ministry of Science of the Republic of Serbia.

References

Appunu, C., Sen, D., Singh, M.K., Dhar, B. 2008. Variation in symbiotic performance of *Bradyrhizobium japonicum* strains and soybean cultivars under field conditions. J. Cent. Eur. Agr. 9:185–189.

Caetano-Anollés, G., Gresshoff, P. 1991. Plant genetic control of nodulation. Annu. Rev. Microbiol. 4:345–382.

Delić, D., Tomić, Z., Miličić, B., Radović, J., Stajković, O., Stanojković, A., Knežević-Vukčević, J. 2006. Evaluation of *Sinorhizobium meliloti* effect on N_2 fixation variability in alfalfa. pp. 369–372. Proceedings of XXVI EUCARPIA Fodder Crops and Amenity Grasses Section and XVI *Medicago* spp. Perugia, Italy.

Hefny, M., Dolinski, R., Malek, W. 2001. Variation in symbiotic characters of alfalfa cultivars inoculated with *Sinorhizobium meliloti* strains. Biol. Fertil. Soil 33:435–437.

Jessen, D.L., Barnes, D.K., Vance, C.P. 1988. Bidirectional selection in alfalfa for activity of nodule nitrogen and carbon-assimilating enzymes. Crop Sci. 28:18–22.

Mytton, L.R., Broockwell, J., Gibson, A.H. 1984. The potential for breeding an improved lucerne – *Rhizobium* symbiosis, 1st assessment of genetic variation. Euphytica 33:401–410.

Mytton, L.R., Skot, L. 1993. Breeding for improved symbiotic nitrogen fixation. In: Hayward, M.D., Bosemark, N.O., Romagosa, I. (eds.), Plant Breeding: Principles and Prospects (pp. 451–472). Chapman and Hall, London.

Phillips, A.D., Teuber, R.L. 1985. Genetic improvement of symbiotic nitrogen fixation in legumes. In: Evans, H.J., Bottomley, P.J., Newton, E.W. (eds.), Nitrogen Fixation Research Progress (pp. 11–17). Martinus Nijhoff, Dordrecht.

Rengel, Z. 2002. Breeding for better symbiosis. Plant Soil 245:147–162.

Vance, C.P. 1998. Legume symbiotic nitrogen fixation: agronomic aspects. In: Spaink, H.P., Kondorosi, A., Hooykaas, P.J.J. (eds.), The Rhizobiaceae (pp. 509–530). Kluwer, Dordrecht.

Vincent, M.J. 1970. A manual for the Practical Study of Root-Nodule Bacteria. In: IBP Handbook (Vol. 15). Blackwell, Oxford.

Vitale, M., Pupilli, F., Scotti, C., Arcioni, S. 1996. Extent of RFLP and RAPD Variability in Tetraploid Populations of Alfalfa (pp. 305–308). Proceedings XVI EGF Meeting. Grado. Italy.

Chapter 10
Physiological and Genetic Diversity in *Rhizobium sullae* from Morocco

Nadia Elboutahiri, Imane Thami-Alami, El-Houssine Zaid, and Sripada M. Udupa

Abstract Sulla is an important forage legume in the northern parts of Morocco and has a significant role in minimizing soil erosion. Sulla consists of two species *Hedysarum coronarium* L. and *H. flexuosum* L and both are well adapted to marginal areas where drought, salinity and alkalinity are the major problems. With the aim to characterize the bacterial symbiotic partner, *Rhizobium sullae* isolates from the nodules were sampled from sulla growing regions of Morocco. A total of 62 isolates were characterized so far, for various physiological characters such as resistance to salinity stress, water stress, high temperature stress, antibiotics, heavy metals, and various pH levels. The results revealed a considerable diversity for various physiological traits. The 62 isolates were divided into 14 clusters which showed a wide range of diversity both within and between the clusters. Genetic diversity of the isolates was analyzed by Amplified Ribosomal DNA Restriction Analysis (ARDRA) using two enzymes *Hinf*I and *Hae*III and data was used for cluster analysis. The results revealed that the individual genetic clusters contain isolates with diverse phenotypic traits, indicating no relationship between physiological and genetic groupings.

Keywords *Rhizobium sullae* · Physiological traits · ARDRA · Genetic diversity

Introduction

Sulla is a perennial legume used as a forage crop in the Mediterranean basin. The genus *Hedysarum* which includes over 100 species, distributed throughout Europe, Africa, Asia and North America (Squartini et al., 2002). In Morocco, sulla is the common name for two similar species, *Hedysarum coronarium* L. and *Hedysarum flexuosum* L. (Glatzle et al., 1986). The plant has been successfully introduced in

I. Thami-Alami (✉)
Institut National de la Recherche Agronomique (INRA), B.P. 415, Rabat, Morocco
e-mail: thamialami_ma@yahoo.fr

C. Huyghe (ed.), *Sustainable Use of Genetic Diversity in Forage and Turf Breeding*,
DOI 10.1007/978-90-481-8706-5_10,

marginal areas including highly calcareous soil (Nuti and Cassella, 1989). Some *Hedysarum* spp. are known to have good forage value in semiarid conditions and are successful crops for soil improvement and honey production in North Africa and the Near-East (Squartini et al., 2002; Kishinevsky et al., 2003).

Nitrogen-fixing bacteria nodulating sulla has been isolated and referred to as *Rhizobium sullae* (Squartini et al., 2002). Strains have been characterized as fast growing rhizobia and highly host-specific for sulla, and no cross-nodulation.

With the aim of providing an adequate characterization of this bacterial symbiotic partner, the *Rhizobium sullae* were isolated from the nodules sampled from sulla growing regions of Morocco, and were characterized, for various physiological characters. The genetic diversity was analyzed by using Amplified Ribosomal DNA Restriction Analysis (ARDRA) using two enzymes *Hin*fI and *Hae*III.

Materials and Methods

Isolate Collection

The 62 rhizobia isolates used in this study were isolated from nodules sampled from sulla growing regions of Morocco. Rhizobia were isolated using standard procedures (Vincent, 1970). All isolates were incubated at 28°C and maintained on Yeast Extract Mannitol (YEM) Agar slants at 4°C, or in 20% (v/v) glycerol at –20°C. All 62 isolates were Gram-negative, fast-growing, and formed single colonies with colorless circular and convex. They differed greatly in the amounts of extracellular gum of 2–3 mm diameters, attained within 2–3 days on YEM plates.

Physiological Characterization

All physiological tests were carried out on YEM plates, except for water stress. Petri dishes containing defined medium were subdivided into squares, which were inoculated with 10 μl of bacterial culture grown for 48 h in YEM broth. The following treatments (with three replications) were applied: salt tolerance (Gao et al., 1994) at 0–10% NaCl at increments of 1%; temperature tolerance at 28, 32, 36, 40 and 44°C; water stress imposed using PEG 6000 (Buss and Bottomley, 1989) in YEM broth at a level of –0.25, –0.5, –1 and –1.5 MPa; pH tolerance (Gao et al., 1994) at pH 3.0, 3.5, 4.5, 5.5, 7.0, 9.0 and 9.5 (Homopipes buffer 25 mM used for pH range of 3–5, and for pH range 9–9.5 [pKa 7.5 at 25°C] and the MES buffer used for pH range 5–7 [pKa 6.1 at 25°C]); and intrinsic antibiotic (Gao et al., 1994) and heavy metal tolerance (Gao et al., 1994) were determined on solid YEM medium containing the following filter-sterilized antibiotics or heavy metals (all μg/ml): chloramphenicol (25 and 100), spectinomycin (15 and 50), streptomycin (10 and 25) and tetracycline (10 and 25); $CdCl_2.2H_2O$ (5 and 20), $MnCl_2$ (300), $HgCl_2$ (20) and $ZnCl_2$ (200).

Genetic Characterization

Bacterial DNA extraction, Amplified rDNA restriction analysis (ARDRA) and data analysis were performed according to Elboutahiri et al. (2009).

Results and Discussion

The phenotypic characterization of the 62 sampled isolates of *R. sullae* for above characters revealed a large degree of variation. There was a wide variability for salinity tolerance at 171–855 mM (1–5%) NaCl; even isolates sampled from the same area showed some variation for NaCl tolerance. Indeed, 35% of isolates collected tolerated a salt concentration of 855 mM NaCl. This concentration was higher than that reported by Kishinevsky et al. (1996), Struffi et al. (1998), Wei et al. (2006) who found that the tolerance to salinity for *Rhizobium sullae* varied between 342 mM NaCl and 513 mM NaCl.

The tolerance to water stress of isolates of *R. sullae* studied varied between 0 MPa to – 1.5 MPa, with 16% of isolates tolerating a water potential of – 1.5 MPa, while more than 19% of isolates showed good growth under a water potential of –1 MPa. Compared to isolates of *S. meliloti*, the tolerance of isolates of *R. sullae* to water stress remained moderate (Elboutahiri et al., 2009). In addition, strains from dry and hot regions were more tolerant to drying than those from cold and humid regions (Hartel and Alexander, 1984).

For most rhizobia, the optimum temperature range for culture growth is 28–31°C, and many cannot grow at 37°C (Graham, 1992). At 28, 32 and 36°C respectively, 100, 88.70 and 82.25% of the isolates grew well. At 40°C, only 1.6% of the isolates grew. These results are similar to those by Struffi et al. (1998) and Kishinevsky et al. (1996, 2003) who reported that the maximum temperature tolerated by *R. sullae* isolated from *H. coronarium* is 37°C.

All the isolates tested could grow at pH values 9 and 9.5. Indeed, *R. sullae* is able to nodulate *H. coronarium* in soils with a pH 9.3, implying that its *nod* genes, unlike those of other rhizobia (Richadson et al., 1988), could be induced and expressed under such conditions. At very low pH (pH3 and pH3.5), none of the isolates were able to grow.

The intrinsic resistance to antibiotics depended on the antibiotics tested. The degree of resistance could vary from one species to another and even between strains of the same species.

The sampled isolates showed good tolerance to heavy metals such as Mn, Zn and Cd. The highest number of isolates grew well in 300 μg/ml Mn (100%), followed by 5 μg/ml Cd (88.70%) and 200 μg/ml Zn and 20 μg/ml Cd (85.48%). But the growth of all isolates was inhibited by Hg.

When the 16S rDNA regions were amplified and electrophoressed on native polyacrylamide gels, 6 polymorphic DNA bands were detected. Further digestion of the PCR products with *Hae*III and *Hin*fI restriction enzymes (ARDRA analysis), resulted in increased number of polymorphic bands. The *Hin*fI endonuclease enzyme revealed the highest level of polymorphism. The cluster analysis based on results of ARDRA analysis, using PowerMarker ver. 3.25 (Liu and Muse 2005), classified the sampled isolates into 9 clusters. However, 8 isolates (out of 62 isolates) did not distribute to any of the 9 clusters (unclustered strains) at 89% of similarity.

References

Buss, M.D., Bottomley, P.J. 1989. Growth and nodulation responses of *Rhizobium meliloti* to water stress induced by permeating and nonpermeating solutes. Appl. Environ. Micrbiol. 55: 2431–2436.

Elboutahiri, N., Thami-Alami, I., Zaïd, E., Udupa, S.M. 2009. Genotypic characterization of indigenous *Sinorhizobium meliloti* and *Rhizobium sullae* by rep-PCR, RAPD and ARDRA analysis. African J. Biotechnol. 8:979–985.

Gao, J.L., Sun, J.G., Li, Y., Wang, E.T., Chen, W.X. 1994. Numerical taxonomy and DNA relatedness of Tropical rhizobia isolated from Hainan Province, China. Int. J. Syst. Bacteriol. 44:151–158.

Glatzle, A., Schulte-Batenbrock, T., Brockwell, J. 1986. Symbiotic incompatibility between two forage species of *Hedysarum*, grown in Morocco, and their homologus rhizobia. FEMS. Microbiol. Lett. 37:39–43.

Graham, P.H. 1992. Stress tolerance in *Rhizobium* and *Bradyrhizobium*, and nodulation under adverse soil conditions. Can. J. Microbiol. 38:475–484.

Hartel, P.G., Alexander, M. 1984. Temperature and desiccation tolerance of cowpea rhizobia. Can. J. Microbiol. 30:820–823.

Kishinevsky, B.D., Nandasena, K.G., Yates, R.J., Nemas, C., Howieson, J.G. 2003. Phenotypic and genetic diversity among rhizobia isolated from three *Hedysarum* species: *H. spinosissimum, H. coronarium* and *H. flexuosum*. Plant and Soil 251:143–153.

Kishinevsky, B.D., Sen, D., Yang, G. 1996. Diversity of Rhizobia isolated from various *Hedysarum* species. Plant and Soil 186:21–28

Liu, K., Muse, S.V. 2005. Powermarker: integrated analysis environment for genetic marker data. Bioinformatics 21:2128–2129.

Nuti, M.P., Casella, S. 1989. Advances in the rhizobia in arid environments. Arid Soil Res. 3: 243–258.

Richadson, A.E., Djordjevic, M.A., Rolfe, B.G., Simpson, R.J. 1988. Effect of pH, Ca, and Al on the exudation from clover seedlings of compounds that induces the expression of nodulation genes in *Rhizobium trifolii*. Plant and Soil 109:37–47.

Squartini, A., Struffi, P., Döring, H., Seleka-Pobell, S., Tola, E., Giacomini, A., Vendramin, E., Velàsquez, E., Mateos, P.F., Martinez-Molina, E., Dazzo, F.B., Casella, S., Nuti, M.P. 2002. *Rhizobium sullae* sp.nov. (formerly "*Rhizobium hedysari* "), the root-nodule microsymbiont of *Hedysarum coronarium* L. Int. J. Syst. Evol. Microbiol. 52:1267–1276.

Struffi, P., Corich, V., Giacomini, A., Benguedouar, A., Squartini, A., Casella, S., Nuti, M.P. 1998. Metabolic proprieties, stress tolerance and macromolecular profiles of rhizobia nodulating *Hedysarum coronarium*. J. Appl. Microbiol. 84:81–89.

Vincent, J.M. 1970. A Manual for the Practical Study of Root Nodule Bacteria. IBP Handbook, No 15. Blackwell Scientific Publications Ltd., Oxford, England.

Wei, G.H., Zhang, Z.X., Chen, C., Chen, W.M., Ju, W.T. 2006. Phenotypic and genetic diversity of rhizobia isolated from nodules of the legume genera *Astragalus*, *Lespedeza* and *Hedysarum* in northwestern China. Microbiological Res. Doi: 10.1016/j.micres.2006.09.005.

Chapter 11
Genetic Diversity in Tall Fescue Using AFLP Markers

Sandrine Flajoulot, Jean-Christophe Caillet, Vincent Béguier, and Philippe Barre

Abstract Despite some data on tall fescue diversity using molecular markers, mainly from United States accessions, there is a lack of knowledge on the global diversity all over the world for this species. The objectives of this study were to evaluate the sub-structure of tall fescue world wide variability and to place French cultivars into this sub-structure using AFLP markers. One plant per accession from 32 ecotypes and 40 plants from six French cultivars were used. A total of 116 polymorphic fragments were scored from two primers combinations. Clustering analysis on 32 ecotypes of tall fescue from all over the world showed two distinct clusters: European and Mediterranean types. In the Mediterranean type two clusters could be distinguished probably illustrating different colonization origins. The analysis of six French cultivars representing the French seed market showed six genetically distinct subsets which were included in the European type.

Keywords AFLP markers · Structuration · Tall fescue

Introduction

Festuca arundinacea Schreb, commonly known as Tall fescue is a cool season grass widely used as a major component of pasture and turf. Tall fescue is a hexaploïde ($2n=6x=42$) and contains 3 genomes (PG1G2). Tall fescue is native from Central Europe and two geographic types can be distinguished: the European type present in a large part of Europe and the Mediterranean type present in North Africa and in South of Spain. In the 19th century, West European ecotypes were introduced in United States and used in breeding program to give the first US tall fescue cultivar: Kentuky31, KY31 (1943). The majority of American cultivars derive from KY31 which has been also largely introduced all over the world (Australia, New Zealand,

S. Flajoulot (✉)
Jouffray-Drillaud, Station expérimentale La Litière, 86 600, Saint Sauvant, France;
Breeder Association of Forage Varieties (ACVF), 7, rue du Coq-Héron,
Paris Cedex 01 75030, France
e-mail: sflajoulot@lusignan.inra.fr

C. Huyghe (ed.), *Sustainable Use of Genetic Diversity in Forage and Turf Breeding*,
DOI 10.1007/978-90-481-8706-5_11,

Japan, South of Africa and South of America). Despite some data on tall fescue diversity using molecular markers (Mian et al., 2002, 2005; Busti et al., 2004), there is a lack of knowledge on the species diversity structure all over the world. The objectives of this study were to evaluate the sub-structure of tall fescue world wide variability using AFLP markers and to locate French cultivars into this sub-structure. The understanding of tall fescue diversity repartition has considerable implications on germplasm conservation and on breeding strategies.

Materials and Methods

One plant per accession from 32 ecotypes (Table 11.1) from all over the world, and 40 plants per cultivar from six French cultivars (A, B, C, D, E and F) were used.

Table 11.1 List and origin of the 32 ecotypes used in this study

Code	Accession	Country
1	191	France
2	198	France
3	340	France
4	400	France
5	446	France
6	572	France
7	PI 235470	Switzerland
8	Fest5	South Africa
9	Fest9	New Zealand
10	Fest14	Italia, Sicily
11	Fest15	Italia, Tuscany
12	Fest-24	Japan
13	Fest29	Mauritania
14	PI 204446	Turkey
15	PI 283290	Portugal
16	PI 283314	Former Soviet Union
17	PI 292851	Germany
18	PI 476285	United States, New York
19	PI 499495	China
20	PI 504538	Greece
21	PI 423120	Spain
22	PI 538931	Kazakhstan
23	PI 577095	Morocco
24	PI 595072	Romania
25	PI 598596	Kazakhstan
26	PI 598836	Morocco
27	PI 598917	Tunisia
28	PI 614890	United States, Washington
29	PI 618976	Spain
30	PI 619005	China
31	PI 632515	United States, California
32	PI 632567	Tunisia

AFLP were chosen because they quickly deliver a high number of markers across the whole genome. For the 32 ecotypes, a total of 116 polymorphic fragments were scored from two primer combinations. For the six French cultivars only 80 fragments among the 116 scored fragments were polymorphic and scored.

Principal Coordinate analysis (PCoA) and cluster analysis based on the unweighted pair-group method of arithmetic averages (UPGMA) were performed using NTSYSpc version 2.1, using a matrix of squared Euclidian distance.

In order to test whether the six French cultivars form separate clusters, we used the software program STRUCTURE version 2.2 (Pritchard et al., 2000; Falush, et al., 2007).

Results and Discussion

Global Diversity in Tall Fescue

In the PCoA on ecotypes (Fig. 11.1) the first two principle coordinates explained 36.9% (Dim 1 = 29% and Dim 2 = 7.9%) of the genetic variation The first axis showed the separation between European and Mediterranean types. The second

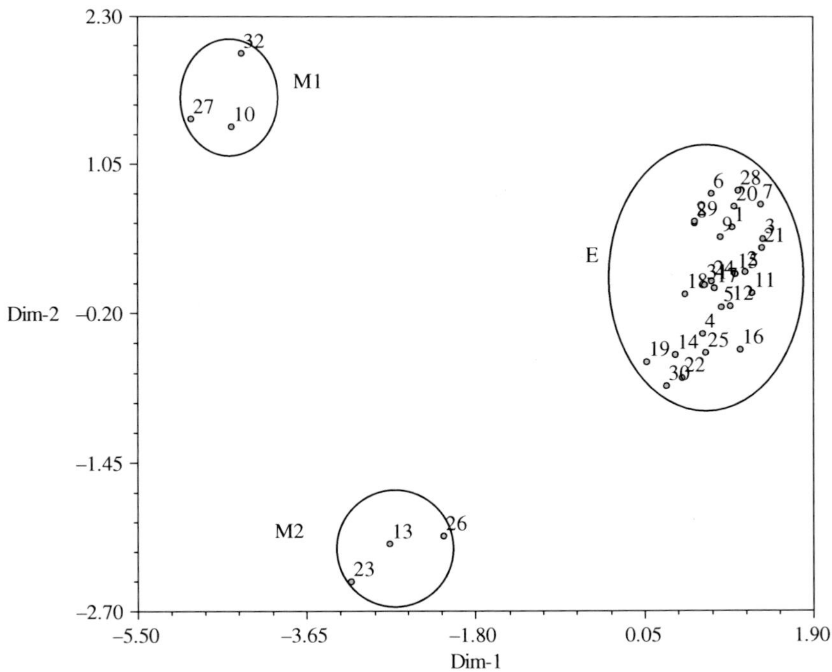

Fig. 11.1 PCoA using distance matrix (squared Euclidian) of 32 ecotypes of tall fescue calculated from 116 AFLP markers

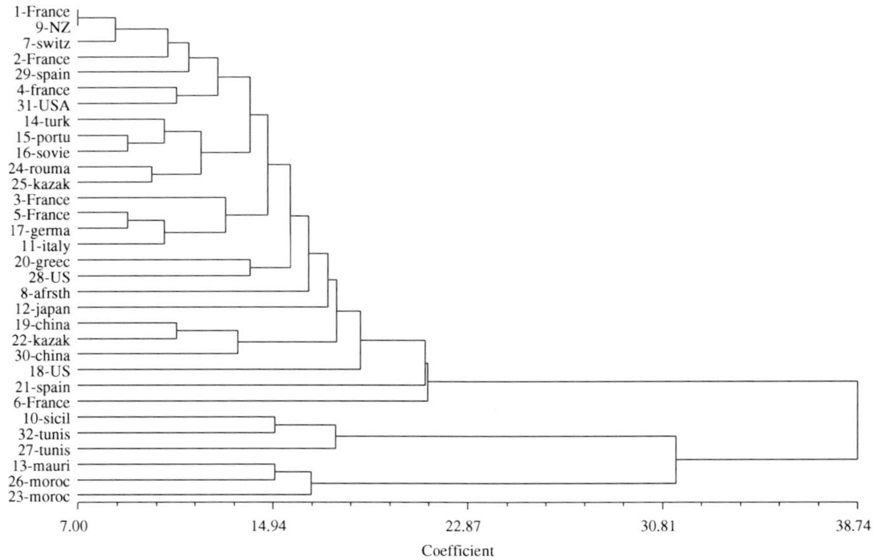

Fig. 11.2 Phylogenetic tree of 32 ecotypes of tall fescue based on distance matrix calculated from 116 AFLP markers

axis showed the separation of the Mediterranean type in two clusters M1 and M2 illustrating probably different colonization origins. The second cluster included 26 ecotypes from all over the world. This large cluster could come from an introduction of West European ecotypes in United States and then from this location further introductions all over the world. This clustering using PCoA was in accordance with the cluster analysis using UPGMA (Fig. 11.2).

Diversity in Six French Cultivars

In the PCoA on both ecotypes and French cultivars, we observed that all cultivar plants were overlapping with the 26 European type ecotypes (data not shown) as expected. In the PCoA on all cultivar plants (Fig. 11.3), the first and second principal coordinates explained 7.9% and 7.4% of the total molecular variation. A partial overlap of all cultivars was observed except for B and E cultivars which were more separated.

The most probable number of cluster (K) inferred by STRUCTURE analysis was 6 showing that all cultivars could be distinguished from each other using AFLP markers. Moreover, when the number of clusters was fixed to three, three distinct clusters were effectively formed, with B and E cultivars in two clusters, and the other cultivars (A, C, D, F) in the third cluster which was in accordance with the PCoA analysis. B and E cultivars appeared genetically more distinct from the others cultivars.

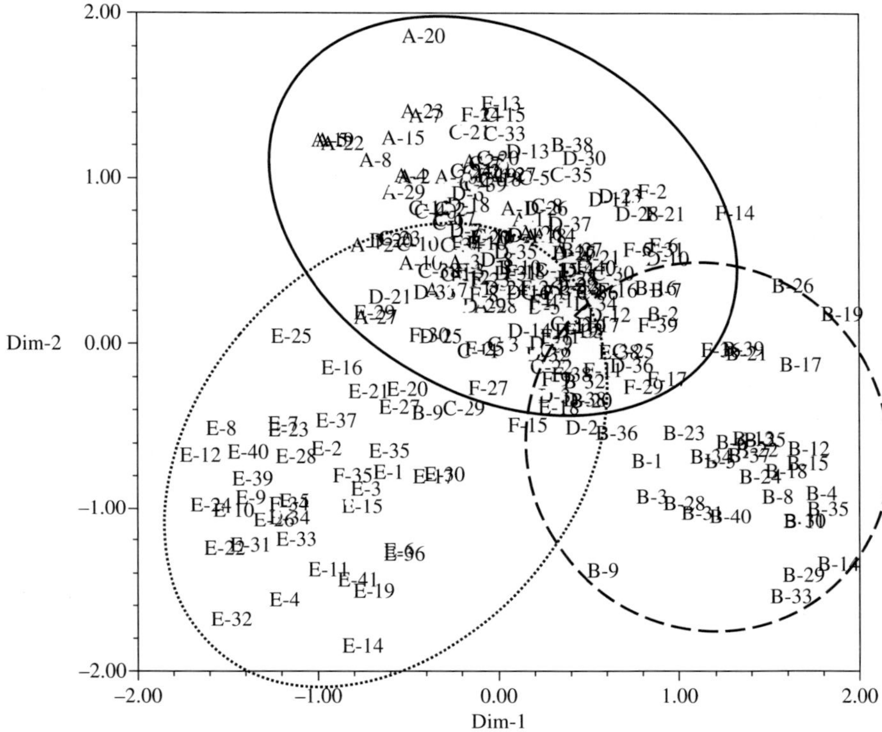

Fig. 11.3 PCoA using distance matrix (squared Euclidian) of 40 individuals from 6 French cultivars of tall fescue calculated from 80 polymorphic AFLP markers: —— *sub cluster with A, C, D and F cultivars;* – – – *sub cluster with B cultivar;* ● ● ● ● *sub cluster with E cultivar*

Clustering analysis on 32 ecotypes of tall fescue from all over the world showed two distinct clusters: European and Mediterranean types. In the Mediterranean type two clusters could be distinguished probably illustrating different colonization origins. The analysis of six French cultivars representing the French seed market with AFLP markers showed six genetically distinct subsets which were included in the European type.

References

Busti, A., Caceres, M.E., Calderini, O., Arcioni, S., Pupilli, F. 2004. RFLP markers for cultivar identification in tall fescue (*Festuca arundinacea* Schreb.). Genet. Resour. Crop Evol. 51: 443–448.

Falush, D., Stephens, M., Pritchard, J.K. 2007. Inference of population structure using multilocus genotype data: dominant markers and null alleles. Mol. Ecol. Notes 7: 574–578.

Mian, R.M.A., Hopkins, A.A., Zwonitzer, J.C. 2002. Determination of genetic diversity in tall fescue with AFLP markers. Crop Sci. 42: 944–950.
Mian, R.M.A., Saha, M.C., Hopkins, A.A., Wang, Z.Y. 2005. Use of tall fescue EST-SSR markers in phylogenetic analysis of cool-season forage grasses. Genome 48: 637–647.
Pritchard, J.K., Stephens, M., Donnelly, P. 2000. Inference of population Structure using multilocus genotype data. Genetics 155: 945–959.

Chapter 12
What are the Cellular Components Underlying the Genetic Diversity of Leaf Growth in *Lolium perenne* L.?

François Gastal, Philippe Barre, and Serge Carré

Abstract Grasses exhibit a large genetic variability in leaf growth and length, with major consequences on productivity and morphology of forage/turf types. The cellular components of the variability in leaf growth were analysed in *Lolium perenne* L., on eight genotypes of contrasting leaf length from a forage type mapping population, and on one turf genotype. Leaf elongation rate, cell number and size, and cellular dynamics in the leaf growth zone were determined on leaf 6. The variability in mature leaf length was due to variability in rate rather than duration of leaf elongation. In the forage type population, variability in leaf growth rate was due to variation in cell production rate and to variation in the number of cells simultaneously elongating in the growth zone, whereas elongation rate and duration of individual cells were similar. In the turf type, the lower leaf elongation rate was due to a decrease in both cell production and cell elongation rates. In conclusion, cell division plays a major role in genetic variation of leaf elongation among forage types, but cell elongation rate also discriminates between turf and forage types.

Keywords Cell division · Cell elongation · Genetic variability · Leaf growth · *Lolium perenne*

Introduction

Perennial grasses exhibit a large genetic variability in leaf growth and length (Sleper and Nelson, 1989; Gastal and Nelson, 1994; Ghesquière et al., 1994). This variability in leaf growth has major consequences on productivity (Horst et al., 1978; Barre et al., 2006), on morphology of the entire plant for example when comparing forage and turf types (Barre et al., Chapter 6), and on competition and dynamics

F. Gastal (✉)
INRA – UR4 Unité de Recherche Pluridisciplinaire Prairies et Plantes Fourragères (UR P3F), F86600 Lusignan, France
e-mail: Francois.Gastal@lusignan.inra.fr

C. Huyghe (ed.), *Sustainable Use of Genetic Diversity in Forage and Turf Breeding*, DOI 10.1007/978-90-481-8706-5_12,

of individual plants within populations (Hazard and Ghesquière, 1995). In order to suggest characteristics for genetic analysis, the cellular components determining the variability in leaf growth were analysed in *Lolium perenne* L.

Material and Methods

The study was conducted on 8 genotypes of contrasting leaf length originating from a forage type mapping population (Barre et al., 2009), and on a genotype from a turf type breeding population from DLF-Trifolium. The mapping population was obtained by a cross between a long-leaved genotype and a short-leaved genotype. Plants were vegetatively propagated, grown in a cabinet under optimal climatic conditions (19°C, 65% RH, 400 μE m^{-2} s^{-1} PPFD) and received a full strength nutrient solution. After 4 weeks of acclimation, plants were defoliated to a height of 3.5 cm. On a first subset of 10 plants per genotype plants, leaf length of leaves 2–6 was measured every 2 days, in order to calculate leaf elongation. After app. 6 weeks of growth, leaf 6 had reached its final size and was sampled to evaluate cell number and cell size on the entire leaf. On a second subset of 10 plants per genotype plants, length of leaf 6 was measured daily immediately after emergence, during 3 consecutive days. This growing leaf, which was then at it maximum elongation rate, was sampled to evaluate epidermal cell length profile along the growth zone, and to derive cellular parameters.

Imprints of leaf surface were taken for subset 1 at 4 positions along the mature leaf 6 (middle sheath, base, middle and upper portion of lamina), and for subset 2 along the entire leaf growth zone (the first 50 mm from leaf base), after removal of the protecting sheath. Epidermal cell length was measured microscopically, using an image analysis system. Leaf growth parameters were derived from leaf length measurements (leaf elongation rate in the linear phase of leaf elongation; mature leaf length; leaf elongation duration, calculated as the time to reach 98% of final leaf length). Cell number and average cell size along a row of epidermal cells over the length of mature leaf 6 were calculated by cumulating and averaging cell number and cell size over the four leaf segments previously defined. Cell growth parameters in the growth zone at the stage of rapid leaf elongation (cell elongation duration; cell elongation rate; number of cells simultaneously elongating in the leaf growth zone) were derived from the profile of epidermal cell length according to Schnyder et al. (1990).

Results and Discussion

Leaf Elongation Rate and Duration

Length of mature leaf 6 (L6) varied in a significant range among the genotypes of the forage population (from 235 to 340 mm), and as expected was shorter in the turf

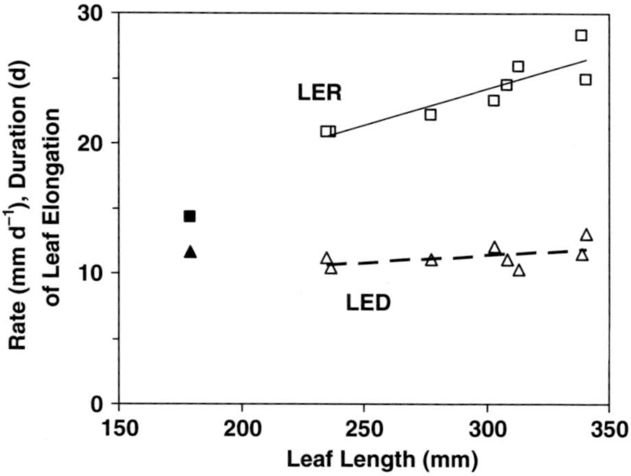

Fig. 12.1 Leaf elongation rate (LER, *square*) and leaf elongation duration (LED, *triangle*) in relation to leaf length of mature leaf 6 in eight forage (*light symbols*) and one turf (*dark symbols*) genotypes

genotype (179 mm). The 8 genotypes chosen in the present study were representative of the population in term of leaf length (not shown). The variation in leaf length was determined by a limited variation in leaf elongation duration (LED), and by a larger variation in leaf elongation rate (LER) (Fig. 12.1). These results are in line with earlier observations on divergent selection lines (Bahmani et al., 2000; Hazard et al., 1996), and on mapping populations (Barre et al., 2009). Nevertheless it is possible to find genotypes differing for both LER and LED (Barre et al., Chapter 6).

Interestingly, this conclusion also applies for the turf genotype. The shorter leaf length of the turf genotype corresponded to a lower LER than any genotype of the forage population, while LED of the turf genotype was close to the LED of forage genotypes.

Cell Number and Cell Size Along the Entire Leaf

In the forage genotypes, the variation in length of mature leaf 6 was mostly determined by variation in the number of epidermal cells per file along the leaf, whereas the average cell size did not vary significantly (Fig. 12.2). In contrast, in the turf genotype, both the number of epidermal cells per row along the leaf, and the average cell size, were decreased in comparison to the cell number and the cell size of the forage genotypes.

Cellular Dynamics

In the grass leaf, the growth zone is located at the base of the leaf and comprises sequentially a zone of cell division, where cells divide and elongate slowly, and

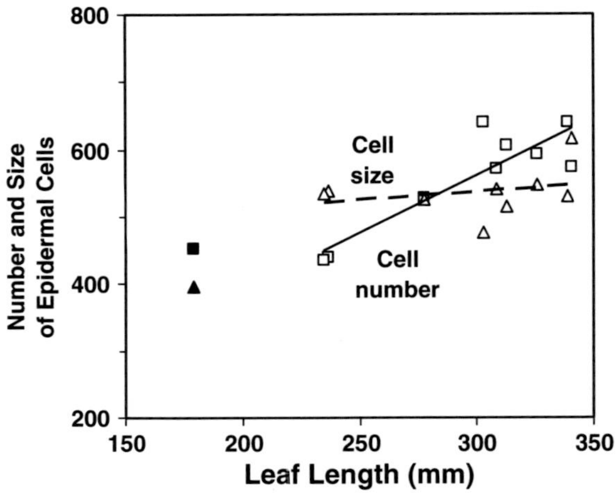

Fig. 12.2 Number (*square*) and average length (μm, *triangle*) of epidermal cells along a cell row over the entire leaf, in relation to leaf length of mature leaf 6 in eight forage (*light symbols*) and in a turf (*dark symbols*) genotypes

a zone of elongation, where cell elongation rate increases progressively, reaches a maximal value and decreases rapidly to stop almost suddenly, at the distal end of the growth zone (MacAdam et al., 1989).

The cell elongation rate (at its maximum rate) only marginally varied between forage genotypes (Fig. 12.3), and cell elongation duration did not vary at all, revealing that in the mapping forage population, the variation in leaf length was not determined by a significant change in cell elongation parameters. However, the turf genotype exhibited a lower rate of cell elongation, compared to the forage

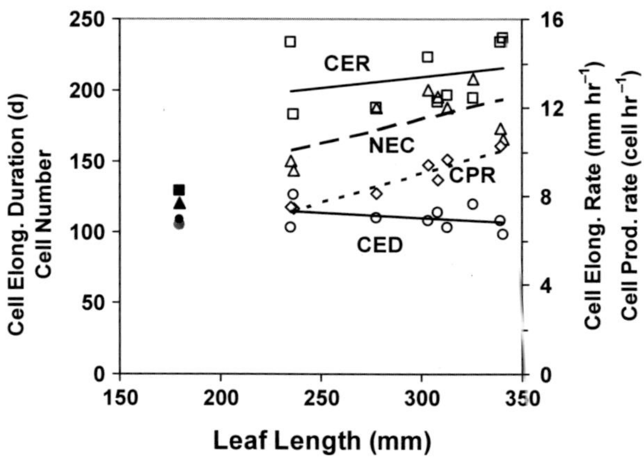

Fig. 12.3 Maximal cell elongation rate (CER, *square*), cell elongation duration (CED, *dots*), cell production rate (CPR, *diamond*) and number of elongating cells (NEC, triangle) within the growth zone of mature leaf 6, in relation to its length in eight forage (*light symbols*) and one turf (*dark symbols*) genotypes

genotypes, therefore indicating that its shorter leaves where partly due to a lower cell elongation rate.

In contrast, the cell production rate was significantly different among the forage genotypes, and between the turf and the forage genotypes. In addition, the number of epidermal cells simultaneously undergoing elongation within the leaf growth zone, increased significantly with final leaf size, both in the forage genotypes and for the turf genotype compared to the forage types. Therefore, the two components, cell production rate and, consequently, number of cells simultaneously elongating, are the main determinants of the differences in leaf length in the forage genotypes. They determine, in conjunction with a lower cell elongation rate, the shorter leaf of the turf type. The results on the forage mapping population are in line with the study from MacAdam et al. (1989) on tall fescue, showing that the variation in leaf elongation rate between populations derived from a divergent selection was mostly determined by cell production rather than cell elongation.

Conclusion

The variability in leaf length appears to be determined differently, at the cellular level, within a forage mapping population and between turf and forage types. In the forage population originating from two highly contrasting genotypes for leaf length, the variability in leaf growth rate was due to variation in the cell production rate and in the number of cells simultaneously elongating in the growth zone, whereas rate and duration of individual cell elongation were similar. In the turf type, the lower leaf elongation rate was due to a decrease in both cell production and cell elongation rates. In conclusion, cell division plays a major role in genetic variation of leaf elongation among forage types, but cell elongation rate is an additional characteristic which discriminates between turf and forage types.

References

Bahmani, I., Hazard, L., Varlet-Grancher, C., Betin, M., Lemaire, G., Matthew, C., Thom, E.R. 2000. Differences in tillering of long – and short- leaved perennial ryegrass genetic lines under full light and shade treatments. Crop Sci. 40(4):1095–1102.

Barre, P., Emile, J.-C., Betin, M., Surault, F., Ghesquière, M., Hazard, L. 2006. Morphological characteristics of perennial ryegrass leaves that influence short-term intake in dairy cows. Agron. J. 98:978–985.

Barre, P., Moreau, L., Mi, F., Turner, L., Gastal, F., Julier, B., Ghesquière, M. 2009. Leaf length QTLs in perennial ryegrass. Grass For. Sci. 64:310–321.

Gastal, F., Nelson, C.J. 1994. Nitrogen use within the growing leaf blade of tall fescue. Plant Physiol. 105: 191–197.

Ghesquière, M., Hazard, L., Betin, M. 1994. Breeding for management adaptation in perennial ryegrass (*Lolium perenne*). II. Genetic variability and heritability of leaf morphogenesis components. Agronomie 14:267–272.

Hazard, L., Ghesquière, M. 1995. Evidence from the use of isozyme markers of competition in swards between short-leaved and long-leaved perenial ryegrass. Grass For. Sci.50:241–248.

Hazard, L., Ghesquière, M., Barraux, C. 1996. Genetic variability for leaf development in perennial ryegrass populations. Canadian J. Plant Sci. 76: 113–118.

Horst, G.L., Nelson, C.J., Asay, K.H. 1978. Relationship of leaf elongation rate to forage yield of tall fescue genotypes. Crop Sci. 18:715–719.

MacAdam, J.W., Volenec, J.J., Nelson, C.J. 1989. Effects of nitrogen on mesophyll cell division and epidermal elongation in tall fescue leaf blades. Plant Physiol. 89:549–556.

Schnyder, H., Seo, S., Rademacher, I.F., Kuhbauch, W. 1990. Spatial distribution of growth rates and of epidermal cell lengths in the elongating zone during leaf development in *Lolium perenne* L. Planta 181:423–431.

Sleper, D.A., Nelson, C.J., 1989. Productivity of Selected High and Low Leaf Area Expansion *Festuca arundinacea* Strains (pp. 379–380). XVI International Grassland Congress, Nice, France.

Chapter 13
Evaluation and Utilization of Morphological Variation in a *Medicago truncatula* Core Collection

Yuanhong Han, Christy M. Motes, and Maria J. Monteros

Abstract *Medicago truncatula* is a model species for legume biology and has been used to develop tools for molecular genetics and genomics. Nested core collections representing the existing genetic diversity from the USDA germplasm collection have previously been identified using molecular markers. The practical value of nested core collections is that they allow implementation of efficient strategies to characterize phenotypic variation compared to random selection of accessions from the whole collection or using geographic stratification to select accessions for evaluation. The goals of this research were to efficiently use the *M. truncatula* core collections to characterize the natural existing variation associated with morphological traits. We assayed variation in pods (length, number of coils, direction of coiling, number of pods per raceme), leaves (morphology and pigmentation patterns), and roots (morphology and biomass) from accessions included in the nested core collections. A remarkable natural variation was identified for the morphological characteristics evaluated in pods, leaves and roots. Accessions with contrasting phenotypes for multiple traits are currently being used to develop mapping populations. The *M. truncatula* core collections are publicly available and should enable researchers to efficiently evaluate genetic variation for additional traits of interest.

Keywords *Medicago truncatula* · Core collection · Morphology

Introduction

Legumes represent an important component of the world's crop production due to their symbiotic nitrogen fixation capabilities, high protein and oil content, and nutritional value. The annual legume *Medicago truncatula* Gaertn., commonly known

Y. Han (✉)
Forage Improvement Division, The Samuel Roberts Noble Foundation, Sam Noble Parkway, Ardmore, OK 73401, USA
e-mail: yhan@noble.org

C. Huyghe (ed.), *Sustainable Use of Genetic Diversity in Forage and Turf Breeding*,
DOI 10.1007/978-90-481-8706-5_13,

as barrel medic, has been used as a model for legume species. The USDA-ARS National Plant Germplasm System maintains one of the three main *M. truncatula* collections. A previous study evaluated the genetic relationships of the *M. truncatula* accessions in the USDA collection using 40 SSR markers (Zhang et al., 2009). The maximization strategy (M strategy), implemented in the "MSTRAT" software (Gouesnard et al., 2001), was used to construct a series of nested core collections from the entire USDA *M. truncatula* collection. Core collections are sub-samples of collections and include accessions chosen to represent the majority of the genetic diversity contained in the larger collection (Frankel, 1984). The goals of this research were to characterize the natural existing variation associated with morphological traits in leaves, pods and roots from *M. truncatula* accessions included in the 64-entry core collection.

Materials and Methods

Plant Materials

The 64-entry *M. truncatula* nested core collection (Zhang et al., 2009) was used to assay morphological variation in pods, leaves and roots.

Leaf and Pod Morphology

Seeds from the *M. truncatula* 32-entry core collection (two generations of single seed descent) and from entries 33–64 (one generation of single seed descent) were surface sterilized, scarified and stored at 4°C for 48 hours. Five germinated seedlings per accession were transferred to soil and plants were grown in a greenhouse at 25°C and 18 h light photoperiod. Data for leaf and plant morphology (Traits 1–6; Table 13.1) were collected when plants were at full bloom. Data for pod morphology (Traits 7–14; Table 13.1) were collected when pods started to dry down.

Root Length and Biomass

Twenty accessions from the 32-entry *M. truncatula* core collection (two generations of single seed descent) were germinated and grown in a greenhouse at 25°C and 18 h light photoperiod. Roots from two separate 10 week old plants were washed, blotted dry and used to measure root length and biomass.

Table 13.1 Morphological traits evaluated in the 64-entry *M. truncatula* core collection

Traits	Description
Growth habit	1 = Erect, 9 = Prostrate
Flower per raceme	Mean of three randomly chosen flowering racemes from different plants
Internode length	Length of third internode from the base of the plant. Mean of three plants
Leaflet length	Length of middle leaflet. Mean of three mature leaflets from different plants
Leaflet width	Width of middle leaflet. Mean of three mature leaflets from different plants
Plant height	Mean from three mature plants
Pod length	Mean of the length of three randomly chosen pods from different plants
Pod spines	1 = Smooth, 9 = Spiny
Pods per raceme	Mean of three randomly chosen racemes with pods from different plants
Pod tightness	LP = Loose pods, MX = Both loose and tight pods, TP = Tight pods
Spine curvature	1 = No curves, 2 = Slightly curved, 3 = Curved, 4 = Slightly hooked, 5 = Hooked
Spine length on pod	Mean length of the three hooks per pod
Coils on pods	Number of coils per pod
Direction of pod tip coil	C = clockwise, CC = counterclockwise

Results and Discussion

A remarkable range in natural variation for morphological characteristics in pods, leaves and roots was identified in the *M. truncatula* core collection (Table 13.2, Fig. 13.1). The existing variation for the morphological characteristics evaluated

Table 13.2 Summary of morphological traits from the 64-entry *M. truncatula* core collection

Trait	Minimum	Maximum	Average
Growth habit	1	9	5.4
Flower per raceme	1	6	2.8
Internode length (mm)	13.8	88.3	38.0
Leaflet length (mm)	10.9	34.8	22.1
Leaflet width (mm)	7.7	26.8	15.6
Plant height (mm)	267	1160	638
Pod length (mm)	3.2	13.5	6.9
Pod spines (No.)	1	9	5.8
Pods per raceme (No.)	1	5	1.8
Spine curvature	1	5	3.2
Spine length on pod (mm)	0.3	5.6	2.6
Coils on pods	3	7	4.2
Root length (mm)	39	283	187
Root biomass (g)	0.2	3.8	1.0

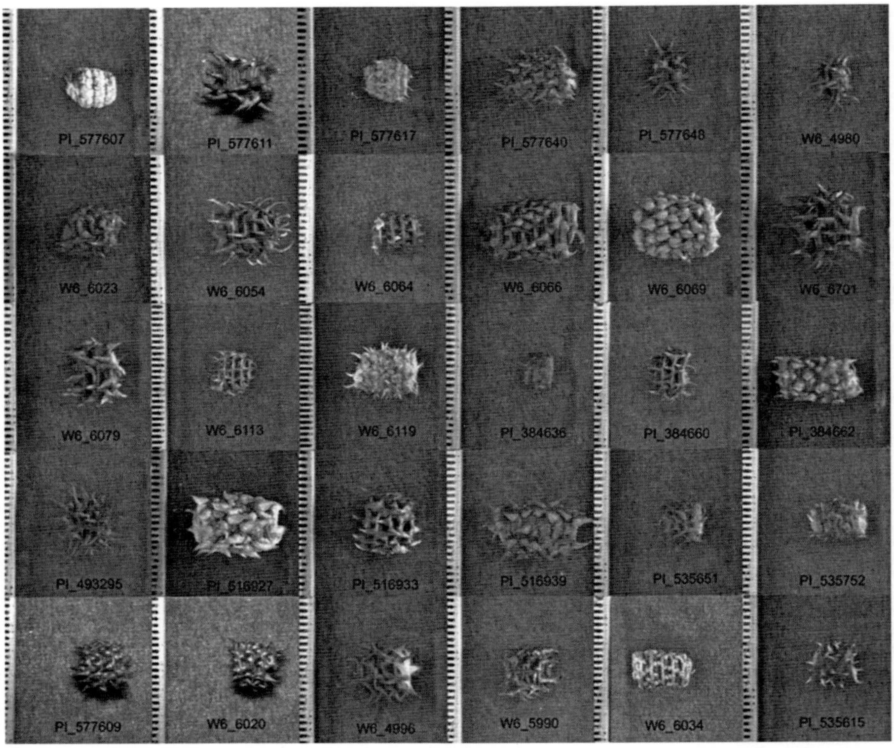

Fig. 13.1 Pod shape variation in the 64-entry *M. truncatula* core collection

can be used to optimize exploitation of these genetic resources for improving traits of agronomic importance. Current efforts include generating *M. truncatula* inbred lines (two generations of single seed descent) from the entire 64-entry core collection, which will be available for distribution through GRIN. The morphological characteristics from the entire core will be re-evaluated with all inbred lines grown simultaneously and the morphological descriptors and pictures from leaves, pods and roots for each accession will be publicly available through the Germplasm Resource Information Network (http://www.ars-grin.gov/) database. Accessions with contrasting morphological characteristics and stress responses are currently being used to develop segregating populations to identify QTLs associated with these traits.

The USDA-ARS *M. truncatula* core collection represents a valuable resource that will enable researchers to: (1) Efficiently evaluate genetic variation for additional traits of interest, such as abiotic stress tolerance, disease resistance, and nodulation, (2) Further understand the underlying mechanisms affecting the traits evaluated, (3) Identify single nucleotide polymorphism (SNP) markers, and (4) Select diverse genotypes to use in breeding programs and in QTL and comparative mapping studies to facilitate subsequent map-based cloning of desirable genes.

References

Frankel, O.H. 1984. Genetic perspectives of germplasm conservation (pp. 161–170). In: Arber, W., Llimensee, K., Peacock, W.J., Starlinger P. (eds.), Genetic manipulation: impact on man and society. Cambridge University Press, Cambridge, UK.

Gouesnard, B., Bataillon, T.M., Decoux, G., Rozale, C., Schoen, D.J., David, J.L. 2001. MSTRAT: An algorithm for building germplasm core collections by maximizing allelic or phenotypic richness. J. Hered. 92:93–94.

Zhang, Y., Han, Y., Jiang, G., Sledge, M.K., Greene, S., Coyne, C.J., Kisha, T., Monteros, M.J. 2009. Nested core collections maximizing genetic diversity and population structure of the model legume *Medicago truncatula*. Genome (In Press).

Chapter 14
Genetic Diversity in a Collection of Lucerne Populations from the Mediterranean Basin Evaluated by SSR Markers

Bernadette Julier, Yasmina Semiani, and Meriem Laouar

Abstract The Mediterranean Basin is a major region of lucerne (*Medicago sativa*) cultivation. We used seven microsatellite markers to evaluate within-population diversity and test differentiation among lucerne populations from the Mediterranean Basin. A set of 16 populations from France, Italy, Tunisia, Algeria, Morocco, Australia and the USA, each represented by 30 individuals, was analysed for seven microsatellite markers. One marker was excluded from the analyses because of presence of null alleles. Number of alleles and expected heterozygozity were calculated to evaluate within-population diversity. F_{IS} was calculated to test departure of populations from panmictic equilibrium. Population structure was analysed by the calculation of F_{ST} index. Within-population genetic diversity was large in all populations, and the populations appeared to be at panmictic equilibrium. Global F_{ST} was significant but very low (0.016). Three groups were identified, one composed by only Gabes, a second group composed by three North African populations, the two Australian cultivars, one American cultivar and one European population and a third group with six European populations, one Moroccan population and one American cultivar. Differentiation among populations was limited, which could be explained by the numerous possibilities of gene flow among populations, such as seed exchange or pollen flux. Reproductive biology (allogamy) and genetics (autotetraploidy) favour the maintenance of large within-population diversity and limits the possibility of population differentiation for neutral markers. This relative lack of differentiation with neutral markers is observed while the populations are known to noticeably differ for agronomic traits.

Keywords *Medicago sativa* · microsatellite · genetic variability · F_{ST}

B. Julier (✉)
INRA, UR 4, BP 6, 86600 Lusignan, France
e-mail: Bernadette.Julier@lusignan.inra.fr

C. Huyghe (ed.), *Sustainable Use of Genetic Diversity in Forage and Turf Breeding*,
DOI 10.1007/978-90-481-8706-5_14, © Springer Science+Business Media B.V. 2010

Introduction

Structure of lucerne (*Medicago sativa*) in the Mediterranean Basin is little described. Populations from the North of Africa mainly are landraces. Traditionnally, each farmer produces its own seeds, but seed exchanges between farmers are frequent. In addition, other landraces or foreign varieties are also cultivated. Each population is adapted to local pedoclimatic conditions. Populations from the South of Europe are mostly varieties bred for improved forage yield and adaptation. Germplasm used by breeders mainly originates from former landraces and other cultivars. Varieties adapted to Mediterranean climates were also selected in the USA and Australia. All these cultivars have a low autumn dormancy, so that winter forage yield is high. However, cultivars from Europe usually are more dormant than those from North Africa (Julier et al., 1995).

Diversity in lucerne populations is usually high, especially when evaluated with molecular markers. Differentiation among populations is difficult, firstly because of the large within population variation and secondly because of small mean differences among populations (Flajoulot et al., 2005).

Sixteen populations, from North Africa, South Europe, Australia and the USA were selected to analyse their diversity and structure with microsatellite markers (SSR). The main questions are (1) Is the within-population diversity similar in traditional landraces and in cultivars? (2) Are the populations from the North and from the South of Mediterranean Basin different? (3) Are cultivars from Australia and North America more similar to one or the other origin?

Material and Methods

A set of 16 populations from France, Italy, Tunisia, Algeria, Morocco, Australia and the USA (Table 14.1) was analysed, each represented by 30 individuals. Seven microsatellite markers were chosen for a codominant interpretation. They had previously been shown to give clear amplification profiles, enabling dosage reading and with a low frequency of null alleles (Flajoulot et al., 2005). Amplification protocols for microsatellite markers were previously described (Flajoulot et al., 2005). PCR products were separated on acrylamide gels in Licor sequencer. For each individual and each marker, the allele dosage was scored, each allele being present in 1–4 doses or absent.

Within-population diversity was calculated using the mean number of alleles per individual and the average over markers of expected heterozygosity (Thrall and Young, 2000). F_{IS} was calculated to test departure of populations from panmictic equilibrium (Thrall and Young, 2000). Population structure was analysed by the calculation of F_{ST} index, using Gene4x (Ronfort et al., 1998) and Spagedi softwares (Hardy and Vekemans, 2002). The matrix of F_{ST} between pairs of populations was used to draw a dendrogram and define groups of related populations.

Table 14.1 Geographical origins of the 16 lucerne populations and estimation of their diversity: Mean number of alleles per individual (A_i) and mean expected heterozygozity (D) over six SSR markers

Population name	Origin	A_i	D	Type of population
North Africa				
Africaine	Morocco	2.62	0.711	variety
Demnat203	Morocco	2.49	0.696	landrace
Erfoud	Morocco	2.47	0.705	landrace
Gabes	Tunisia	2.45	0.674	landrace
Rich2	Morocco	2.64	0.719	landrace
Tamantit	Algeria	2.31	0.694	landrace
South Europe				
Ecotipo siciliano	Italy	2.63	0.721	landrace
Coussouls	France	2.45	0.673	variety
Mamuntanas	Italy	2.75	0.760	landrace
Magali	France	2.75	0.735	variety
Melissa	France	2.68	0.730	variety
Prosementi	Italy	2.65	0.707	variety
Australia and USA				
ABT805	USA	2.74	0.730	variety
Ameristand	USA	2.57	0.710	variety
Sardi Ten	Australia	2.54	0.718	variety
Siriver	Australia	2.54	0.694	variety

Results and Discussion

Diversity of Populations

Number of alleles ranged from 12 to 37 for the seven markers. Number of alleles was similar for each population. Mean number of alleles per individual ranged from 2.3 for MAA660456 to 3.0 for MTIC451. The test for panmictic equilibrium, carried out for each population and each marker, suggested that MAA660456 was not at equilibrium in the populations (F_{IS} ranging from 0.18 to 0.31, $P < 0.01$). This led us to conclude that null alleles at the locus MAA660456 disturbed the analysis causing an apparent disequilibrium and therefore, this marker was excluded from further analysis. The other six markers were at equilibrium in most cases so that global F_{IS} for each population over these six markers were not significant. As the populations were at equilibrium for all other markers, we concluded that the locus MAA660456 by itself caused the apparent disequilibrium. The presence of null alleles at this locus is likely to produce this effect. Therefore, MAA660456 was excluded from further analysis. The panmictic equilibrium of the populations was expected considering the mode of conservation of landraces or the mode of obtention of varieties through several generations of panmixia.

The two criteria describing within-population diversity (mean number of alleles per individual and average of expected heterozygozity) indicated a large diversity.

The number of alleles per individual averaged 2.58 and showed little differences among populations (Table 14.1). Depending on the markers, expected heterozygozity varied from 0.621 for MTIC432 to 0.818 for MTIC451. On average over the six SSR markers, it ranged from 0.673 for the French variety Coussouls to 0.760 for the Italian landrace Mamuntanas (Table 14.1), but again, differences among populations were limited. All populations thus showed a large within population diversity, both at the within and among individual levels. No difference for the level of diversity was noticed between varietys and landraces, nor among geographical origins (South Europe, North Africa, USA-Australia).

Genetic Structure of the Populations

Over all six markers, a global F_{ST} of 0.016 was obtained over the 16 populations. Taken individually, each marker generated global F_{ST} that ranged from 0.010 to 0.026. All these values were significant but very low. However, they were similar to those calculated in other studies on lucerne populations (Flajoulot et al., 2005; Herrmann et al., 2008). F_{ST} between pairs of populations varied from 0.0 for the populations Sardi10 and Demnat203 to 0.049 for the populations ABT805 and Gabes.

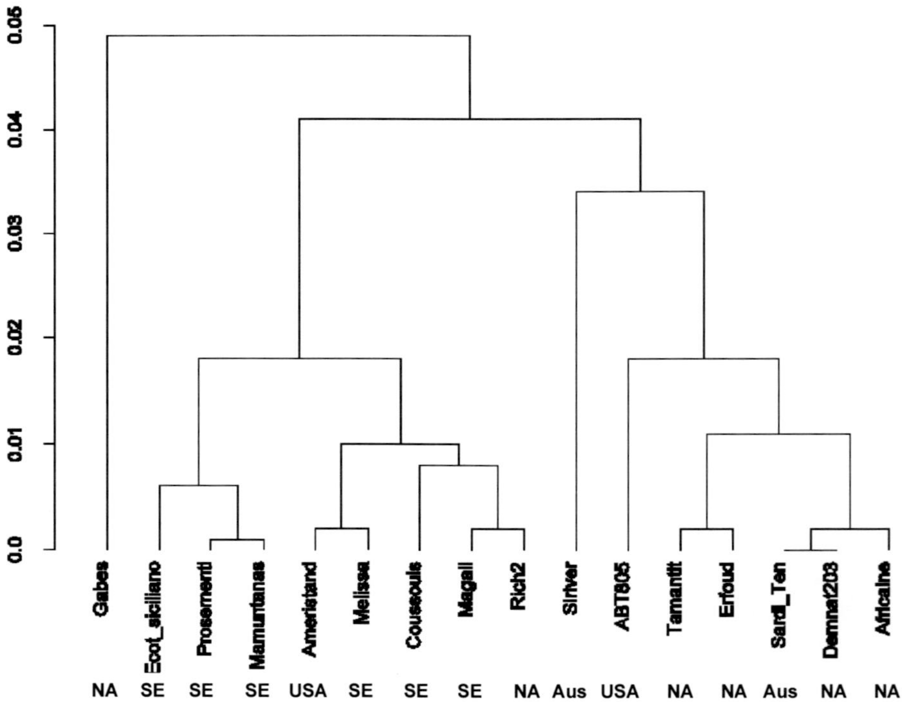

Fig. 14.1 Dendrogram obtained from a clustering procedure on the F_{ST} matrix among populations. Geographic origin is indicated: NA: North of Africa, SE: South of Europe, USA and Aus: Australia

The dendrogram calculated from the F_{ST} values among pairs of populations contained three groups (Fig. 14.1): one composed by only Gabes, a second group with the populations Ecotipo Siciliano, Prosementi, Mamuntanas, Ameristand, Melissa, Coussouls, Magali, Rich2 and a third group composed by Siriver, ABT805, Tamantit, Erfoud, Sardi Ten, Demnat203, Africaine. The second group gathered almost all varieties of South Europe but also a Moroccan landrace and an American variety. The third group gathered four populations from North Africa, the varieties from Australia and the USA and no European population. This structure did not exactly fit that expected from the geographical origins. However, populations from North Africa tended to be separated from the European populations, and among North African populations, Gabes was the most different. The two Australian varieties of this study (Sardi Ten and Siriver) and the American variety ABT805 might have been bred from genetic resources originating from North Africa. Contrastingly, the American variety Ameristand likely originates from genetic resources of Europe.

Large within-population diversity was observed in Mediterranean populations of lucerne as in other populations (Flajoulot et al., 2005). Differentiation among populations is weak. This could be explained by the numerous possibilities of gene flows among populations, such as exchanges or pollen fluxes. Reproductive biology (allogamy) and genetics (autotetraploidy) favour the maintenance of large within-population diversity and limit the possibility of population differentiation with neutral markers. However, Mediterranean populations differ for agronomic traits (Julier et al., 1995). Breeding thus consisted in increasing the frequency of favourable alleles at specific genes while keeping the diversity of the genetic background. However, a differentiation between populations from the North and the South of the Mediterranean Basin was observed with neutral markers. This could be related to a slight modification of the genetic background through traditional and modern breeding.

Acknowledgements This study was supported by European Union, program INCO PERMED (Improvement of native PERennial forage plants for sustainibility of MEDiterranean farming systems), 2004–2009. We thank Mounawer Badri (CBBC, Tunisia) for DNA extraction, Chrystel Gibelin and Denise Cadier for genotyping.

References

Flajoulot, S., Ronfort, J., Baudouin, P., Barre, P., Huguet, T., Huyghe, C., Julier, B. 2008. Genetic diversity among alfalfa (*Medicago sativa*) cultivars coming from a breeding program, using SSR markers. Theor. Appl. Genet. 111:1420–1429.

Hardy, O.J., Vekemans X. 2002. SPAGeDi: A versatile computer program to analyse spatial genetic structure at the individual or population levels. Mol. Ecol. Notes 2:618–620.

Herrmann, D., Flajoulot, S., Barre, P., Huyghe, C., Ronfort, J., Julier, B. 2008. Comparison of morphological traits and SSR to analyse diversity and structure of alfalfa cultivars. North American Alfalfa Improvement Conference, Dallas.

Julier, B., Porcheron, A., Ecalle, C., Guy, P. 1995. Genetic variability for morphology, growth and forage yield among perennial diploid and tetraploid lucerne populations (*Medicago sativa* L.). Agronomie 15:295–304.

Ronfort, J., Jenczewski, E., Bataillon, T., Rousset, F. 1998. Analysis of population structure in autotetraploid species. Genetics 150:921–930.

Thrall, P.H., Young, A. 2000. AUTOTET: A program for analysis of autotetraploid genotypic data. J. Hered. 91:348–349.

Chapter 15
Drought Survival of Some Perennial Grasses in Moroccan Rainfed Conditions: Agronomic Traits

Rajae Kallida, Naima Shaimi, and Chaouki Al Faiz

Abstract A 4 year experiment was carried out to evaluate perenniality and some adaptive responses to drought of *Dactylis glomerata*, *Festuca arundinacea* and *Phalaris aquatica*, within the multi-site activity of the EU-PERMED project. The trial held in Merchouch experimental INRA station at 68 km west of Rabat/Morocco, in a vertic deep silty clay soil (> 1 m 80 deep) was sown on 20 October 2005 to compare 16 grass accessions. Measurements included: sward establishment, dry matter production, average row cover, sward senescence and phenology. Plant emergence and establishment were good and regular for almost all cultivars. In all seasons, fescue yielded significantly more than cocksfoot. The harsh conditions and summer droughts have affected significantly production and persistence of grasses, mostly cocksfoot cultivars which disappeared completely at the end of the third year except the summer dormant cultivar Kasbah. Four groups of grass cultivars have been distinguished according to their persistence under extreme drought conditions.

Keywords Perennial grasses · Perenniality · Summer dormancy

Introduction

Perennial grasses are an important source of high quality forage which could complement grazed fallow, and contribute to the agro ecosystems sustainability by reducing soil erosion and conserving soil water. These forage species may present a valuable alternative to annual forage crops. They will improve environmental and economic sustainability of Mediterranean agro pastoral farming system. However, previous studies had shown the difficulties to find perennial grasses with the ability

---

R. Kallida (✉)
Institut National de la Recherche Agronomique (INRA), Rabat, Morocco
e-mail: rkallida@yahoo.fr

C. Huyghe (ed.), *Sustainable Use of Genetic Diversity in Forage and Turf Breeding*,
DOI 10.1007/978-90-481-8706-5_15,

to overcome severe Mediterranean summer droughts. Summer dormancy, induced by photoperiod and temperature, is the key-trait for drought avoidance, which improve plant persistence and allows the achievement of a range of agronomic and environmental goals (Volaire and Norton, 2006).

It is important to study the performance of elite grass material from recent plant improvement work under arid Mediterranean climate and to define its potential of production under severe drought conditions.

Material and Methods

The trial was located in north western Morocco in the INRA experimental station of Merchouch (33°47′ N, 6° 42′ E, 85 m a.s.l.) on a vertic deep silty clay soil about 180 cm deep. The average annual rainfall is 407 mm. The tested material belong to three perennial grass species: (i) *Dactylis glomerata*: Currie, Delta, Jana, Kasbah, Medly, Ottava and Porto; (ii) *Festuca arundinacea*: Centurion, Flecha, Fraydo, Lutine, Sisa, Tanit and (iii) *Phalaris aquatica*: Australian, Atlas, Partenope and Sirolan. Seedbed preparation included ploughing, disk harrowing and cultivation. Nitrogen and P_2O_5 at 40 and 80 kg ha^{-1}, respectively, were incorporated into the soil before sowing. We provided about 40 kg ha^{-1} of N after each harvest.

The design was based on 4 randomised blocks. All cultivars were sown on 20 October 2005. We secured the establishment of the swards by providing an irrigation of 90 mm in autumn of the first year. Kasbah, Ottava and Lutine cultivars have been resown. Trial was kept weed free by hand weeding all over the growing seasons. No disease or pest damage was observed through seasons.

Observations and measurements included plant emergence by visual scoring, flowering date and visual assessment of linear density rate of living plants (LDR) at starting of the regrowth following autumn rehydrations (scale 0–100). The difference 100-LDR indicates the proportion of empty segments of lines (absence or apparently dead plants, or weeds). Variation between average values of spring LDR and following autumn LDR is a good indicator of mortality/perenniality.

Table 15.1 Meteorological conditions during the 3 years of trial at Merchouch station

	2005–2006	2006–2007	2007–2008
Rainfall Oct-March (mm)	378	119.7	189.8
Rainfall Apr-June (mm)	33	51.6	90
Rainfall July-August (mm)	2.18	0	0
Total rainfall (mm)	413.18	171.3	279.8
Mean daily min. temp. Jan–Feb (°C)	4.9	6.05	5.84
Absolute min. temp. winter (°C)	0.14	0.55	0.54
Mean daily max. temp. July–August (°C)	27.8	26.5	27.5

Row cover was recorded in autumn and spring on the first year and at each autumn after regrowth. Dry matter yield was estimated by cutting when 50% plants reached flowering. Plots were harvested at 5 cm above ground level on a sample area of 3 m^2 per plot. Meteorological data were daily recorded (Table 15.1).

Results and Discussion

Flowering

Within cocksfoot, Kasbah was the earliest cultivar and could be taken as reference. The other cocksfoot flowered 5–11 days after Kasbah. Porto cv was the latest flowering with 3 weeks of delay from Kasbah (Table 15.2).

The cultivars of tall fescue Fraydo, Flecha, Sisa and Centurion flowered earlier (3–7 days from Kasbah) than Tanit, Lutine and FlechaE which flowered 12 days after Kasbah. For phalaris, Sirolan and Australian cultivars were the latest.

Table 15.2 Flowering date (from 1st January) and dry matter production of 18 perennial grasses cultivars sown in Marchouch location /Morocco

Species	Cultivars	Flowering date (from 1st January)	DM 1st year (t ha^{-1})	DM 2nd year (t ha^{-1})	DM 3rd year (t ha^{-1})
Dactylis glomerata	Currie	94	5.77	1.14	0.55
	Delta	89	1.58	1.31	0.43
	Jana	93	6.41	1.09	0.55
	Kasbah	82	3.41	1.26	2.68
	Medly	89	5.5	0.79	0.49
	Ottava	88	3.94	0.79	0.83
	Porto	105	4.21	0.12	0.06
Festuca arundinacea	Centurion	86	9.01	1.9	3.41
	FlechaE	86	9.12	2.55	5.79
	FlechaN	83	14.26	3.13	5.58
	Fraydo	78	10.06	2.65	5.30
	Lutine	90	8.21	1.28	0.98
	Sisa	84	10.15	1.45	2.37
	Tanit	88	10.31	1.41	2.40
Phalaris aquatica	Atlas	93	9.01	1.51	4.79
	Australian	100	13.67	1.66	4.37
	Parténope	84	12.19		3.05
	Sirolan	100	11.56	1.32	4.97
	Mean	90	8.24	1.49	2.70
	LSD 0.05	4.4	2.51	0.74	1.52

Dry Matter Production

Under rainfed Moroccan conditions, we did not make more than two cuttings during each growing season. In the first year, when the climatic conditions were favorable (429 mm), yields were high and reached 7.7 t ha^{-1} in average. Indeed, we noted a good crop establishment and an important winter growth. In the second year, after experiencing a very severe summer drought as well as during the growth season (only 180 mm), the yields were affected and the average did not exceed 1.5 t DM ha^{-1}.

Autumn and winter productions were low and without great differences between cultivars (Table 15.2). Cultivars Flecha and Fraydo accumulated more dry matter in winter (0.8 and 1.06 t ha^{-1}). The regeneration of the cocksfoot cv. Kasbah was better than the other Mediterranean cultivars Medly, Jana, Curie and Delta. Cultivars Porto and Ottawa had no production.

The relative advantage of fescue cvs Flecha and Fraydo was due to the winter and spring growth (1.9–2.1 t ha^{-1}).

The drought scenario was still active in the third year (281 mm). Autumn rainfall did not allow good grass regeneration. Yields were of 5.1 t ha^{-1} for the best fescue with stands of hardly 75% of that sown in the first year.

Phalaris cultivars had been penalized by the drought and produced 4.7 t ha^{-1} with little differences with the productive fescue cultivars. However, cocksfoot yields did not exceed half a ton of dry matter/ha for most cultivars. Kasbah performed well and produced 2.7 t ha^{-1}. Kasbah seems to be the most adapted because selected from Moroccan material. Indeed, the production of cocksfoot was significantly affected by the poor stand density during the third year and their poor persistence.

Grasses Persistence

Plant survival through summer droughts is one of the most important adaptive responses. It determines persistence, long-term production and water use efficiency during the next seasons. Persistence was approached mainly by the changes in the linear row density.

The severe drought had great effects on sward density. After 2 years, the reduction was on average 55% for cocksfoot, with variability between cultivars. Cultivar Kasbah kept dense swards in spite of the severe droughts benefiting of its two main characteristics which are drought resistance and summer dormancy. Cultivar Porto was the least persistent under our conditions. Another severe drought (180 mm) further reduced the LDR down to less than 10%, except for cultivar Kasbah which confirmed its adaptation to the drought and its good perenniality.

Fescue cultivars were more persistent than cocksfoot. LDR decreased by 23% on average. The decrease was more important for Sisa and Lutine which had disappeared completely by the end of the third year (Fig. 15.1).

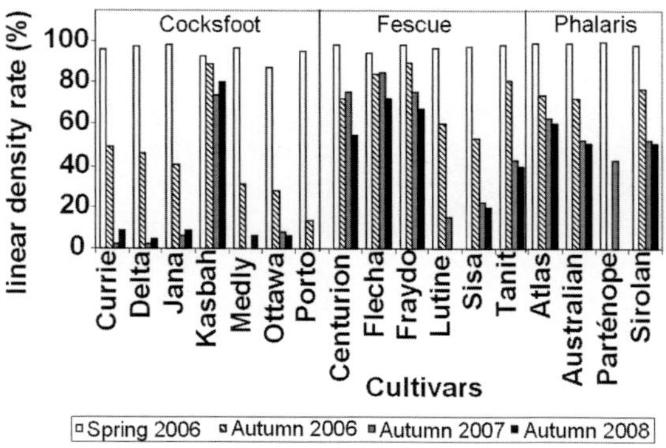

Fig. 15.1 Variation in the linear density rating (LDR, scale 0–100) in Merchouch location/ Morocco for spring 2006, autumn 2006, 2007 and 2008

Phalaris cultivars showed similar decreasing trend of LDR. Sward density reduction reached 25%, comparable to fescue cultivars. Cultivar Parténope behaved as an annual crop.

In Morocco, with drastic summer droughts, the changes in linear density rate allowed to select 4 groups of cultivars according to their persistence to severe drought in summer and even during the growing seasons:

(i) High level of drought resistance and high perenniality: This group contains Flecha, which kept optimal plant densities in the second year. The recovery after the third summer drought maintained a good plant density.

(ii) Intermediate level of drought resistance and perenniality: It includes cultivars Fraydo, Centurion and Kasbah, having started with optimal sward density and survived the first drought with LDR superior to 85%. The third year of drought had reduced their LDR of about 25%.

(iii) Limited level of drought resistance and perenniality: This group gathers Phalaris cultivars Atlas, Australian, Sirolan and the fescue Tanit. They had an initial optimal sward density but LDR decreased after the first drought and dropped by half after the third year.

(iiii) Low level of resistance and perenniality: This group includes all cocksfoot cultivars of Mediterranean origin except Kasbah and the temperate tall fescue Lutine and Sisa which has a survival level much lower than Mediterranean fescues as already demonstrated for the other cultivars (Demeter, Clarine,. . .), even in deep soils like in this trial (Norton et al., 2006).

Conclusion

Perennial grasses cultivars tested under Moroccan climate were characterized by very low yields especially in the stressful conditions during the second and third years. This was particularly useful to detect varieties able to stand severe drought stress levels. Probably perennial grasses have a decreasing interest when aridity increases, where annual rainfall is less than 350–400 mm.

Tall fescue cultivars were distinctly better yielding than Phalaris and cocksfoot varieties during this 3 years evaluation, showing advantages in yield and in persistence. Cultivar Flecha was the top-yielding one, while Fraydo and Centurion were intermediately persistent in our environment. Cultivar Flecha, selected in Argentina (apparently from Tunisian germplasm) and largely grown in Australia, could be introduced for cultivation into Moroccan growing environments. In cocksfoot, Kasbah (the only summer dormant variety) tended to outperform the other cultivars for forage yield and persistence. All the other cocksfoot cultivars were consistently low yielding and proved to be poorly persistent.

Acknowledgements This work was in part supported by the European Commission (PERMED project, INCO-CT-2004-509140) for financial support.

References

Volaire, F., Norton, M. 2006. Summer dormancy in perennial temperate grasses. Annal. of Botany 98:927–933.

Norton, M., Volaire, F. et Lelievre F. 2006. Summer dormancy in *Festuca arundinacea* Schreb., the influence of season of sowing and a simulated mid-summer storm on two contrasting cultivars. Aus. J. Agri. Res. 57:1267–1277.

Chapter 16
Drought Survival of Some Perennial Grasses in Moroccan Rainfed Conditions: Eco-physiological Traits

Rajae Kallida, Naima Shaimi, and Chaouki Al Faiz

Abstract A 4 year experiment was carried out to evaluate perenniality and some adaptative responses to drought of *Dactylis glomerata*, *Lolium arundinacea* and *Phalaris aquatica*, The trial was held in Merchouch experimental INRA station at 68 km west of Rabat/Morocco, in a vertic deep silty clay soil (>1m80 deep) was sown on 20 october 2005 to compare 18 grass accessions. Measurements included: establishment, dry matter production, sward senescence, water content and water soluble carbohydrates in survival organs.

The potential of production was important for tall fescue and phalaris compared to cocksfoot cultivars. Drought progression was different through varieties. Summer dormant cocksfoot became totally senescent at onset of drought, while non dormant cocksfoot had more green tissue. Festuca cultivars had slightly slow senescence along summer and behave identically against drought progression. Senescence levels reached more than 87% at the middle of summer drought. Water content in survival organs decreased gradually under summer drought. Festuca cultivars had maintained higher levels at the end of summer (48%) than cocksfoot cultivars (39%), which did not show any significant differences between varieties. Inversely, soluble water carbohydrates in survival organs increased with drought in the same pattern.

The harsh conditions and summer droughts have affected significantly production and persistence of grasses. Four groups of grass cultivars have been distinguished according to their persistence under extreme drought conditions.

Keywords Perennial grasses · Perenniality · Summer dormancy · Sward senescence · Water content and Water soluble carbohydrates

---

R. Kallida (✉)
Institut National de la Recherche Agronomique (INRA), Rabat, Morocco
e-mail: rkallida@yahoo.fr

C. Huyghe (ed.), *Sustainable Use of Genetic Diversity in Forage and Turf Breeding*, DOI 10.1007/978-90-481-8706-5_16, © Springer Science+Business Media B.V. 2010

Introduction

Perennial grasses are an important source of high quality forage which could complement grazed fallow, and contribute to the agro ecosystems sustainability by reducing soil erosion and conserving soil water. These forage species may present a valuable alternative to annual forage crops. They will improve environmental and economic sustainability of Mediterranean agro pastoral farming system. However, for forage plants growing in areas subject to prolonged and severe summer drought, the most important agronomic character is not the ability to produce during drought but the ability to survive, recover in autumn and grow actively during the rainy seasons (Volaire et al., 1998). Different adaptations have evolved among the grasses of the Mediterranean region ensuring their survival during the arid summer. Therefore, this study aimed to identify the main eco-physiological adaptations of an elite grasses material from recent plant improvement work from Mediterranean basin under harsh conditions.

Material and Methods

The trial was located in north western Morocco in the INRA experimental station of Merchouch (33°47′ N, 6° 42′ E, 399 m a.s.l.) on a vertic deep silty clay soil about 180 cm deep. The average annual rainfall is 407 mm. The tested material belong to three perennial grass species: (i) *Dactylis glomerata*: Currie, Delta, Jana, Kasbah, Medly, Ottava and Porto; (ii) *Festuca arundinacea*: Centurion, Flecha E, Flecha NE, Fraydo, Lutine, Sisa and Tanit and (iii) *Phalaris aquatica*: Australian, Atlas, Partenope and Sirolan. Seedbed preparation included ploughing, disk harrowing and cultivation. Nitrogen and P_2O_5 at 40 and 80 kg ha^{-1}, respectively, were incorporated into the soil before sowing. We provided 40 kg ha^{-1} of N after each harvest.

The design was based on 4 randomised blocks. All cultivars were sown on 20 October 2005.

Dry matter yield was estimated by regular mowing when plants reached 50% of flowering. A visual rating of sward senescence, assessed in the onset, middle and end of summer drought and visual assessment of linear density rate of living plants (LDR) at starting of the regrowth following autumn rehydration (scale 0−100) were performed. The difference 100-LDR indicates the proportion of empty segments of lines (absence or apparently dead plants, or weeds). Variation between average values of spring LDR and following autumn LDR is a good indicator of mortality/perenniality.

Leaf bases were dissected for determination of water content and water soluble carbohydrates (Dubois et al., 1956). Total rainfall was 413, 180 and 280 mm respectively for 2005/2006, 2006/2007 and 2007/2008.

Results and Discussion

Yield

Under rainfed Moroccan conditions, we were not able to make more than two cuttings each growing season. In first year, when the climatic conditions were favourable (429 mm), yields were highs and reached 7.7 t ha^{-1} in average. Indeed, we noted a good crop establishment and an important winter growth rate. In the second year, a very severe summer drought as well as during the growth season (only 180 mm) drastically reduced the yields, which did not exceed 1.5 t ha^{-1} on average (Fig. 16.1).

Autumn and winter production after regeneration was weak and without any significant difference between cultivars. Flecha cv and Fraydo cv accumulated more dry matter in winter compared to the other cultivars (0.8 and 1.06 t ha^{-1}). The recovery of the cocksfoot Kasbah was better compared with the other Mediterranean cultivars Medly, Jana, Curie and Delta. Porto and Ottava cultivars had an insignificant production.

The yields of the tall fescue cvs. Flecha and Fraydo were higher in winter and spring (1.86 – 2.07 t ha^{-1}).

In the third year, although with 281 mm, autumn rainfall did not allow good grass recovery. With stands of hardly 75% of that sown in the first year, yields were 5.1 t ha^{-1} only for the best fescue.

Phalaris cultivars had been penalized by the drought and produced 4.71 t ha^{-1} without great differences from productive fescue cultivars. Cocksfoot yields did not exceed half a ton of dry matter/ha for almost all cultivars; Kasbah performed relatively better and produced 2.68 t ha^{-1}, likely because selected from Moroccan material. In general, the production of cocksfoot was significantly affected by the poor stand density during the third year.

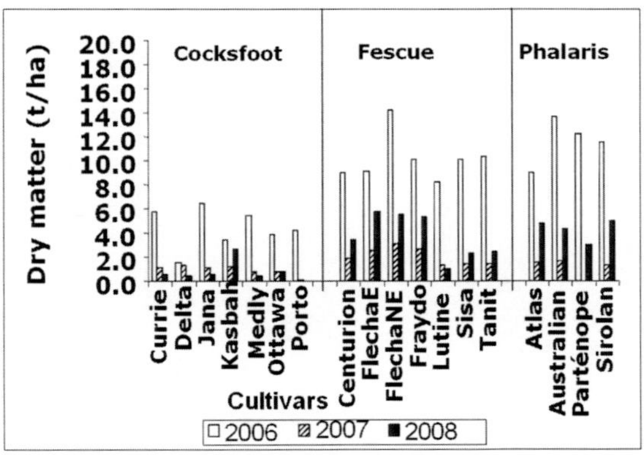

Fig. 16.1 Dry matter production (t ha^{-1}) of 18 perennial grasses cultivars sown in Marchouch /Morocco during 3 growing seasons

Responses to Summer Drought

Sward Senescence

During the driest years, senescence process had started more quickly for all cultivars than in 2006, when more rainfall was available. Senescence scores were two fold those reached in 2006 already at the beginning of summer. Volaire et al. (2007) reported a good correlation between the level of senescence and the residual soil water at the end of the summer. All cultivars of tall fescue exhibited a similar pattern with a regular progression of senescence from May to September before rehydration. These levels were very high at the middle of drought and reached more than 85%. In contrast, all cocksfoot exhibited a senescence that progressed more quickly than the tall fescues and therefore less than 5% of green material remained in the aerial tissues in September. Kasbah was the only cultivar to exhibit a complete endogenous dormancy with a senescence of almost 100% from June. The phalaris cultivars were also dormant although the progression of senescence was not as quick as for Kasbah (Fig. 16.2).

The progress of the senescence with drought is considered as being an answer to summer drought. The tested cultivars differed significantly in their senescence, whether at the beginning, middle or at the end of drought, suggesting different potential for drought tolerance. The less persistent fescue cultivars kept a high photosynthetic activity compared to other cultivars. The use of carbohydrates could be associated with their poor perenniality. Senescence of Kasbah was inferred in the beginning of drought which confirms its potential of drought tolerance.

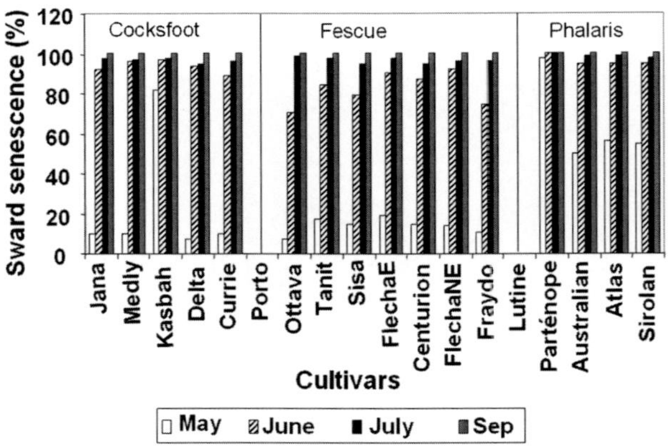

Fig. 16.2 Changes in aerial tissues senescence of perennial grasses sown at Merchouch (Morocco) in four dates during summer 2008

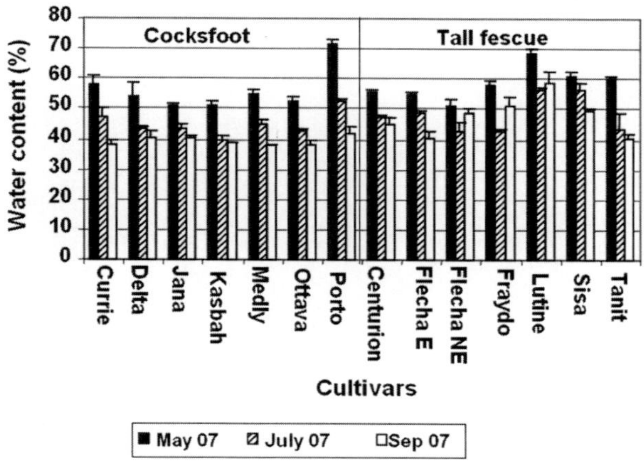

Fig. 16.3 Changes in water content of survival organs from perennial grasses tested at Merchouch (Morocco) in three dates during summer 2007

Water Content of Survival Organs

The water content of enclosed leaf bases decreased gradually under drought for all tested cultivars, but recovered after autumn rehydration. Indeed the moisture contents reached levels of 70–80% in autumn of the years of regeneration. It followed the progression of aerial senescence. The meristems of tall fescue remained more hydrated than those of cocksfoot (48% in September). Although more hydrated at onset of drought, the meristems of cocksfoot reached final values of 39% (Fig. 16.3).

In July, meristems of Kasbah had the lowest hydration due to its dormant status. In summer, water content in fescue cultivars ranged in wider interval than in cocksfoot cultivars. The meristems sampled for the phalaris were not representative since they were attached to the very few remaining vegetative tillers. In contrast to the reproductive tillers that formed basal bulbs, these vegetative tillers remained well hydrated.

Under dry seasons, levels of meristems hydration were lower than in the wettest year, when differences between cultivars had been revealed. Fescue meristems maintained a higher level of hydration than cocksfoot. Indeed, persistent cultivars were less hydrated at the end of drought for both cocksfoot and fescue.

Water Soluble Carbohydrates

Meristems water soluble carbohydrates content ranged between 15.5 and 38 mg /g of fresh matter at the beginning of drought, with higher contents for fescue cultivars. Drought progression increased total water soluble sugars in the surviving parts by 26 and 52% respectively for cocksfoot and fescue.

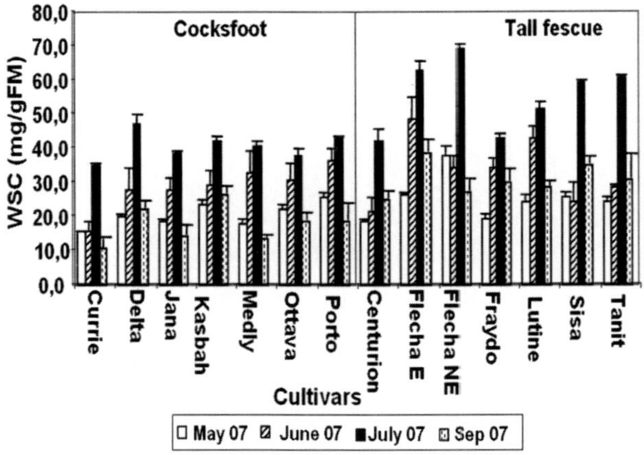

Fig. 16.4 Changes in water soluble carbohydrate of survival organs from perennial grasses tested at Merchouch (Morocco) in 3 dates during summer 2007 and after autumn rehydration

After rehydration, plants recovered active growth and metabolic activity. Sugars contents dropped to approximately 24% (Fig. 16.4). Sugar contents were higher for fescue cultivars in particular Flecha cv. It was reported that survival to drought was correlated to a large pool of water soluble sugars, in particular fructans in the organs of survival.

Persistence

The evolution of linear density rate (Fig. 16.5) had permitted to select four groups of cultivars according to their persistence to severe drought in summer and even during the growing season.

G1: The first group only contained the fescue cv. Flecha which kept dense sward density. Plant disappearance did not exceed 20% after severe droughts.

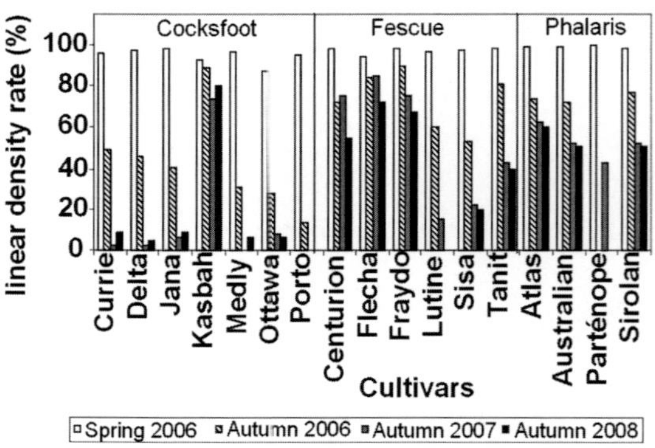

Fig. 16.5 Changes in the linear density rating (LDR, scale 0–100) in Merchouch location/Morocco for spring 2006, autumn 2006, 2007 and 2008

G2: The second group consisted of fescue cultivars Fraydo, Centurion and cocksfoot Kasbah. It started with optimal sward density and survived the first drought with dense swards (85%). The third year of drought had reduced their density of about 25%. This group was considered to have intermediate persistence.

G3: This group started with an optimal density of populating but the LDR decreased from the first drought and dropped by a half after the third year. This group contains the fescue cultivar Tanit and Phalaris Atlas PG, Australian and Sirolan. The varieties of Phalaris (cvs Australian Atlas PG and Sirolan) recovered slowly but significantly all along autumn through progressive emergence of summer dormant underground buds. They probably progressed along the autumn, but too slowly which constituted an important limit to early autumn production for Phalaris cultivars.

G4: This group consisted of material with limited drought resistance and low perenniality. It included all cocksfoot cultivars of Mediterranean origin apart from Kasbah, which belonged to the second group. The temperate tall fescue Lutine and Sisa had a much lower survival level than Mediterranean types (belonging to G2). This was already demonstrated for the other cvs (Demeter, Clarine,...), even in deep soils like in this trial (Norton et al., 2006).

The persistence of the cultivars after four summer droughts showed that the most affected cultivars are Mediterranean cocksfoot, except Kasbah and the fescue with oceanic origin. Phalaris seems to be intermediate. The level of dormancy exhibited in the field cannot explain the differences in drought sensitivity.

Conclusion

Perennial grasses cultivars tested under Moroccan climate were characterized by very low yields especially in the stressful conditions during the second and the third year. This was particularly useful to detect varieties able to stand severe drought stress levels. Probably this material has decreasing interest when aridity increases, where annual rainfall is less than 350–400 mm. Tall fescue varieties were distinctly higher yielding than Phalaris and cocksfoot cultivars during this 3-years evaluation, showing advantages in yield and persistence. Flecha cultivar was the top-yielding and could be widely introduced for cultivation into Moroccan growing environments. In cocksfoot Kasbah (the only summer dormant variety) tended to outperform the other cultivars for forage yield and persistence.

Acknowledgements This work was in part supported by the European Commission (PERMED project, INCO-CT-2004-509140) for financial support.

References

Dubois, M., Gilles, K.A., Hamilton, J.K., Robers, P.A., Smith, F. 1956. Colorimetric method of determination of sugars and related substances. Analy. Chem. 28:350–356.

Volaire, F., Norton, M., Lelievre, F. 2007. What is summer dormancy in temperate perennial grasses? Breeding and seed production for conventional and organic agriculture. Proceedings of the XXVI meeting of the EUCARPIA fodder crops and amenity grasses section, XVI meeting of the EUCARPIA *Medicago* spp group, Perugia, Italy, 2–7 September 2006, pp. 348–352.

Volaire, F., Thomas, H., Bertagne, N., Bourgeois, E., Gautier, M.F., Lelievre, F. 1998. Survival and recovery of perennial forage grasses under prolonged Mediterranean drought. II. Water status, solute accumulation of dehydrin transcripts in bases of immature leaves. New. Phytol. 140:451–460.

Norton, M., Volaire, F. et Lelievre F. 2006. Summer dormancy in *Festuca arundinacea* Schreb., the influence of season of sowing and a simulated mid-summer storm on two contrasting cultivars. Aust.. J. Agric. Res. 57:1267–1277.

Chapter 17
Forage and Seed Yield Components in Four French Landraces of Grass Pea (*Lathyrus sativus* L.)

Aleksandar Mikić, Vojislav Mihailović, Branko Ćupina, Đorđe Krstić, Sanja Vasiljević, and Dragan Milić

Abstract A small-plot trial was carried out in 2006, 2007 and 2008 at the Experimental Field of the Institute of Field and Vegetable Crops at Rimski Šančevi, including four grass pea local landraces from southern France, Le Cambou from Dordogne and Fléchou, Parranquet and Bon Encontre from Lot-et-Garonne. Each trial was composed of two identical parts, one for forage and another for seed. The plants in the first part were cut at the stage of full flowering, while in the second part the plants were harvested at the stage of full maturity of seeds in the oldest pods. Main forage and seed yield components were evaluated. The landrace Bon Encontre had the highest forage yields, with 42.08 g plant^{-1} of green forage and 8.18 g plant^{-1} of forage dry matter, while the landrace Le Cambou had the lowest forage yields, with 22.82 g plant^{-1} of green forage and 4.51 g plant^{-1} of forage dry matter. However, Le Cambou had the greatest average values of number of fertile nodes (17.7 plant^{-1}), number of pods (18.3 plant^{-1}), number of seeds (36.9 plant^{-1}) and seed yield (7.20 g plant^{-1}). The landrace Bon Encontre was the most promising for the development of dual-purpose grass pea cultivars.

Keywords Forage dry matter yield · Green forage yield · Grass pea · *Lathyrus sativus* · Seed yield · Yield components

Introduction

Grass pea (*Lathyrus sativus* L.) is one of the oldest European crops: its seeds were used by the hunter-gatherers as early as 12,000 BC, while it seems to have been cultivated in the Iberian Peninsula since 7,500 BC (Ljuština and Mikić, 2008). Similarly to other botanically related species, such as pea (*Pisum sativum* L.) or

A. Mikić (✉)
Forage Crops Department, Institute of Field and Vegetable Crops, Maksima Gorkog 30, Novi Sad, Serbia
e-mail: mikic@ifvcns.ns.ac.yu

C. Huyghe (ed.), *Sustainable Use of Genetic Diversity in Forage and Turf Breeding*,
DOI 10.1007/978-90-481-8706-5_17,

vetches (*Vicia* spp.), grass pea is a multifunctional crop that may be used for grain, forage, biomass and green manure (Campbell, 1997).

Grass pea in Serbia today is an almost completely forgotten crop and has been used extremely locally, with no official data (Mihailović et al., 2005). One of the aims of a concerted research between the Institute of Field and Vegetable Crops and the Faculty of Agriculture in Novi Sad is the re-introduction of legume species such as grass pea in Serbian agriculture. This required the establishment of a collection of grass pea accessions, with a primary aim of serving as a basis for the first Serbian breeding programmes and as a source for other research.

A grass pea collection in Novi Sad is a part of the Annual Forage Legumes Collection (AFLCNS) and is maintained in the Forage Crops Department of the Institute of Field and Vegetable Crops. The first grass pea accessions that entered AFLCNS were four local landraces from southern France, kindly donated by the Laboratoire d'Ecologie Moléculaire, IBEAS, University of Pau, France, in 2002. Today, the grass pea collection in Novi Sad has about 60 accessions of diverse geographical origin and status.

The aim of the study was to evaluate forage and seed yields and yield components in four French landraces of grass pea under growing conditions of Serbia.

Materials and Methods

A small-plot trial was carried out in 2006, 2007 and 2008 at the Experimental Field of the Institute of Field and Vegetable Crops at Rimski Šančevi. It included the first four grass pea accessions in AFLCNS, originating from Southern France, namely Le Cambou from Dordogne and Fléchou, Parranquet and Bon Encontre from Lot-et-Garonne.

The trial had two identical parts, one for forage and another for seed. In both parts and in each year, all four accessions were sown in early March, with a plot size of 5 m^2 and three replications. The sowing rate for all four accessions was 120 seeds m^{-2}, being greater than for grass pea grain production and in accordance with a widely used practice that forage production requires higher sowing rates than grin production (Mikić et al., 2007). In the first part, the plants were cut at the stage of full flowering. On the basis of a sample taken during cutting, the most important forage yield components were determined, such as plant height (cm), number of stems (plant^{-1}) and number of internodes (plant^{-1}), as well as green forage yield (g plant^{-1}). Forage dry matter yield (g plant^{-1}) was determined on the basis of a green forage sample after the drying at room temperature.

The plants of all four cultivars in the second part were harvested at the stage of full maturity of seeds in the oldest pods. On the basis of a sample taken during harvest, the most important seed yield components were determined, such as number of fertile nodes (plant^{-1}), number of pods (plant^{-1}), number of seeds (plant^{-1}) and thousand seed mass (g), as well as seed yield (g plant^{-1}).

Results were processed by analysis of variance (ANOVA) applying the Least Significant Difference (LSD) test using the computer software MSTAT-C.

Results and Discussion

There were significant differences in characteristics related to forage yields among the examined four French grass pea landraces (Table 17.1).

The average plant height ranged from 68 cm in Fléchou to 91 cm in Bon Encontre, which was more than in the preliminary results in the same environment (Mihailović et al., 2008). The greatest average values of both number of stems and number of internodes were found for Bon Encontre (8.3 plant^{-1} and 98.7 plant^{-1}), while the smallest average values of both number of stems and number of internodes were found for Le Cambou (4.4 plant^{-1} and 61.9 plant^{-1}). The landrace Bon Encontre also had the highest forage yields, with 42.08 g plant^{-1} of green forage and 8.18 g plant^{-1} of forage dry matter, while the landrace Le Cambou had the lowest forage yields, with 22.82 g plant^{-1} of green forage and 4.51 g plant^{-1} of forage dry matter.

Among the four French grass pea landraces, there were also significant differences in the characteristics related to seed yield (Table 17.2).

The landrace Le Cambou had the greatest average values of number of fertile nodes (17.7 plant^{-1}), number of pods (18.3 plant^{-1}) and number of seeds (36.9 plant^{-1}), while the landrace Fléchou had the lowest average values of number of fertile nodes (10.6 plant^{-1}), number of pods (11.1 plant^{-1}) and number of seeds

Table 17.1 Average values of the characteristics related to forage yield in the four French grass pea landraces for 2006–2008 at Rimski Šančevi

Landrace	Plant height (cm)	Number of stems (plant^{-1})	Number of internodes (plant^{-1})	Green forage yield (g plant^{-1})	Forage dry matter yield (g plant^{-1})
Le Cambou	78	4.4	61.9	22.82	4.51
Fléchou	68	6.4	75.1	28.16	4.91
Parranquet	78	7.3	98.1	26.73	5.81
Bon Encontre	91	8.3	98.7	42.08	8.18
$LSD_{0.05}$	8	2.4	27.4	9.48	1.44
$LSD_{0.01}$	11	3.7	38.5	13.16	2.62

Table 17.2 The average values of the characteristics related to seed yield in the four French grass pea landraces for 2006–2008 at Rimski Šančevi

Landrace	Number of fertile nodes (plant^{-1})	Number of pods (plant^{-1})	Number of seeds (plant^{-1})	Thousand seed mass (g)	Seed yield (g plant^{-1})
Le Cambou	17.7	18.3	36.9	210	7.20
Fléchou	10.6	11.1	24.3	186	4.37
Parranquet	13.7	14.9	25.3	220	5.30
Bon Encontre	15.7	17.6	33.3	163	5.47
$LSD_{0.05}$	5.2	6.2	5.6	54	3.01
$LSD_{0.01}$	7.0	8.8	8.2	71	4.48

(24.3 plant^{-1}). The average values of thousand seed mass ranged from 163 g in Bon Encontre to 220 g in Parranquet. With an average seed yield of 7.20 g plant^{-1}, the landrace Le Cambou confirmed its great potential for seed production (Mihailović et al., 2007).

Conclusion

Among the four tested landraces, some with good potential for either forage or seed production were identified which may be successfully used as sources of genetic variation for breeding or conservation. Landraces such as Bon Encontre may be useful in the development of dual-purpose grass pea cultivars, being able to answer the needs of one part of the market for a cultivar that can provide both forage and grain.

Acknowledgements The curators of AFLCNS would like to acknowledge Dr. Daniel Combes for his donation of its first grass pea accessions.

This research provided the basis for the Project 20090 (2008–2010) of the Ministry of Science and Technological Development of the Republic of Serbia.

References

Campbell C.G. 1997. Grass Pea. *Lathyrus sativus* L. Institute of Plant Genetics and Crop Plant Research (IPK), Gatersleben, Germany – International Plant Genetic Resources Institute (IPGRI), Rome, Italy, p. 91.

Ljuština, M., Mikić, A. 2008. Grain legumes technology transfer in Old Europe – archaeological evidence. Book of Abstracts of the Second GL-TTP Workshop *Integrating Legume Science and Crop Breeding*, Novi Sad, Serbia, 27 and 28 November 2008, pp. 42–43.

Mihailović, V., Mikić, A., Ćupina, B., Erić, P. 2005. Field pea and vetches in Serbia and Montenegro. Grain Legumes 44:25–26.

Mihailović, V., Mikić, A., Ćupina, B., Matić, R., Katić, S., Karagić, Đ., Milošević, M., Vujaković, M. 2007. Comparing yields of feed pea, common vetch, lentil and grasspea. Proceedings of the First Joint Prince of Songkhla University (PSU) – University of Novi Sad (UNS) International Conference on BioScience: Food, Agriculture, and the Environment, Hat Yai, Songkhla, Thailand, 17–19 August 2006, pp. 188–197.

Mihailović, V., Mikić, A., Ćupina, B., Vasiljević, S., Milić, D. 2008. Forage yields and forage yield components in grass pea (*Lathyrus sativus* L.). Lathyrus Lathyrism Newsletter 5 (in press).

Mikić, A., Mihailović, V., Ćupina, B., Vasić, M., Đorđević, V., Balešević-Tubić, S., Zdravković, M. 2007. Potential of grass pea for protein yield. Book of Abstracts of the 6th European Conference on Grain Legumes *Integrating Legume Biology for Sustainable Agriculture*, Lisbon, Portugal, 12–16 November 2007, pp. 115–116.

Chapter 18
Development of *Melilotus siculus* – A New Salt and Waterlogging-tolerant Annual Fodder Legume Species for Mediterranean-type Climates

Phillip Nichols, Andrew Craig, Amanda Bonython, Mary-Jane Rogers, Ross Ballard, Nigel Charman, Stephen Hughes, Timothy Colmer, Darryl McClements, and Ed Barrett-Lennard

Abstract In Australia large areas of agricultural land are currently affected by dryland salinity, with this area expected to reach 17 million hectares by 2050. Many of these soils are also subject to periods of waterlogging. The commonly sown pasture and fodder legumes in southern Australia are sensitive to these conditions. A recent series of field experiments across southern Australia found that of 33 self-regenerating annual legumes, the undomesticated species, *Melilotus siculus* (Turra) Vitman ex B.D. Jacks, was the only one productive and persistent beyond the first year on waterlogged, saline (EC_e levels in summer > 8 dS/m in the top 0–10 cm) sites. The salinity and waterlogging tolerance of *M. siculus* in the vegetative phase has been confirmed by glasshouse experiments, while recent studies have shown mechanisms for salinity tolerance and avoidance in germinating seedlings. Recent work has identified suitable rhizobia able to nodulate regenerating plants on saline soils. Evaluation of *M. siculus* genotypes will now commence, with the aim of developing a new fodder legume cultivar suitable for saline soils prone to waterlogging.

Keywords Fodder legumes · *Melilotus siculus* · Nodulation · Rhizobia · Salt tolerance · Waterlogging tolerance

Introduction

Large areas of southern Australia have become seriously affected by dryland salinity, due to rising watertables resulting from the clearing of native perennial vegetation. The National Land and Water Resources Audit (2001) estimated that

---

P. Nichols (✉)
Cooperative Research Centre for Future Farm Industries, Crawley, WA 6009, Australia
e-mail: phil.nichols@agric.wa.gov.au

C. Huyghe (ed.), *Sustainable Use of Genetic Diversity in Forage and Turf Breeding*,
DOI 10.1007/978-90-481-8706-5_18,

5.7 million hectares of agricultural land in Australia are currently affected by dryland salinity or at risk from shallow watertables, with this area expected to rise to 17 million hectares by 2050. Areas affected by dryland salinity are also often subjected to waterlogging as a consequence of the groundwater rising close to, or intercepting, the plant root-zone. Plants growing in such environments are subject to the additional challenge of hypoxia (low oxygen concentration), making plants more susceptible to high shoot concentrations of Na^+ and Cl^-. These adverse conditions have severe effects on plant growth and survival (Barrett-Lennard, 2003a). The combined effects of salt and waterlogging render such salt-affected areas unsuitable for cropping and they are used principally for livestock production. Severely affected areas are unable to support plant growth and are abandoned from agriculture.

Over the past 10 years there has been growing interest by farmers in increasing the productivity of salt-affected land (Barrett-Lennard, 2003b). Saltland pastures are generally based on saltbush (*Atriplex* species) in low rainfall areas or salt-tolerant grasses, such as puccinellia (*Puccinellia ciliata* Bor) or tall wheat grass (*Thinopyrum ponticum* (Podp.) Z.-W. Liu & R.-C. Wang) in higher rainfall areas (Barrett-Lennard, 2003b). However, saline landscapes are generally nitrogen deficient, resulting from denitrification and the lack of a legume base (Rogers et al., 2005). Companion legumes with salt and waterlogging tolerance are, therefore, required to sustain productivity of these systems.

Self-regenerating annual pasture legumes, originating from the Mediterranean basin, are widely used in the farming systems of southern Australia. However, currently used legumes are among the most sensitive plants to salinity (Rogers et al., 2005). Subterranean clover (*Trifolium subterraneum* L.), the most commonly sown pasture legume in Australia (Nichols et al., 2007), is particularly sensitive (West and Taylor, 1981). Of particular importance to self-regenerating annual legumes is the ability to germinate and establish in the years following sowing. This occurs following the opening autumn rains, when soil surface salinity levels are generally at their highest (Nichols et al., 2008). There is clearly a need to identify annual legumes with increased adaptation to the combined stresses of salinity and waterlogging on saline soils.

Field and glasshouse studies over the past 6 years have identified *Melilotus siculus* (Turra) Vitman ex B.D. Jacks (syn. *M. messanensis* (L.) Mill.) as a very promising annual fodder legume for saline, waterlogged soils (Nichols et al., 2008, 2009; Rogers et al., 2008). This upright, vigorous species is native to saline, marshy areas of the Mediterranean basin and western Asia (Marañon et al., 1989). It is also naturalised in similar environments in the Australian States of Victoria (Jeanes, 1996) and Western Australia (Paczkowska and Chapman, 2000). This paper summarises the results of field and glasshouse studies on *M. siculus* and discusses work being undertaken to enable its commercialisation as a new plant for agriculture.

Annual Legume Productivity and Survival at Saline Sites

Herbage production and persistence of 42 annual pasture legumes from 33 species were measured over 3 years at five sites across southern Australia that varied in extent of both salinity and waterlogging (Nichols et al., 2008). *Melilotus siculus* was the only species able to persist beyond the first year on sites with prolonged waterlogging that also had surface (0–10 cm) saturated paste extract electrical conductivities (EC_e) in summer greater > 8 dS/m. Although *M. siculus* plants were able to set seed and regenerate on these soils, their productivity in the years after pasture establishment was severely constrained, due to inability of the commercial annual medic rhizobia (*Sinorhizobium medicae* strain WSM1115) to persist and nodulate regenerating seedlings (Charman et al., 2006; Nichols et al., 2008).

Glasshouse Studies of Salinity and Waterlogging Tolerance

The salt and waterlogging tolerances of *M. siculus* have been confirmed in glasshouse experiments (Rogers et al., 2008). Nineteen *Melilotus* and three control species (*Trifolium fragiferum* L. cv. Palestine, *T. michelianum* Savi cv. Paradana and *Medicago sativa* L. cv. Sceptre) were screened as mature plants in a glasshouse with four NaCl treatments (0, 80, 160 and 240 mM NaCl). After 28 days, shoots of *Melilotus siculus* in 240 mM solution (equivalent to 24 dS/m) had 89% the biomass of non-saline controls, compared to *T. michelianum* (31%), *T. fragiferum* (68%) and *Medicago sativa* (75%). The same species were grown in non-saline solution in a controlled environment room under aerated or stagnant (designed to emulate waterlogging) conditions (Rogers et al., 2008). Shoot growth of *Melilotus siculus* after 28 days under stagnant conditions was 102% that of aerated controls, compared to 99% for the waterlogging tolerant species *T. michelianum* and *T. fragiferum* and 29% for *Medicago sativa*. Recent studies have also shown that *Melilotus siculus* has tolerance superior to *T. michelanium* cv. Paradana and *Medicago polymorpha* L. cv. Scimitar under the combined stresses of both salinity and waterlogging (S.M. Bowman et al.; N.L. Teakle et al., unpublished data).

Salinity Tolerance and Avoidance Mechanisms at Germination

Melilotus siculus demonstrates a range of salinity tolerance and avoidance mechanisms at germination (Nichols et al., 2009). Their experiments showed the maximum NaCl concentrations, for which no reduction in germination percentage occurs, was 300 mM for *M. siculus*, compared to 240 mM for *Medicago polymorpha* cv. Scimitar and 120 mM for *T. michelianum* cv. Frontier and *T. subterraneum* cv. Dalkeith. Following 21 days exposure to 600 mM NaCl, *Melilotus siculus* was able to recover 31% germinability upon transfer to non-saline solution. K^+ levels in imbibed seed tissues were preserved in *M. siculus* at NaCl concentrations up to 600 mM (equivalent to 60 dS/m), indicating an ability to maintain intact cell membranes under high osmotic stress.

Seed coat impermeability ('hard seeds') was shown to act as a protection against the toxic effects of salinity (Nichols et al., 2009). This study also showed *M. siculus* had a delay in hard seed breakdown ('seed softening') over the summer–autumn period, compared to *T. subterraneum* and *T. michelianum*, which were ready to germinate by early autumn. This delay acts as a salinity avoidance mechanism to defer germination until late autumn-early winter, when reliable rainfall, capable of flushing salts from the surface, is more likely to occur.

Other Desirable Characters of *M. siculus*

Melilotus siculus has a major advantage over other *Melilotus* species by having negligible levels of the chemical coumarin (Nair et al., unpublished data, Stevenson, 1969), which can be converted into dicoumarol and cause a haemorrhagic condition in livestock, known as sweetclover disease, when hay becomes mouldy (Masters et al., 2001). Coumarin can also taint the flavour of meat, milk and eggs if ingested by livestock, and of flour, if contaminated with grain. Seeds of *M. siculus* are larger than other annual fodder legumes ($\sim$ 8–12 mg), pods are non-shattering and are retained on upright stems, all of which should make seed harvesting feasible with conventional cereal headers. Preliminary studies also indicate *M. siculus* suffers little damage from redlegged earth mites (*Halotydeus destructor* Tucker), a major legume pest in southern Australia.

Current and Future Work

The major limitation to commercialising *M. siculus* as a new fodder legume for saline, waterlogged soils has been the inability of commercially available rhizobia to nodulate regenerating *M. siculus* plants. Field testing of 80 strains of rhizobia, mainly sourced from naturalised *Medicago* and *Melilotus* plants in southern Australia, has recently shown that several are better able to survive over summer and form effective nodules than the current annual *Medicago* strain WSM1115. A trip to southern Spain was also recently undertaken to collect *M. siculus* rhizobia from saline, waterlogged soils. Field evaluation of different *M. siculus* accessions will now commence with the aim of selecting a cultivar.

Acknowledgements Funding for attendance at the Eucarpia conference was provided by the AW Howard Memorial Trust Inc.

References

Barrett-Lennard, E.G. 2003a. The interaction between waterlogging and salinity in higher plants: causes, consequences and implications. Plant Soil 253:35–54.

Barrett-Lennard, E.G. 2003b. Saltland pastures in Australia – A practical guide (2nd ed.). Department of Agriculture Western Australia, South Perth, Australia.

Charman, N., Ballard, R., Craig, A. 2006. *Melilotus siculus (*syn *messanensis)* is constrained by a lack of suitable rhizobia. In: Turner, N.C., Acuna, T., Johnson, R.C. (eds), Proceedings of the 13th Agronomy Conference, Perth, Western Australia.

Jeanes, J.A. 1996. Fabaceae, *Melilotus* (pp. 719–721). In: Walsh, N.G., Entwhistle, T.J. (eds), Flora of Victoria, Volume 3, Dicotyledons: Winteraceae to Myrtaceae. Inkata Press, Melbourne, Australia.

Marañon, T., Garcia, L.V., Troncoso, A. 1989. Salinity and germination of annual *Melilotus* from the Guadalquivir delta (SW Spain). Plant Soil 119:223–228.

Masters, D.G., Norman, H.C., Dynes, R.A. 2001. Opportunities and limitations for animal production from saline land. Asian-Aust. J. Anim. Sci. 14:119–211.

National Land and Water Resources Audit 2001. Australian dryland salinity assessment 2000. Extent, impacts, processes, monitoring and management options. National Land and Water Resources Audit, Canberra, Australia.

Nichols, P.G.H., Craig, A.D., Rogers, M.E., Albertsen, T.O., Miller, S., McClements, D.R., Hughes, S.J., D'Antuono, M.F., Dear, B.S. 2008. Production and persistence of annual legumes at five saline sites in southern Australia. Aust. J. Exp. Agric. 48:518–535.

Nichols, P.G.H., Loi, A., Nutt, B.J., Evans, P.M., Craig, A.D., Pengelly, B.C., Dear, B.S., Lloyd, D.L., Revell, C.K., Nair, R.M., Ewing, M.A., Howieson, J.G., Auricht, G.A., Howie, J.H., Sandral, G.A., Carr, S.J., de Koning, C.T., Hackney, B.F., Crocker, G.J., Snowball, R., Hughes, S.J., Hall, E.J., Foster, K.J., Skinner, P.W., Barbetti, M.J., You, M.P. 2007. New annual and short-lived perennial pasture legumes for Australian agriculture – 15 years of revolution. Field Crops Res. 104:10–23.

Nichols, P.G.H., Malik, A.I., Stockdale, M., Colmer, T.D. 2009. Salt tolerance and avoidance mechanisms at germination of annual pasture legumes and their importance for adaptation to saline environments. Plant Soil 315:241–255.

Paczkowska, G., Chapman, A.R. 2000. The Western Australian flora: A descriptive catalogue. Wildflower society of Western Australia (Inc.), Western Australian Herbarium, CALM, Botanic Gardens and Parks Authority, Perth, Australia.

Rogers, M.E., Craig, A.D., Munns, R., Colmer, T.D., Nichols, P.G.H., Malcolm, C.V., Brown, A.J., Semple, W.S., Evans, P.M., Cowley, K., Hughes, S.J., Snowball, R.S., Bennett, S.J., Sweeney, G.C., Dear, B.S., Ewing, M.E. 2005. The development of fodder plants for the salt-affected areas of southern and eastern Australia: An overview. Aust. J. Exp. Agric. 45:301–329.

Rogers, M.E., Colmer, T.D., Frost, K., Henry, D., Cornwall, D., Hulm, E., Deretic, J., Hughes, S.R., Craig, A.D. 2008. Diversity in the genus *Melilotus* for tolerance to salinity and waterlogging. Plant Soil 304:89–101.

Stevenson, G.A. 1969. An agronomic and taxonomic review of the genus *Melilotus* Mill. Can. J. Plant Sci. 49:1–20.

West, D.W., Taylor, J.A. 1981. Germination and growth of cultivars of *Trifolium subterraneum* L. in the presence of sodium chloride salinity. Plant Soil 62:221–230.

Chapter 19
Domestication of New Mediterranean Annual Pasture Legumes

Phillip Nichols, Angelo Loi, Bradley Nutt, Richard Snowball, and Clinton Revell

Abstract Fifteen years ago subterranean clover (*Trifolium subterraneum* L.) and annual medics (*Medicago* spp.) dominated annual pasture legume sowings in the Mediterranean-like climate of southern Australia. Since then a number of sustainability and economic challenges to existing farming systems have emerged, exposing shortcomings in these species and a lack of legume biodiversity. A selection program, largely based in Western Australia, with testing sites across southern Australia, has responded to these challenges by domesticating new annual pasture legume species native to the Mediterranean basin to overcome the deficiencies in traditional species (Nichols et al., 2007). Seven new species to agriculture have been commercialised (*Ornithopus sativus*, *Biserrula pelecinus*, *Trifolium glanduliferum*, *T. dasyurum*, *T. spumosum*, *T. purpureum* and *Medicago sphaerocarpos*), while *Lotus ornithopodioides* and *Melilotus siculus* are under evaluation. Traits incorporated include deeper root systems, protection from false breaks (germination-inducing rainfall events followed by death from drought), a range of hardseed levels, acid-soil tolerant root nodule symbioses, tolerance to pests, diseases and salinity and provision of less expensive seed through ease of seed harvesting and processing. The contributions of genetic resources, rhizobiology, pasture ecology and agronomy, plant pathology, physiology, entomology, plant chemistry and animal science have been paramount to this success.

Keywords Pasture legumes · Cultivars · Domestication · *Trifolium* · *Medicago* · *Ornithopus* · *Biserrula*

P. Nichols (✉)
Cooperative Research Centre for Future Farm Industries, Crawley, WA 6009, Australia
e-mail: phil.nichols@agric.wa.gov.au

C. Huyghe (ed.), *Sustainable Use of Genetic Diversity in Forage and Turf Breeding*,
DOI 10.1007/978-90-481-8706-5_19, © Springer Science+Business Media B.V. 2010

Introduction

The agricultural area of south-western Australia has a Mediterranean-type climate. It also has inherently low soil fertility and no native legumes suited to grazing by sheep and cattle. This has resulted in development of farming systems incorporating self-regenerating annual pasture legumes native to the Mediterranean basin (Howieson et al., 2000). In combination with the use of superphosphate, pasture legumes have led to improved soil fertility and increased crop yields (Puckridge and French, 1983) and to greater animal production (Doyle et al., 1993). Howieson et al. (2000) listed a range of benefits for incorporating pasture legumes into farming systems. These include their ability to fix atmospheric nitrogen (with benefits for both companion pasture species and subsequent crops), ability to increase soil fertility and structure, and capacity to break disease and pest life cycles of crops when grown in rotation.

Prior to the early 1990s annual pasture legume options in Western Australia were largely confined to subterranean clover (*Trifolium subterraneum*) on acid soils and annual medics (*Medicago* spp.) on neutral-alkaline soils. The two keys to widespread use of subterranean clover have been its tolerance of heavy grazing and a suite of cultivars differing in flowering time, enabling it to be grown in environments with a wide range of growing season lengths. Annual medics have been widely sown in low-medium rainfall areas. Their high levels of hardseededness make them well adapted to crop rotation systems. The main annual medic species sown have been *Medicago truncatula*, *M. littoralis* and *M. polymorpha* var. *brevispina*, with limited sowing of *M. scutellata*, *M. rugosa*, and *M. tornata*. The acid-tolerant *M. murex* was also commercialised in the mid 1980s (Oram, 1990). Other commercially available species with cultivars registered at the time were yellow serradella (*Ornithopus compressus*), balansa clover (*T. michelianum*), Persian clover (*T. resupinatum*), rose clover (*T. hirtum*), cupped clover (*T. cherleri*) and slender serradella (*O. pinnatus*) (Oram, 1990).

The Drivers for Change

Since the early 1990s a number of sustainability and economic challenges to existing farming systems have emerged, exposing shortcomings in the traditional species and a lack of legume biodiversity. Some of the drivers for change include: (1) poor adaptation of subterranean clover and annual medics to difficult soils, particularly deep, acid sands and soils subject to salinity; (2) seed bank depletion of soft-seeded pasture legumes from increased cropping frequencies; (3) environmental concerns from soil erosion caused by vacuum harvesting subterranean clover and annual medic seeds; (4) the need for cheaper seed for re-sowing pastures; (5) the need for specialist fodder legumes; (6) the need for deeper-rooted plants to reduce groundwater recharge and the potential for dryland salinity; and (7) the need for greater annual legume diversity within paddocks to stabilise productivity. These new challenges demanded an expansion in the range of pasture legume options with traits to meet the needs of current and prospective farming systems. This paper discusses the processes involved in domesticating new annual legumes and describes the release of seven new species for agriculture in Western Australia. A more complete description of this work is given in Nichols et al. (2007).

Processes Involved in Domestication of New Species

The methodology used in the germplasm acquisition, selection and cultivar release processes is summarised in Fig. 19.1. Of paramount importance is the initial stage of problem definition to identify agronomic problems that require new plant-based solutions or the potential for new productivity opportunities. The next stage involves acquiring diverse germplasm of a range of potential species and genera either from Genetic Resource Centres or, where limited options are available, from targeted collection missions to countries with regions that have matching eco-geographic parameters to target environments. Germplasm is characterised and examined as spaced plants or in short rows under irrigation for important traits required for commerce. It is important to note that seed traits lending themselves to production of inexpensive, high quality seed are given a high priority, as this is what largely drives commercial uptake. Characters include those that allow harvesting by conventional headers, such as good pod retention, non-shattering pods and ability of pods to flow through the machine, and those that allow ease of seed processing, particularly extraction of seed from pods. Traits for high productivity are also included to determine species with the most promise.

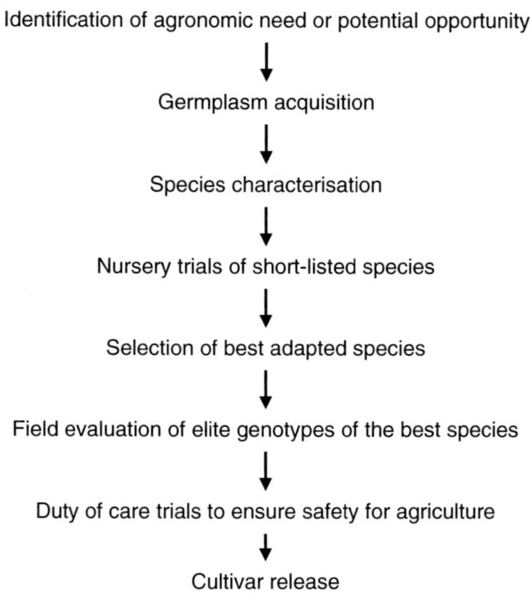

Fig. 19.1 Steps involved in the domestication of new annual pasture legume species

Nursery trials of species with desirable traits are then conducted in small plots in a limited number of target environments for two or three years. Species entries can consist of individual genotypes that represent the species or as bulks of a few genotypes sown together. The main aim of these trials is to determine the general adaptation of the species to target environments over two or more seasons. Survival and persistence, in comparison with existing commercial options, is of key importance. The next process is to examine a wider range of genotypes of the species identified as having the most promise. This involves characterising them for productivity, seed production and phenological traits. Measurement of flowering time is particularly important to enable identification of genotypes adapted to environments of particular growing season lengths. The levels of any anti-nutritional or toxic chemicals that are believed could be present in the species are also measured; unacceptable genotypes are eliminated from further work. Parallel research into identifying compatible, robust root nodule bacteria is also crucial for success.

Elite genotypes with the most promising productivity and seed production traits, that match the phenological requirements of the intended target environment, then enter intensive field evaluation trials at several sites in the target zone in multi-replicated trials under appropriate simulated farming conditions. Measurements include herbage and seed production, seedling regeneration in the years after sowing, and reaction to local pests and diseases. Trials generally run for at least three years and comparisons of productivity and persistence are made with commercial pasture legumes recommended for the district. The best performing genotypes are then selected for potential cultivar release.

The final step in the domestication process is to conduct ‘Duty of care’ trials (Revell and Revell, 2007), in which grazed animals on large plots of the candidate cultivar are monitored to ensure safety for agriculture. This involves measuring animal performance (wool, meat or dairy production) against matched animals grazing large plots of existing pasture options, monitoring health by a qualified veterinarian, conducting a post-mortem on slaughtered animals to detect any liver damage from build-up of any toxic compounds, and taste-testing of meat taken from the animals to identify any tainting or negative effects on meat quality. The rate of spread of regenerating seedlings is also monitored to determine any weed risk potential. The duty of care process is discussed in more detail by Revell and Revell (2007). If the candidate cultivar passes these criteria, it is released to the seed industry.

Post-release agronomic research of new species is also important to optimise grazing management and seed production practices and to gain confidence in its use by farmers and agri-business.

Cultivar Releases of New Species

Seven new species to agriculture have been commercialised since 1993 (*Ornithopus sativus*, *Biserrula pelecinus*, *Trifolium glanduliferum*, *T. dasyurum*, *T. spumosum*, *T. purpureum* and *Medicago sphaerocarpos*) (Table 19.1), while *Lotus*

Table 19.1 Cultivars of new annual pasture legume species domesticated in Western Australia since 1993

Species	Common name	Cultivar	Release year
Biserrula pelecinus	Biserrula	Casbah	1997
		Mauro(b)	2002
Medicago sphaerocarpos	Sphere medic	Orion	1993
Ornithopus sativus	French serradella	Cadiz(b)	1996
		Erica(b)	2003
		Margurita(b)	2003
Trifolium dasyurum	Eastern star clover	AGWEST® Sothis	2007
T. glanduliferum	Gland clover	Prima	2001
T. purpureum	Purple clover	ELECTRA™	2006
T. spumosum	Bladder clover	AGWEST® Bartolo	2008

ornithopodioides and *Melilotus siculus* are under evaluation. Traits incorporated include deeper root systems, protection from false breaks (germination-inducing rainfall events followed by death from drought), a range of hardseed levels, acid-soil tolerant root nodule symbioses, tolerance to pests, diseases and salinity and provision of less expensive seed through ease of seed harvesting and processing (Nichols et al., 2007). A farmer survey in Western Australia has shown widespread adoption of the new pasture legumes (Nichols et al., 2007) and their use is expected to increase with greater farmer experience.

Acknowledgements Funding for attendance at the Eucarpia conference was provided by the AW Howard Memorial Trust Inc.

References

Doyle, P.T., Grimm, M., Thompson, A.N. 1993. Grazing for pasture and sheep management in the annual pasture zone (pp. 71–90). In: Kemp, D.R., Michalk, D.L.(eds), Pasture management technology for the 21st century. CSIRO Australia, Melbourne, Australia.

Howieson, J.G., O'Hara, G.W., Carr, S.J. 2000. Changing roles for legumes in Mediterranean agriculture: Developments from an Australian perspective. Field Crops Res. 65:107–122.

Nichols, P.G.H., Loi, A., Nutt, B.J., Evans, P.M., Craig, A.D., Pengelly, B.C., Dear, B.S., Lloyd, D.L., Revell, C.K., Nair, R.M., Ewing, M.A., Howieson, J.G., Auricht, G.A., Howie, J.H., Sandral, G.A., Carr, S.J., de Koning, C.T., Hackney, B.F., Crocker, G.J., Snowball, R., Hughes, S.J., Hall, E.J., Foster, K.J., Skinner, P.W., Barbetti, M.J., You, M.P. 2007. New annual and short-lived perennial pasture legumes for Australian agriculture – 15 years of revolution. Field Crops Res. 104:10–23.

Oram, R.N. 1990. Register of Australian herbage plant cultivars (3rd ed.). CSIRO Australia, Canberra, Australia.

Puckridge, D.W., French, R.J. 1983. The annual legume pasture in cereal-ley farming systems of southern Australia: A review. Agric, Ecosys. Environ. 9:229–267.

Revell, C.K., Revell, D. K. 2007. Meeting 'duty of care' obligations when developing new pasture species. Field Crops Res. 104:95–102.

Chapter 20
The Population Genetic Structure of Diploid *Medicago sativa* L. Germplasm

Muhammet Sakiroglu, Jeffrey J. Doyle, and E. Charles Brummer

Abstract The three subspecies *Medicago sativa* subsp. *caerulea* (syn. *coerulea*), *M. sativa* subsp. *falcata,* and *M. sativa* subsp. *hemicycla* are considered to form the diploid gene pool of cultivated alfalfa (*M. sativa* subsp. *sativa*). The diploid gene pool is underutilized in breeding programs despite extensive morphological variation and the simplicity of disomic inheritance. Population structure and the genetic basis of the current morphologically-based classification of diploid germplasm are not known. We analyzed the population genetic structure of wild diploid alfalfa germplasm by evaluating 374 individual genotypes from 120 accessions, representing the broad natural variation of the three subspecies, with 89 microsatellite markers. We found that the three subspecies formed distinct clusters, with evidence of further subdivision of *falcata* and *caerulea* into two subclusters each. The genetic distinction between the two *falcata* subclusters is more definitive than that of the two *caerulea* groups. Genome composition suggests extensive gene flow where subspecies and/or groups within subspecies grow sympatrically. The results will help breeders identify appropriate diploid accessions to maximize diversity in applied germplasm development.

Keywords Genetic diversity · *Medicago sativa* · Population structure · SSR markers

Introduction

Alfalfa is one of the main forage legumes cultivated around the globe (Michaud et al., 1988). Modern alfalfa cultivars are the product of the intense selection from the taxonomic group called the *Medicago sativa-falcata* complex. The complex

E.C. Brummer (✉)
Institute for Plant Breeding, Genetics, and Genomics, University of Georgia, 111 Riverbend Rd., Athens, GA 30602, USA
e-mail: brummer@uga.edu

C. Huyghe (ed.), *Sustainable Use of Genetic Diversity in Forage and Turf Breeding*,
DOI 10.1007/978-90-481-8706-5_20,

consists of a number of interfertile taxa mainly classified based on ploidy and morphology (e.g., flower color, pod shape, and pollen shape). Tetraploid taxa are purple flowered *M. sativa* subsp. *sativa* with coiled pods, yellow flowered *M. sativa* subsp. *falcata* with sickle shaped pods, and *M. sativa* subsp. *varia*, the natural hybrid of *sativa* and *falcata*, which has variegated flowers and pods with approximately one coil. Analogous to the tetraploids, the diploid taxa are purple flowered *M. sativa* subsp. *caerulea, M. sativa* subsp. *falcata*, and hybrid *M. sativa* subsp. *hemicycla* which has variegated flowers and pods of one or few coils (Quiros and Bauchan, 1988, Lesins and Lesins, 1979). Although the vast majority of cultivars are tetraploid, diploid germplasm can be used to enhance breeding germplasm because direct hybridization across ploidy levels is possible. Breeding at the diploid level offers the possibility to avoid complex tetrasomic inheritance (McCoy and Bingham, 1988; Kaló et al., 2000).

Evaluation of population structure in diploid germplasm can be used to test whether the current morphology-based taxonomy reflects patterns of genetic variation in the complex. This is an important first step toward effective utilization of the diploid germplasm for future breeding efforts and for use in association mapping. In this study, we used SSR markers to evaluate a wide range of wild diploid accessions of the *M. sativa-falcata* complex from throughout their natural range.

Material and Methods

Plant Materials

We selected 374 diploid individuals from 120 accessions from the USDA National Plant Germplasm System (NPGS) collections to represent the entire geographical range of subspecies *caerulea*, *falcata*, and *hemicycla*. All the plants were grown in the University of Georgia greenhouses. We recorded flower color and pod shape for these genotypes, because they have been the main criteria used to assign genotypes into subspecies. Accessions with yellow, purple, and variegated flowers were considered *falcata*, *caerulea*, and *hemicycla*, respectively. Similarly, accessions with sickle-shaped (falcate), coiled, and intermediate pods were considered *falcata*, *caerulea*, and *hemicycla*, respectively. We harvested pods from 121 genotypes, recorded the coil number for 10 individual pods per genotype, and computed a mean coiling value. Pods were scored in ¼ coil intervals for the number of coils, ranging from completely straight (0 coils) to four coils.

Genotyping

DNA was extracted from all 374 genotypes using young leaves from greenhouse grown plants, following the CTAB method (Doyle and Doyle 1987). Eighty-nine polymorphic SSR markers were selected from previous studies in alfalfa (Diwan et al., 2000; Julier et al., 2003; Sledge et al., 2005; Robins et al., 2007). PCR was conducted following Schuelke (2000). An automated ABI3730 sequencer at the UGA DNA Sequencing Facility was used for genotyping and allele scoring was performed using GENEMARKER software (SoftGenetics, State College, PA).

Data Analyses

In order to infer the population structure of the entire set of genotypes without regard to the preexisting subspecies classification or geographical information, we used the software STRUCTURE (Pritchard et al., 2000). To deduce the optimal number of clusters (K) present in the entire population, we evaluated $K = 1$–10 assuming admixture and correlated allele frequencies with 5 replicate runs performed for all K values. The length of burn-in Markov Chain Monte Carlo (MCMC) replications was set to 10,000 and data were collected over 100°000 MCMC replications in each run. The best estimation of the true number of K was obtained using the method developed by Evanno et al. (2005). Further experimental details are available in Şakiroğlu et al. (2009).

Results and Discussion

Assuming $K = 2$, two large clusters corresponding to *falcata* and *caerulea* were identified. Genotypes denoted as *hemicycla* based on morphology showed a hybrid genome composition consisting of roughly equal parts of *sativa* and *falcata*, as expected. This result showed that flower color and genome composition data were in agreement. The population structure analysis suggested that the optimal number of clusters is five ($K = 5$). *Falcata* and *caerulea* groups were further structured into two distinct subclusters indicating a hierarchical population structure, and subspecies *hemicycla* formed a separate cluster (Fig. 20.1). Passport information on the collection location for each accession suggested that the separation within *caerulea* germplasm clusters is based on geography (southern vs. northern subclusters), but that the separation of *falcata* germplasm is based on ecogeography (upland vs. lowland ecotypes).

The genome composition of several genotypes suggested that they were hybrids between clusters. We categorized those hybrid genotypes into three main hybrid types. The first group, hybrids between the two *caerulea* populations, included 21 genotypes from six accessions. Of these, 12 genotypes from three accessions collected from the former Soviet Union (PI283640) and Russia (PI577548 and PI315466) did not have precise collection information, so we could not hypothesize further regarding their putative hybridity. The remaining hybrid accessions, PI464723 and PI464726 from Turkey and PI621924 from Armenia, were collected from locations geographically at the border of separation between the southern vs. northern *caerulea* subgroups.

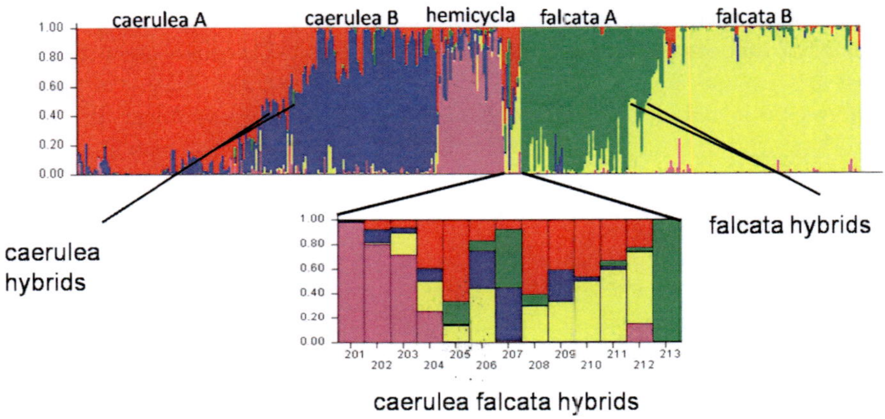

Fig. 20.1 Structure output of 374 diploid alfalfa accessions assuming the number of populations to be five. Hybrid accessions among five groups are indicated

The second set of hybrids were genotypes with significant portions of their genome from each *falcata* subcluster (Fig. 20.1). Of the three hybrid accessions, PI577558 and PI538987 were collected from eastern Russia, about 900 km apart. Both accessions were collected from the plains but very close to river basins where admixture between the lowland and upland ecotypes could occur. The third hybrid accession, PI631568 from Italy, was collected from the Tagliamento River basin at the base of the Italian Alpines, again a potential hybrid zone.

The third group of hybrids includes accessions in subsp. *hemicycla*. By definition, *hemicycla* is a hybrid subspecies between *falcata* and *caerulea*, but STRUCTURE indicates that it forms its own separate group. This suggests that after initial hybridization, *hemicycla* individuals tend to intercross within the subspecies, thus leading toward a distinct population. Interestingly, a number of *caerulea* × *falcata* hybrid genotypes had a different genome composition than the rest of *hemicycla* populations (Fig. 20.1). These genotypes are admixtures of *falcata* and *caerulea* populations in different degrees, implying recent hybridization since the progenitor populations could be inferred. Nine genotypes from six *hemicycla* accessions indicated a recent admixture pattern of genome composition. Genome composition suggests that recent hybrids are mainly between southern *caerulea* and upland *falcata* with the exception of two Russian genotypes, which are hybrids between either upland or lowland *falcata* and northern *caerulea*. The Georgian accession of PI577543 had purple flowers but a mean pod coil of 1.375, within the expected range for *hemicycla*. The two Turkish accessions PI464727 and PI464728 were collected from Kars province in northeast Turkey. All three of these accessions were collected from locations where *falcata*, *caerulea*, and *hemicycla* are sympatric; thus, these accessions could be recent hybrids.

Conclusions

Our results clearly show that diploid alfalfa germplasm is structured both within and among subspecies. Although we detected evidence of gene flow among groups in some accessions from regions where the subspecies grow sympatrically, the majority of accessions – including some that are in regions of sympatry with other groups – are clearly distinct from the others based on molecular marker data. This suggests that reproductive and/or geographic mechanisms prevent completely free genetic exchange among taxa included in the *M. sativa-falcata* complex. We have noted that *falcata* genotypes tend to flower later than *caerulea* when grown in a common garden in Georgia, USA (34°N latitude), which could play a role if similar differences are seen in native regions (the *caveat* being that these are typically higher latitudes than our common garden). Pollinator preferences for flower color are well known, and may also impact the gene flow among groups.

Acknowledgements We would like to thank Yanling Wei for technical assistance; the Turkish Government for funding Muhammet Şakiroğlu; and the National Research Initiative (NRI) Plant Feedstock Genomics for Bioenergy Program for funding this project.

References

Diwan, N., Bouton, J.H., Kochert, G., Cregan, P.B. 2000. Mapping of Simple Sequence Repeat (SSR) DNA Markers in Diploid and Tetraploid Alfalfa. Theor. Appl. Genet. 101:165–172.

Doyle, J.J., Doyle, J.L. 1987. A rapid DNA isolation for small quantities of fresh leaf tissue. Phytochem. Bull. 19:11–15.

Evanno, G., Regnaut, S., Goudet, J. 2005. Detecting the number of clusters of individuals using the software structure: A simulation study. Mol. Ecol. 14:2611–2620.

Julier, B., Flajoulot, S., Barre, P., Cardinet, G., Santoni, S., Huguet, T., Huyghe C. 2003. Construction of two genetic linkage maps in cultivated tetraploid alfalfa (*Medicago sativa*) using microsatellite and AFLP markers. BMC Plant Biol. 3:9.

Kaló, P., Endre, G., Zimányi, L., Csanádi, G., Kiss, G.B. 2000. Construction of an Improved Linkage Map of Diploid Alfalfa (*Medicago sativa*). Theor. Appl. Genet. 100:641–657.

Lesins, K., Lesins, I. 1979. Genus *Medicago* (Leguminasae): A taxogenetic study. Kluwer, Dordrecht, Netherlands.

McCoy, T.J., Bingham E.T. 1988. Cytology and cytogenetics of Alfalfa (pp. 739–776). In: Hanson, A.A., et al. (eds.), Alfalfa and alfalfa improvement. ASA-CSSA-SSSA, Madison, WI, USA.

Michaud, R., Lehman, W.F., Rumbaugh, M.D. 1988. World distribution and historical development (pp.25–91). In: Hanson, A.A., Barnes, D.K., Hill, R.R. (eds.), Alfalfa and Alfalfa Improvement. ASA-CSSA-SSSA, Madison, WI, USA.

Pritchard, J.K., Stevens, M., Donnelly, P. 2000. Inference of population structure using multilocus genotype data. Genetics 155:945–959.

Quiros, C.F., Bauchan, G.R. 1988. The genus *Medicago* and the origin of the *Medicago sativa* complex (pp. 93–124). In: Hanson, A.A., et al. (eds.), Alfalfa and alfalfa improvement. ASA-CSSA-SSSA, Madison, WI, USA.

Robins, J.G., Luth, D., Campbell, T.A., Bauchan, G.R., He, C., Viands, D.R., Hansen, J.L., Brummer, E.C. 2007. Genetic mapping of biomass production in tetraploid alfalfa (*Medicago sativa* L.). Crop Sci. 47:1–10.

Şakiroğlu, M., Doyle, J.J., Brummer, E.C. 2009. Inferring population structure and genetic diversity of a broad range of wild diploid alfalfa (*Medicago sativa* L.) accessions using SSR markers. Theor. Appl. Genet. [in review].

Schuelke, M. 2000. An economic method for the fluorescent labeling of PCR fragments. Nat. Biotechnol. 18:233–234.

Sledge, M.K., Ray, I.M.,. Jiang, G. 2005. An expressed sequence tag SSR map of tetraploid alfalfa (*Medicago sativa* L.). Theor. Appl. Genet. 111:980–992.

Chapter 21
Characterization and Preliminary Evaluation of *Hedysarum coronarium* L. Ecotypes in Mediterranean Environment

Mauro Salis, Antonio M. Carroni, Alessandro Longu, Patrizia Manunza, and Maurizio Pitzalis

Abstract The changes of cropping systems in the last decades have caused a strong reduction of perennial forage genetic resources, especially for landraces of legume species and where the agriculture is more intensive.

The aim of our research was to characterize and preliminarily evaluate Italian genetic resources of sulla (*Hedysarum coronarium* L.). The trial was carried out in south Sardinia (Italy) during 2007–2008. Thirteen ecotypes of sulla representative of the main regions of cultivation (Sardinia, Sicily, Tuscany, Abruzzo, Marche) and three control varieties ('Carmen', 'Grimaldi' and 'Sparacia') were evaluated in a trial designed as a randomized complete block with three replicates. In each plot (1.26 × 1.80 m) 70 plants were transplanted at the end of November, and on 40 plants the following variables were observed: winter mortality; growth habit in spring; time of beginning of flowering; dry matter yield; leaf/stem ratio. These variables were analyzed using univariate and multivariate methods, and the results were used to identify genetic resources useful for local breeding of sulla.

Keywords Breeding strategy · Genetic resources · Sulla

Introduction

The changes of cropping systems in the last decades have caused a strong reduction of perennial forage genetic resources, especially for landraces of legume species and where the agriculture is more intensive.

There are several basic reasons which lead to consider Mediterranean plant germplasm as a breeding material of high value. Particularly, with a regard to

M. Salis (✉)
Unità di ricerca per i sistemi agropastorali in ambiente mediterraneo, Consiglio per la Ricerca e sperimentazione in Agricoltura (CRA), Podere Ortigara Loc. Sanluri Stato, Sanluri (VS) 09025, Italy
e-mail: mauro.salis@entecra.it

C. Huyghe (ed.), *Sustainable Use of Genetic Diversity in Forage and Turf Breeding*, DOI 10.1007/978-90-481-8706-5_21, © Springer Science+Business Media B.V. 2010

Italian genetic resources of forage species, there are different microclimates and soil situations that increase the genetic variability.

Sulla (*Hedysarum coronarium* L.) is a member of *Leguminosae* family originating and growing in the semi-arid Mediterranean environments with an exceptional adaptability, a great supply of nitrogen produced by symbiotic fixation, a satisfactory seed production (Roggero et al., 1996), a high nutritive value (Bullitta et al., 1999). It is used for hay, silage and pasture; moreover, it is an interesting example of a multiple-use species (Bullitta and Sulas, 1998). For these reasons, the sulla can improve its role in Mediterranean forage systems. A better knowledge of the development morphology and quality changes of sulla in different environments is necessary to optimize its potential for livestock production (Borreani et al., 2003).

The aim of our research was to characterize and preliminarily evaluate Italian genetic resources of sulla.

Materials and Methods

The trial was carried out in South Sardinia (Italy) during 2007–2008 in a site characterized by clay-loam soil (pH 7.1) and an annual average temperature of 16.5°C and 563 mm of rainfall, distributed in 72 days. Thirteen ecotypes of sulla representative of the main regions of cultivation (Sardinia, Sicily, Tuscany, Abruzzo, Marche) and three control varieties ('Carmen', 'Grimaldi' and 'Sparacia') were evaluated in a trial.

Ecotype/Variety	Origin	Material	Ecotype/Variety	Origin	Material
Alberese	Tuscany	Nat.Pop.	Offida	Marche	Nat.Pop.
Aragona	Sicily	Nat.Pop.	S.Giovanni Lipioni	Abruzzo	Landrace
Cerami	Sicily	Nat.Pop.	Sardara	Sardinia	Nat.Pop.
Gangi	Sicily	Nat.Pop.	Simala	Sardinia	Nat.Pop.
Gesturi	Sardinia	Nat.Pop.	Torrebruna	Abruzzo	Landrace
Guasila-Pimentel	Sardinia	Nat.Pop.	Carmen	Tuscany	Variety
Melia	Sicily	Nat.Pop.	Grimaldi	Umbria	Variety
Monteroni d'Arbia	Tuscany	Nat.Pop.	Sparacia	Sicily	Variety

On October 2007, seeds were sown into containers and, at the beginning of December, 70 plants were transplanted in plots (1.26 × 1.80 m) arranged in a randomized complete block with three replicates. On 40 plants the following variables were observed: winter mortality, as number of plants at the beginning of spring; growth habit and height of plants in mid spring; time of beginning of flowering; mid and end spring dry matter yield; leaf/stem ratio.

These variables were analyzed using univariate and multivariate methods. The multivariate statistical analysis was performed using average values for the ecotypes and included linear correlation coefficient analysis, principal component analysis and cluster analysis.

Results and Discussion

The analysis of variance among ecotypes showed “Aragona” as the ecotype with a lower winter mortality (Table 21.1). About the growth habit, “Gangi”, “Sparacia”

Table 21.1 Number of plants m^{-2}, growth habit, height (cm) at the beginning of spring and number of days for flowering from transplant

Ecotypes	N. plants m^{-2}	Growth habit	Height (cm)	Flowering
Alberese	29.49a	60.0b–d	23.7d–f	133.7a–c
Aragona	24.62b	60.0b–d	26.7c–e	124.0f
Carmen	29.23a	75.0a–c	26.3c–e	131.3a–d
Cerami	28.72a	60.0b–d	20.2d–g	126.0df
Gangi	28.20ab	90.0a	38.2a	128.0c–f
Gesturi	27.18ab	45.0de	13.4gh	130.7a–e
Grimaldi	28.97a	80.0ab	32.4a–c	134.0ab
Guasila	29.49a	65.0b–d	23.0d–f	128.3b–f
Melia	27.69ab	50.0de	19.8e–g	126.0d–f
Monteroni	27.95ab	45.0de	17.1 fg	135.0a
Offida	28.72a	65.0b–d	22.0d–f	130.3a–e
S.Giovi.Lip	27.95ab	80.0ab	27.4 cd	134.3a
Sardara	28.97a	30.0f	8.7 h	126.3d–f
Simala	27.30ab	52.5c–e	19.6e–g	125.0ef
Sparacia	27.69ab	90.0a	31.1bc	126.0d–f
Torrebruna	28.20ab	90.0a	34.8ab	132.7a–c
CV	6.6	19.4	15.6	2.3

Means with the same letter are not different per $P \leq 0.05$

Table 21.2 Averange dry matter production (t ha^{-1}) (DM 1) and leaf/steam ratio at the mid of spring, dry matter production at the end of spring (DM 2), total dry matter production

Ecotypes	DM 1 (t ha^{-1})	Leaf/Steam ratio	DM 2 (t ha^{-1})	Total DM (t ha^{-1})
Alberese	3.67a–c	1.73b–e	0.88a–e	4.55a–d
Aragona	2.99a–d	1.53c–e	0.35hi	3.35de
Carmen	4.06a–c	1.66b–e	0.89a–e	4.94ab
Cerami	3.71a–c	1.98bc	0.37 g–i	4.94ab
Gangi	4.29a	1.30e	0.61e–h	4.08a–d
Gesturi	2.76 cd	2.07ab	0.75b–f	4.90a–c
Grimaldi	4.11ab	1.66b–e	0.90a–d	5.02ab
Guasila	3.29a–d	1.69b–e	0.70c–f	3.99a–d
Melia	3.23a–d	2.07ab	0.57f–i	3.79b–e
Monteroni	2.92b–d	2.48a	0.98a–c	3.90a–e
Offida	3.54a–d	1.79b–d	1.06a	4.60a–d
S.Giovi.Lip	4.04a–c	1.73b–e	0.98a–c	5.02ab
Sardara	2.24d	1.69b–e	0.32i	2.57e
Simala	3.30a–d	1.65b–e	0.63d–g	3.93a–e
Sparacia	3.44a–d	1.65b–e	0.34hi	3.78b–e
Torrebruna	4.23a–b	1.35d–e	0.99a–b	5.22a
CV	19.0	14.0	20.7	16.9

Means with the same letter are not different per $P \leq 0.05$

and “Torrebruna” showed an erect growth habit, while “Sardara” was the only one with a prostrate growth habit, as confirmed by height analysis. The ecotype “Aragona” is resulted earlier, while “Monteroni” later (Table 21.2).

“Sardara” showed lower mid Spring dry matter production. “Monteroni” showed the highest leaf/stem ratio, “Torrebruna” and “Gangi” the lowest. Higher productions were observed in the ecotypes “Carmen”, “Grimaldi”, “S. Giovanni Lipioni” and “Torrebruna”, with about 5 t/ha.

Principal components analysis (not reported data) showed that the first two components (PC1 and PC2) explained over 80% of the variance. PC1 was more correlated to the variable “growth habit”, “height”, “mid spring dry matter” and “total dry matter”.

Figure 21.1 shows the position of the ecotypes in relation to the two principal components.

The populations on the right, for positive values of PC1, include the tall and highly productive ecotypes, as confirmed by the analysis of variance among clusters

Table 21.3 Matrix of the first two principal components (PC) of some of variables observed in the 16 ecotypes compared

Variables	PC 1	PC 2
N. plants m^{-2}	0.145	0.340
Growth habit	0.424	–0.253
Height	0.413	–0.307
Flowering	0.257	0.515
DM 1	0.455	–0.065
Leaf/Steam	–0.253	0.452
DM 2	0.286	0.488
Total DM	0.461	0.121

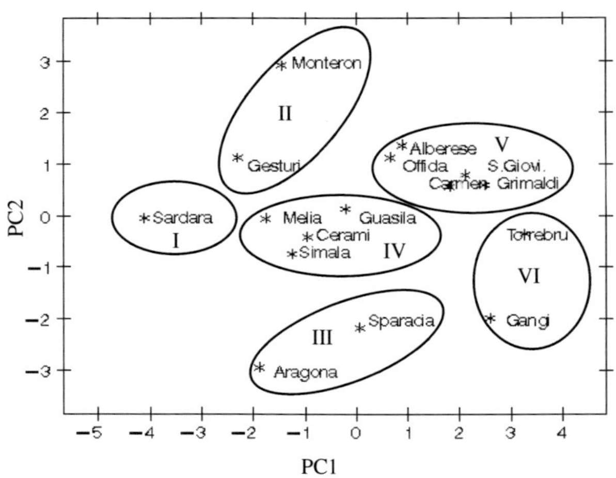

Fig. 21.1 Plot of the 16 ecotypes for the first two principal components

Table 21.4 Means and analysis of variance of the six cluster for each of the eight variables

Clusters	N. plants m-2	Growth habit	Height	Flowering	DM 1	Leaf/Steam	DM 2	Total DM
I (1)	28.97a	30.0d	8.7e	126.3b	2.24d	1.69b	0.32c	2.57c
II (2)	27.56a	45.0 cd	15.2d	132.8a	2.84 cd	2.27a	0.86a	3.71b
III (2)	17.44b	50.0c	19.2 cd	83.3c	2.14d	1.06c	0.23c	2.38c
IV (4)	28.52a	57.5bc	21.1c	126.3b	3.39bc	1.88b	0.54b	3.93b
V (5)	28.87a	72.0b	26.3b	132.7a	3.88ab	1.72b	0.94a	4.83a
VI (2)	28.21a	90.0a	36.5a	130.3a	4.25a	1.32c	0.80a	5.06a
F-ratio	42.84	17.87	37.03	251.26	12.34	20.07	25.14	20.30
P	≤0.0001	0.0001	≤0.0001	≤0.0001	0.0005	≤0.0001	≤0.0001	≤0.0001

The number of populations contained in each cluster are shown in parentheses

(Table 21.4). Table 21.3 shows the matrix of the first two principal components (PC) of some of variables observed in the 16 ecotypes compared.

These results are preliminary and the evaluation of the genetic resources of sulla will continue in the future in the Unit of Research in Sanluri.

Acknowledgements The work was funded by the Italian Ministry for Agricultural, Food and Forestry Policies within the project Plant Genetic Resources – FAO Treaty.

References

Borreani, G., Roggero, P.P., Sulas, L., Valente, M.E. 2003. Quantifying morphological stage to predict the nutritive value in sulla (*Hedysarum coronarium* L.). Agron. J. 95:1608–1617.

Bullitta, P., Sulas, L., Porqueddu, C., Caredda, S. 1996. Sistemi pascolivi della Sardegna (pp. 269–290). In: *Attualità e prospettive della foraggi coltura da prato e da pascolo*, Lodi, Italy.

Bullitta, P., Sulas, L. 1998. Le possibilità di utilizzazione della sulla con il pascolamento e il suo inserimento nei sistemi pascolivi sardi (pp.53–71). I Georgofili, Quaderni, 1, Firenze.

Roggero, P.P., Santilocchi, R., Sargenti, P. 1996. Integration of forage and seed production of sulla (*Hedysarum coronarium* L.) in Central Italy (pp. 297–300). In: Proc. of 16th General Meeting of EGF, Grado, Italy.

Chapter 22
Phenotypic Assessment of Variability in Tillering and Early Development in Ryegrass (*Lolium* spp.)

Oana Saracutu, Gerda Cnops, Isabel Roldán-Ruiz, and Antje Rohde

Abstract Tillering is enormously variable in the genus *Lolium*. Exploitation of tillering characteristics in breeding programs requires a systematic characterization of this trait at the between-species, within-species, and within-cultivar levels. We have analyzed tillering in forage and turf cultivars of *L. perenne,* forage cultivars of *L. multiflorum* and annual *L. temulentum* genotypes. The collection was also enriched with wild *L. perenne* populations.

The tillering phenotype is investigated based on total tiller number, tiller orders, regrowth capacity after cutting, plant height and diameter. All of these traits showed great heterogeneity between and within accessions. As expected, turf grasses had the most tillers and *L. temulentum* the least. Regrowth after cutting was not linked to the general tillering capacity of an accession, suggesting that different genetic mechanisms control tillering and re-growth. These results, which are based on the evaluation of individual plants, provide a first set of observations for understanding tillering behavior in *Lolium*.

Keywords Plant architecture · Ryegrass · Tillering

Introduction

Plant architecture has a great influence on crop yield and quality. A complex interaction of genetic and environmental factors determine plant architecture. Changes in plant architecture made a major contribution to the evolution of the various plant species we know today. The genetic diversity in plant architecture also played an important role in the domestication of crop plants. One of the best-known examples of this is the *TEOSINTE BRANCHED 1* gene, which explains the striking difference between maize and its ancestor teosinte (Doebley et al., 1997). Despite large

O. Saracutu (✉)
Plant Sciences Unit – Growth and Development, Institute for Agricultural and Fisheries Research (ILVO), Caritasstraat 21, Melle 9090, Belgium
e-mail: oana.saracutu@ilvo.vlaanderen.be

C. Huyghe (ed.), *Sustainable Use of Genetic Diversity in Forage and Turf Breeding*, DOI 10.1007/978-90-481-8706-5_22,

architectural changes being controlled by only one or few genes and the wealth of knowledge about model plants, the genetic control of plant architecture remains poorly understood in most crop species.

However, the exploration of plant architecture offers new traits and concepts for the improvement of yield and production methods. Important traits are influenced by plant architecture. In forage grasses, these traits include persistence, sward density, and yield. Modifications to plant architecture can improve yield stability throughout the years, which is essential in permanent grassland. The importance of plant architecture as a yield component is widely accepted, but plant architecture is seldom a noted trait in production or breeding. One possible reason is that the available knowledge is not readily applicable to breeding. However, extensive knowledge exists on the genetic regulation of branching in model plants such as Arabidopsis, pea and maize (Schmitz and Theres, 2005). Branching is determined by the developmental state of the apical meristem, the branching pattern and the ability of axillary meristems to grow out. Genes regulating these three steps have been revealed (McSteen and Leyser, 2005).

Tillering in ryegrass influences ground cover, and is correlated with important processes such as senescence in summer or persistence in winter. Persistence includes tolerance to stresses, such as defoliation, trampling, flooding, drought and cold. Rapid re-growth after these stresses prevents the introduction of unwanted grasses and herbs into the sward. The success of an individual genotype in a sward also depends on its competing ability. Competing ability is linked to the height and width of an individual genotype as it spreads and "overgrows" other genotypes in the stand.

The genetic constitution of a genotype to tiller is strongly modified by management practices. In sward conditions, often only 10% of all axillary meristems will form a tiller (Parsons and Chapman, 2000). There is also a compensatory mechanism between leaf area and number of tillers as yield components, called the "self-thinning rule" (Parsons and Chapman, 2000). Because of this, breeding for long- or short-leafed cultivars may have led to a selection of low- and high-tillering plants, respectively. However, some cultivars that combine long leaves and high tillering have been described (Barre et al., 2006).

Tillering is enormously variable in the genus *Lolium*. Tillering characteristics can only be exploited in breeding programs if a detailed phenotypic description of tillering is developed. Based on phenotypic variation in tillering traits, a suitable collection of genotypes is being established and will be used for association genetics.

Material and Methods

A collection of genotypes of different *Lolium* accessions was grown; first as individual spaced plants in a greenhouse during one growing season, then potted and placed outside during the second growing season. Plants were exposed to natural light conditions. The following groups were included: forage (mostly diploid, but also some tetraploids) and turf cultivars of *L. perenne*, forage cultivars of *L. multiflorum, L. temulentum,* and wild *L. perenne* populations. These wild populations are of particular relevance, as they are potential sources of genetic diversity in branching genes not yet exploited in breeding programs.

In the first growing season, 22 accessions (15 cultivars with 10 individual genotypes each, 7 wild populations with 15 individual genotypes each) were investigated for total tiller number and number of tillers per tiller order. In the second growing season, vegetative growth was measured in 24 accessions (14 cultivars with 10 individual genotypes each and 10 wild accessions with 15 individuals each). Genetic diversity was investigated in 5 cultivars (n=10) and 5 wild populations (n=20) with 12 SSR markers, distributed across all chromosomes.

Results and Discussion

Distribution of Tillers over Different Positions in the Plant

An important criterion during establishment of a sward is a rapid tillering response, *i.e.*, tillers arising from all possible positions on the plant. First-order tillers emerge in the sheath of the leaves from the main stem. Tillers that emerge from the sheath of leaves from the first order tillers are considered second order tillers, and so on. Two months after sowing, first and second order tillers were present in all investigated accessions (Fig. 22.1). However, third order tillers appeared more frequently in turf grasses. Notably, the wild populations showed a tendency to produce many tillers of second order. The variability in number of tillers per tiller order was high in the investigated categories – between and within species, between and within cultivars.

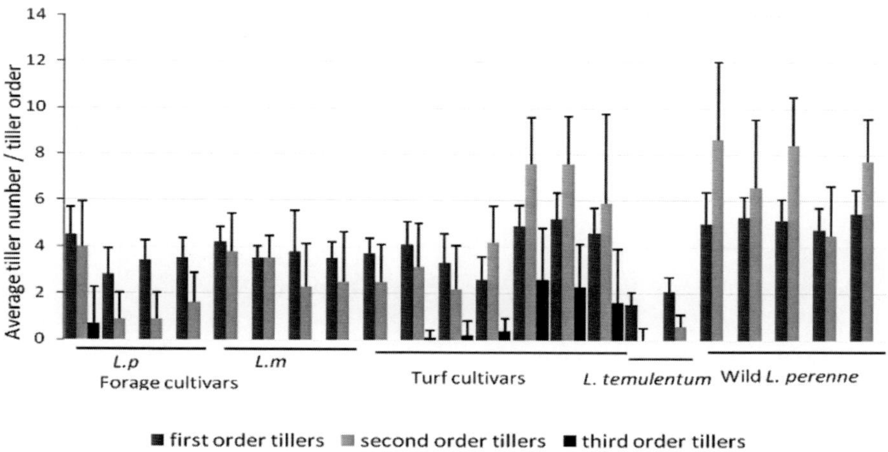

Fig. 22.1 Average tiller number per tiller order ± standard deviation in 2-month old plants of 15 cultivars (n=10), 2 L. temulentum populations (n=15) and 5 L. perenne populations (n=15)

Total Tiller Number

Total tiller number increases rapidly during early growth. Remarkable differences in total tiller number appear over time (Fig. 22.2). Turf type grasses develop nearly twice as many tillers as forage types. In *L. multiflorum*, the total tiller number increased more slowly, probably due to the significantly higher leaf area production than *L. perenne*. After winter, the total tiller number had decreased slightly in all accessions. Again, total tiller number varied tremendously between and within accessions (Fig. 22.2).

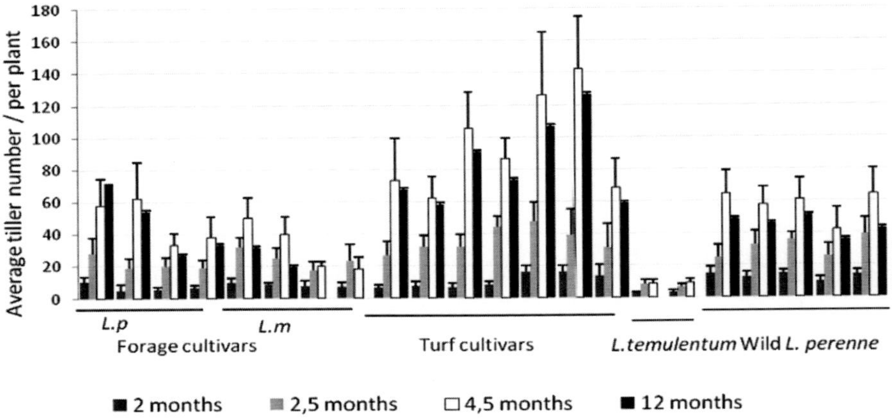

Fig. 22.2 Average tiller number ± standard deviation in 2-, 2.5-, 4.5- and 12-month-old plants of 15 cultivars ($n=10$), 2 L. temulentum populations ($n=15$) and 5 L. perenne populations ($n=15$)

Plant Height and Diameter

In the second growing season, surrogate traits were used for phenotypic assessment of the individual plants in addition to discrete descriptions of the tillering phenotype. Height and diameter showed no specific correlation, but *L. multiflorum* is represented at high height and low diameter values, whereas turf accession showed the opposite relationship (Fig. 22.3). Notably, the wild *L. perenne* populations cover the complete phenotypic range.

Genetic Diversity Supports Phenotypic Diversity

Similar to phenotypic diversity, there was a high degree of genetic diversity. Generally, the total number of alleles, as well as the number of alleles per locus, was lower in the cultivars than in wild populations, but was still considerable (Table 22.1). The observed heterozygozity is comparable for cultivars and wild populations, but in both cases the values are lower than the expected heterozygozity.

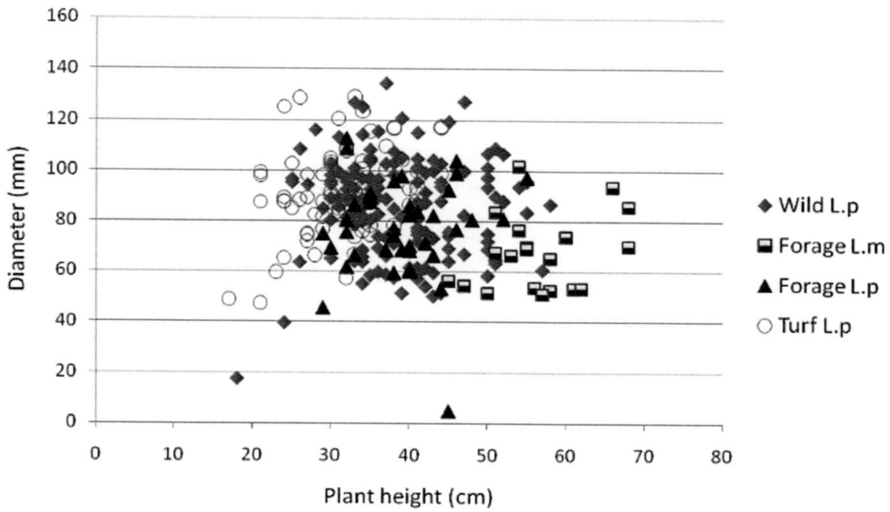

Fig. 22.3 Correlation between plant height (cm) and diameter (mm) based on individual data in 14 cultivars (n=10) and 10 wild populations (n=15)

Table 22.1 Allele number and heterozygosity in 5 cultivars (n=10) and 5 wild populations (n=20), based on 12 SSR markers

	Cultivars	Populations
total allele number / all SSR	50	84
average allele number / SSR and accession	4.2	7
average observed heterozygosity	0.52	0.56
average expected heterozygosity	0.59	0.71

Conclusion

A large heterogeneity in tillering was observed in *Lolium* ssp., between and within species as well as between and within cultivars. Re-growth capacity was low in all investigated accessions. Interestingly, accessions other than those with the highest branching exhibited a high re-growth capacity. These observations suggest that different genetic mechanisms control tillering and re-growth. In light of this phenotypic and genotypic diversity, further work with individual genotypes is required.

References

Barre P., Emile J.C., Betin, M., Surault, F., Ghesquière, M., Hazard, L. 2006. Morphological characteristics of perennial ryegrass leaves that influence short-term intake in dairy cows. Agron. J. 98:978–985.

Doebley, J, Stec, A., Hubbard, L. 1997. The evolution of apical dominance in maize. Nature 386:485–488.
McSteen, P., Leyser, O. 2005. Shoot branching. Annu. Rev. Plant Biol. 56:353–374.
Parsons, A.J., Chapman, D.F. 2000. The principles of pasture growth and utilization. In: Hopkins A (ed.), Grass – its production and utilization. (3rd ed., pp.31–89). Blackwell Science, Oxford.
Schmitz, G., Theres, K. 1999. Genetic control of branching in Arabidopsis and tomato. Curr. Opin. Plant Biol. 2:51–55.

Chapter 23
Agronomic Evaluation of Moroccan Ecotypes of Tall Fescue

Naima Shaimi, Rajae Kallida, and Chaouki Al Faiz

Abstract Drought or reduced water availability is the main factor limiting crop production in Mediterranean area. The aim of this study was to select ecotypes of *Festuca arundinacea* Schreb. adapted to the Moroccan environment. During 3 years, eleven ecotypes of tall fescue, collected from different regions of Morocco, and eight commercial cultivars were evaluated for the following agronomical traits: dry matter yield, summer growth, heading date, survival rate, leaf area index (LAI) and plant height.

Principal Component Analysis showed that 3 components explain 80% of total variability. Perenniality, plant height, ground cover, heading date and summer dormancy were associated with the 1st component. Ground cover and dry matter yield were associated with the 2nd component, while resistance to disease was associated with the 3rd component. Cluster analysis yielded three groups of tall fescue accessions and one unclustered ecotype.

Dry matter yield was significantly correlated with plant height, number of tillers and summer growth ($r = 0.68$, $r = 0.61$ and $r = 0.56$ respectively).

In terms of production potential and perenniality, some local ecotypes present certain superiority in comparison with the commercial varieties. These ecotypes could be eventually exploited to create new Moroccan varieties.

Keywords Tall fescue · Evaluation · Perenniality · Ecotypes

Introduction

Morocco has one of the richest flora in the western Mediterranean area and offers great diversity of soil types, winter temperatures and rainfalls. These features together with Morocco being a main centre of diversity for *Phalaris*,

C. Al Faiz (✉)
Institut National de la Recherche Agronomique (INRA), Rabat B.P. 415, Morocco
e-mail: faizchawki@yahoo.fr

C. Huyghe (ed.), *Sustainable Use of Genetic Diversity in Forage and Turf Breeding*,
DOI 10.1007/978-90-481-8706-5_23,

Festuca and *Dactylis*, have provided a wide range of genetic variation within these species (Saidi et al., 2005). Valuable attributes of North African perennial grass germplasm include summer dormancy (Silsbury, 1961) and vigorous winter growth (Reed et al., 1980). Compared to European tall fescue, Moroccan ecotypes have shown good winter growth, good level of summer dormancy, resistance to disease and presence of non-toxic endophyte (West and Piper, 2008).

Drought or reduced water availability is the main factor limiting crop production in Mediterranean area. The aim of this study was to select ecotypes of *Festuca arundinacea* Schreb. well adapted to the Moroccan environment.

Materials and Methods

Eighteen ecotypes were evaluated under field conditions. These included seven commercial cultivars (FlechaNE, Pastelle, Tanit, Sisa, Centurion, Lutine, Demeter and Fraydo), all originating from Mediterranean area. Seeds of local ecotypes were collected during the summer of 2004 in different regions of Morocco. The field experiment was conducted at the Guich experimental station of INRA, Rabat/Morocco (Latitude 34°03′N, Longitude 06°46′W, Elevation 10.5 m) on an alfic xeropsamment soil, having similar texture throughout (pH 6.8 and organic matter content 1.2%). The soil was without depth limitation for root growth and was fertilized with 28 kg/ha of N, 56 kg/ha of P and 28 kg/ha of K at sowing. We provided extra nitrogen (about 40 kg/ha of N) at tillering and after each harvest (except after the last harvest of each year). Seeds of the ecotypes were pre-germinated in Petri dishes and transplanted into the field on 24 November 2004 at the two tiller stage. Eight plants per accession were planted in 2 rows of 1 meter and were replicated twice. Row spacing was 0.2 m. The total rainfall was 373 mm in 2004/2005, 636 mm in 2005/2006, 323 in 2006/2007 and 350 mm in 2007/2008. The trial was irrigated in the summer of the first year in order to evaluate summer dormancy. Dry matter yields were estimated at each harvest (g DM/m). Flowering time was recorded at 50% of heading (days from 1st January). Plant height and leaf area index (LAI) were recorded at heading time. LAI was measured using the area meters (LAI 2000, LI-COR). Disease (mildew and rust) was scored 0 when there are no symptoms, 1–3 when slight symptoms, 4–6 when moderate symptoms and 7–9 with high symptoms. Plant height and leaf area index (LAI) at heading date were measured. Herbage senescence (visual scoring with 0 = all tissues are green and 100= no visible green tissue) was scored on 15th July and perenniality was assessed as the percentage of plants that resumed growth 10 d after rehydration of summer 2006, 2007 and 2008. The data analysis was based on Principal Component Analysis (PCA) and cluster analysis.

Results and Discussion

Dry Matter (g DM/m), LAI and Plant Height

In the first year, Pastelle accumulated the highest biomass followed by Demeter and Fraydo. Among tested ecotypes, F19, F21 and F28 had the highest production (Fig. 23.1). In the second year, production was three-fold higher than in the first year on average. The cultivar Fraydo accumulated the highest biomass followed by the ecotypes F21 and F25. The severe drought of the third year plus a reduced survival rate caused a considerable decrease of annual production. Indeed, production decreased by 52% in average in comparison of the second year. The ecotype F28 accumulated the highest production followed by Fraydo.

Leaf area index scores were 7.4 on average. Centurion was estimated to have the highest LAI (9.0), while the ecotype F18 had the lowest LAI (5.4).

In general, ecotypes were shorter than cultivars, with 116 cm and 132 cm respectively for ecotypes and cultivars, with Fraydo being the highest (157 cm). Among ecotypes, F22 was the shortest (83 cm) while F25 was the highest (143 cm).

Dry matter yield was significantly ($p < 0.05$) correlated with plant height, number of tillers and summer growth ($r = 0.68$, $r = 0.61$ and $r = 0.56$ respectively).

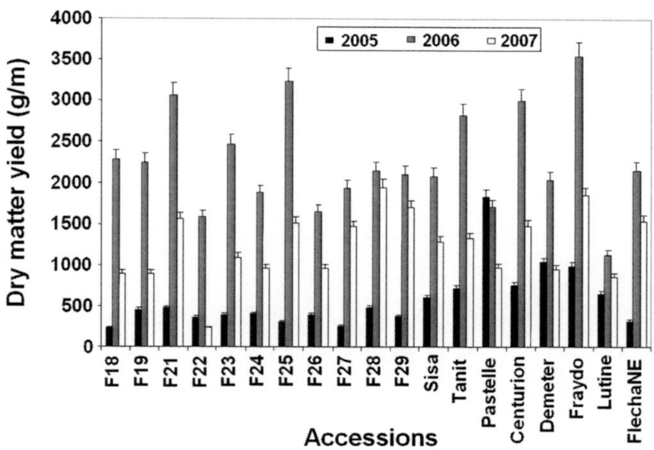

Fig. 23.1 Total yield (DM g/m) for Moroccan ecotypes and cultivars of tall fescue. Bars on the lines represent the standard deviation (SD)

Flowering Time

Fraydo and Tanit were the earliest to flower (Day 75 in average). Pastelle was the latest cultivar to flower (Day 114), while the ecotypes were intermediate to very late (flowering at 87–116 days).

Disease Resistance

All cultivars and ecotypes were slightly to moderately susceptible to both mildew and rust, except the ecotype (D8), which was very susceptible to disease.

Summer Dormancy

It is only under summer irrigation that summer dormancy can be measured to avoid any confounding between the response to water deficit and the expression of dormancy (Norton et al., 2008). Summer dormancy, assessed as percentage of senescence tissues, was very low (5% in average). However, growth seemed to be restricted during summer. Norton et al (2008) reported that tall fescue is incompletely dormant.

Drought Survival and Perenniality

After the summer drought of the second year, survival rate was at least 88% (seven out of eight plants surviving) for all accessions, with the exception of the ecotypes F22 and F23 with 38 and 63%, respectively (Fig. 23.2). The severe and long summer drought of the third and fourth year caused a great decrease of persistence and survival rate dropped to 62% on average. Cultivar Flecha NE and ecotypes F27, F28 and F29 were the most persistent (88%) followed by ecotype F21 and cultivars Centurion, Demeter and Fraydo (75%). The ecotype F22 was the least persistent (13%).

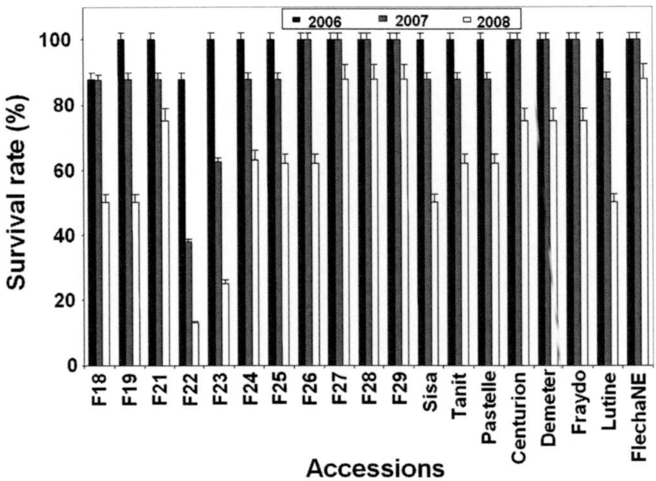

Fig. 23.2 Survival rate (%) for Moroccan ecotypes and cultivars of tall fescue. Bars on the lines represent the standard deviation (SD)

Principal Component Analysis (PCA) and Cluster Analysis

PCA showed that 3 components explain 80% of total variability. Perenniality, plant height, LAI, heading date and summer dormancy were associated with the 1st component. LAI and dry matter yield were associated with the 2nd component, while resistance to disease was associated with the 3rd component.

A dendrogram based on the matrices of the first two principal components (Fig. 23.3) showed, at 80% of similarity, three clusters plus an unclustered accession. The ecotype F21 was unclustered and was characterised by a very good biomass production and perenniality.

The cluster (A) contains Tanit, Demeter, Lutine and Sisa cvs. These cultivars showed very good biomass production and good level of perenniality.

The cluster (B) is the largest one and includes cultivars Pastelle, Fraydo and FlechaNE and the ecotypes F19, F23, F24, F25, F26, F27, F28 and F29. This group was characterised by good to intermediate level of biomass production and perenniality.

The cluster (C) includes the two ecotypes F18 and F22, which presented low level of biomass production and perenniality.

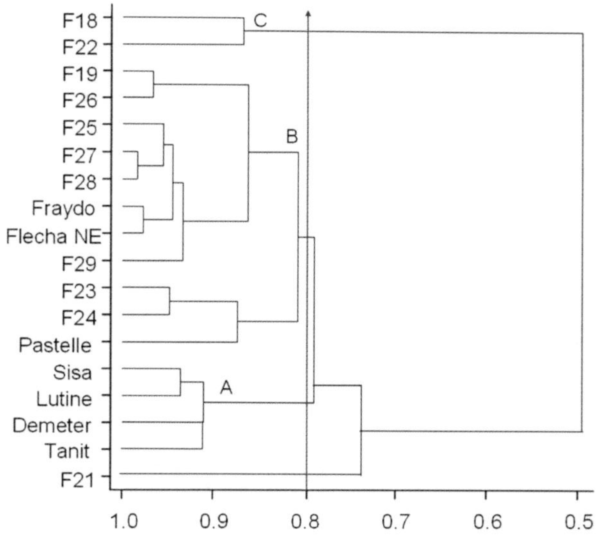

Fig. 23.3 Dendrogram of ecotypes and cultivars of tall fescue studied after cluster analysis on the basis of two principal component mean value

Conclusion

In terms of production potential and perenniality, some local ecotypes presented certain superiority in comparison with commercial varieties. The most productive ecotypes should be included in a breeding and seed production program in order to create Moroccan cultivars. There is a demand for germplasm for forage production adapted to severe summer drought areas, such as in southern Mediterranean areas.

References

Norton, M.R., Volaire, F., Fukai, S., Lelievre, F. 2008. Measurement of summer dormancy in temperate perennial pasture grasses. Aust. J. Agric. Res. 59:498–509.

Reed, K.F.M., Cade, J.W., Williams, A.E. 1980. The significance of Mediterranean plant introduction for increasing the winter growth of pasture plants in the high rainfall aareas of Victoria. Proc. 1st Aust. Agron. Conf., Gatton, Qld. P. 238.

Saidi, N., Al Faiz, C., Thami Alami, I, Ouabbou, H. 2005. Management of biodiversity of forage, pasture and medicinal plant species in Morocco. International conference on promoting community-driven conservation and sustainable use of dryland agrobiodiversity. ICARDA, Aleppo, Syria, April 2005. p. 63.

Silsbury, J.H. 1961. A study of dormancy, survival and other characteristics in *Lolium perenne* L. at Adelaide, S.A. Aust. J. Agric. Res. 12:1–9.

West, C.P., and Piper, E.L. 2008. Non-toxic endophytes, plants injected therewith and methods for injecting plants. U.S. Patent 7465855. 16 December.

Chapter 24
The Czech Core Collection of *Trifolium repens* L.

Tomáš Vymyslický, Jan Pelikán, Pavlína Gottwaldová, and Jan Nedělník

Abstract Altogether 32 characters were evaluated in the set of 161 accessions (varieties, newly bred varieties and wild forms collected in the nature) within world collection of the *Trifolium repens* L. stored in the Czech national gene bank. Thirty plants of each accession were planted on the field; ten of them were evaluated in the years 2007 and 2008. All the evaluated characters were included into the analyses. Missing values were replaced by mean value. Cluster analysis was performed in the software Statistica for Windows for all the accessions together. Complete linkage method was used for clustering and Euclidean distance was used as the measure of distance. Established core collection of the white clover consists of 41 accessions from 17 countries.

Keywords Cluster analysis · Core collection · Evaluation · Morphological characters · Plant individuals · White clover · Yield characters

Introduction

The studies of genetic resources and their evaluation have a long tradition in the Czech Republic (Vacek, 1963; Zapletalová, 1986, Pelikán et al., 2005; Vymyslický et al., 2006). During last 50 years, about 2000 accessions have been evaluated, described and stored in the Gene bank, but many characters have not yet been evaluated at the plant individuals. A large attention is devoted to detailed evaluation of the world assortment of fodder crops in the last years. At present, the evaluation of morphological characters only is not sufficient.

T. Vymyslický (✉)
Agricultural research, Ltd., Zahradní 1, Troubsko 664 41, Czech Republic
e-mail: vymyslicky@vupt.cz

C. Huyghe (ed.), *Sustainable Use of Genetic Diversity in Forage and Turf Breeding*,
DOI 10.1007/978-90-481-8706-5_24,

So evaluation of disease resistance and molecular analyses are performed. While the "core collections" are created, the importance of wild species/forms is increasing as input materials for breeding programmes and for use in organic farming and nature protection.

Drobná and Žáková (2001) evaluated the set of 15 characters on 36 genotypes of *Trifolium repens*. After statistic evaluation by cluster analysis, they found that varieties clustered with accessions of similar geographic origin or with similar qualitative traits.

Boller et al. (2003) studied the collection of old Swiss landraces of the red clover (*Trifolium pratense*). They evaluated set of 20 varieties and a large variability among and within varieties was found. The authors pointed out the old cultivars and landraces in breeding process, as potential sources of valuable traits.

Gubiš (2001) studied the resistance of alfalfa to the Bacterial wilt. He found the most resistant materials among the wild forms.

The study of variability in the collections is mostly based on morphological, phenological and agronomical characteristics of individual genotypes. By application of the multidimensional statistics, the varieties could be differentiated or aggregated according to evaluated traits/characters (Užík and Žofajová, 1997). Kouamé and Quesenberry (1993) used the cluster analysis for classification of the red clover collection as the basis for creating the core collection.

Material and Methods

All the available 161 seed samples of *Trifolium repens* were included into detailed evaluation of morphological and yield characters. They were varieties, newly bred varieties and wild forms collected in the nature and had been stored at the Czech national gene bank. Seeds were picked up from the gene bank storage; small seedlings were planted from the original seeds and then were transplanted into growth boxes. Plants that had been cultivated in growth boxes were planted on the field in the spring 2006 at a spacing of 100 × 80 cm. Thirty individuals of each accession were planted. The evaluation was made on ten individuals of each accession during the years 2007 and 2008. Thirty two morphological, phytopathological and yield characters were evaluated both in the field and in the laboratory. NIRS analyses were also performed to determine the chemical composition of dry matter samples and these results were included into the evaluation together with other characters as a whole.

Quantitative data obtained from 10 individuals were statistically analysed (point estimations of the mean value, estimations of correlation coefficients among measured characters). Point estimations of the mean values were transformed into nine point scale. Other non-quantitative characters were directly evaluated by the nine point scale according to Czech national descriptor's list for genus *Trifolium* L. (Užík et al., 1985).

Cluster analysis was performed in the software Statistica for Windows (STATSOFT, INC., 2003) for all the accessions together. The complete linkage method was used for clustering and Euclidean distance as the measure of distance.

Finally, we established the Czech core collection of the white clover, consisting 41 accessions from 17 countries. This Czech core collection is a subset of 161 accessions available in the Czech national gene bank collection.

Results and Discussion

Based on the cluster analysis, all 161 evaluated accessions were divided into the 39 groups (Fig. 24.1). Good predictive ability of cluster analysis was confirmed. Most of the New Zealand accessions were concentrated in the right side of the dendrogram, while Czech varieties and newly bred materials (Slavice breeding station) were in the middle part of the dendrogram. The most widespread type *hollandicum* was concentrated in clusters 1, 2, 4, 6, 10, 21, 33 and 35. Clusters 17, 18, 38 and 40 were formed by representatives of this type and some of two other types. Cluster no. 19 consisted of type *giganteum*. Clusters 11 and 12 contained members of the type *giganteum* plus one member of the type *hollandicum*. Only members of *sylvestre* type were spread through all the clusters with no significant concentration. Results of cluster analysis would tend to show the fact, that only the morphological descriptors are poor to define the varieties of white clover. This is most probably caused by the fact, that the taxonomical differences among varieties are unclear thanks to intensive breeding process within the different ecotypes, forms, types of the white clover.

Established core collection of the white clover consists of 41 accessions from 17 countries (for details see Table 24.1). Czech Republic has 8 accessions; United Kingdom has 5 accessions; New Zealand, Netherlands and Denmark have 4 accessions; France has 3 accessions; Sweden and Poland have 2 accessions and Finland, USA, Germany, Japan, Ireland, Estonia, South African Republic, Latvia and Hungary have only one representative. Form *hollandicum* is in the collection 17 times, form *gigantem* 9 times, form *sylvestre* 8 times and by 7 accessions we did not find the form. Altogether, 39 subclusters for selection of representatives for the core collection were determined. Accessions of the most widespread form *hollandicum*, are concentrated in 12 subclusters. Accessions of the form *giganteum* are concentrated in 3 subclusters. Accessions of form *sylvestre* are scattered through other subclusters.

The closest pairs found by this clustering method were linked together and consisted of six accessions: Romena (POL) and SE-123 (CZE – breeding line), Klondike (DNK) and KARP 101/93 (CZE – wild accession), AberDai and AberEndura (GBR).

Interesting cluster was formed by the New Zealand accessions (Huia, Challenge, Emerald, Milton, Tribute, Bounty and Sustain) and by the Czech accessions (TROU 191/95, TROU 87/95, Jura, SE-124, SE-117, SE-119 and SE-115) – wild forms, one variety and breeding lines (from the Slavice breeding station).

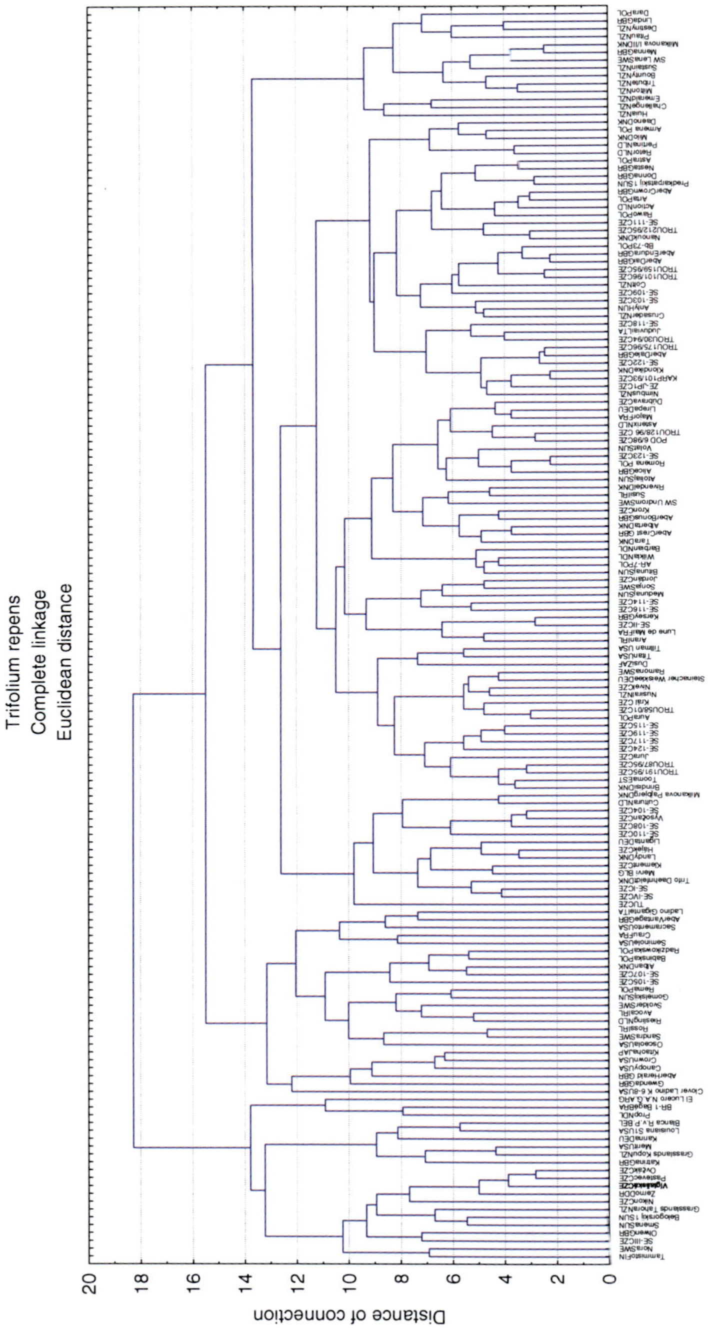

Fig. 24.1 Dendrogram of 161 accessions of *Trifolium repens*

Table 24.1 The Czech core collection of *Trifolium repens* accessions

No.	Accession	Country	No.	Accession	Country
1	Tammisto	FIN	22	Lune de Mai	FRA
2	Olwen	GBR	23	Jordán	CZE
3	Grassland Tahora	NZL	24	Barbian	NDL
4	Ovčák	CZE	25	AberCrest (AC 50)	GBR
5	Merit	USA	26	Rivendel	DNK
6	Karina	DEU	27	Alice	GBR
7	Prop	NDL	28	Major	FRA
8	Kitaoha	JAP	29	AberDale	GBR
9	Ross	IRL	30	Juduviai	LTA
10	Rema	POL	31	Anly	HUN
11	Alban	DNK	32	Colt	NZL
12	Crau	FRA	33	Nanouk	DNK
13	TU	CZE	34	Action	NDL
14	Klement	CZE	35	Astra	POL
15	Hájek	CZE	36	Pertina	NDL
16	Vysočan	CZE	37	Milo	DNK
17	Tooma	EST	38	Grassland Challenge	NZL
18	SE-119	CZE	39	Tribute	NZL
19	Král	CZE	40	SW Lena	SWE
20	Ramona	SWE	41	Linda	GBR
21	Dusi	ZAF			

Conclusions

Within all the 161 evaluated accessions available in the Czech national gene bank we established core collection of the white clover, which consists of 41 accessions from 17 countries. Altogether, 39 subclusters for selection of representatives for the core collection were determined. *Hollandicum* type is in the collection 17 times, *gigantem* type 9 times, *sylvestre* type 8 times and by 7 accessions we did not find the detailed taxonomic level (form, type). It is a pity, that by the varieties and breeding lines we do not know detailed pedigree of the varieties and breeding lines for determination of their joint progenitors.

Acknowledgement The results were obtained in the frame of institutional project, number MSM 2629608001, financed by the Ministry of Education, Youth and Sports of the Czech Republic.

References

Boller, B., Tanner, P., Gűnter, S., et Schubiger F.X. 2003. Description and Evaluation of a Collection of former Swiss Red Clover Landraces. Czech J. Genet. Plant Breed. 39:31–37.

Drobná, J., Žáková M. 2001. The evaluation of the collection of the White clover (*Trifolium repens* L.) genetic resource by Cluster analysis. Scientific studies Research Institute of Crop Production, Piešťany, 30:187–192. [in Slovak].

Gubiš, V. 2001 Predisposition of Alfalfa genetic resources to the Bacterial wilt. Scientific studies Research Institute of Crop Production, Piešťany, 30:119–123. [in Slovak].

Kouamé, C.N., Quesenberry, K.H. 1993. Cluster analysis of a world collection of red clover germplasm. Gen. Res. and Crop Evol. 40:39–47.

Pelikán, J., Vymyslický, T., Gottwaldová, P. 2005. The evaluation of genetic resources of fodder plants the species of family *Fabaceae* and their relatives in the Czech Republic. Proceedings of the international conference: Hodnotenie genetických zdrojov rastlín, Research Institute of Crop Production, Piešťany: 73–76. [in Czech].

STATSOFT, INC. 2003. STATISTICA Cz [Software system for data analysis], version 6. www.StatSoft.cz

Užík, M., Vacek V., Tomašovičová A., Bareš I., Sehnalová J., Blahout J. 1985. Descriptors List. Genus *Trifolium* L. – Genetic resources 23, Research Institute of Crop Production Praha-Ruzyně.

Užík M., Žofajová, A. 1997. The classification of characters and varieties of barley (*Hordeum vulgare* L.). – In: Genetické zdroje rastlín, SPU Nitra: 7–16. [in Slovak].

Vacek, V. 1963. Study, Maintaining and use of the world assortment of fodder crops. I Wild flora, A. family *Fabaceae.* – Ms., depon. Research Institute for Fodder Crops, Ltd. Troubsko. [in Czech].

Vymyslický, T., Pelikán, J., Gottwaldová, P., Nedělník, J. 2006. The Czech national core collection of *Trifolium pratense.* – In: Proceedings of the Eucarpia conference "Breeding and seed production for conventional and organic agriculture", Perugia, Italy, 3–7.9.2006: 200–205. ISBN-978-88-87652-12-3.

Zapletalová, I. 1986. Study of the among-varieties and within-varieties variability of genetic resources of the species *Trifolium repens* L. Ph. D. Thesis, depon. Research Institute for Fodder Crops, Ltd. Troubsko.

Chapter 25
Towards an Enhanced Utilization of Plant Genetic Resources in Grass Breeding by Characterization and Evaluation Trials

Evelin Willner, Susanne Hünmörder, and Klaus J. Dehmer

Abstract Characterizations and evaluations of the most important grass species maintained there belong to the fundamental tasks of the Malchow Satellite Collection of the IPK Genebank. In recent years and in cooperation with universities, breeding companies and research institutions, several trials were performed in this respect.

The most comprehensive analyses were performed on *Lolium perenne* L. (number of accessions maintained at Malchow: more than 2,900), with a special focus on material from a collection trip to Romania. Primary evaluations of 455 ecotypes showed e.g. large variations in traits like heading date and growth type, partially outperforming standard varieties. Based on these results, individual plants of ecotypes of the same origin and with identical traits were unified to subgroups and introduced as 85 additional accessions into the genebank. These were then subjected to secondary evaluations. Here again, some accessions did as well as the standard varieties or even better.

In summary it could be demonstrated that the collected ecotypes constitute a suitable basic material for ryegrass breeding.

Keywords *Lolium perenne* L · Perennial ryegrass · Plant genetic resources · Characterization · Evaluation · Phenotypic variation of traits

Introduction

Besides oil seed crops and other fodder crops, the Malchow Satellite Collection of the IPK Genebank is holding a large grass collection with more than 10,000 accessions. Besides maintenance, characterization and evaluation of the most important

E. Willner (✉)
Leibniz Institute of Plant Genetics and Crop Plant Research (IPK), Genebank, Satellite Collections North, Inselstrasse 9, Malchow/Poel 23999, Germany
e-mail: willner@ipk-gatersleben.de

C. Huyghe (ed.), *Sustainable Use of Genetic Diversity in Forage and Turf Breeding*,
DOI 10.1007/978-90-481-8706-5_25, © Springer Science+Business Media B.V. 2010

grass species are the fundamental tasks of the genebank work here in order to e.g. provide better described material for breeding and research.

In recent years and in cooperation with universities, breeding companies and research institutions, we have performed several trials in this respect.

The most comprehensive analyses were performed on accessions of the genus *Lolium*, with special focus on material from a collection trip to Romania in 1993. This paper outlines the results of characterizations and evaluations conducted between 1994 and 1997.

Material and Methods

In 1993, a joined plant exploration for grass species (in cooperation with the Eurograss breeding company) was carried out in central Romania in the four regions of Crisana, Transylvania, Carpathians (partly) and Subcarpathians (Fig. 25.1), with a total of 29 areas being covered (Tab. 1). As a result, 455 populations of *Lolium perenne* L. were collected (as seed or living material), besides other important grass genera like *Dactylis, Festuca* and *Poa*.

Between 1994 and 1997, the material was characterized in a single plant trial. From each of these 455 populations, 24 single plants were observed and compared to eight standard varieties (Bardonna, Castle, Citadel, Fennema, Limes, Liprinta, Livre, Vigor), representing the range of variability of variety material for the traits of interest (score units 1–9). The most important traits like ear emergence, growth rate at different stages, reaction to diseases, leaf width, growth habit and persistency was assessed at Malchow/Poel, Northern Germany (Longitude 11°28′26″ E, Latitude

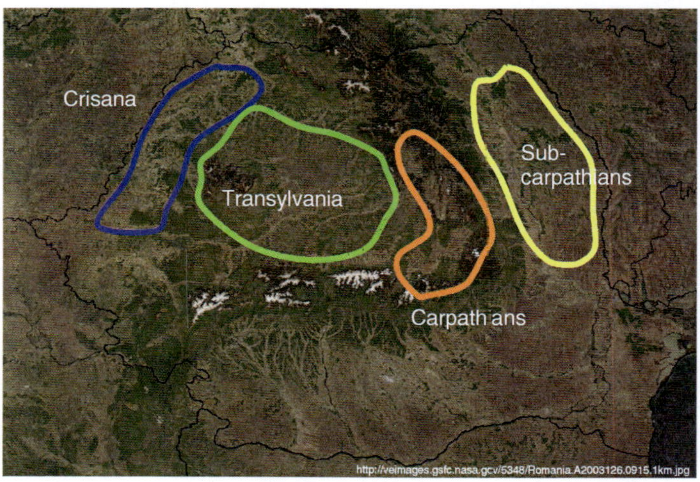

Fig. 25.1 Collecting regions of the 1993 plant exploration in Romania

53°59′40″ N, 2 m ASL, mean annual precipitation 562 mm, annual temperature 8.9°C, sandy loam) over a 3 year period. This trial was carried out as block design without replication for the 455 populations because of the large number of entries, while the standard varieties were repeated five times.

Statistical analyses were carried out only as descriptive statistics, because the data basis concerning variance, sample size, homogeneity of variances and normal distribution was very heterogeneous. Being an initial study, these analyses did not aim at testing neither averages nor variances for significant differences. As a primary goal of collecting trips is looking for new variability and closing gaps in existing genebank collections, they should rather elucidate whether material from plant exploration regions or areas in Romania exhibited a wider range of variability than *Lolium perenne* standard varieties.

Results and Discussion

Table 25.1 shows the results for the five of the traits most relevant for breeding, ear emergence, growth habit, leaf width, bulk before first cut and rust susceptibility for the different collection areas in the four Romanian collection regions in comparison to standard varieties.

No material from the 29 collection areas was on average of all populations earlier or later than the early standard variety Bardonna and the late one, Vigor, respectively. Areas vary between scores 5–7, indicating an intermediate to late heading date. They did not differ significantly from one area to the other. In respect to regions, there was no large difference in variation either. However, there were individual populations with very early or late ear emergence, e.g. from the areas Crisul valley or Crisana plain (Table 25.1). Some populations were earlier than Bardonna, but no population was later than Vigor. In addition, there seems to be a tendency that material from mountainous areas is less variable with regard to this trait compared to populations from the plains.

Regarding growth habit, areas varied within the range of the two standard varieties around a score of 6 to 8 with a tendency towards a prostrate growth habit (Table 25.1). Material from no area reached the level of Bardonna or below (erect growth). However, accessions from some areas showed a growth habit nearly as prostrate as Vigor (Crisana plain, Cimpia, Somes hills).

Nevertheless, some individual populations of these areas exhibited the maximum trait score and outcompete the variety, Vigor. In Crisul valley, a very wide variation could be observed.

For leaf width, no large variation could be observed between varieties, material from different areas and/or regions, scores ranging around 5 (Table 25.1). In some areas only, like Crisana plain, Crisul valley, Moldavia or Cimpia, individual populations showed variation of three to four units of score within the same area. Narrow-leafed populations could be found in one area only, Siret valley.

Table 25.1 Mean, minimum and maximum values for five traits of collected populations, grouped by regions (*bold*) or areas

		ear emergence 1 very early 9 very late		growth habit 1 erect 9 prostrate		leaf width 1 very narrow 9 very wide		bulk before first cut* 1 very low 9 very high		rust susceptibility 1 very low 9 very high	
regions / areas	no. of pops.	MW	Min/Max	MW	Min/Max	MW	Min/Max	MW	Min/Max	MW	Min/Max
standard varieties	**8**	**6.1**	**3.8/8.6**	**6.4**	**4.4/8.3**	**5.1**	**4.6/5.8**	**4.9**	**4.0/5.7**	**4.0**	**3.5/4.2**
collection material	**442** +	**5.6**	**1.7/7.8**	**6.8**	**1.8/9.0**	**5.0**	**2.5/7.8**	**5.7**	**4.1/7.4**	**4.2**	**1.9/5.7**
Crisana	**139**	**5.6**	**4.0/6.5**	**6.7**	**4.5/8.0**	**5.0**	**4.3/6.3**	**5.2**	**4.6/5.8**	**3.6**	**2.8/4.5**
Crisana plain	87	5.7	3.7/6.8	7.8	5.0/9.0	5.1	4.0/7.5	5.3	4.2/6.3	3.7	2.9/4.9
Crisul valley	32	4.9	1.7/6.6	5.8	1.8/8.3	5.5	4.8/7.7	5.5	4.6/6.2	3.5	2.8/4.8
Marghita depression	2	5.8	5.3/6.3	7.0	7.0/7.0	4.7	4.6/4.8	5.1	5.0/5.2	3.7	3.7/3.8
Satu-Mare plain	18	5.8	5.3/6.4	6.3	4.0/7.8	4.8	3.9/5.0	5.0	4.4/5.5	3.5	1.9/4.5
Transylvania	**209**	**5.6**	**4.9/6.4**	**6.9**	**5.7/8.2**	**5.0**	**4.6/5.7**	**5.8**	**5.2/6.3**	**4.1**	**3.6/4.6**
Cimpia	63	5.3	4.8/6.6	7.5	5.7/9.0	5.1	4.2/7.8	5.8	4.8/7.0	4.4	3.7/4.9
Cimpia hill country	21	5.6	4.8/6.6	6.4	4.8/8.7	5.4	4.7/6.8	5.5	4.6/5.9	3.5	3.0/4.3
Hirtibaciu hills	2	5.3	5.3/5.4	7.1	6.3/7.8	4.9	4.8/5.0	5.7	5.5/5.8	4.4	4.0/4.8
Hirtibaciu valley	2	4.7	4.0/5.3	7.1	6.3/7.8	4.9	4.7/5.2	5.9	5.5/6.3	4.6	4.4/4.7
Meses mountains	11	5.6	5.0/6.6	7.6	6.2/9.0	5.0	4.5/5.3	5.5	5.0/6.2	3.7	3.3/4.1
Mures valley	7	5.6	5.0/6.3	6.8	6.5/7.0	5.0	4.7/5.2	6.5	6.3/6.8	3.9	3.5/4.2
Sebes valley	2	5.0	4.3/5.6	6.6	6.2/7.1	4.9	4.8/5.0	6.8	6.7/6.9	4.0	3.5/4.4
Secas hills	9	6.2	5.4/7.0	7.5	6.4/8.7	4.9	4.5/5.5	6.3	5.5/7.2	4.6	4.1/4.8
Silvana hills	8	7.1	6.7/7.8	6.0	5.3/6.8	5.0	4.7/5.3	4.7	4.1/5.4	3.6	3.4/3.8
Somes hills	25	6.2	5.3/6.8	8.0	6.3/9.0	4.9	4.5/5.5	5.5	4.3/6.3	4.0	3.3/4.8
Subcarpathian Transylvania	36	5.3	4.4/6.8	6.8	4.3/8.7	5.0	4.5/5.5	6.0	5.0/6.5	4.2	3.2/5.0
Tirnava hills	23	4.8	3.4/5.5	5.8	4.3/8.2	5.3	4.7/6.2	5.5	4.8/5.8	4.1	3.3/4.9

Table 25.1 (continued)

regions / areas	no. of pops.	ear emergence 1 very early 9 very late		growth habit 1 erect 9 prostrate		leaf width 1 very narrow 9 very wide		bulk before first cut* 1 very low 9 very high		rust susceptibility 1 very low 9 very high	
		MW	Min/Max	MW	Min/Max	MW	Min/Max	MW	Min/Max	MW	Min/Max
Carpathians	**34**	**5.1**	**4.4/5.9**	**6.6**	**5.3/8.1**	**5.0**	**4.7/5.3**	**5.8**	**5.4/6.3**	**4.3**	**3.9/4.8**
Brasov depression	12	5.4	4.3/6.3	6.9	5.0/8.6	5.0	4.7/5.3	6.1	5.3/6.6	4.5	4.3/4.8
Ciuc depression	6	5.6	5.3/6.0	7.2	5.5/8.7	5.2	4.8/5.5	6.0	5.7/6.2	4.1	3.7/4.8
Gurghiu mountain	5	4.9	4.4/5.2	6.2	5.8/7.0	4.8	4.5/5.0	5.5	5.1/5.9	4.3	3.8/4.6
Harghita mountain	11	4.6	3.8/5.9	6.0	4.8/8.0	5.1	4.8/5.5	5.5	5.3/6.3	4.2	3.8/4.8
Subcarpathians	**60**	**5.9**	**5.6/6.3**	**7.1**	**6.2/8.1**	**4.9**	**4.2/5.4**	**6.0**	**5.5/6.6**	**4.6**	**4.1/4.9**
Bahlui plain	3	6.3	6.2/6.3	7.2	6.5/7.8	4.6	4.3/5.0	5.9	5.1/6.8	4.6	4.3/4.8
Birlad plateau	8	5.5	5.2/5.8	6.3	5.3/7.7	5.2	5.0/5.3	6.1	5.7/6.4	4.6	3.8/4.9
Falciu hills	8	5.9	5.5/6.6	6.9	5.7/8.5	4.8	4.1/5.2	5.2	4.6/6.4	4.1	3.7/4.8
Hirlau hills	3	6.1	5.9/6.3	7.4	7.0/8.0	4.6	4.0/5.0	6.7	6.6/6.8	4.9	4.8/5.2
Jijia plain	5	6.2	5.9/6.7	7.8	7.3/8.3	4.7	4.2/5.2	6.1	5.3/6.7	4.4	4.1/4.8
Moldavia	17	6.0	5.2/6.9	7.1	5.0/8.2	5.5	4.5/7.3	6.3	5.6/7.4	5.1	4.2/5.7
Prut valley	3	6.0	5.8/6.2	6.7	5.7/7.5	5.0	4.8/5.2	5.7	5.5/5.9	4.4	4.0/4.5
Siret valley	8	5.7	5.5/6.3	7.0	6.2/8.3	4.3	2.5/5.2	6.0	5.6/6.6	4.5	4.2/4.9
Tutova valley	5	5.5	5.3/5.8	7.5	7.0/8.3	5.0	4.7/5.3	5.9	5.7/6.1	4.6	4.2/4.8

Underlined: minimum and maximum values for respective trait; * visual estimate; + 13 additional populations not listed (one pop. per area, no means)

The best variety in respect to bulk before first cut (only as visual estimate, not measured), Fennema, with a score of 6, was outperformed by populations from various areas (Hirlau hills, Mures and Sebes valleys) from regions like the Subcarpathians or Transylvania (Table 25.1). Here, the best populations were scored with a 7 for high yield. Remarkably, the populations with the lowest yield did not fall below the score for standard variety Castle. Populations of areas within the Crisana region and the Carpathians were rather uniform, with neither the mean nor the extremes varying by more than one unit of score. On the contrary, Transylvanian populations showed the widest range of variation between and within areas.

Concerning rust susceptibility, there were no big differences between and within the areas or regions (Table 25.1). The Crisana region appeared to be very interesting, as all its four areas had an average below the mean of this trial. The best populations with a score of 2 were found in the Satu-Mare plain. The Moldavia area showed the highest average score at 5, i.e. intermediate susceptible.

When employing box plot analyses for a visualization of the results, traits ear emergence and growth habit show the largest variability compared to the other three traits evaluated (Fig. 25.2). In comparison to the standard varieties, the latters' range of variability was only surpassed by a few populations of the Crisana und Transylvania region (earlier ear emergence, prostrate or erect growth habit, respectively; Fig. 25.2a,b).

Leaf width (Fig. 25.2c) was the least variable trait, both when comparing between regions and to the standard varieties (mean value and 50% of population means ranged around score 5). However, the rather large variability within

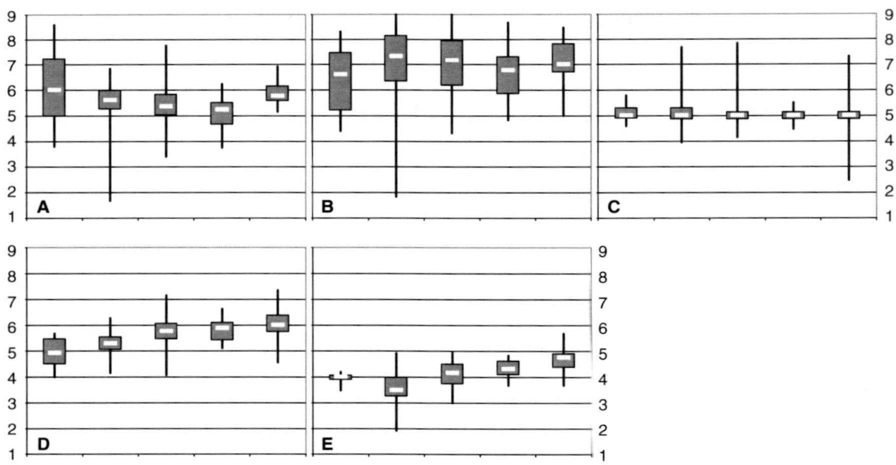

Fig. 25.2 Box plots displaying the variability found for the five evaluated traits, shaded boxes indicating the central 50% quantile, black vertical lines the total variation per trait and white horizontal lines the mean values: (**a**) ear emergence, (**b**) growth habit, (**c**) leaf width, (**d**) bulk before first cut, (**e**) rust susceptibility; material evaluated: standard varieties, populations from Crisana, Transylvania, Carpathians and Subcarpathians, respectively (from left to right in each graph)

the Subcarpathians region with narrow and broad leaved individuals is striking. It is higher than within the three other regions, while also being the most variable trait of the Subcarpathians. Bulk before first cut (Fig. 25.2d) and rust susceptibility (Fig. 25.2e) hardly vary between and within regions. Only Crisana region with 75% of the population means below the variety means raise expectations for an increased rust resistance.

These results on rust susceptibility are relevant with regard to a second evaluation by H. Lellbach, Julius Kuehn Institute (Lellbach and Willner, 2002), who applied the rust leaf segment test to progenies of the populations presented in this paper. 23 (27%) of 85 subgroups analysed were scored as resistant. These subgroups originated mainly from the Crisana region (Crisana, Satu-Mare plain) or from the Transylvania region (Cimpia, Tirnava hills). Susceptible subgroups were located mostly in Moldavia and Subcarpathian Transylvania (data not shown).

As large numbers of additional accessions from a geographically defined region – like the 455 Romanian ecotypes/populations presented here – would mean a too excessive amount of additional work load of maintaining them in a genebank for the future, 18 individual plants from one or more populations per area with similar traits in ear emergence (early, medium, late), bulk (highest scores), and rust susceptibility (low) were multiplied together. The resulting progenies were integrated as new accessions into the IPK genebank collection. This resulted in 85 instead of 455 new entries (cf. the 85 subgroups from the Lellbach rust evaluations), permitting a more effective genebank management of the Romanian material. These accessions can be viewed as "Pre-Breeding" material as described by Paul et al. (1994). At the same time, this procedure led to the formation of a core collection of the Romanian exploration material, as the subgroups/the new accessions cover the maximum of variability of the five traits of interest (see also Oetmann, 1994 for *Lolium* and van Hintum, 1994 for core collections in general).

Conclusions

Characterization and evaluation trials lead to better described plant genetic resources collections (see also Casler, 1995 for a geographically broader collection, scored for a more limited number of traits). The data from such trials assist external users in deciding which accession(s) to request from large and diverse germplasm collections. For genebank curators, they furthermore permit the selection of individual plants for uniting them to subgroups with similar traits, which then can be accepted as well characterized, new accessions into a genebank. The respective accessions contribute to demonstrate the desirable variation for specific traits as postulated by Connolly (2001). The gathered data are in addition available for the EPGRIS3 project, aiming at making all European plant genetice resources data publicly available (ECPGR working group documentation and information, see: http://www.epgris3.eu/).

References

Casler, M.D. 1995. Patterns of variation in a collection of perennial ryegrass accessions. Crop Sci. 35:1169–1177.

Connolly, V. 2001. Breeding improved varieties of perennial ryegrass. Crops Research Centre Oak Park, Carlow, End-of-project report, ISBN 1 84170 199 8.

Hintum, Th.J.L. van 1994. Drowning in the genepool – managing genetic diversity in genebank collections. Ph.D. Thesis, Swedish University of Agricultural Sciences, Svalöv, Sweden.

Lellbach, H., Willner, E. 2002. Genetische Ressourcen bei Weidelgräsern – Kronenrostresistenz von Ökotypen in Bezug zum geografischen Ursprung und zu anderen züchterischen Merkmalen. Biologische Vielfalt mit der Land- und Forstwirtschaft, Schriftenreihe des BMVEL, Reihe A: Angew. Wissensch. 494:286–287.

Oetmann, A. 1994. Untersuchungen zur intraspezifischen phänotypischen Variabilität autochthoner Weidelgrasherkünfte (*Lolium perenne* L.) und ihre Bedeutung für die Erhaltung wertvoller Standorte vor Ort (*in situ*). Diss. Universität/Gesamthochschule Kassel.

Paul, C., Posselt, U.K., Scheller, H. 1994. "Pre-Breeding" genetischer Ressourcen des Deutschen Weidelgrases (*Lolium perenne* L.). Vortr. Pflanzenzücht. 27:200–204.

Part III
Genetic Changes in Grassland and Turf Communities

Chapter 26
Ecological and Population Genetic Concepts for Creating New Varieties

Isabelle Litrico, Philippe Barre, and Christian Huyghe

Abstract The agronomic value of grasslands tends to decrease over time, leading to the need for repeated ploughing and resowing, which cause long term environmental damage. In order to extend the time during which grasslands are productive, we must understand the causes for this decline. Changes in population genetic structure of agronomic grasses due to the interaction of selection, migration and drift may provide part of the answer.

Selection pressures can be mediated by abiotic factors (e.g., cultural practices, soil, climate) or biotic factors (e.g., competitors, diseases). Gene flow into sown grasslands may occur via pollen or seed from the surrounding landscape, and small effective population size may exacerbate genetic drift. Given that grasslands are composed of multiple species and cultivars, similar mechanisms may control their species composition. Different pedo-climatic conditions, different species and genetic grassland compositions and different cultural practices lead to different adaptations of grassland community.

Productivity in sown grasslands will be determined initially by the sown composition of species and genotypes within species. Productivity and resistance to invasion can be enhanced by complementary resource use among species or genotypes. Declines in productivity may arise if the traits that determine the initial success of species or genotypes are negatively correlated with traits that determine productivity over the long term. If so, taking measures to enhance the maintenance of diversity over time may extend the period of high productivity in sown grasslands, and developing cultivars or mixtures for which the productivity- success trade off is unlikely, are important future goals.

Keywords Agronomic value · Competition · Facilitation · Functional diversity · Genetic diversity · Specie diversity

---

I. Litrico (✉)
Institut National de la Recherche Agronomique – Unité de Recherche Pluridisciplinaire Prairies et Plantes Fourragères, Route de Saintes, Lusignan 86600, France
e-mail: isabelle.litrico@lusignan.inra.fr

C. Huyghe (ed.), *Sustainable Use of Genetic Diversity in Forage and Turf Breeding*,
DOI 10.1007/978-90-481-8706-5_26, © Springer Science+Business Media B.V. 2010

From Fixed Pure Lines to Grassland Communities

For many years, we have considered cultivated populations of perennial forage species as stable populations without genetic changes over time. This assumption is coming from most annual crops where varieties are fixed pure lines (wheat, bean...) or F1 hybrids (maize...). This assumption of fixity and homogeneity deeply influenced the view of perennial forage species, by considering that their varieties were also stable over time. But, when considering meadows, the situation is very different. Indeed, unlike annual crops, varieties used in meadows present a high level of genetic diversity. Indeed, varieties are synthetics obtained from the free intermating of variable number of individuals. For example, the number of alleles for SSR markers is higher in forage varieties compared to annual crops in which no more than two alleles are observed (Fig. 26.1). In addition to this genetic diversity, specific diversity is often observed and used in meadows. Indeed, farmers sow several species in mixtures where the number of species ranges from two to ten. Mixtures such as – cocksfoot, fescues, lotus, alfalfa -, – fescue, timothy, red clover, sainfoin -, – ryegrass, fescue, red clover, lotus – or – ryegrass, timothy, red clover, sainfoin – are recommended depending on location and expected exploitation regime (Grenier, 2004). Moreover, meadows are exploited during several years, with variable agricultural management and under variable soil and climate conditions. The high species and genetic diversity with a long period of exploitation generates obvious possibilities of changes over time.

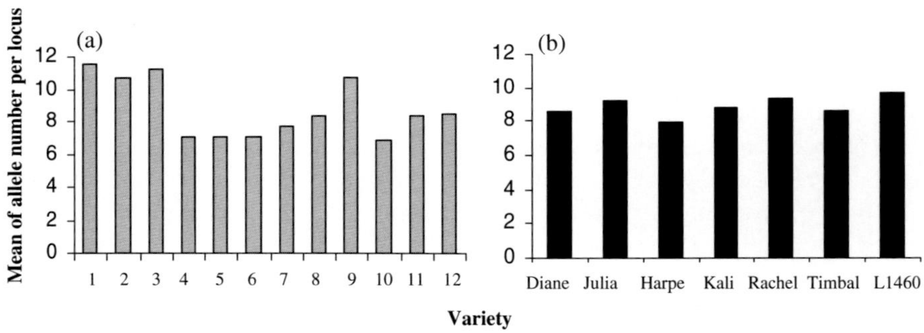

Fig. 26.1 Mean of number of allele per locus for (**a**) 12 varieties of *Lolium perenne* with seven loci SSR (unpub. and private data) and (**b**) seven varieties of *Medicago sativa* with eight loci SRR (Flajoulot et al., 2005)

Ecology and Population Genetics and Evolution of Communities

Like natural heterogeneous populations and communities under environmental pressures, species and genetic composition of meadows is likely to change over time via adaptation mechanisms. Concepts of ecology and population genetics should lead to a better understanding of meadow dynamics and its possible consequences on agronomic value changes.

Some definitions are necessary for further understanding. A population is defined as a group of individuals (genotypes) belonging to the same species which intermate with the same probability at a given time and space. At the upper level, a community is defined as a group of populations of different species which interact at a given time and space. Considering a species, its fundamental niche (Hutchinson, 1957) is defined as all resources potentially available for this species taking into account its biology, physiology, and ecology when no competition occurs. In natural conditions, the occupied niche for a species is narrower than fundamental niche and it is called the real or realized niche. Indeed, the realized niche results from competition for resources.

Within communities, presence of species and their respective abundance depend on factors such as edaphic and climatic conditions, which will make it possible for a species to survive and produce progenies in this environment. It will also depend on biotic factors. Interactions among species within a community will shape this community. The theory of competitive exclusion (Hardin, 1960) or Gause's theory (Gause, 1934) reported that two species occupying simultaneously the same niche will not live together for long. As a consequence, segregation of niches and functional divergence of individuals only will make it possible for several species to coexist in a community, these species having a similar fundamental niche but exhibiting different real niches. This was well illustrated on finches by Darwin (Lack, 1945) or on warblers by MacArthur (1958). Competition is not the only interaction existing in a community; facilitation (Bruno et al., 2003) may be extremely advantageous for species coexistence. Indeed, beyond segregation and complementary of experienced real niches, some species may facilitate settlement, survival and reproduction of other species. In meadows, this is clearly the case for legumes which facilitate grasses through contribution to the N budget of the community.

As said earlier, a community is composed of populations of several species and these populations may be genetically heterogeneous. For most perennial forage species used in sown grasslands, natural populations as well as varieties show a broad genetic diversity. This diversity, as well as the species diversity, will be shaped depending on abiotic and biotic factors. Factors and mechanisms influencing community structure may similarly influence genetic structure of populations whose the unit is the allele.

Within a population, when considering that each genotype possesses a fundamental niche, the real niche of this genotype may depend on competition and facilitation that this given genotype has with the other genotypes of the population and the other species of the community. As a consequence, the genetic and allelic structure of the population and its equilibrium will also depend on the existing segregation and complementary generated by the interactions within the population and the community.

On the same basis, it is possible to draw a parallel analysis of the factors and mechanisms explaining the population structure, using the concepts of population genetics, to explain the community structure. To do so, it is necessary to change scale and to apply the same principles. When interested in selection mechanism at population level, genetic selection modifies allelic frequency over time. Individuals carrying favourable alleles in a given environment survive and efficiently breed. Frequencies of their alleles will increase in the subsequent generations while the frequencies of unfavourable alleles will decrease.

Considering the same mechanism at community scale with species as the elementary unit, under selection pressure, it may be expected that species with a low fit will decline or disappear. Thus, selection tends to reduce both genetic and species diversity. At the population level, selection directly acts on non-neutral loci while frequency of neutral alleles is expected to vary at random (i.e. independently of selection effect). However, it is important to notice that neutral loci may be influenced by selected loci, through a so-called "hitch hiking mechanism" (Maynard Smith and Haigh, 1974; Aguade et al., 1989; Stephan and Langley, 1989; Begun and Aquadro, 1992; Stephan et al., 2006). But we will not go in detail into this mechanism.

The random variation of neutral alleles in a population is called genetic drift. An example of such as situation of significant drift of allele frequency is observed when population size suddenly decreases (bottleneck), as a consequence of a random event. The drift effect is much stronger on small sized populations. Similarly, if a stochastic phenomenon induces a shrink of community size, species abundance and even their presence may be affected. This will generate a random variation of species diversity of the community. When several steps of drift occur, species and alleles are likely to disappear.

Migration is another mechanism influencing allele frequency in a population, as arrival of new alleles and emission of pollen and propagule to neighbouring populations will change allele frequencies over generation. Fluxes of pollen and to a lesser extent of seeds are common in perennial forage species populations due to their anemogamy and entomogamy. Similarly, arrival of seeds of a species in a community will change species abundance. It may also generate an increase in species number of the community. Migration of seeds may occur in meadows as a consequence of wind, but also as a consequence of animal behaviour. Seeds may be transported on animal skin (exozoochory) or through their gut (endozoochory), where seeds of perennial forage species may successfully survive. This was documented in permanent grasslands (Pakeman et al., 2002) or in experimental design with animal in barns (Huyghe et al., 2008).

The main underlying process of migration is the sexual reproduction. Sexual reproduction within a population has little effect on allele frequency of the population if all individuals similarly contribute to pollen and seed pools. At a population scale, individual plants will breed via self or cross-pollination. In the case of cross-pollination, individuals may randomly intermate (equal probability of crosses). This is panmixy. In a panmictic situation and in absence of selection, allele and genotype frequencies do not vary over generations. However, if preferential crosses occur as a consequence of homogamy or heterogamy, genotype frequency will change over generations, while allele frequency will be stable. Thus despite stable allele frequencies, genotype frequency may change. Sexual reproduction will thus influence frequency of heterozygous genotypes as well as the consequence of selection mechanism. If heretozygous genotypes are counter-selected, sexual reproduction will balance against selection effect. At each generation, heterozygous genotypes will be produced and this will reduce the response intensity to population selection. A similarity could be found at a community level through species hybridization. However, the analogy of effect is far less obvious than the previously analysed mechanisms. Altogether, sexual reproduction, with or without migration is likely to modify allele, genotype and species frequency.

The last evolution force that may theoretically influence allele frequency is the mutation, i.e. a spontaneous modification of allele DNA sequence. We will not discuss here this mechanism as mutation occurs at a very low rate and will not act on the duration of a sown grassland sward.

Species and Genetic Composition of Temporary Grasslands over Time

Temporary grasslands are increasingly sown with mixtures of several varieties and species. At sowing and seedling emergence, the sward has a given species and genetic composition, which confers a given agronomic value. Species composition will vary over growing seasons. For instance, in swards sown with a mixture of Italian ryegrass and tall fescue will have a high proportion of ryegrass in the first cycle, while fescue will become predominant later on. It is also obvious that temporary grasslands may become invaded by weeds, the origin of weed seeds being either the migration or the emergence from the resident soil seed bank (Pakeman and Small, 2005). Changes in species composition have been heavily documented.

On the opposite, changes in allele and genotype frequencies have been little documented. This may be simply explained by the fact that it is difficult to easily measure the occurrence this phenomenon. Straub (2006) published one of these studies. She analysed the genetic composition of 12 grasslands sown with a single perennial ryegrass variety, Herbie. These grasslands were located in a wide range of soil and climate and experienced a wide range of management regimes. Differences among sampled grasslands were evidenced as well as with the control samples of the initial seeds (Fig. 26.2). Among the underlying mechanisms, Straub (2006) showed occurrence of migration with new alleles and selection/drift leading to loss of alleles (Fig.26.3).

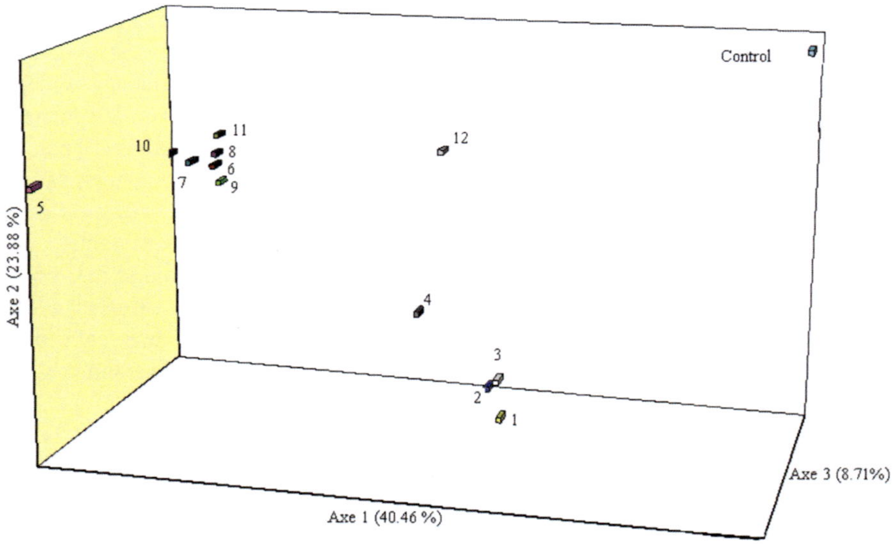

Fig. 26.2 Differentiation among sampled grassland from Straub (2006) based on F_{st} values from 9 SSR markers. The global F_{st} value is 0.09 ($p < 0.0001$). Population of "St Ferréol": one cattle grazing grassland (1); population of "Ordiap": one sheep grazing grassland (2); populations of "Le pin au Haras": one cutting grassland (3), two cattle grazing grasslands (4 and 5); population of "Theix": one cutting grassland (6); populations of "Lusignan": two cutting grasslands (7 and 11), three cattle grazing grasslands (8, 9 and 12), one sheep grazing grassland (10)

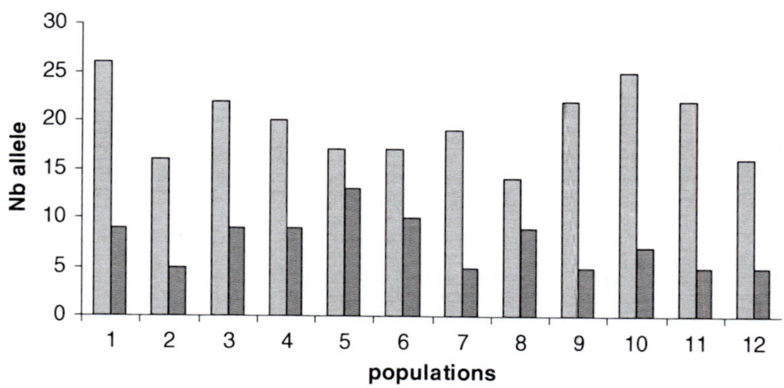

Fig. 26.3 Number of absent alleles and new alleles and ☐ new alleles ■ in sampled populations relative to control (Straub, 2006). Population of "St Ferréol": one cattle grazing grassland (1); population of "Ordiap": one sheep grazing grassland (2); populations of "Le pin au Haras": one cutting grassland (3), two cattle grazing grasslands (4 and 5); population of "Theix": one cutting grassland (6); populations of "Lusignan": two cutting grasslands (7 and 11), three cattle grazing grasslands (8, 9 and 12), one sheep grazing grassland (10)

Managements, Genetic/Species Diversity and Agronomic Value

Management practices strongly condition the environment that grassland communities have to face. From an ecological point of view, practices are perturbations inducing selection pressures generating mechanisms which will lead to changes in species, genotype and allele frequencies.

Defoliation (cutting and grazing) is a main selection pressure. Morphological and physiological traits may confer avoiding ability. This is the case of a prostrate growth habit and short leaves. Under a frequent defoliation regime (continuous grazing at a high stocking rate, mowing), individuals with such functional traits will be positively selected. On the opposite, infrequent cutting will generate a strong competition for light. Plants with long leaves will better survive and out yield their neighbours (Hazard and Ghesquière, 1995, Hazard et al., 2001).

Beyond selection, grazing may induce drift. Indeed, treading may kill individual's plants at random, with random variation of allele and species frequencies. As a consequence, cutting and grazing reduce frequencies or may even induce allele/species disappearance. But grazing will generate recruitment sites, where seedling recruitments may occur in case of migration and sexual reproduction of the sward. Indeed, animal treading or animal dungs will open gaps in the swards which will be sites for such seedling emergence and possibly adult plants if competition with the existing sward is not too strong.

Alongside with defoliation and treading, neighbourhood is an important biotic factor for community structure and its changes. Sowing of various varieties and/or species in mixtures conditions a diversity of possible neighbourhood for a genotype. Interactions within communities are complex including competition and facilitation. Depending on the main mechanism, diversity may or may not be preserved. Competition, even when it is not exclusive, will lead to a loss of diversity through selection. At the opposite, facilitation will favour a preservation of diversity. However, each individual or species has an effect on its close environment as a consequence of nutrient uptake and release, light capture, space occupancy, effect on parasite or pollinator populations. Consequently, it modifies the environment of its neighbours. Variability of genetic and species composition will generate a heterogeneity of local environments and interactions and thus a heterogeneity of available niches. This heterogeneity may favour diversity through diversifying selection. Through this mechanism, complementary of genotypes/species in resources exploitation and effect on biotic environment will have a positive effect on diversity.

As a conclusion, agronomic practices as well as the initial composition of grassland community will condition pressures exerted on grassland communities and this will partly determine changes in species and genetic diversity.

Management of Agronomic Value Decline and New Varieties

When a variety is registered, it has a given genetic composition and is characterised by a given agronomic value in the registration trials. Farmers will then use it alone or in mixtures with other varieties and species. During the first cycles and the first year, it may be anticipated that the species and genetic composition corresponds to the sown mixture and the expected agronomic value, even though the establishment phase and the early emergence phase may be critical (Korner et al., 2008). Over time, species (Mosimann and Charles, 1996) and genotype (Straub, 2006) composition will change under the pressure previously described. Selection pressure will lead to an increase in the proportion of the genotype/species the most adapted to the environmental conditions which will be defined by the practices, the weather and soil conditions and the composition of the community. And it does not exist a relationship between the agronomic value of genotype and their adaptive value to environmental conditions. This relationship seems to be stronger at the species level. It would suggest that the most competitive species are also the most productive, explaining why in presence of competition only, the yield of mixtures of species is similar to the best yielding species (Tilman et al., 1997; Hector et al., 1999).

Recruitment of seedlings from other meadows via migration may generate an increased diversity, but the agronomic value of the migrants may be poor.

As a consequence, selection, drift and migration leading to changes in species and genotype frequency may also induce changes in agronomic value of sown grasslands, especially if the duration of the sward is long.

Such a decrease over time of both biomass production and forage quality has been documented by Elgersma and Schlepers (1997) and Schils et al. (1999). Once this decrease is too strong, grasslands are ploughed and have to be re-sown. The anticipated decline of agronomic value has led to recommendations for temporary grasslands, with optimum duration being thought to be 3 or 4 years. This fairly short duration may have detrimental environmental impacts, as grassland ploughing and the subsequent soil perturbation induces a strong nitrogen mineralization and possible nitrogen leachings (Benoît et al., 1995, Benoît and Simon, 2004), soil erosion (Chisci and Zanchi, 1981). It also may cause a loss of biodiversity (Wingerden et al., 1992), temporary grasslands being the only permanent landscape elements in mixed crop-livestock farming systems. In a context of sustainable agriculture with a compulsory economic and environmental performance, more persistent grasslands are necessary, the persistency being defined by the preservation of a high agronomic value over a high number of growth-defoliation cycles.

The choice of species and variety composition is determined by production and feeding value objectives for the first years of exploitation. However, in a context of sustainable agriculture, it is necessary to produce more and better.

Some mutispecies seed mixtures take into account these objectives and include species with different life durations. It is anticipated that their respective abundance will change over years and thus increase the persistency of the swards. The understanding of the mechanisms is critical to define the best initial compositions, as it was shown in the previous sections that both agronomic practices and species composition will determine the changes in composition and agronomic value. Another option would be to try preserving the initial genetic and species composition which was associated to a high agronomic value. This could be achieved by a reduction of the selection pressure and a balance of the selection effect by re-injecting the diversity which was lost (through overseeding or through sexual reproduction of the existing swards). However, this would also mean that the soil fertility does not change over cycles and years. And this is very unlikely.

Species richness and functional diversity with grasslands communities strongly influence their agronomic value, especially the biomass production (Hector et al., 1999; Wilsey and Polley, 2004). Considering the hypothesis that species diversity could act as a diversifying selection (Whittaker, 1975; Harper, 1977), a sound diversity of genotypes or species could reduce the competition among genotypes or species, increase facilitation processes and thus improve persistency of the plant community and of its agronomic value. If the individuals exhibit complementary for functional traits involved in the acquisition of a given resource, the competition for this resource will be reduced via a complementary of niche (for instance plants from the same species or from different species with different root depth will not compete for water and mineral nitrogen uptake). As a consequence, the community will be stabilized by the saturation of niches (in the above example, all soil horizons will be explored). Similarly, the process of facilitation for acquisition of a resource between individuals of the same species or of different species will improve the coexistence of the different plants. This implies local adaptation to the identity of the neighbours (Turkington and Aarssen, 1984; Turkington, 1989; Vavrek, 1998). A positive correlation could exist between the functional diversity of the individuals (degree of dissimilarity for functional traits) and the persistency of agronomic value.

Broad genetic diversity within species does not imply that it is necessary to broaden the genetic diversity in varieties. First of all, it is necessary to investigate whether the diversity already present within the present varieties of perennial forage species is broad enough. This diversity was shown to be very large, especially when investigated with neutral molecular markers as shown by Flajoulot et al. (2005) for alfalfa or Auzanneau et al. (2007) for perennial ryegrass. In a second step, broadening the diversity would be more easily achieved by mixing varieties. Indeed, mixing variety with different morphological and physiological traits is likely to contribute more efficiently to broadening functional diversity among individuals. Moreover, this would avoid difficulties with DUS tests at variety registration.

To balance the loss of diversity due to selection, it could be possible to use the naturally occurring sexual reproduction through adaptation of the agronomic practices at the adequate period. This would require a compromise with the instantaneous agronomic value, but could be beneficial on the long term. Indeed, sexual reproduction which is associated with stem production and stem lignification will induce a lower feeding value at the precise growth cycle when reproduction is sought. But this loss of agronomic value could be compensated for by an increased persistency. Another option could be the use of over-seeding, without prior destruction of the existing sward. This makes it possible to control the species and genotype identity of the seeds and does not require looking for a compromise between persistency and agronomic value. Over-seeding technologies are under quick improvement, but the rate of failure is still high, especially when the existing swards are highly degraded and include highly competitive species, such as *Agrostis* (Huguenin-Elie et al., 2007).

This innovative view of the way of defining and managing diversity of sown grasslands requires theoretical and experimental approaches in order to

(i) better determine and understand mechanisms involved in the variations of agronomic value over time under various practices,
(ii) define the optimum level of complementary and facilitation within and among species which will make it possible to tune the succession of species and to stabilize the agronomic value,
(iii) define the key traits of plants involved in those mechanisms. These will have to be taken into account in breeding programs. The assemblage of variety and species in grassland communities will be based upon the complementary and thus the appropriate functional diversity for traits,
(iv) establish the optimum management of grasslands in order to exploit this diversity.

Considering the importance of biotic and abiotic environmental factors for the future of plant communities, it is obvious that registration trials and decision should take the ability of a variety to contribute to a high agronomic value of grassland communities. However, it is not possible to run registration trials in mixtures. Indeed, it would simply be impossible to investigate all possible combinations of species. Thus, the option is to collect data for functional traits which are relevant for the performance in condition of mixtures. The wide range of data collected for the DUS as well as the VCU trials provides an outstanding source of information which has to be valorised, once the key traits have been defined.

Finally, advice to farmers on the optimum variety and species composition could thus be based on knowledge of the relevant mechanisms and would be validated in a few conditions by extension services or seed companies.

References

Aguade, M., Miyashita, N., Langley, C.H. 1989. Reduced variation in the yellow-achaete- scute region in natural populations of *Drosophila melanogaster*. Genetics 122:607–615.

Auzanneau, J., Huyghe, C., Julier, B., Barre, P. 2007. Linkage disequilibrium in synthetic varieties of perennial ryegrass. Theor. Appl. Genet. 115:837–847.

Begun, D.J., Aquadro, C.F. 1992. Levels of naturally occurring DNA polymorphism correlate with recombination rates in *D. melanogaster*. Nature 356:519–520.

Benoît, M., Simon, J.C. 2004. Graslands and water resources: Recent findings and challenges in Europe. pp 117–128. In: Lüscher, A., Jeangros, B., Kessler, W., Huguenin, O., Lobsiger, M., Millar, N., Suter, D. Proceedings 20th General Meeting European Grassland Federation, Luzern, Suisse, 21–24 Juin 2004.

Benoît, M., Saintôt, D., Gaury, F. 1995. Mesures en parcelles d'agriculteurs des pertes en nitrates. Variabilité sous divers systèmes de culture et modélisation de la qualité de l'eau d'un bassin d'alimentation. C.R. Academie d'Agriculture 81:175–188.

Bruno, J.F., Stachowicz, J.J., Bertness, M.D. 2003. Inclusion of facilitation into theory. Trends Ecol. Evol.. 18:119–125.

Chisci, G., Zanchi, C. 1981. The influence of different tillage systems and different crops on soil losses on hilly silty-clayed soil (pp 52, 134–146, 211–217). In: Morgan R.P.C. (eds.), Soil conservation: problems and perspectives. Wiley.Cutting. Grass and Forage Science, Chichester, UK.

Elgersma, A., Schlepers, H. 1997. Performance of white clover/perennial ryegrass mixtures under cutting. Grass Forage Sci. 52:134–146

Flajoulot, S., Ronfort, J., Baudoin, P., Barre, P., Huguet, T., Huyghe, C., Julier, B. 2005. Genetic diversity among alfalfa (*Medicago sativa*) cultivars coming from a breeding program, using SSR markers. Theor. Appl. Genet. 111:1420–1429.

Gause, G.F. 1934. The struggle for existence. Hafner, New York.

Grenier, J. 2004. Les prairies multi-espèces. In: Bulletin du pôle scientifique bio du massif central, May 2004, N°2.

Hardin, G. 1960. The competitive exclusion principle. Science 131:1292–1297.

Harper, J.L. 1977. The population biology of plants. Academic Press, New York, NY.

Hazard, L., Barker, D., Easton, S. 2001. Morphogenetic adaptation to defoliation and soil fertility in perennial ryegrass. New Zeal. J. Agr. Res. 44:1–12.

Hazard, L., Ghesquière, M. 1995. Evidence from the use of isozyme markers of competition in swards between long-leaved and short-leaved perennial ryegrass. Grass Forage Sci. 50: 241–248.

Hector, A., Schmid, B., Beierkuhnlein, C., Caldeira, M.C., Diemer, M., Dimitrakopoulos, P., Finn, J.A., Freitas, H., Giller, P.S., Good, J., Harris, R., Högberg, P., Huss-Danell, K., Joshi, J., Jumpponen, A., Körner, C., Leadley, P.W., Loreau, M., Minns, A., Mulder, C.P.H., O'Donovan, G., Otway, S.J., Pereira, J.S., Prinz, A., Read, D.J., Scherer-Lorenzen, M., Schulze E.D., Siamantziouras, A.S.D., Spehn, E.M., Terry, A.C., Troumbis, A.Y., Woodward, F.I., Yachi, S., Lawton, J.H. 1999. Plant diversity and productivity experiments in European grasslands. Science 286:1123–1127.

Huguenin-Elie, O., Stutz, J., Lüscher, A. 2007. Amélioration des prairies par le sursemis, Revue Suisse d'Agriculture 39:25–29.

Hutchinson, G.E. 1957. Concluding remarks. In: *Population Studies: Animal Ecology and Demography. Cold Spring Harbor Symposia on Quantitative Biology*, (Vol. XXII, pp. 415–427). The Biological Laboratory, Cold Spring Harbor, L.I., New York.

Huyghe, C., Litrico, I., Martinez, F.I., Rolston, P. 2008. Effect of sexual reproduction and seedling recruitment on vegetation dynamics in grasslands (pp 544–550). In: Xie, H., Huang, J. (eds), Multifunctional grasslands in a changing word, proc. XXI Int. Grassland congress, VIII Int. Rangeland Congress, Vol II.

Körner, C., Stöcklin, J., Reuther-Thiebaud, L., Pelaez-Riedl, S. 2008. Small differences in arrival time influence composition and productivity of plant communities. New Phytol. 177 : 698–705.

Lack, D.L. 1945. The Galapagos finches (Geospizinae); a study in variation. Occasional Papers of the California Academy of Sciences 21:36–49.

MacArthur, R. 1958. Population ecology of some warblers of northeastern coniferous forests. Ecology 39(4):599–619.

Maynard Smith, J., Haigh, J. 1974. The hitch-hiking effect of a favourable gene. Genet. Res. 23**:** 23–35.

Mosimann, E., Charles, J. 1996. La conception des mélanges fourragers en Suisse Fourrages 145:17–31.

Pakeman, R.J., Digneffe, G., Small, J.L. 2002. Ecological correlates of endozoochory by herbivores. Funct. Ecol. 16:296–304.

Pakeman, R.J., Small, J.L. 2005. The role of seed bank, seed rain and the timing of disturbance in gap regeneration. J. Veg. Sci. 16:121–130.

Schils, R.L.M., Vellinga, T.V., Kraak, T. 1999. Dry-matter yield and herbage quality of a perennial ryegrass/white clover sward in a rotational grazing and cutting system. Grass Forage Sci. 54:19–29.

Stephan, W., Langley, C.H. 1989. Molecular genetic variation in the centromeric region of the X chromosome in three *Drosophila ananasae* populations. I. Contrasts between the vermilion and forked loci. Genetics 121:89–99.

Stephan, W., Song, Y., Langley, C.H. 2006. The Hitchhiking effect on linkage disequilibrium neutral loci. Genetics 172:2647–2663.

Straub, C. 2006. Evolution génétique de prairies monovariétales de ray-grass anglais. Phd, Université de Poitiers, p. 180.

Tilman, D., Lehman, C.L., Thomson, K.T. 1997. Plant diversity and ecosystem productivity: theoretical considerations. Proc. Nat. Acad. Sci. 94:1857–1861.

Turkington, R. 1989. The growth, distribution and neighbour relationships of *Trifolium repens* in permanent pasture V. the coevolution of competitors. J. Ecol. 77:717–733.

Turkington, R., Aarssen, L.W. 1984. Local-scale differentiation as a result of competitive interactions (pp. 107–127). In: Dirzo, R., Sarukhàn, J. (eds.), Perspectives on plant population ecology. Sinauer Associates Inc. Sunderland, MA.

Vavrek, M.C. 1998. Within – population genetic diversity of *Taraxacum officinale* (Asteraceae): differential genotype response and effect on interspecific competition. Am. J. Bot. 85:947–954.

Whittaker, R.H. 1975. Communities and ecosystems. MacMillan Publishing Compagny, NewYork, NY.

Wilsey, B.J.; Polley, H.W. 2004. Realistically low species evenness does not alter grassland species-richness-productivity relationships. Ecology 85, 10, 2693–2700.

Wingerden, W.K.R E. van, Musters, J.C.M., Cannemeijer, F., Bongers, W. 1992. Simulation of hatching dates in three *Chorthippus* species (Orthoptera: Acrididae) in unfertilized and lightly fertilized grasslands (pp. 86–93). Proceedings of the Section Experimental and Applied Entomology of the Netherlands Entomological Society (3).

Chapter 27
Selection of Spaced Plants of Perennial Ryegrass in Association with White Clover

Joost Baert and An Ghesquiere

Abstract In conventional breeding programs of perennial ryegrass (*Lolium perenne* L.) spaced plants are usually evaluated on bare soil with the application of mineral nitrogen fertilizer. In organic farming, perennial ryegrass is mostly sown in association with white clover without any dressing of mineral N fertilizer. Will the selection under these organic conditions lead to varieties that are different from varieties obtained by a conventional breeding management?

In 2007 we established 48 seedlings of each of 12 diploid and 9 tetraploid populations of different heading dates as spaced plants on bare soil under a conventional management with the application of 220 kg N/ha/year. 48 seedlings of each of the same populations were planted in a freshly sown stand of white clover without any N fertilizer application. Growth and crown rust infection of these plants were observed in 2008.

Almost no crown rust occurred on the plants in the white clover sward. We selected plants under both managements. Although there were differences in competition and N availability the percentage of plants selected per population was nearly the same under both managements. The selected plants were put in isolated polycrosses for seed production and further progeny testing.

Keywords Perennial ryegrass · Spaced plants · Selection · White clover

Introduction

Most perennial ryegrass breeding schemes consist of the following phases: observation and selection of individual plants, recombination in polycrosses, test of progenies and production of breeders seed. In conventional breeding programs

---

J. Baert (✉)
ILVO Plant, Caritasstraat 21, Melle 9090, Belgium
e-mail: joost.baert@ilvo.vlaanderen.be

C. Huyghe (ed.), *Sustainable Use of Genetic Diversity in Forage and Turf Breeding*,
DOI 10.1007/978-90-481-8706-5_27,

of perennial ryegrass, spaced plants are usually evaluated on bare soil with the application of mineral nitrogen fertilizer. In organic farming perennial ryegrass is mostly sown in association with white clover without any dressing of mineral N fertilizer. These two managements differ in competition pressure and time and amount of available nitrogen for the perennial ryegrass. There is a lot of controversy about the most suitable perennial ryegrass cultivar in mixtures with white clover (Baert et al., 2006). Will the selection under conventional or organic management lead to different varieties? We tried to answer this question in an empirical way.

Material and Methods

Plant Material and Growing Conditions

12 diploid and 9 tetraploid populations of different heading dates (from 16th of May until 22nd of June) were used in the experiment. In 2007 we established 48 seedlings of each of these populations as spaced plants on bare soil under a conventional management. We applied 220 kg of mineral nitrogen per ha per year degressively distributed over the 4 cuts (70, 60, 50 and 40 kg N/ha respectively) and kept the soil weed free. 48 seedlings of each of the same populations were planted in a freshly sown stand of the medium leaved white clover cv. Merwi. Here, no N fertilizer or herbicide were applied. Plants were spaced at 0.5 m $\times$ 0.5 m intervals and 250 kg K_2O/ha/year was applied in both managements.

Observation and Selection

After the establishment year, we scored in 2008 the spring growth, regrowth after each of the four cuts and the crown rust resistance of the spaced plants. We applied a scale from 1 to 5 (5 being the best). Based on these scores we selected the plants with an overall average score of 4 and a minimum score of 3 for spring growth and rust resistance. We determined the percentage of selected plants per population under both managements.

Results and Discussion

We observed almost no crown rust on the plants in the white clover sward. It is well known that in mixtures with white clover perennial ryegrass is less infected by crown rust, if the clover content is high enough and well distributed in the pasture.

The percentages of the selected plants in the diploid and tetraploid populations under both managements were similar (Fig. 27.1). Their correlation coefficient was

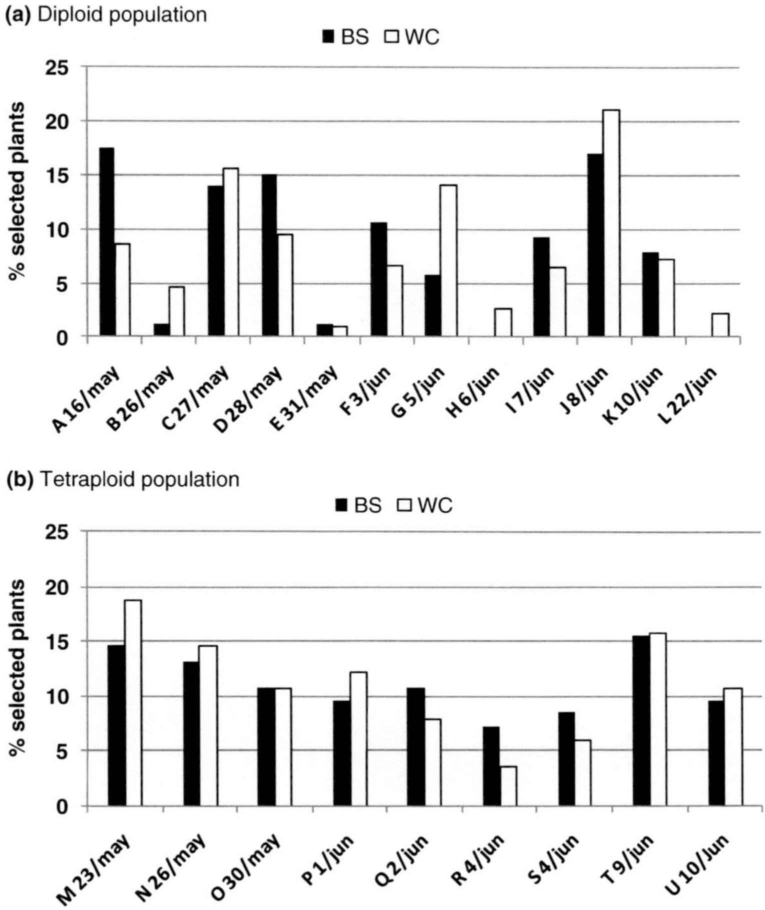

Fig. 27.1 Relative percentage of plants selected on bare soil (*BS*) and in association with white clover (*WC*) for diploid (**a**: A–L) and tetraploid (**b**: M–U) populations (in order of ascending heading date)

0.73 and 0.90 respectively for the diploid and the tetraploid populations. Under both managements plants were selected in all maturity groups. There was no shift towards a specific maturity group while selecting the perennial ryegrass in association with white clover. Van Eekeren and Wagenaar (2002) found that early varieties were more competitive to white clover.

Although the differences in competition and N availability under both managements the percentage of plants selected per population was nearly the same. The growth habit of the selected plants will further be evaluated. They were also put in isolated polycrosses for seed production and further progeny testing.

References

Baert, J., Van Eekeren, N., Ghesquiere, A. 2006. Breeding fodder grass and clover for low input/organic conditions in N.W. Europe. Proc. of the 26th Eucarpia Fodder Crops and Amenity Grasses section, Perugia, 31–37.

Van Eekeren, N., Wagenaar, J. 2002. Finding suitable perennial ryegrass cultivars for optimal grass/white clover management under cutting/grazing regime. In: Multi-functional grasslands. EGF, Grassland Sci. in Europe 7:482–483.

Chapter 28
Determination of Botanical Composition in Multispecies Forage Mixtures by Near Infrared Reflectance Spectroscopy

Fabienne Chataigner, Fabien Surault, Christian Huyghe, and Bernadette Julier

Abstract To improve on the time-consuming manual method often used to determine botanical composition of grasslands, near infrared reflectance spectroscopy (NIRS) was used to predict the proportion of each family (grass vs. legume) and also the proportion of each species in grass or grass-legume forage mixtures. Samples were collected in multispecies sown swards ("real mixtures") and artificial mixtures were produced from samples collected in monospecific swards. The mixtures were composed of perennial ryegrass (*Lolium perenne*), cocksfoot (*Dactylis glomerata*), tall fescue (*Festuca arundinacea*) with or without white clover (*Trifolium repens*). Plant samples were dried and ground, and a total of 3607 spectra were collected on a NIRS Foss 6500 machine. An independent validation set was available. The proportion of white clover in the mixtures was very well predicted (SEPC = 3.3), as a consequence of differences in biochemical composition between grasses and legumes. Prediction of each grass species was not as good (SEPC between 5.4 and 9.1) but still satisfactory, particularly considering that their morphological and biochemical characteristics are close. These equations can be used to rapidly determine species composition in multispecies grasslands.

Keywords Botanical composition · Grass · Legume · NIRS · Prediction

Introduction

In studies carried out on temporary multispecies grasslands, it is often interesting to follow the botanical evolution through time. Indeed, species frequency, in terms of contribution to total biomass, is one of the main factors explaining variation for quantity and quality of biomass production. The most commonly used method is

---

F. Chataigner (✉)
INRA, UR 4, Unité de Recherche Pluridisciplinaire Prairies et Plantes Fourragères, Lusignan BP6, 86600, France
e-mail: fabienne.chataigner@lusignan.inra.fr

C. Huyghe (ed.), *Sustainable Use of Genetic Diversity in Forage and Turf Breeding*, DOI 10.1007/978-90-481-8706-5_28,

manual sorting, but it is time-consuming. Near Infra-Red Reflectance Spectroscopy (NIRS) is a promising tool for estimating the proportion of the various species in a mixture. It has been shown that it is possible to predict the proportion of legume species in the mixtures using NIRS (Coleman et al., 1985; Petersen et al.,1987; Pitman et al.,1991; Shaffer et al.,1990; Wachendorf et al., 1998, Locher et al., 2005). To our knowledge, prediction of proportions of different grasses in mixtures was not reported.

Several strategies can be used to build a NIRS calibration, depending on the origin of the samples. They come either from monospecific swards, from artificial mixtures made from samples of monospecific swards or from multispecies swards ("real mixtures"). Artificial mixtures are easier to handle so that an equation of prediction can be built more rapidly. Pitman et al., (1991) showed that equations developed from artificial mixtures gave excellent statistical parameters but were less precise in predicting the specific composition of mixtures exploited by grazing. Equations built from monospecific swards are more suited to the analysis of samples composed by only grasses or legumes. Although this type of equation is able to predict the composition of grasses vs. legumes, no studies have been reported predicting the proportion of different grass species in a sward.

We have worked on mixtures made from two or three grass species with or without a legume species. Both artificial and real mixtures were used to build the calibration that was further tested on real mixtures.

Material and Methods

Microplots sown at Lusignan in 2003 with four species (perennial ryegrass – *Lolium perenne*, cocksfoot – *Dactylis glomerata*, tall fescue – *Festuca arundinacea* and white clover – *Trifolium repens*) were used (Chapter 35). Seven different types of mixtures (Table 28.1) and monospecific plots for each of the four species were sown. The design included three replicates and was conducted with two defoliation frequencies (25 and 45 days) and two levels of nitrogen fertilization. On monospecific plots, samples were harvested in several cuts in 2004 and 2005, dried and ground. Artificial mixtures were created from samples harvested in monospecific swards, with a species proportion varying from 0 to 100% by 5% steps. Samples from multispecies stands were also harvested in 2004 and 2005 and the species were manually separated, dried and weighed to calculate the proportion of each species. Each original sample was brought together and then ground.

All the samples were analysed with a NIRS Foss 6500. Reflectance data were collected between 1100 and 2500 nm and a partial least square regression (PLS) was used. The calibration equation was built using real and artificial samples. A few samples were collected twice. A total of 2930 and 677 spectra were collected for artificial and real mixtures, respectively (Table 28.2). The set thus contained 3607 spectra. Validation was carried out on 80 real mixtures obtained in 2005 and independent to those included in the calibration set.

Table 28.1 The seven real mixtures used to build the NIRS calibration set

Mixture	*Lolium perenne*	*Dactylis glomerata*	*Festuca arundinacea*	*Trifolium repens*
2.1	X	X	–	–
2.2	–	X	X	–
2.3	X	–	X	–
2.4	X	–	–	X
2.5			X	X
3.1	X	–	X	X
3.2	X	X	–	X
4.2	X	X	X	X

Table 28.2 Number of spectra collected on each type of mixtures

	Real mixtures								
Type of samples	2.1	2.2	2.3	2.4	2.5	3.1	3.2	4.2	Total
Artificial samples (2004)	156	157	156	156	106	192	192	–	2930
Artificial samples (2005)	263	270	257	257	110	331	327	–	
Real samples (2004)	48	48	48	48	–	48	48	–	677
Real samples (2005)	71	38	48	116	–	48	60	8	
TOTAL	538	513	509	577	216	619	627	8	3607

Results and Discussion

Calibration gave the best adjustment for *T. repens* with a SECV of 1.7% and a R^2 of 1. Similarly, Locher et al. (2005) obtained a SECV of 2.4–3.3% for the prediction of the legume proportion in multispecies legume-grass mixtures. The proportion of each grass was less accurately predicted with SECVs of 4.5, 5.2 and 7.5% for *D. glomerata*, *F. arundinacea* and *L. perenne,* respectively (Table 28.3).

Validation on external real mixtures is presented in Table 28.4. These real mixtures were independent from those used for calibration but they originated from the same experimental design. *Trifolium repens* was the species for which the proportion

Table 28.3 Calibration statistics for NIRS determination of species proportion in artificial and real mixtures

Species	*N*	Mean %	SD	SEC	R^2	SECV
Lolium perenne	3569	33.68	29.45	7.42	0.94	7.55
Dactylis glomerata	3553	25.62	28.78	4.47	0.98	4.54
Festuca arundinacea	3532	21.81	28.89	5.14	0.97	5.24
Trifolium repens	3541	18.09	28.18	1.68	1.00	1.71

N Number of samples; SD Standard deviation; SEC Standard error of Calibration; SECV Standard Error of Cross Validation.

Table 28.4 Validation statistics for NIRS prediction of the percentage of species in real mixtures

Species	N	Mean %	SD	bias	SEP	SEPC	R^2	RPD
Lolium perenne	80	34.08	27.03	3.99	9.85	9.01	0.89	2.7
Dactylis glomerata	80	29.11	27.03	3.13	7.91	7.31	0.96	3.4
Festuca arundinacea	80	12.32	15.51	−8.01	9.63	5.38	0.91	1.6
Trifolium repens	80	24.50	35.16	1.23	3.53	3.33	1.00	10.0

N Number of samples; SD Standard deviation, SEP Standard Error of Prediction; SEPC Standard error of Prediction corrected of mean bias; RPD Ration between SD and SEP.

in the mixtures was best predicted (SEP = 3.5%). This can be related to the different biochemical composition of a legume compared to a grass, leading to particular spectral signatures. Proportions of *L. perenne, D. glomerata* and *F. arundinacea* were predicted with SEP of 9.9, 7.9 and 9.6%, respectively. Taking into account the important bias for *F. arundinacea* between the predictions and manual values in the validation set, the SEPC was calculated as 5.4%. This bias can be explained by the small proportion of *F. arundinacea* in mixtures of the external validation set. Indeed the average *F. arundinacea* content was only 12.3% compared to 34.1%, 29.1% and 24.5% for *L. perenne, D. glomerata* and *T. repens*, respectively. On 12 samples, the error on manual determination by two technicians was estimated to be 20.7%, 7.0%, 4.5% and 5.3% for *L. perenne, D. glomerata, F. arundinacea and Trifolium repens.* We can thus consider that results of NIRS prediction are satisfactory, especially considering the similar morphological and biochemical characteristics of the three grass species.

The potential of predicting legume proportion in mixtures was confirmed. More interestingly, we demonstrated that the prediction of the proportion of a given grass species in a mixture composed of two or three grass species was feasible. The calibration set still needs to be enriched by additional samples. Samples collected in the same field trial harvested in 2006, 2007 and 2008, and samples collected in grass – lucerne (*Medicago sativa*) mixtures will be added to the calibration set in order to improve the quality of prediction. Experimental errors also occur in manual determination of species proportion and these could limit further progress in prediction accuracy. However, these equations can already be used to rapidly determine species composition in multispecies grasslands.

Acknowledgements We thank Rodrigue Véron for forage sampling and manual sorting of species and Catherine Levêque for collecting NIRS spectra.

References

Coleman, S.W., Barton, F.E., Meyer, R.D. 1985. The use of near-infrared reflectance spectroscopy to predict species composition of forage mixtures. Crop Sci. 25:834–837.

Locher, F., Heuwinkel, H., Guster, R., Schmidhalter, U. 2005. Development of near infrared reflectance spectroscopy calibrations to estimate legume content of multispecies legume-grass mixtures. Agr. J. 97:11–17.

Petersen, J.C., Barton, F.E., Windham, W.R., Hoveland, C.S. 1987. Botanical composition definition of tall fescue-white clover mixtures by near infrared reflectance spectroscopy. Crop Sci. 27:1077–1080.

Pitman., W.D., Piacitelli, C.K., Aiken, G.E., Barton, F.E.1991. Botanical composition of tropical grass-legume pastures estimated with near-infrared reflectance spectroscopy. Agron. J. 83 103–107.

Shaffer, J.A., Jung, G.A., Shenk, J.S.,Abrams, S.M. 1990. Estimation of Botanical Composition in Alfalfa/Ryegrass Mixtures by Near Infrared Spectroscopy. Agron. J. 82:669–673.

Wachendorf, M., Ingwersen, B., Taube, F. 1998. Prediction of the clover content of red clover and white clover-grass mixtures by near-infrared reflectance spectroscopy. Grass For. Sci. 54:87–90.

Chapter 29
Investigating the Competition for Water and the Depth of Water Extraction in Multispecies Grasslands Using ^{18}O Natural Abundance

Jean-Louis Durand, Thierry Bariac, Youri Rothfuss, Patricia Richard, Philippe Biron, and François Gastal

Abstract Depending on root depth and the vertical distribution of water, the soil contributes to satisfy the transpiration demand of grass crops. Using soil humidity measurements repeated in time at different depth, the depth of water extraction (DWE) of a species can be measured in pure swards. However, that may be irrelevant in mixed swards because (i) roots of all species can be found in all horizons and (ii) the relationship between root density, soil humidity and water extraction is complex. To date, the use of natural abundance of ^{18}O is the only way to precisely compare the DWE of plants sharing the same ground. The work reported here describes the results obtained in a mixture of *Dactylis glomerata, Lolium perenne, Festuca arundinacea* grown in dense sward and believed to exhibit various abilities to exploit soil resources. Pure stands were also analysed and compared to the mixed crop. The results showed that when the soil profile exhibited a monotonous gradient of natural ^{18}O abundance in water, the ranking of the DWE of three species was possible and explained the differences in plant water status observed in summer.

Keywords Perennial ryegrass · Tall fescue · Cocksfoot · Water use · Isotopic signature · ^{18}O

Introduction

Many cropped grasslands consist in mixtures of commercial varieties of pure species. Practical conditions often induce a limitation in water resources because they rarely are irrigated. Better knowledge of the processes involved in water use during summer is important in order to lead breeders toward further progress more efficiently (Sampoux et al., 2010). The use of cultivars in multispecies grasslands

J.-L. Durand (✉)
INRA – URP3F Unité de Recherche Pluridisciplinaire Prairies et Plantes Fourragères, BP6, Lusignan 86600, France
e-mail: jean-louis.durand@lusignan.inra.fr

C. Huyghe (ed.), *Sustainable Use of Genetic Diversity in Forage and Turf Breeding*, DOI 10.1007/978-90-481-8706-5_29, © Springer Science+Business Media B.V. 2010

also raises questions concerning the optimal balance and the putative competition for resources between species. In mixed sown grasslands (*i.e.* typically two to four species) as in any vegetation, the competition for any resource occurs when that resource is shared by several species and is lacking for at least one of them (Casper and Jackson, 1997). The latter is largely dependent on the demand of the species, which may vary in space and time. The depth of water extraction (DWE) is a synthetic variable that depicts the use of water in soil during a particular period of time (Durand et al., 2007). It can be computed following equation (29.1)

$$\mathrm{DWE} = \frac{\sum_{z} z.\Delta_z}{\sum_{z} \Delta_z} \tag{29.1}$$

where Δ_z is the variation in humidity during the time period. Water availability for each species mainly depends on the soil volume explored by roots and on the distribution of water. Whereas water availability mainly varies with depth and can be easily measured using several methods, root depth and density remains extremely difficult to measure with any method accurate enough to catch the competitive ability of closely related species. Root biomass, root specific length and the proportion of living vs dead roots are only three of the important variables required to depict the ability of a root system to access to soil water resources. Recently however, the measurement of natural abundance both in soil and in the tiller base of plants was precise enough to distinguish between genotypes of *Festuca* and *Lolium* (Durand et al., 2007). Rainfall water has a general isotopic signature of *circa* –7 ‰. When evaporation occurs over several days discrimination against ^{18}O in the vapour phase brings about an enrichment of the water in ^{18}O at the surface of the soil. Subsequent downward diffusion of isotopes in soil generates a basically one dimensional gradient in natural abundance of ^{18}O with soil depth. As plants use water from all horizons, the natural abundance of ^{18}O in their non transpiring tissues hence depends on the depth of water extraction (DWE). The comparison of DWE of several plants therefore is possible if they grow on the same soil, showing the same ^{18}O gradient. The DWE in pure stands in tall fescue cocksfoot and perennial ryegrass was studied earlier (Lemaire and Denoix, 1987; Durand et al., 1997). How they share water in mixed swards remains unknown. The work reported here shows how the use of the variation of natural abundance of ^{18}O can allow for ranking of the depth of water extraction of these closely related grass species.

Material and Methods

Swards (2.5 × 5 m^2) of *Lolium perenne* (*Lp*, cv Milca), *Dactylis glomerata* (*Dg*, cv Ludac) and *Festuca arundinacea* (*Fa,* cv Soni) were sown in pure stands and mixtures (three species at same initial density) on 11 May 2005 with optimal fertilisers supply. The herbage was cut four times each year. Access tubes for a neutron probe for measuring the volumetric humidity were set in each plot down to *circa* 1.70 m depth. Measurements were made every 2 weeks and every 15 cm. On 13 July 2007, soil samples were collected to measure the isotopic signature ($\delta^{18}O$) of water. On the same date, the base of tillers growing less than one meter away from the soil profile were collected between 11 am and 15 pm in pure and mixed swards (4 replicates.). The outer sheath, from the oldest leaf of the tiller was discarded to avoid any possible evaporation-induced enrichment in ^{18}O of the plant material. The water was extracted from the soil and the plant samples and analysed following the procedure described in Durand et al. (2007).

Results

On day of sampling in 2007, the swards had been submitted to no serious water deficit. The relative water content of the soil was always higher than 40% of field capacity (data not shown). During that period, the measurements of soil humidity were used to derive the mean depth of water extraction of the swards with the equation (1). For the period between 11 and 19 July, all crops used water from the same layers of soil (Fig. 29.1). Both in pure and mixed stands, the average depth were approximately 57 cm during that period of time.

The absence of rainfall during the week before 13 July was sufficient to generate a deep gradient of ^{18}O in the soil (Fig. 29.2) in all situations. The isotopic profiles were very similar. The deep soil water had a isotopic signature ranging between –6‰ in *Lp*, and –6.8 ‰ for *Fa* and the mixture, Dg exhibiting intermediate values. In the surface, the mixture had a higher enrichment in ^{18}O with an $\delta^{18}O$ value of approximately –0.8 ‰ all pure swards exhibiting similar values (close to –2.3 ‰). The gradient extended down to 90 cm and was especially high between 0 and 20 cm. (close to 0.10 ‰ cm^{-1} for the mixed sward) and became lower under that depth (0.03 ‰ cm^{-1} for the mixed sward.)

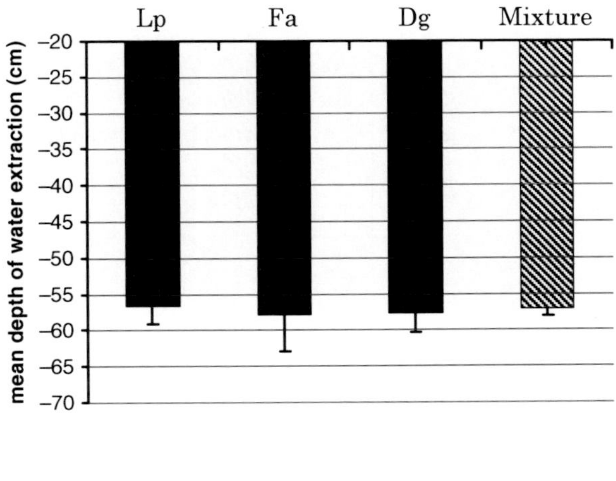

Fig. 29.1 Depth of water extraction between the 11 and 19 July measured using a neutron probe in pure (*plain bars*) or mixed (*hatched bars*) swards of perennial ryegrass (*Lp*), cocksfoot (*Gg*) or tall fescue (*Fa*)

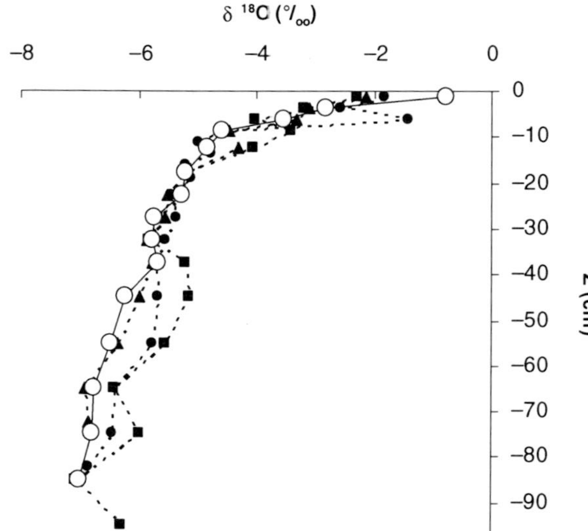

Fig. 29.2 Isotopic composition ($\delta^{18}O$ °/oo) of soil water in the plots of pure swards of perennial ryegrass (squares), cocksfoot (closed circles), tall fescue (triangles) or in the mixture of the three species (open circles) on 13 July

In pure swards, the values of isotopic composition of the water extracted from the tillers were similar, on average, for the three species (Fig. 29.3). The mean value was close to –3.7 ‰, *i.e.* within the range of values found for the soil samples. The cocksfoot tended to exhibit lower, but not significantly, values than *Fa* and *Lp*. In the mixture by contrast there were larger differences between species, $\delta^{18}O$ ranging between –3 and –4.5 ‰. The *Lp* $\delta^{18}O$ exhibited higher values than *Dg* and *Fa*, which did not differ significantly.

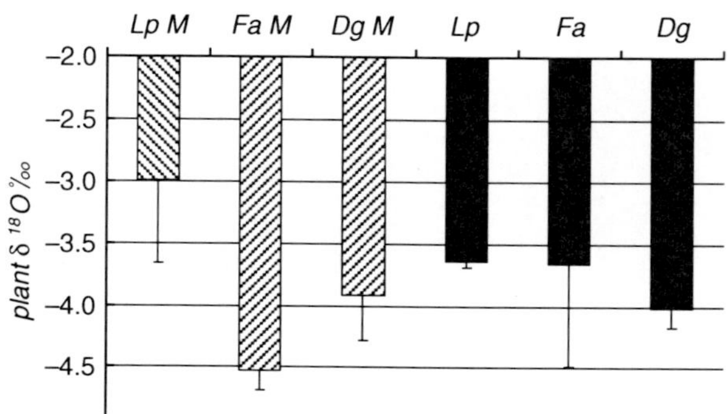

Fig. 29.3 Isotopic signature of water extracted from the base of tillers of perennial ryegrass (*Lp*), tall fescue (*Fa*) and cocksfoot (*Dg*) sampled either in pure stands (plain bars) or mixture (hatched bars) on 13 July 2007

Discussion and Conclusion

Despite different depth of rooting when grown in pure stands (Durand et al., 1997) the three species studied in this experiment exhibited similar depth of water extraction between 11 and 19 July. This was confirmed by the use of isotopic signature. It is noteworthy that the soil layer where the water had the same $\delta^{18}O$ value as the tillers, *i.e* approximately –15 cm, was much higher than the actual depth of water extraction found to be close to circa –56 cm. This shows that the $\delta^{18}O$ value can only be used to rank the plants between each other and not to derive any absolute depth of water extraction. Furthermore, because the soil profiles of $\delta^{18}O$ differed slightly, it could be deduced that *Lp* likely was even closer to the two other species. The values found in pure stands hence could not be used to derive any information on the relative situation of the different species of the mixed sward where a clear difference appeared between *Lp* and the others. Two years after sowing, *Lp* was dominated by *Dg* and *Fa* in the canopy (data not shown). With time this probably induced a reduction of the rooting depth of Lp in comparison with the two others. Also, its leaves being often in the shade of the others, it likely exhibited a lower transpiration rate. This is known to reduce the depth of water extraction during the day (Durand et al., 2007).

To further investigate the resource partitioning between grass species in mixed swards, more work on the identification of species in root biomass and on the energy budget of species leaf area in complex canopies (Sinoquet et al., 2000) will be necessary.

Aknowledgements This work was supported by the ANR and the Région Poitou Charentes.

References

Casper, B.B, Jackson, R.B, 1997. Plant competition underground. Ann. Rev. Ecol. Syst. 28: 545–570.

Durand, J.L., Bariac, T., Ghesquière, M., Biron, P., Richard, P., Humphreys, M., Zwierzykovski, Z. 2007. Ranking of the depth of water extraction by individual grass plants, using natural ^{18}O isotope abundance. Environ. Exp. Bot. 60:137–144.

Durand, J.L., Gastal, F., Etchebest, S., Bonnet, A.C., Ghesquière, M., 1997. Interspecific variability of plant water status and leaf morphogenesis in temperate forage grasses under summer water deficit. Eur. J. Agron. 7:99–107

Lemaire, G., Denoix, A. 1987. Croissance estivale en matière sèche de peuplements de fétuque élevée (*Festuca arundinacea* Schreb.) et de dactyle (*Dactylis glomerata* L.) dans l'Ouest de la France. II. Interaction entre les niveaux d'alimentation hydrique et de nutrition azotée. Agronomie 7:381–389

Sampoux, J.P., Métral, R., Ghesquière, M., Baudouin, P., Bayle, B., Béguier, V., Bourdon, P., Chosson, J.F., de Bruijn, K., Deneufbourg, F., Galbrun, C., Pietraszek, W., Tarel, B., Viguié, A., 2010. Genetic improvement in rye-grass (*Lolium perenne*) from turf and forage breeding over the four past decades. In: C. Huyghe (ed.), Sustainable use of genetic diversity in forage and turf breeding, Springer, Dordrecht, The Netherlands, pp. 325–330.

Sinoquet H., Rakocevic M., Varlet-Grancher C., 2000. Comparison of models for daily light partitioning in multispecies canopies. Agr. Forest Meteorol. 101:251–263.

Chapter 30
Botanical and Genetic Change in Grass-clover Based Systems

An Ghesquiere, Joost Baert, Marianne Malengier, and Jan De Riek

Abstract Swards based on grass-clover mixtures are regaining importance in grassland production in Flanders. The population structure in these swards develops in response to abiotic and biotic stresses. In this study we analysed botanical and genetic change in grass-clover mixtures.

We have screened 15 varieties of white clover (*Trifolium repens* L.) and 15 varieties of red clover (*Trifolium pratense* L.) by amplified fragment length polymorphism (AFLP) markers. We selected five varieties of each of the species with high genetic distances between them. We set up a field experiment with these clover varieties in mixture with perennial ryegrass. The plots were mown during the growing seasons of 2006, 2007 and 2008 and the botanical composition was analysed at the last cut of each year. The proportion of the grass was low in the mixtures with red clover. The percentage white clover was low in the mixtures with both legumes in the first years. Since red clover is not persistent white clover was expanding in these plots in 2008.

In August 2008 we took samples to test the genetic shift in the white and red clover varieties growing in association with perennial ryegrass. The genetic diversity decreased in all varieties.

Keywords Botanical change · Genetic shift · Grass-clover mixture · Molecular markers

Introduction

Changing from a conventional livestock production system, based on fertilized grass swards, to an organic management system requires the establishment of legume-based swards. Pasture mixtures that contain legumes offer many

---

A. Ghesquiere (✉)
Institute for Agricultural and Fisheries Research, Applied Genetics and Breeding, Plant Science Unit, Caritasstraat 21, Melle B 9090, Belgium
e-mail: an.ghesquiere@ilvo.vlaanderen.be

C. Huyghe (ed.), *Sustainable Use of Genetic Diversity in Forage and Turf Breeding*,
DOI 10.1007/978-90-481-8706-5_30,

advantages but also require more careful management than pure grass swards. The advantages of having legumes in a pasture are improved forage quality and reduced nitrogen fertilization. The population structure of the mixture develops in response to abiotic and biotic stresses. In this study, we analysed botanical and genetic change in the clover components of grass-clover mixtures.

Material and Methods

We sowed 15 varieties of red clover and 15 of white clover in the greenhouse on November 18, 2004. The varieties were chosen from the most important ones for Western Europe developed by different breeders. After approximately 6 weeks, leaf material of each variety (30 individual plants from each) was harvested for DNA extraction. DNA was isolated based on the CTAB procedure described by Doyle and Doyle (1987). The AFLP reactions were performed according to Vos et al. (1995) with a commercial kit. DNA was digested with two restriction enzymes (EcoRI and MseI) and adaptors were ligated to the DNA-fragments. Each individual was profiled with six primer combinations to generate polymorphic bands. All primer combinations were labelled with different fluorescent dyes. The separation of the fragments was carried out on a capillary ABI 3130 Genetic Analyser. Each band was treated as a separate putative locus, and scored as present (1) or absent (0) in each individual plant by the Genemapper software. For estimating genetic distances between varieties, Euclidean distance was used. For comparing pairs of individual plants, the Jaccard coefficient was applied. Cluster analysis was based on the unweighted paired-group method using arithmetic averages (UPGMA) of the SPSS software. Principle component analysis was carried out using the SPSS software and the entire set of polymorphic markers obtained through AFLP analysis.

Based on the results of the AFLP study we set up a field experiment that was sown on June 1, 2005 in three replicates. We selected five varieties of each of the clover species with high genetic distances between each other. The selected clover varieties were Aberherald, Alice, Barbian, Crusader and Klondike for white clover and Lemmon, Merula, Milvus, Mistral and Ruttinova for red clover. The clover was sown in different types of mixtures in association with perennial ryegrass. (Table 30.1).

We applied a moderate nitrogen fertilization on the plots (180 N per year). The plots were cut three to four times per year. The dry matter yield was measured at each cut in 2006, 2007 and 2008. We sampled the plots at the last cut of each year for botanical analyses. In August 2008, we took samples on the plots of type A and B to test the genetic shift of one white or red clover variety growing in association

Table 30.1 Different types of grass clover mixtures used to analyse the botanical and genetic change

Type	Number of entries	Description
A	5	each of the RC
B	5	each of the WC
C	1	mixture of the 5 RC
D	1	mixture of the 5 WC
E	1	mixture of the 5 RC and the 5 WC
F	5	each of the RC with a mixture of the 5 WC
G	5	each of the WC with a mixture of the 5 RC

RC: Red clover variety; WC: White clover variety

with perennial ryegrass. Using the same six AFLP primer combinations 30 plants per variety were fingerprinted. We scored respectively 208 and 235 polymorphic markers in the red and white clover dataset. The results were compared with the results of the originally sown seed lots.

Results and Discussion

Botanical Change

On average the total dry matter yield of the plots with red clover (type A and C) was not significantly different from the yield of the plots with both legumes (E, F, G). In 2008 the plots with the red clover varieties Lemmon and Milvus were yielding significantly less than the other mixtures in trial. In 2006 and 2007, most white clover varieties (B, D) were yielding less than the mixtures with red clover (A, C, E, F, G). In the last year the dry matter yield of the plots with white clover was similar to the yield of the plots with both legumes.

Red clover was overgrowing the grass in the sowing year which resulted in a low proportion of grass in the plots with red clover in November 2005 (24% in the plots with only red clover (A, C) and 28% in the plots with both legumes). In the plots with white clover (B, D) the percentage perennial ryegrass was 46%. In the two following growing seasons the proportion of red clover was even increasing (Fig. 30.1). Since the red clover plants were not persistent the white clover was expanding in the plots with both legumes in 2008. Although there was only a minor decrease in the proportion of the red clover in the red clover plots (from 82% in 2007 to 71% in 2008) there was a large reduction in the dry matter yield of the last cut (Fig. 30.2).

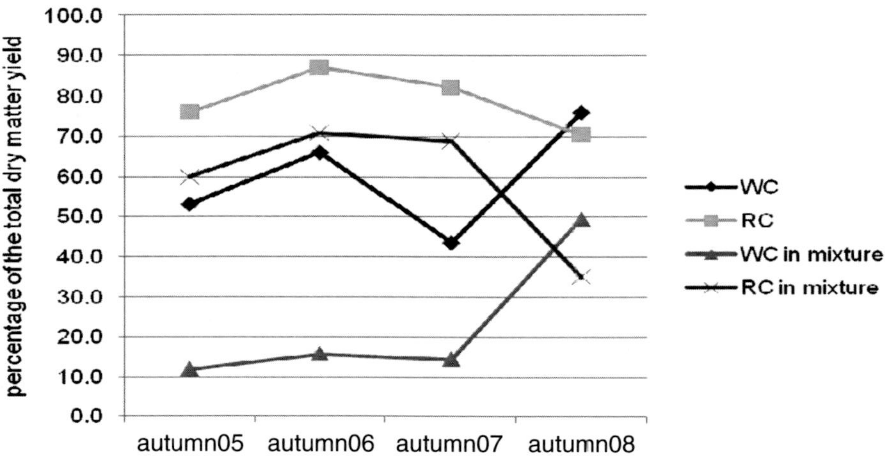

Fig. 30.1 Proportion of the dry matter yield of the white (*WC*) and red (*RC*) clover growing in association with perennial ryegrass and in association with the other clover species (WC and RC in mixture)

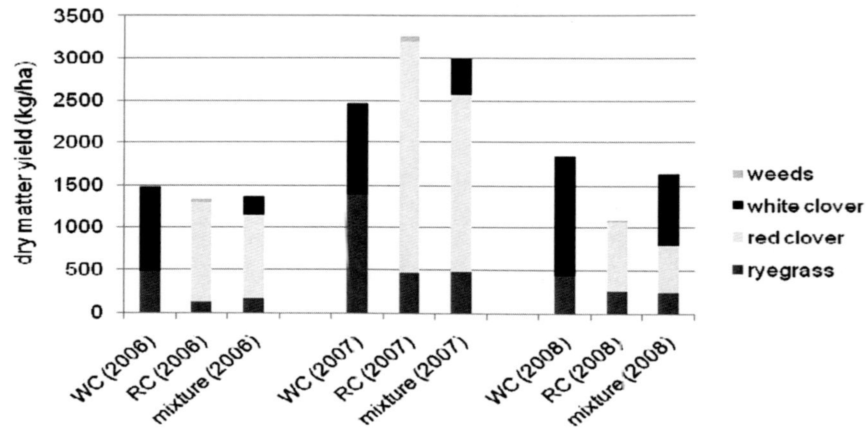

Fig. 30.2 Dry matter yield (in kg/ha) of the different components in the grass-clover mixtures at the last cut of the three growing seasons (white (*WC*) or red (*RC*) clover or a mixture of both species growing in association with perennial ryegrass)

Genetic Change

Table 30.2 shows the genetic similarity between individual plants of the original seed lots of the red and white clover varieties and of the samples taken in

Table 30.2 Similarity within the original seed lots and the field samples taken after three growing seasons (mean Jaccard coefficient between pairs of plants of the same variety)

Variety	Species	Original seed lot	Field samples
Lemmon	red clover	0.3800	0.4419
Merula	red clover	0.3732	0.4046
Milvus	red clover	0.3862	0.4030
Mistral	red clover	0.3707	0.3982
Ruttinova	red clover	0.3820	0.3873
Aberherald	white clover	0.3896	0.4227
Alice	white clover	0.3758	0.4524
Barbian	white clover	0.3831	0.4198
Crusader	white clover	0.3780	0.4906
Klondike	white clover	0.3376	0.4134

the plots in the summer of 2008. The diversity decreased in all varieties. There was a selection in the field that resulted in more similar plants. In the PCO plots (Figs. 30.3 and 30.4) there is a clear shift of the samples taken after three growing seasons when compared to the initial situation. Varieties can be clearly distinguished from each other but at the end of the trial plants tend to become more alike.

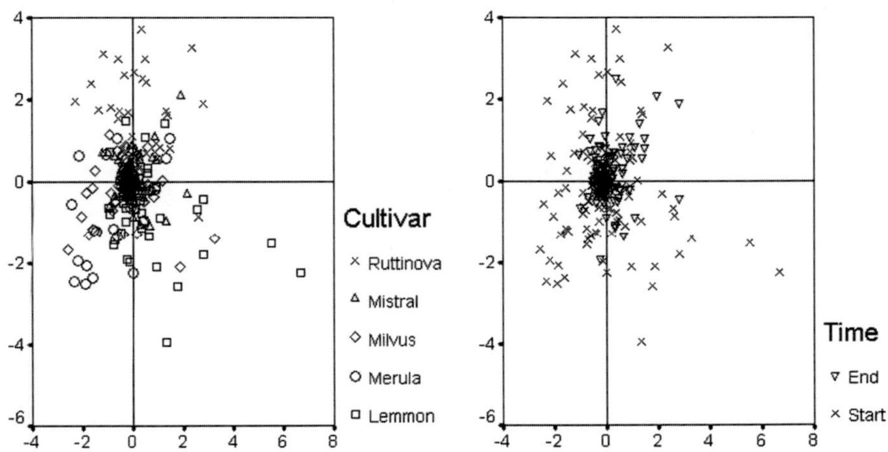

Fig. 30.3 PCO plot showing the genetic diversity (*left*) and changes (*right*) of the red clover varieties over the trial. PCO axes 1 and 2 explain 14.4% and 8.3% of the variation

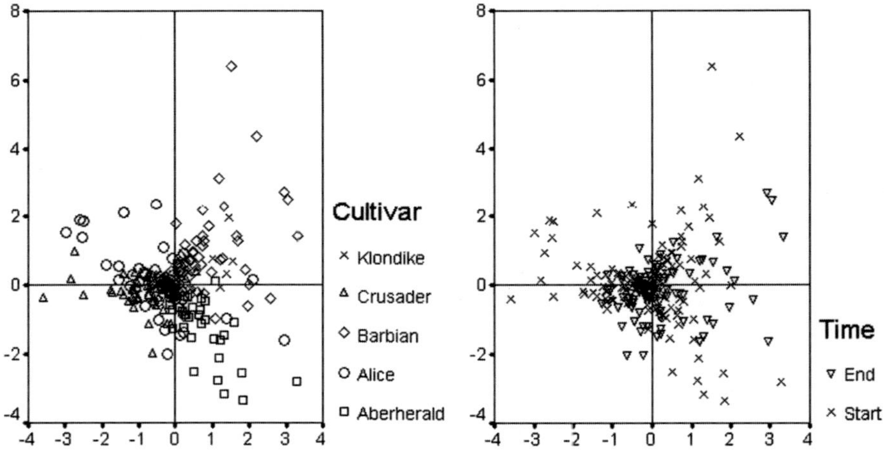

Fig. 30.4 PCO plot showing the genetic diversity (*left*) and changes (*right*) of the white clover varieties over the trial. PCO axes 1 and 2 explain 14.4% and 10.4% of the variation

References

Doyle, J., Doyle, J .1987. A rapid DNA isolation procedure for small quantities of fresh leaf tissue. Phytochem. Bull. 19:11–15.

Vos, P., Hogers, R., Bleeker, M., Reijans, M., van de Lee, T., Hornes, M., Frijters, A., Pot, J., Peleman, J. Kuiper, M., Zabeau, M. 1995. AFLP: A new technique for DNA fingerprinting. Nucleic Acids Res. 23:4407–4414.

Chapter 31
Plasticity Versus Selection in Morphological Evolution of Populations. A Case Study of *Lolium perenne* L. Mini-swards

Stéphane Grenier and Isabelle Litrico

Abstract Yields of sown meadows decrease over years leading to the necessity of ploughing and re-sowing new meadows which are not without environmental impacts. Therefore, a challenge is to understand how meadows evolve in order to be able to increase their persistency. This evolution could come from both plants selection and plasticity. Indeed, the decrease of yield in a meadow could be due to selection for the most persistent genotypes which would be the less productive and/or an adaptation to defoliation of all the genotypes by limiting their size. The objective of this study was to quantify the part of both selection and plasticity in perennial ryegrass mini-sward evolution under two evolutive pressures defined by two defoliation frequencies. The population used was a pseudo F2 from a cross between a forage and a turf genotype. The part of selection and plasticity will be evaluated over two generations of selection pressure. This will be achieved using morphological traits and fitness value. Genetic evolution will be followed using molecular markers identified in QTL analyses on adaptive traits.

The results from morphological data after the first year of selection pressures highlight plasticity between two rhythms of defoliation and significant differences between genotypes.

Keywords Adaptation · *Lolium perenne* · Phenotypic plasticity · Selection

Introduction

Sown meadows are highly diverse with respect to both species and genetic diversity. Yields of sown meadows decrease over years leading to the necessity of ploughing and re-sowing new meadows. This may have negative environmental impacts. So,

S. Grenier (✉)
INRA – Unité de Recherche Pluridisciplinaire Prairies et Plantes Fourragères, Route de Saintes, Lusignan BP 6, 86600, France
e-mail: stephane.grenier@lusignan.inra.fr

C. Huyghe (ed.), *Sustainable Use of Genetic Diversity in Forage and Turf Breeding*, DOI 10.1007/978-90-481-8706-5_31, © Springer Science+Business Media B.V. 2010

a challenge is to understand how meadows change over time in order to be able to increase their persistency. This evolution could come from both plants selection and phenotypic plasticity. Indeed, the decrease of yield in a meadow could be due to a selection for the most persistent genotypes which could be the less productive ones and/or an adaptation to defoliation of all genotypes by limiting their size (Sibly, 1995; Hazard and Ghesquiere, 1997). The objective of this study was to quantify the part of both selection and plasticity in a perennial ryegrass mini-sward evolution under two evolution pressures defined by two defoliation frequencies.

Material and Methods

The population used was a pseudo F2 from a cross between one forage and one turf heterozygous genotypes. It consists of 240 genotypes. The part of selection and plasticity will be evaluated over two generations of selection pressures. This will be achieved using firstly morphological traits including plant height, leaf length, dry matter and number of tillers and secondly fitness value evaluated in terms of reproductive effort (number of seeds) and mortality. Three clones per genotype were used in each treatment. During the treatments, plants were cut at 3 cm after a growing period of 6 weeks in infrequent defoliation (ID) and 2 weeks in frequent defoliation (FD). Plants were sown in October 2007 and treatments were applied between January and October 2008. The next generation will be studied in 2010 for the same period of treatment.

Plasticity is the morphology difference between the two treatments. It is calculated by taking the absolute value of the difference between the two treatment values.

Genetic evolution will be followed using "neutral" and "selected" molecular markers identified in QTL analyses on adaptive traits.

Results

Evolution of Population Morphology

- As expected a difference of plant height was observed between treatments (Fig. 31.1). This difference decreased over the time and was not significant when the treatments were stopped.
- No significant difference of plant diameter was observed between treatments during their application but a significant difference was observed when the treatments were stopped.

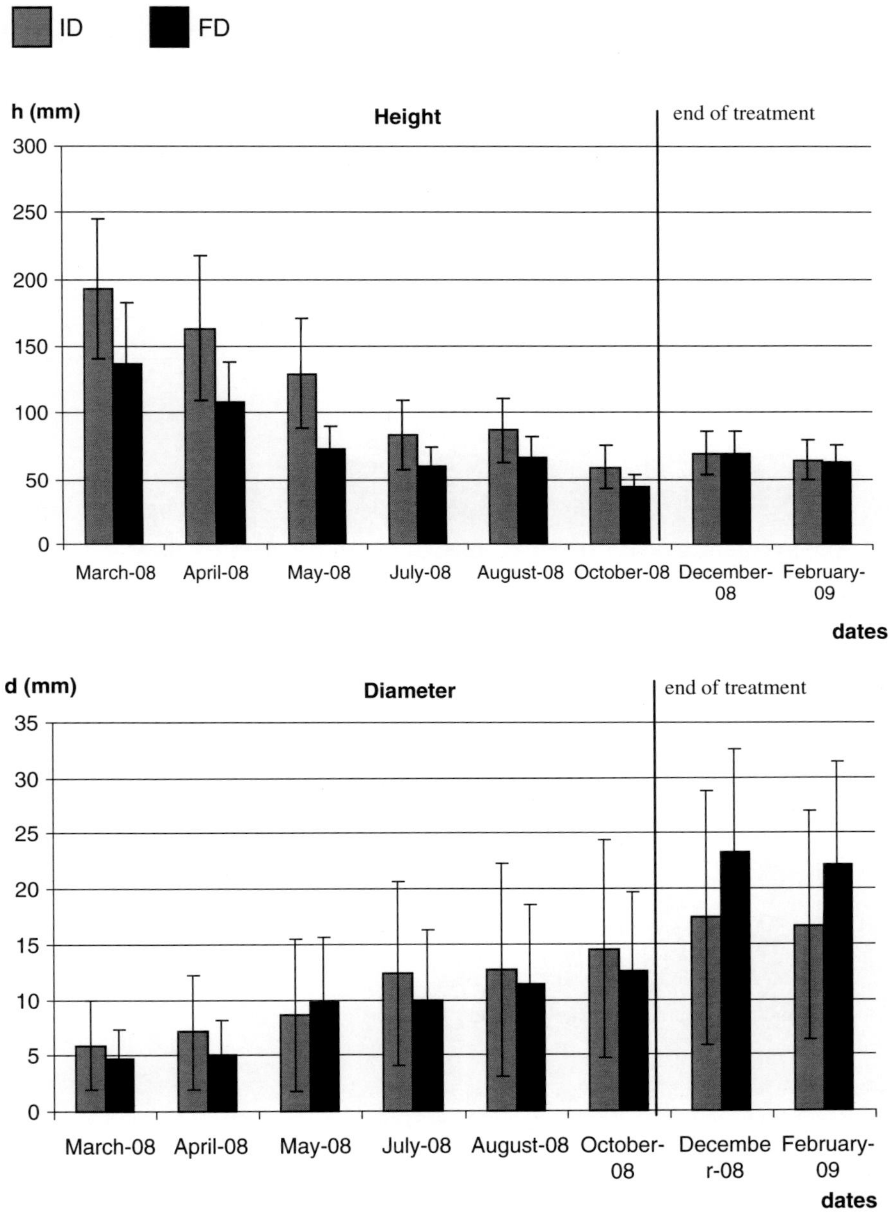

Fig. 31.1 Changes in height and diameter in each treatment

Evolution of Phenotypic Plasticity

Plasticity decreased with time for plant height and increased for diameter (Fig. 31.2). The genotypes were significantly different, many presented little plasticity whereas some presented a high plasticity (Fig. 31.3).

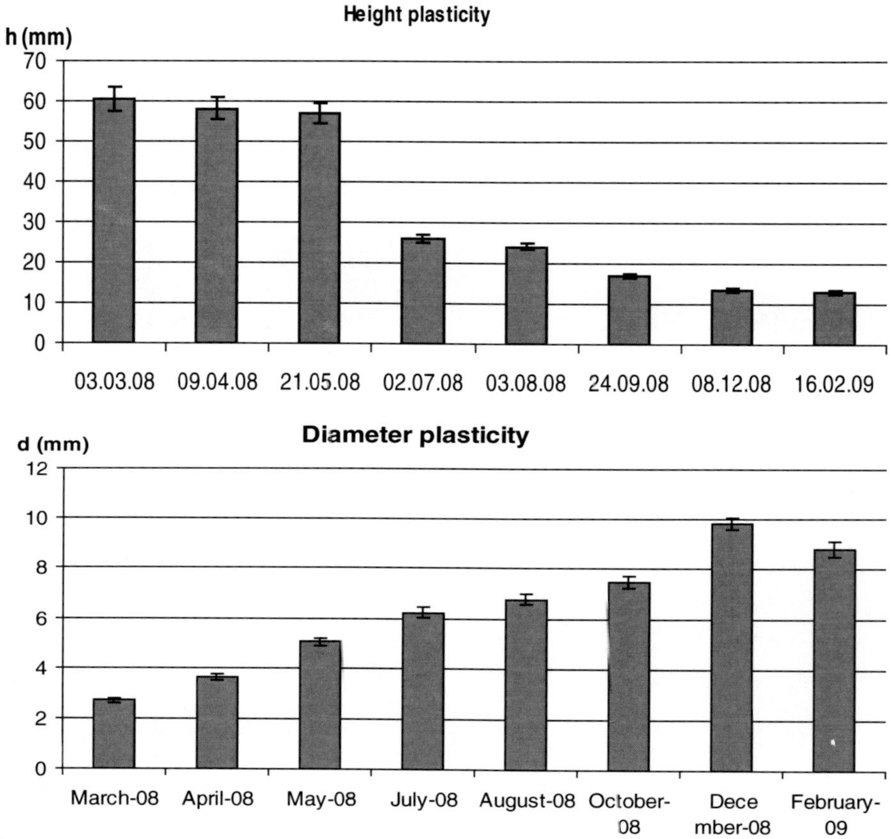

Fig. 31.2 Changes in means for height and diameter

Selection

Mortality represented 13.5% of the total population which is lower than expected considering the selection pressures. We counted 92 dead plants in Infrequent Defoliation and 106 in Frequent Defoliation on a total of 720 plants in each treatment (Fig. 31.4).

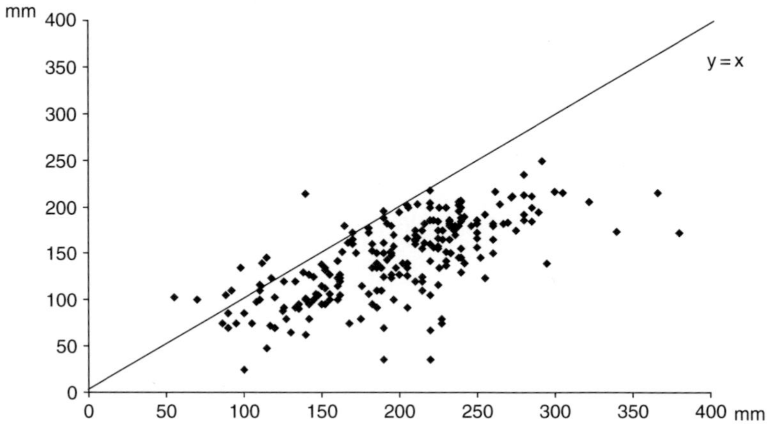

Fig. 31.3 Height in ID as a function of the height in FD (for March 08)

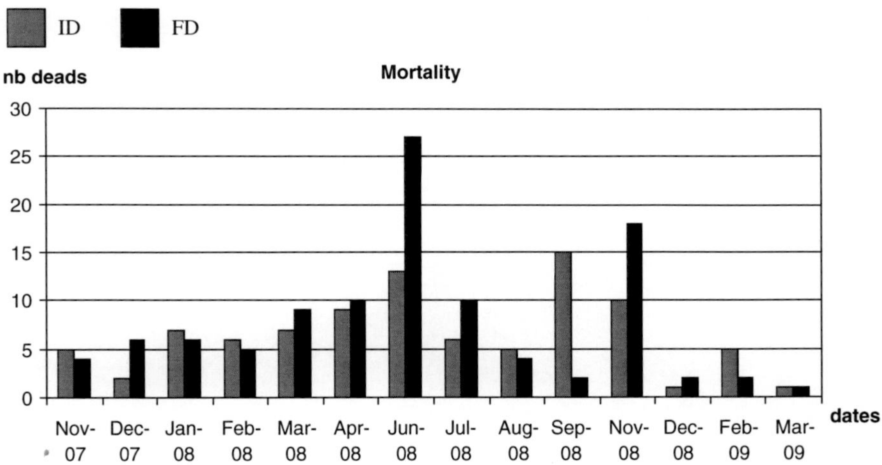

Fig. 31.4 Dynamics of mortality in each treatment

Mortality could have been caused by a treatment effect, genotype effect or more likely by a neighbour effect but, except for several genotypes which were very weak, no clear causal effect could be identified.

Conclusions and Perspectives

- After the first year of selection pressures we highlighted plasticity for leaf length, dry matter and the number of tillers.

  The genotypes are significantly different for all traits and for their phenotypic plasticity.

- Mortality is weak and no obvious causal effect was found.
- Selection cannot be assessed yet. The genetic data are being collected and the population is currently multiplied to obtain the next generation.

  The data on the two generations will provide an estimate of the selection intensity.

Acknowledgements All the technical staff and students involved in this experimentation for their patient help and the French Poitou-Charentes Regional Council for funding.

References

Hazard, L., Ghesquiere, M., 1997. Productivity under contrasting cutting regimes of perennial ryegrass selected for short- and long-leaves. Euphytica 95:295–299

Sibly, R.M., 1995. Life-history evolution in spatially heterogeneous environments, with and without phenotypic plasticity. Evol. Ecol. 9:242–257

Chapter 32
The Contribution of Hybrid Ryegrass (*Lolium x hybridum* Hausskn) to Dry Matter Yield in Mixtures with Perennial Ryegrass

J. Alan Lovatt, Richard Hayes, Ruth Sanderson, and Sheena Duller

Abstract In UK agriculture, forage grasses are invariably grown in mixtures composed of different varieties and species. A study was carried out to estimate the contribution of varying amounts of hybrid ryegrass to the performance of a typical silage ley mixture comprising hybrid ryegrass, intermediate flowering perennial ryegrass, and late flowering perennial ryegrass (*Lolium perenne* L.). In a plot experiment, it was established that dry matter yield responded linearly to the percentage hybrid in each harvest year confirming that the hybrid ryegrass had a significant effect on the yield of the mixture even in the third harvest year.

Keywords Dry matter yield · *Lolium x hybridum* · Mixtures · Ryegrass

Introduction

In UK agriculture, forage grasses are invariably grown in mixtures composed of different varieties and species, usually with white clover. It is generally thought that having several varieties and/or species in a sward confers enhanced flexibility (Ingram, 1997). Some decades ago grass seed mixtures could comprise many varied components which would later be reduced to six species in the 'Cockle Park' mixture (Lazenby, 1981) but examination of modern seed catalogues shows the majority of mixtures are comprised of various combinations of ryegrasses with or without clover.

A common mixture type comprises of mainly perennial ryegrasses of both intermediate and late flowering types to provide yield and ground cover respectively and there has been some research carried out on perennial ryegrass mixtures (Humphreys and O'Kiely, 2007; Jones and Roberts, 1994). Very often a hybrid

J.A. Lovatt (✉)
Institute of Biological, Environmental and Rural Science, Aberystwyth University, Gogerddan, Aberystwyth SY23 3EB, Wales, UK
e-mail: alan.lovatt@aber.ac.uk

C. Huyghe (ed.), *Sustainable Use of Genetic Diversity in Forage and Turf Breeding*, DOI 10.1007/978-90-481-8706-5_32, © Springer Science+Business Media B.V. 2010

ryegrass will be added to provide additional dry matter yield especially at the main silage cut but very few investigations have been carried out on mixtures including hybrid ryegrass. The aim of this study was to use mixtures containing varying amounts of hybrid ryegrass to provide information on the contribution of the hybrid ryegrass to the dry matter yielding ability of such a sward over three harvest years.

Materials and Methods

A typical silage ley mixture comprising 23% AberEcho hybrid ryegrass, 31% AberDart intermediate flowering perennial ryegrass, 23% AberAvon late flowering perennial ryegrass and 23% AberZest late flowering perennial ryegrass was chosen as the basis for the experiment. A further ten seed mixtures were made up comprising identical proportions of the perennial ryegrasses but the proportion of the hybrid ryegrass ranged from 0 to 37% of the sward (Table 32.1). Plots of each of the component varieties of the mixture were also sown as monocultures in this experiment. White clover would normally be included in this mixture but was not included in this trial to enable the contribution of the grass component to be clearly determined.

The ten mixtures, the original mixture and the component varieties were compared in a plot experiment carried out at the Plas Gogerddan site (52° 27'N, 4° 01'W) of IBERS on Rheidol series soil, pH 6.8. The plots (3 m × 1 m) were broadcast sown in August 2005 and arranged in four randomized replicate blocks (64 experimental plots in total). The plots were harvested over 3 years with a Haldrup forage plot harvester at an approximate cutting height of 50 mm. The harvesting regime consisted of an initial spring harvest in early April followed by a silage cut in late May and a second silage cut at between 5–6 weeks later

Table 32.1 Details of mixture components

Mixture	AberEcho	%AberAvon	AberDart	AberZest
1	0	30	40	30
2	4	29	38	29
3	8	28	37	28
4	12	26	35	26
5	16	25	34	25
6	19	24	32	24
7	27	22	29	22
8	30	21	28	21
9	33	20	27	20
10	37	19	25	19
Original mixture	23	23	31	23
AberDart			100	
AberAvon		100		
AberZest				100
AberEcho	100			

and 4 further cuts at monthly intervals except for the third harvest year where a total of only 6 cuts were completed. Fresh herbage samples of approximately 300 g were retained and oven-dried at 80°C. The dried samples were then weighed to allow the determination of the dry matter yield. Equal amounts of fertilizer (22:4:14:7.5 N:P_2O_5:K_2O:SO_3) were applied to all the plots at the beginning of March and after each cut to give an annual rate of 428 kg N/ha. Percentage ground cover estimates were performed visually at the end of each harvest year.

Results and Discussion

There was significant variation between the mixtures and components in all three harvest years (2006 $P < 0.001$, 2007 $P < 0.001$ and 2008 $P < 0.05$, Table 32.2). In each of the three harvest years, AberEcho was the highest yielding component being significantly higher yielding than any of the mixtures or components in harvest years 1 and 2. The mean ground cover in the trial declined significantly from 2006 to 2007 and also from 2007 to 2008 ($P < 0.001$). AberEcho had a significantly lower ground cover than all the mixtures and varieties in the trial ($P < 0.001$). While the mixtures did not yield as well as AberEcho, they did exhibit a significantly better ground cover.

A regression analysis was carried out for the yields in each harvest year against the percentage of hybrid in the sward. Dry matter yield responded linearly to the

Table 32.2 Mean annual dry matter yield and ground cover in November over 3 harvest years of mixtures and component varieties

	Total annual dry matter yield (t/ha)						% Ground cover		
Mixture	2006		2007		2008		2006	2007	2008
1	18.3	(102)	10.8	(95)	10.7	(100)	36	44	36
2	17.6	(98)	11.2	(98)	10.8	(100)	41	46	40
3	18.8	(105)	11.7	(102)	11.4	(106)	40	44	40
4	18.3	(102)	11.4	(99)	10.8	(101)	40	39	34
5	18.1	(101)	11.4	(100)	10.9	(102)	38	40	35
6	19.0	(106)	11.6	(101)	11.3	(105)	35	40	37
7	19.3	(108)	11.8	(103)	10.7	(99)	35	39	34
8	19.7	(110)	11.7	(103)	11.1	(103)	34	36	34
9	18.7	(105)	11.7	(103)	11.4	(106)	36	39	35
10	19.2	(107)	11.9	(104)	11.6	(108)	34	38	35
Original mixture	17.9	(100)	11.6	(102)	10.9	(102)	35	36	33
AberDart	17.9	(100)	11.4	(100)	10.7	(100)	40	45	38
AberAvon	17.5	(98)	11.1	(97)	11.1	(103)	40	41	35
AberZest	16.8	(94)	10.8	(94)	10.7	(100)	36	40	34
AberEcho	21.6	(121)	13.5	(118)	11.7	(109)	34	34	26
LSD ($P = 0.05$)	1.12		0.67		0.69		4.7	6.0	4.9

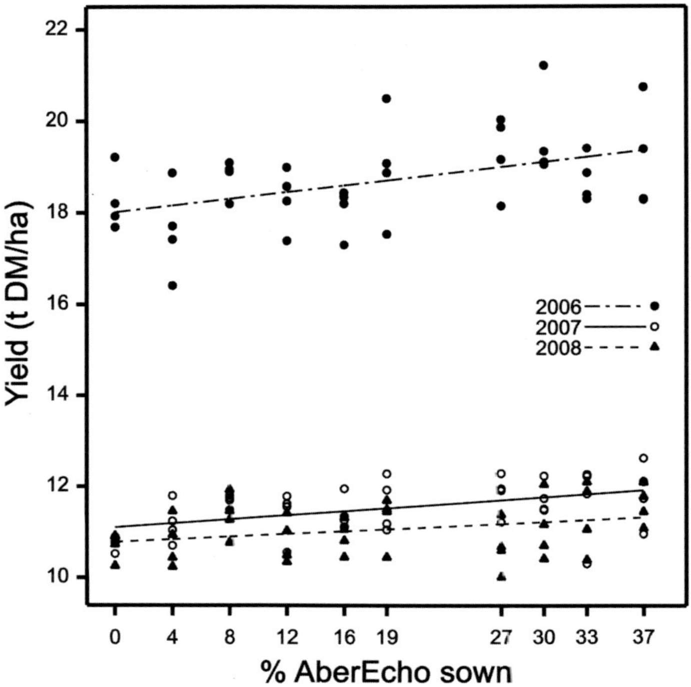

• – *2006* $y = 18.015$ (s.e. 0.190) + 0.03649 (s.e. 0.00857)
○ – *2007* $y = 11.104$ (s.e. 0.190) + 0.02194 (s.e. 0.00857)
▲ – *2008* $y = 10.784$ (s.e. 0.190) + 0.01454 (s.e. 0.00857)

Fig. 32.1 Influence of percentage hybrid ryegrass on annual dry matter yield over 3 harvest years

percentage hybrid in each harvest year (2006 $p < 0.01$; 2007 $p < 0.01$; 2008 $p < 0.05$) confirming that the hybrid ryegrass still had a significant effect on the yield of the mixture even in the third harvest year (Fig. 32.1). No significant differences were observed in the slopes of the lines ($p = 0.188$) indicating a similar response to the percentage of hybrid ryegrass in the mixture in all 3 harvest years.

Further experimentation is required to determine if this response is reproducible under differing environmental conditions.

References

Humphreys, J., O'Kiely, P. 2007. Effects of two mixtures of perennial ryegrass cultivars with contrasting heading dates, and differing in spring grazing frequency and silage harvest date, on characteristics of silage from first-cut swards. Grass and Forage Science 62:389–404.

Ingram, J. 1997. Some theoretical aspects of mixtures (pp. 176–188). In: Weddell, J.R. (ed.), Seeds of progress. BGS Occasional Symposium no. 31.

Jones, E.L., Roberts, J.E. 1994. Herbage quality and production of perennial ryegrass cultivars in monoculture and mixtures. Ir. J. Agr. Food Res. 33:169–176.

Lazenby, A. 1981. British grasslands: Past, present and future. Grass and Forage Sci. 36:243–66.

Chapter 33
Characterisation of Variation in Condensed Tannin Levels and Persistence in *Lotus* spp.

Athole Marshall, Michael Fothergill, Elaine Rees, and Ellen Sizer-Coverdale

Abstract *Lotus corniculatus* (bird's-foot trefoil) and *L. uliginosus* syn. *L. pedunculatus* (greater bird's- foot trefoil) have potential benefits to UK grassland because of the presence of proanthocyanidins, also known as condensed tannins (CTs) in the herbage. We have begun a programme of research to quantify the variation in CT levels and persistence within a number of varieties and selection lines as the first step in the identification of material which can be used in a crossing programme aimed at the development of varieties. We have used a high throughput method of quantifying the level of CTs in herbage of these *Lotus* species which has revealed significant variation in CT content within and between varieties. A CT content of up to 35 mg/g DM was common among the 20 varieties analysed with some plants showing values of 70 mg/g DM. One variety had a CT content of 150 mg/g DM, which is at a level that could have a negative effect on animal performance. Significant variation in plant persistence was also found with a rhizomatous line (Highgrove) exhibiting high levels of persistence.

Keywords Lotus · Proanthocyanidins · Condensed tannins · Persistence

Introduction

Lotus corniculatus (bird's-foot trefoil) ($2n = 12$ and 24) and *L. uliginosus* syn. *L. pedunculatus* (greater bird's- foot trefoil) ($2n = 12$) are relatively minor perennial forage legume species within UK grassland. The varieties of both *Lotus* species that are commercially available have not been bred for UK conditions and tend to lack persistence in mixed swards (Hopkins et al., 1996). However both species have important characteristics with potential benefits for grassland agriculture. Their herbage contains proanthocyanidins, also known as condensed tannins (CTs), which

A. Marshall (✉)
Institute of Biological, Environmental and Rural Sciences (IBERS), Aberystwyth University, SY233EB, Aberystwyth, Ceredigion, UK
e-mail: thm@aber.ac.uk

C. Huyghe (ed.), *Sustainable Use of Genetic Diversity in Forage and Turf Breeding*,
DOI 10.1007/978-90-481-8706-5_33,

help to reduce bloat, have anthelminthic properties and can protect protein in the rumen (Waghorn and Shelton, 1992) while reducing the rate of protein degradation and nitrogenous losses to the environment. The levels of CTs in the herbage of *Lotus* species are critical as they can have a beneficial or detrimental effect on ruminant livestock depending upon their concentration. A high throughput system of analysing proanthocyanidins has recently been developed (Marshall et al., 2008) which will enable genetic variation in CT content to be quantified and when combined with information on variation in agronomic performance enable identification of germplasm for use in future genetic improvement programmes. In this paper we describe field experiments to quantify variation in CT content and agronomic performance within varieties of *L. corniculatus* and *L. uliginosus*.

Materials and Methods

Plant material and experimental (1)- Twenty seeds of each of 18 varieties of *L. corniculatus* and 2 varieties of *L. uliginosus* were sown in pots and when established 10 plants of each were transplanted into bare soil of the Nercwys series at Aberystwyth University. In May of the following year, 3 stems were collected from each plant and freeze-dried for tannin assay of leaf and stem. Each plant was then cut to a height of 3 cm with hand held shears and dry weight quantified after oven drying in a preheated forced-draft oven at 80°C for 24 h. Individual plant persistence was scored in the spring of the following year. Tannin content was quantified by placing between 3 and 4 mg of ground freeze dried leaf samples into a 2 ml well in an Eppendorf 96 well plate and using the high throughput assay described by Marshall et al. (2008).

Plant material and experimental (2)- Planted plots of a selection of 9 varieties from the first experiment were established on a low fertility site at Bronydd Mawr Research farm. The plots were sown in 2005 with *Festuca pratensis* and with or without the large leaved white clover variety Katrina. Dry matter yield of the Lotus component was quantified in 2006 and 2007. Prior to harvest a subsample of the Lotus in each plot was harvested and freeze dried for tannin analysis as per described in the spaced plant analysis.

Results and Discussion

Leaf tannin content was greatest in A1 0528 (Table 33.1), and was signficantly greater in this than in other varieties. It was lowest in Dawn, though not significantly lower than Upstart, inbred line, AU-Dewey, Empire and Norcen. There was considerable variation between spaced plants as in A1 0528, tannin content ranged from 30.9 to 159.6 mg CT/g DM and from 5.1 to 34.7 mg CT/g DM in Dawn. The rhizomatous varieties Steadfast and Highgrove had a CT content of 28.9 and 32.6 mg CT/g DM respectively although the variation in Highgrove was considerabley greater than in Steadfast.

Table 33.1 CT content (mg/g dry weight) of the leaf, dry matter yield (g/plant) and % surviving plants of varieties of *L. corniculatus* and *L. uliginosus**. Mean, min and max tannin content is derived from 20 plants per variety. Data is presented as the mean plus s.e.m

	CT (mg/g dry weight)				
Variety	mean	min	max	DM yield (g/plant)	Surviving plants
Georgea	35.0±3.72	20.8	54.3	199.3±20.56	5
Lotar	31.3±2.48	21.9	47.6	305.3±19.62	45
Viking	29.9±5.30	16.4	43.8	252.0±25.81	30
Leo	28.7±2.87	17.6	49.9	262.6±30.73	5
AU-Dewey	17.1±4.95	3.8	52.3	251.2±44.77	25
Norcen	16.6±2.96	5.7	32.4	284.8±18.83	15
GSG	25.4±2.68	10.4	39.1	303.3±22.84	0
Emlyn	42.5±3.54	26.6	59.3	348.9±27.58	10
Upstart	18.3±2.42	7.3	32.9	235.1±19.43	15
G. Goldie	44.6±4.76	25.6	70.9	320.9±22.83	10
A1 0528	111.0±15.42	30.9	159.6	118.2±10.74	90
Dawn	14.9±2.4	5.1	34.7	226.3±28.68	30
Terre	22.9±2.47	13.4	36.8	102.6±13.91	15
Ober	39.1±7.06	14.3	70.3	324.9±21.77	10
Empire	18.5±2.12	5.8	28.5	172.0±15.46	0
Inia Draco	23.7±1.51	17.1	29.1	301.3±34.54	0
Highgrove	32.6±6.40	10.4	64.6	330.1±25.28	60
Steadfast	28.9±3.02	15.5	34.5	220.7±25.49	15
G. Maku*	17.7±3.11	3.1	34.5	220.7±25.49	85
Sunrise*	18.9±0.95	14.4	24.2	296.0±31.05	95

Persistence was greatest in the *L. uliginosus* variety Sunrise (95% plants surviving over winter). A1 0528 was the best performing *L. corniculatus* variety (90%) and Highgrove the best of the two rhizomatous lines (60%).

The CT content required to control bloat is considered to be 5 mg CT/g DM (Li et al., 1996) whereas 20–40 mg CT/g DM optimal for ruminant production (Aerts et al., 1999). CT levels greater than 60 mg CT/g DM can reduce voluntary intake and depress digestion efficiency (Barry and Manley, 1986). Considerable variation was seen within all varieties for CT content though generally within the acceptable range for ruminant production. One variety (A1 0528) had a CT content of 150 mg CT/g DM and although this was the most persistent variety CT levels of this order would have a negative effect on animal performance.

In the second experiment varieties differed in CT content in each year however generally the ranking of varieties in terms of CT content was consistent (Table 33.2) between treatments and years. Sunrise had the highest CT content and Leo the lowest and this was evident in both years and treatments. There was no effect of white clover in 2006 but CT content was significantly lower in the presence of white clover in 2007.

Dry matter yield of Lotus was greatest in Ober and Lotar and lowest in Leo. There was a significant effect of white clover on Lotus yield as it generally increased Lotus

Table 33.2 CT content (mg/g dry weight) and dry matter yield (g/plot) of seven varieties of *L. corniculatus* and two varieties of *L. uliginosus** in two harvest years with and without a white clover companion. Numbers in brackets indicate rank in each year

	CT (mg/g dry weight)				Lotus DM yield (g/m^2)			
	2006		2007		2006		2007	
Variety	Control	+Clover	Control	+Clover	Control	+Clover	Control	+Clover
Lotar	19.1(6)	22.4(5)	19.5(7)	15.7(6)	58.4	128.5	233.2	497.6
ARS2620	17.7(7)	16.9(8)	22.8(4)	12.8(7)	66.8	183.6	121.4	152.8
Steadfast	17.5(8)	18.2(7)	21.7(5)	9.9(8)	40.7	100.0	92.8	190.5
Ober	21.6(4)	24.5(3)	18.9(8)	16.7(5)	87.7	122.3	277.9	446.8
GSG	24.1(3)	24.3(4)	24.8(3)	17.3(4)	18.8	60.3	59.6	168.7
Leo	11.7(9)	11.8(9)	11.9(9)	7.1(9)	39.7	84.1	34.0	173.7
Maku*	56.9(2)	65.9(1)	32.4(2)	37.8(1)	35.8	23.5	157.3	202.9
Sunrise*	66.5(1)	65.0(2)	34.3(1)	29.7(2)	32.9	34.6	194.1	263.8
Inia Draco	20.6(5)	21.9(6)	21.0(6)	21.1(3)	41.9	73.2	112.7	193.2
Mean	28.4	30.1	23.0	18.7	46.9	90.0	142.6	254.4
Sig.	***	***	***	***	***	***	***	***

***Significant at $P < 0.001$

by nearly 100%. This was a surprising finding as it was generally expected that the white clover would suppress the growth of Lotus. There are several potential reasons for this. The experimental area was of low fertility with a low P content and the white clover was sown at a rate of 1.5 kg/ha therefore the contribution of white clover to the total dry matter yield was generally low throughout the experiment. There is some evidence that white clover can have beneficial effects on soil structure and this may have had a positive effect on the growth of Lotus.

Dry matter yield had a significant impact on the CT content delivered from the sward (Table 33.2). Ober and Lotar had a much lower CT content than the two *L. uliginosus* varieties however they produced comparable CT/m^2 (Table 33.3) due to their superior yield. Maku despite high CT levels yields poorly in the first year in the presence of white clover. Leo was the least succesful as it had a low CT content and a low yield The presence of white clover in these low fertility plots significantly enhanced yield of lotus and thus the deliverable CT, although this was more evident in *L. corniculatus* than in the 2 varieties of *L. uliginosus*. This experiment was carried out on a low fertility site therefore further studies are required on higher fertility soils to determine to what extent fertility influences Lotus content and CT levels.

From this experiment it is clear that the development of high yielding varieties of Lotus with reasonable levels of CT content are essential to deliver the benefits of Lotus in mixed swards. Currently we are carrying out a crossing programme using selected material from these experiments with the objective of developing advanced selection lines that combine high dry matter yield, persistence and appropriate levels of CT.

Table 33.3 CT content (mg/m^2) of 7 varieties of *L. corniculatus* and two varieties of *L. uliginosus** in 2 harvest years with and without a white clover companion

	CT (mg/g dry weight)			
	2006		2007	
Variety	Control	+Clover	Control	+Clover
Lotar	1101	2900	3474	7868
ARS2620	1137	3049	2583	1576
Steadfast	737	1856	1907	1886
Ober	1817	3119	4429	7748
GSG	481	1410	1204	2783
Leo	489	979	345	1174
Maku*	2028	1509	5121	8106
Sunrise*	2194	2109	5998	8151
Inia Draco	836	1740	2238	3813
Mean	1202	2075	3033	4789
Sig.	***	***	***	***

***Significant at $P < 0.001$

Acknowledgements This research was funded by the Department for the Environment, Food and Rural Affairs (Defra).

References

Aerts, R.J., Barry, T.N., McNabb, W.C. 1999. Polyphenols and beneficial effects of proanthocyanidins in forages. Agric. Ecosyst. Environ. 75:1–12.

Barry, T.N., Manley, T.R. 1986. Interrelationships between the concentration of total condensed tannin, free condensed tannin and lignin in Lotus spp. and their possible consequences in ruminant nutrition. J. Sci. Food Agric.37:248–254.

Hopkins, A., Martyn, T.M., Johnson, R.H., Sheldrick, R.D., Lavender, R.L. 1996. Forage production by two Lotus species as influenced by companion grass. Grass Forage Sci. 51:343–349.

Li, Y.-G., Tanner, G., Larkin, P.1996. The DMACA-HCl protocol and the threshold proanthocyanidins content for bloat safety in forage legumes. J. Sci. Food Agric. 70:89–101.

Marshall, A.H., Bryant, D., Latypova, G., Hauck, B., Olyott, P., Morris, P., Robbins, M. 2008. A high-throughput method for the quantification of proanthocyanidins in forage crops and its application in assessing variation in condensed tannin content in breeding programmes for *Lotus corniculatus* and *L. uliginosus*. J. Agric. Food Chem. 56:974–981.

Waghorn, G.C., Shelton, I.D. 1992. The nutritive value of *Lotus* for sheep. Proc. NZ Soc. Anim. Prod. 57:89–92.

Chapter 34
Test and Selection of *Festuca/Lolium* Specific Markers for *Festulolium* Genetic Investigations

Nadjette Missi-Bakour, Jean-Louis Durand, and Marc Ghesquière

Abstract *Festulolium* are intergeneric hybrids developed to combine the agronomic advantages of ryegrass (*Lolium*) with the persistence and abiotic stress tolerance of fescues (*Festuca*). To survey genetic evolution of tetraploid *Festulolium* populations bred from *F. glaucescens (Fg)* $\times$ *L. multiflorum (Lm)* cross, we examined 47 EST-SSR and STS public markers for: their amplification efficiency in *Fg* and *Lm*; the genome specificity of their PCR products; and ability for high throughput genotyping (band size $\leq$ 500 bp). A total of 322 bands were scored among which 93 were *Fg*-specific, 173, *Lm*-specific and 56 common to both species. *F. arundinacea* originating EST-SSR amplified *Fg* DNA at a significant higher rate than *L. perenne* EST-SSR and showed a significant higher number of *Fg*-specific bands. *In fine*, we selected 12 markers among which two primer pairs need to be redefined for use in ABI sequencer. We also present the genotypic distribution of 7 markers in a population of tetraploid *Lm* $\times$ *Fg* F1 hybrids.

Keywords STS · EST-SSR · Interspecific hybrids · Tetraploidy · *Festulolium*

Introduction

F. arundinacea var glaucescens Boiss. (*Fg*) and *L. multiflorum* L. (*Lm*) *Festulolium* (4x) were developed to combine drought resistance of *F. glaucescens* with high productivity, rapid establishment and excellent digestibility of *L. multiflorum* (Ghesquière et al., 1996). However, cytogenetic observations in amphiploid *Festulolium* showed that *Lolium* genome tends to be more frequent than *Festuca*

N. Missi-Bakour (✉)
Centre Poitou-Charentes, Institut National de Recherche Agronomique, Route de Saintes, Lusignan 86600, France
e-mail: Nadjette.Missi@lusignan.inra.fr

C. Huyghe (ed.), *Sustainable Use of Genetic Diversity in Forage and Turf Breeding*, DOI 10.1007/978-90-481-8706-5_34,

genome (Kopecky et al., 2006). This disequilibrium is even more pronounced in *F. pratensis* × *Lolium* sp. *Festulolium*. *Festuca* genome frequency decreased from 0.50 in F1 hybrids down to 0.45 in a F6 generation of *F. pratensis x L. perenne* hybrids (Zwierzykowski et al., 2006). In order to better understand *Festuca* genome loss in *Festulolium* populations and, more generally, to assess the possible impact of environment on those genetic variations, molecular markers seem to be more adequate for large scale survey than genomic in situ hybridization (GISH) which requires chromosome spreads. Indeed, co-dominant markers give access to genotype and opportunity to be processed by high throughput devices. Therefore, we selected among public sources the most adequate molecular markers for *Festulolium* studies. These markers have to amplify at a maximum rate both *Festuca* and *Lolium* genomes, to be as polymorphic as possible in producing many *Fg*-specific and *Lm*-specific bands. They should also enable an accurate estimation of *Festuca* and *Lolium* alleles frequencies at any given locus despite tetraploid structure of the genome.

Materials and Methods

Choice of Markers and DNA Sampling

First, we chose for testing 47 published markers isolated from cDNA (Table 34.1). The Noble Foundation (NFFA) and STS markers were chosen for amplification of both *Festuca* and *Lolium* genomes (Saha et al., 2004; Mian et al., 2005; Lem and Lallemand, 2003). ESTDIAS markers were selected for availability of mapping on linkage groups (LG) (Studer et al., 2008). These markers were tested on a DNA kit composed of 33 *Fg* DNAs ($2n = 4x = 28$) including 5 different populations (3715, 3723, 3733, 3740 and Bn354) and 36 *Lm* DNAs belonging to 3 diploid ($2n = 2x = 14$)

Table 34.1 Mean amplification rate (AR) and mean number of *F* (*Fg specific*), L (*Lm* specific) and C (Common to the two species) bands per marker type in *Fg* and *Lm*. N: Mean of total number of bands per marker, Nf: Mean number of *Fg* specific bands per marker, Nl: Mean number of *Lm* specific bands per marker, Nc; Mean number of common bands per marker

			Mean values					
Marker Type	Source	Number	AR_*Lm*	AR_*Fg*	N	Nf	Nl	Nc
Fa EST-SSR	Saha et al., 2004; Mian et al., 2005	24	0.80	0.88	7.04	2.88	2.92	1.25
Lp EST-SSR	Studer et al., 2008	19	0.81	0.57[a]	6.80	1.00[a]	4.65	1.15
STS	Lem and Lallemand, 2003	4	0.90	0.96	8.33	2.17	5.00	1.17
Total mean		47	0.81	0.75	6.96	2.02	3.77	1.17

[a]significantly different from *Lolium* values

cultivars (*Fastyl*, *Shoot* and *Spark*) and to one tetraploid ($2n = 4x = 28$) cultivar (*Lipo*). DNA was isolated according to CTAB method (Doyle and Doyle, 1987).

Polymerase Chain Reaction and PCR Product Separation

PCR-products were run in a 10 μL reaction volume including 25 ng of genomic DNA and 8 μl of PCR mix (1X PCR Buffer including $MgCl_2$ from Qiagen; 0.2 μM Reverse Primer; 0.1 μM Forward Primer; 0.1 μM M13 fluorescent extension; 0.033 U/μL of Qiagen Taq Polymerase and 2 mM of each dNTP). The PCR program included: a touchdown to decrease the annealing temperature from 62°C down to 55°C (7 cycles, 30 s/cycle), an amplification step at 55°C (28 cycles, 30 s/cycle), in addition to denaturation and elongation steps held at 95°C and 72°C resp. (5 mn at the beginning and the end of the program, 30 s/cycle). The PCR products were separated on a 6.5% polyacrylamide gel and image acquisition was performed by LI-COR *IR*2 automated sequencer. Gel analysis was performed with *GeneProfiler* software. Band size was determined with a precision of 0.5–2 bp according to gel quality. *Festuca* and *Lolium* profiles were compared to assign a genomic origin to every band (F for *Festuca glaucescens*, L for *Lolium multiflorum* and C for common bands).

The markers of interest were tested with 25–47 F1 hybrids. Forward primers were marked directly by FAM or VIC fluorochrome and the PCR-products were separated by the mean of ABI Prism 3100 sequencer (standard conditions of Clermont-Ferrand high throughput genotyping platform).

Results and Discussion

Amplification Rate and Allele Diversity

The 47 tested primer pairs positively amplified DNA at a 74.8 % rate on average among the 33 genotypes of *F. glaucescens* and a 80.9% rate among the 36 genotypes of *L. multiflorum* (Table 34.1). Fa EST-SSR, Lp EST-SSR and STS equally amplified the *Lm* genome ($F_{\mathrm{df}=2} = 0.23$) while Lp EST-SSR showed a significantly lower amplification rate in the *Fg* genome ($F_{\mathrm{df}=2} = 7.16$). Six markers amplified exclusively the *Lm* genome (5 EST-DIAS and one NFFA) while only two markers amplified exclusively the *Fg* genome (NFFA131 and NFFA 150) (Fig. 34.1). A total of 322 different bands were scored, including 93 F bands, 173 L bands and 56 common bands. Mean number of F bands per marker was significantly higher in Fa EST-SSR and STS markers than in Lp EST-SSR ($F_{\mathrm{df}=2} = 5.10$). These results were expected for two reasons. First, Fa EST-SSR were previously selected on the base of amplification scores in *Lolium* and *Fg* cited by the authors. Secondly, *F. arundinacea* (Fa) EST are expected to have high homology with both *Lolium*

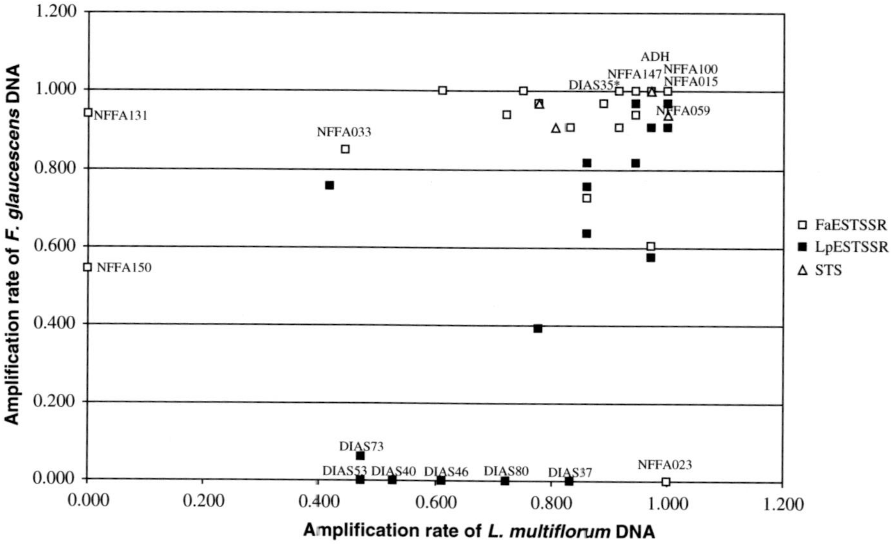

Fig. 34.1 Amplification rate of *Lm* and *Fg* DNAs. DIAS35* represents the new EST DIAS35 primers redesigned on the basis of *Fg* and *Lm* DNA sequences

and *Fg* genomes since the hexaploid Fa genome is composed of the *F. pratensis* diploid genome (P), which is close to the *Lolium* genome, and from the tetraploid *Fg* genome (G1G2).

Markers of Interest

33 markers reached the amplification rate of 60% in both *Fg* and *Lm*. 21 markers produced at least one specific band for both *Fg* and *Lm* and less than three common bands. Furthermore, 11 markers presented an F frequency superior to 0.5 and an L frequency superior to 0.4 (Table 34.2). In addition, NFFA033 which was retained though its *Lm* amplification rate was lower than the threshold, due to its high specificity. ADH and EST DIASG01_044 primer pairs need to be re-designed in order to reduce band size above high throughput threshold (<= 500 bp).

Use on F1 Hybrids

92.1% of F1 genotypes observed across the 7 markers showed expected bi-specific genotypes (LF, LC, FC) (Table 34.3). According to the number of segregating alleles, di-, tri- or tetraallelic genotypes were also clearly distinguished. In comparison with the number of specific alleles in parental species, the number of segregating alleles in the F1 population decreased for most markers. Monospecific (F, FF and L) genotypes indicated the presence of null alleles originating from *Lm* (NFFA100, NFFA033 and NFFA147) or from *Fg* respectively (PHOS).

Table 34.2 Selected markers characteristics. LG : linkage group, Nf: number of *Fg* specific bands, Nl: number of *Lm* specific bands, Nc: number of common bands between *Fg* and *Lm*

		In parental species			In F1 hybrids			
Markers	LG	Nf	Nl	Nc	Nf	Nl	Nc	Comments
NFFA015	6	4	3	1	2	3	1	
NFFA033	NA	3	2	0	2	1	0	AR_*Lm* < Threshold, probable null alleles in *Lm*
NFFA059	5	7	6	2	6	4	2	
NFFA098	NA	4	7	1				
NFFA100	NA	2	1	0	2	1	0	probable null alleles in *Lm*
NFFA110	NA	8	2	0				
NFFA147	NA	2	5	0	2	2	0	probable null alleles in *Lm*
G01_035	6	2	6	0				
G01_035*	6	2	5	0	2	4	0	redesigned primers
G01_044	3	4	6	0	1	6	0	
ADH	4	4	10	0	4	8	0	
CYP	5	2	4	1				
PHOS	3	1	4	0	1	4	0	probable null alleles in *Fg*

Table 34.3 Genotype distribution and allelic frequencies in F1 hybrids, calculated according to the observed genotypes

Obs. Genotype	ESTDIAS35*	NFFA015	NFFA033	NFFA059	NFFA100	NFFA147	PHOS
F			10		2	3	
FF						1	
FC		12		4			
FFC		1					
FFLC				1			
FLC		14		19			
FLLC		3		19			
FFLL	5			1		1	
FL	12		14		16	15	16
FLL	5						8
FFL	15	1	1	2	29	8	
L							3
N F1 hybrids	37	31	25	46	47	28	27
F freq	0.55	0.41	0.51	0.34	0.60	0.55	0.40
L freq	0.45	0.21	0.29	0.37	0.38	0.38	0.55
C freq	0	0.38	0	0.29	0	0	0

In all cases, ability to quantify interspecific genetic variability in *Festulolium* will depend on marker polymorphism. However, the set of markers we isolated herein enables yet the distinction between amphiploid and introgressed *Festulolium* cultivars, all confounded in the new official definition of *Festulolium*.

Acknowledgments The authors would like to thank all colleagues who contributed to this study and La Région Poitou-Charentes for its financial support.

References

Doyle, J.J., Doyle, J.L. 1987. A rapid DNA isolation procedure for small quantities of fresh leaf tissue. Phytochem. Bul. 19:11–15.

Ghesquière, M., Emile, J.C., Jadas-Hécart, J., et al. 1996. First *in vivo* assesment of feeding value of *Festulolium* hybrids derived from *Festuca arundinacea* var. *glaucescens* and selection for palatability. Plant Breed. 115:238–244.

Kopecky, D., Loureiro, J., Zwierzykowski, Z., Ghesquiere, M., Dolezel, J. 2006. Genome constitution and evolution in *Lolium* × *Festuca* hybrid cultivars (*Festulolium*). Theor. Appl. Genet. 113:731–742.

Lem, P., Lallemand, J. 2003. Grass consensus STS markers: An efficient approach for detecting polymorphism in *Lolium*. Theor. Appl. Genet. 107:1113–1122.

Mian, M., Saha, M., Hopkins, A., Wang, Z. 2005. Use of tall fescue EST-SSR markers in phylogenetic analysis of cool-season forage grasses. Genome. 48:637–647.

Saha, M., Mian, M., Eujayl, I., et al. 2004. Tall fescue EST-SSR markers with transferability across several grass species. Theor. Appl. Genet. 109:781–791.

Studer, B., Asp, T., Frei, U., et al. 2008. Expressed sequence tag-derived microsatellite markers of perennial ryegrass (*Lolium perenne* L.). Mol. Breeding. 21:533–548.

Zwierzykowski, Z., Kosmala, A., Zwierzykowska, E., et al. 2006. Genome balance in six successive generations of the allotetraploid *Festuca pratensis x Lolium perenne*. Theor. Appl. Genet. 113:539–547.

Chapter 35
Forage Production of Grasslands Composed by One, Two or Three Varieties of Perennial Ryegrass

Fabien Surault, Bernadette Julier, and Christian Huyghe

Abstract The present study aimed at evaluating the effect of an increased genetic diversity on forage production of perennial ryegrass (*Lolium perenne* L.), increasing genetic diversity being achieved by mixtures of varieties. Mixtures of varieties with similar or different heading dates were sown in 2003 in a micro-plot experiment conducted at INRA of Lusignan. Nine varieties grown in pure stand and 10 mixtures of two or three varieties were analysed. Two defoliation regimes were applied (fast or slow). During 5 years, dry matter production was mesured in each cut. Considering plots composed by one, two or three varieties of the same heading date, no difference in annual forage production was noticed. Increased genetic diversity within a heading date group provided neither advantage nor disadvantage. The effect of heading date was highly significant on forage production. Mixtures of varieties of different heading dates had an intermediate forage production and never produced more than the highest yielding variety. Compared to positive effects of increased specific diversity on grassland production, increased genetic diversity was not related to any increased yield, but can be seen as a method to secure forage production while avoiding a poor production of a particular variety in a given environment.

Keywords Forage production · Heading date · Mixtures of varieties · Pasture · *Lolium perenne*

Introduction

Numerous varieties of perennial ryegrass (*Lolium perenne* L.) are registered to the French Official Catalogue. Varieties are classified in earliness groups, according to their heading date, ranging from very early with a heading date in early May to very

F. Surault (✉)
INRA, UR 4, URP3F, Lusignan 86600, France
e-mail: fabien.surault@lusignan.inra.fr

C. Huyghe (ed.), *Sustainable Use of Genetic Diversity in Forage and Turf Breeding*,
DOI 10.1007/978-90-481-8706-5_35,

late with a heading date in mid-June. Ryegrass varieties are usually used as pure stands or in mixtures with legumes and/or other grasses. Some farmers mix several varieties of ryegrass of different ploidy or earliness (Humphreys and O'Kiely, 2006; 2007). Increasing species diversity in grasslands is often described as leading to an increased forage production (Hector et al., 1999; Roscher et al., 2007). In a mixture of varieties of the same species, genetic diversity increases but no study documented whether the agronomic value was modified. Heading date determining production peak, a mixture of varieties with different heading dates could result in a better yield production than a single variety. To test this hypothesis, mixtures of two or three varieties of perennial ryegrass were compared to pure varieties in an agronomic trial harvested during 5 years.

Material and Methods

In spring 2003, 19 swards of perennial ryegrass were sown in 9 m^2 plots at INRA Lusignan (France). The swards (Table 35.1) were composed by a single variety (numbered 1.1–1.9) or mixtures of two or three varieties with similar or different heading date (numbered 2.1–2.6 and 3.1–3.3, respectively). The trial included three blocks and was arranged in a split-plot design with two defoliation regimes: a cut every 25–30 days to simulate grazing or every 45–50 days to simulate harvest as hay. All plots received an average of 160 kg of nitrogen/ha/year, with 40 and 60 kg/ha after each cut in the frequent and the infrequent regimes, respectively. Plots have been exploited for 5 years. Cutting was mechanically conducted, the harvested forage was weighed and a forage sample was collected from each plot at each cut. Samples were dried at 60°C for 72 h to determine the percentage of dry matter.

Analyses of variance were carried out. Firstly, within a heading date group and for each defoliation regime, effect of number of varieties on forage yield was tested. Secondly, for the nine varieties sown as pure stands, effects of heading date and variety within heading date were evaluated. Thirdly, forage yield of mixtures composed by varieties of different heading dates was compared to the highest and lowest yielding varieties grown as pure stands.

Table 35.1 Description of perennial ryegrass plots sown in 2003

Grassland No	Early varieties			Intermediate varieties			Late varieties		
	1.1 Hamilton	1.2 Vital	1.3 Belramo	1.4 Herbie	1.5 Brest	1.6 Milca	1.7 Ohio	1.8 Kerval	1.9 Barlatan
2.1	o	o							
2.2				o	o				
2.3							o	o	
2.4	o			o					
2.5	o						o		
2.6				o			o		
3.1	o			o			o		
3.2	o	o	o						
3.3				o	o	o			
3.4							o	o	o

Results and Discussion

Forage Production of Mono and Multi-variety Swards of Similar Heading Date

The swards of a given heading date composed by one, two or three varieties were compared for annual and total forage yield over 5 years. No significant difference (Table 35.2), was noticed except in 2007 for the intermediate heading date under the frequent cutting regime (the variety 1.4 and the three-variety mixture 3.3 were the least yielding swards) and in 2004 for the late heading date under infrequent cutting regime (the variety 1.9 yielded significantly less forage). In these two cases, one of the variety grown as pure stand produced significantly less than the other swards, either mono or pluri-variety.

In the conditions of this trial, mixture of two or three varieties did not induce a benefit to annual forage production or total forage production over 5 years, whatever the defoliation frequency.

Table 35.2 Effect of the number of varieties (1, 2 or 3) comprised in the sward in each heading date group on annual yield and 5-year total yield (in t/ha) under frequent defoliation regime

	Heading date											
	Early				Intermediate				Late			
Year	1[a]	2	3	Effect	1[a]	2	3	Effect	1[a]	2	3	Effect
2004	10.5	10.2	10.3	*ns*	9.5	8.3	8.7	*ns*	8.4	8.0	7.8	*ns*
2005	5.9	5.2	5.8	*ns*	5.3	5.2	5.3	*ns*	4.9	4.4	4.6	*ns*
2006	8.9	9.0	9.0	*ns*	9.0	8.6	8.2	*ns*	7.6	7.0	6.8	*ns*
2007	8.1	7.5	8.1	*ns*	8.5	8.2	7.5	**	7.4	7.5	6.8	*ns*
2008	7.4	7.0	8.0	*ns*	7.5	7.5	7.5	*ns*	6.8	6.8	5.9	*ns*
5 year total	40.7	38.9	41.3	*ns*	39.0	37.8	37.1	*ns*	34.2	33.8	31.9	*ns*

*$P > 0.05$; **$P > 0.01$; ***$P > 0.001$; ns non significant
[a]Yield of the best variety grown as pure stand over the three varieties composing the mixture

Effect of Heading Date of the Varieties on Forage Yield of Mono-variety Grasslands

Heading date had a very significant effect of the annual forage yield of the varieties and the 5-year total yield (Table 35.3). Depending on the years, difference in dry matter yield between late and early or intermediate varieties ranged from 0.7 to 2.2 t DM/ha/year in frequent defoliation and from 1.1 to 2.7 t DM/ha/year in infrequent defoliation. For the 5-year total dry matter yield, differences reached 6.4 and 8.8 t DM/ha for the frequent and the infrequent regimes.

Table 35.3 Effect of heading date of the varieties on the annual forage yield and the 5-year total forage yield in the nine mono-variety grasslands

		Effect of	
Defoliation regime	Year	Earliness	Variety within earliness
Frequent	2004	***	ns
	2005	***	ns
	2006	***	ns
	2007	***	*
	2008	*	ns
	5-year total	***	ns
Infrequent	2004	***	**
	2005	***	ns
	2006	***	*
	2007	***	ns
	2008	***	ns
	5-year total	***	***

* $P > 0.05$; ** $P > 0.01$; *** $P > 0.001$; ns non significant

The three late varieties produced always less forage than early or intermediate ones (Fig. 35.1). These two types of varieties had the same dry matter yield over 5 years in frequent (39.4 vs. 38.6 t DM/ha) and infrequent (50.7 vs. 50.6 t/ha) defoliation regimes.

When defoliations were frequent, effect of varieties was not significant except in 2007, and there was no effect over the 5 years. Contrastingly, in infrequent defoliation regime, there was a significant cumulative effect over the 5 years, but difference among varieties within heading date was evidenced only in 2004 and 2006.

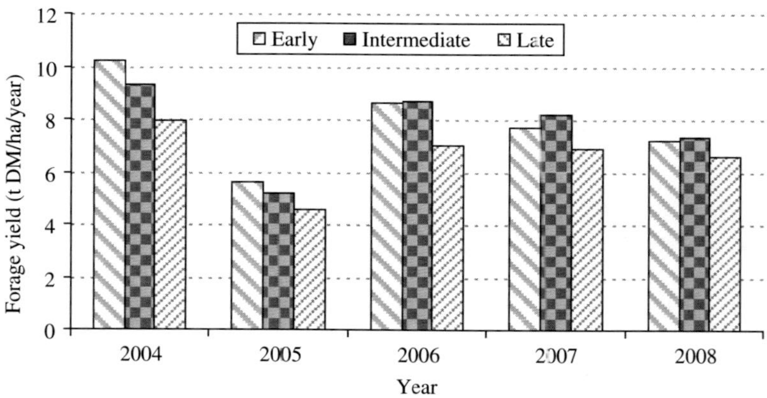

Fig. 35.1 Annual forage yield (t DM/ha/year) of the three types of heading date (means of the three varieties for each year)

Forage Yield of the Mixtures Composed by Varieties of Different Heading Dates

Except for mixture 2.4 in frequent regime, variety mixtures always produced more than the least yielding variety composing the mixtures (Table 35.4). This difference between mixtures and the least yielding variety was generally significant. However, mixtures never produced significantly more than the most productive variety included in the mixture.

In most situations, there was no difference between the yield of the mixture and the most productive variety grown in pure stand, but the best variety yielded significantly more than the mixture in three cases (Table 35.4).

Mixtures of varieties with different heading dates, a trait related to a difference in the date of the forage production peak, did not yield more than the highest yielding variety of each mixture. In the conditions of this trial, this strategy was not a solution to increase dry matter production.

Table 35.4 Five-year total forage yield (t DM/ha) of mixtures of different heading date varieties and of the highest yielding and the least yielding varieties composing the mixtures. (Test of mean with comparison between the mixture yield and the highest yielding or the least yielding varieties)

		Mixtures		Pure stand variety	
Defoliation regime	Heading date[a]	No	Yield	The most yielding	The least yielding
Frequent	E+I	2.4	49.69	50.91 ns	50.76 ns
	E+L	2.5	48.98	50.91 ns	44.73 ***
	I+L	2.6	47.38	50.76 ***	44.73 *
	E+I+L	3.1	48.66	50.91 ns	44.73 ***
Infrequent	E+I	2.4	38.77	39.26 ns	37.84 ns
	E+L	2.5	36.96	39.26 *	34.23 **
	I+L	2.6	36.09	37.84 ns	34.23 ns
	E+I+L	3.1	36.25	39.26 ***	34.23 *

* $P > 0.05$; ** $P > 0.01$; ***; $P > 0.001$; ns non significant
[a]E: Early, I: Intermediate, L: Late heading date

Conclusion

Swards composed with mixtures of perennial ryegrass varieties did not produce more than swards composed by a single variety. This result would indicate that the positive impact of an increased diversity in grasslands achieved through a mixture of species is not observed at the within-species level. However, within-species mixture can be a method to secure forage production under conditions that could favour either early or late varieties. It could also be hypothesized that the genetic diversity within each variety was already large enough to achieve the maximum overyielding.

To better understand mechanisms involved in yield elaboration in variety mixtures, it is important to know whether genetic changes could have occurred during the 5 years. Assessment of the frequency of each variety in the mixtures will be conducted by using microsatellite markers that differentiate the varieties. In addition, yield distribution through the year has to be considered. Mixtures composed by different heading date varieties could have a production more adapted to the requirements of the animals.

Acknowledgements Authors thank Rodrigue Veron for his help in the management of this trial.

References

Hector, A., Schmid, B., Beierkuhnlein, C., Caldeira, M.C., Diemer, M., Dimitrakopoulos, P.G., Finn, J.A., Freitas, H., Giller, P.S., Good, J., Harris, R., Hogberg, P., Huss-Danell, K., Joshi, J., Jumpponen, A., Korner, C., Leadley, P.W., Loreau, M., Minns, A., Mulder, C.P.H., O'Donovan, G., Otway, S.J., Pereira, J.S., Prinz, A., Read, D.J., Scherer-Lorenzen, M., Schulze, E.-D., Siamantziouras, A.-S.D., Spehn, E.M., Terry, A.C., Troumbis, A.Y., Woodward, F.I., Yachi, S., Lawton, J.H. 1999. Plant diversity and productivity experiments in European grasslands. Science 286:1123–1127.

Humphreys, J., O'Kiely P. 2006. Amount and quality of grass harvested for first-cut silage for differing spring-grazing frequencies of two mixtures of perennial ryegrass cultivars with contrasting heading date. Grass Forage Sci. 61:77–88.

Humphreys, J., O'Kiely P. 2007. Effects of two mixtures of perennial ryegrass cultivars with contrasting heading dates, and differing in spring-grazing frequency and silage harvest date, on characteristics of silage from first-cut swards. Grass Forage Sci. 62:389–404.

Roscher, C., Schumacher, J., Weisser, W.W., Schmid, B., Schulze, E.-D. 2007. Detecting the role of individual species for overyielding in experimental grassland communities composed of potentially dominant species. Oecologia 154:535–549.

Part IV
Genetic Progresses to Meet End-Users' Expectations

Chapter 36
Genetic Gain in Agronomic Value of Forage Crops and Turf: A Review

Stephan A.G. van der Heijden and Niels Roulund

Abstract Plant breeding has shown to be an effective process to design and develop varieties to serve the requirements of end users. However, in that respect grasses are very different from other species due to there perennial characteristic and the possibility to produce mixtures of species and varieties. Besides that there are many traits that are genetically negatively correlated so the varieties are compromises to suit the customers. In this presentation an overview will be given of available literature on quantitative characteristics like dry matter yield, nitrogen recovery or nutritional value, and the results will be compared with genetic gains in other crops based on recent quantitative-genetics tools. Besides that examples of genetic improvements for many other traits will be given as disease resistance to invertebrate pests and fungal, bacterial and viral diseases and abiotic stresses.

For the turf grasses a substantial improvement in important agronomic traits has been obtained in the last 30 years. This improvement will be illustrated using data from official lists and literature. Improvements has in particularly been related to traits like sod density, fineness of leafs and tolerance to wear – whereas improvements in respect to diseases resistance like snow mould (*Fusarium nivale*), red threat (*Laetisarium fuciformis*) and abiotic stress like drought have been more restricted due to low heritability of the traits. In order to meet the requirements to a world with more emphasis on low input which already is seen in some countries, focus in the breeding for the future will be on the more complex traits like drought and disease resistance than it has been in the past. Future research is proposed to focus even more on grass root development in order to allow for development for efficient breeding tools since good and efficient roots are crucial under low input.

A very important issue, the importance of good and stable seed yields, for the economic feasibility of grass products will be discussed.

Finally the potential of genomic selection approaches will briefly addressed.

S.A.G. van der Heijden (✉)
Barenbrug Holding, Nijmegen NL-6515 AB, The Netherlands
e-mail: svdheijden@barenbrug.com

C. Huyghe (ed.), *Sustainable Use of Genetic Diversity in Forage and Turf Breeding*,
DOI 10.1007/978-90-481-8706-5_36, © Springer Science+Business Media B.V. 2010

Keywords Genetic gain · Agronomic value · Forage grasses · Turf grasses · Forage legumes · Disease resistance · Microclover · Genetic color · Seed yield · Low input · Dry matter yield · Nitrogen recovery · Persistence · Forage quality · Selection response · Molecular markers · Genomic selection · Economic value

Introduction

Grass is so common in the world around us that it can be easily overlooked, although it plays a very important role in many aspects of human well being ranging from very basic elements as feed and food but also recreation. Grasslands cover about 70% of the world's agricultural area (Soussana and Luscher, 2007). However, there is still al lot of improvement possible in the use of grassland and a major challenge will be an economically, environmentally and sustainable management of this area. Besides the importance of grasslands in the food production, they should also be recognized from the viewpoint of 'multi-functionality' i.e. they are also important as sources of pollution, particular from livestock production, but are also delivering important ecosystems services and play an important role in the tourism, amenity and leisure industry in many parts of the world. There is also a growing interest for the production of bioenergy based on grass based systems (Abberton et al., 2007).

The biological nature of the grassland species in combination with its broad adaptation, its multifunctionality and the possibility to use it in mixtures of varieties or species for many different utilities make the generally used one-dimensional approach such as measuring yield over time as a unreliable tool to compare genetic progress in grasses with many other species. This is also hampered by the fact that in most forage and turf species are perennial with very interesting negative genetic correlation between traits of interest (forage yield, slow growth, quality) and economic aspects as seed yield. Besides that higher yield or more aggressiveness is not always the selection target due to the aspect of final use of the varieties in mixtures.

Genetic Gain of Turf Grasses

The genetic gain in respect to turf quality has for all major cool season grasses been substantial during the last 30 years by breeders at Rutgers University USA and by breeders in Europe. This is illustrated very clearly by Sampoux et al. (2010, in press), in which performance of 31 turf varieties of perennial ryegrass (*Lolium perenne*) listed in France in the period 1975–2005 was compared. A summary of a part of the results are given in Table 36.1.

From the result, it is clear that the main aesthetic components of turf grasses namely turf quality, turf density and fineness of leafs have been improved significantly during the last 30 years. Even for wear resistance, for which improvements are considerably more difficult to obtain due to technical difficulties in performing good wear trials, significant improvements have been made during the investigated period. Genetic colour has become darker by 0.78 points/decade probably due to introduction of American varieties in the test or use of US germplasm for the

Table 36.1 Gradual increase in major turf traits in the period 1975–2005

Trait	Improvement in score of trait over 10 years	Significance level
Turf Quality	0.77 points	***
Turf Density	0.91 points	***
Fineness of Leafs	0.76 points	***
Wear resistance	0.54 points	***
Genetic colour	0.78 points	***
Crown rust (*Puccinia coronata*) resistance	0.55 points	***
Red thread (*Corticium fuciforme*)		Ns
Other disease resistance		Ns

*** = significant effect at 1% level – Ns not significance effect

breeding of new varieties. Crown rust (*Puccinia coronata*) resistance has also been improved during the assessed period – which seems logical because crown rust is one of the few diseases showing a good correlation between scoring on single spaced plants and swards. However, for resistance against red thread (*Laetisarium fuciformis*), another important disease, no significant improvement has been achieved throughout the last 30 years of breeding. This is in accordance with the fact that the disease has only a low heritability and selection for the disease is only possible for grass sown in swards and not on single spaced plants of the breeding population. The substantial progress obtained in turf quality during the last 30 years for ryegrass is well illustrated in Fig. 36.1 in which a turf plot of a new turf cultivar is sown next to a 30 years old variety.

Besides the turf quality turf breeders have been forced to improve the seed yield potential of their varieties in order meet the demands for competitive commercial production of the variety. This has been a considerable challenge, since for most

Fig. 36.1 Difference in turf quality between a new turf cultivar and a 30 years old cultivar of *Lolium perenne*

turf species a negative correlation between shoot density and seed yield is observed. The results of Sampoux et al. (2010, in press) are however showing no progress in seed yield over the years and interestingly a turf variety is being one of the best producers (results not shown).

The future will bring new challenges to the turf breeders. It is clear that one future focus will be on response to lower input for turf grasses used in Europe. The use of pesticides for example is already banned on golf courses in Denmark, a fact which easily could become the case for more countries in the future.

The use of artificial irrigation and fertilizers will surely also become restricted for many users of turf grasses in the future.

Unfortunately, breeding for varieties adapted to low input conditions with better disease resistance is hampered by low heritability of the traits and by no or little correlation between single space plants and the behaviour in the turf.

Genetic gains for these traits have therefore been restricted. One interesting alternative solution for some of the problems is the use of specially selected small leaved white clovers (*Trifolium repens*) in mixture with the turf grass. The white clovers have been selected to be able to grow homogeneously in the turf sward in balance with the grass instead of being too aggressive or getting out competed by the grass. As the white clover will supply the grass with sufficient nitrogen for optimal growth, the turf grass seems almost always to have an advantage of growing with the turf clover. This is manifested by an earlier start of green appearance of the turf, better performance during drought, better disease tolerance and lower rate of weed invasion. Fig. 36.2 shows the different aspect of the same grass mixture with and without the turf clover in trials at Bingley in end of April 2009. The grass-clover mixture is growing vigorously whereas the pure grass mixture still is dormant.

The beneficial effect of the turf clover is also clearly seen in Fig. 36.3 which has been taken on a golf course with sandy soil in Denmark. The clover has been sown

Fig. 36.2 Comparison of the same grass mixture with and without turf clover end of April in trial at STRI Bingley. Clover grass mixture in left side

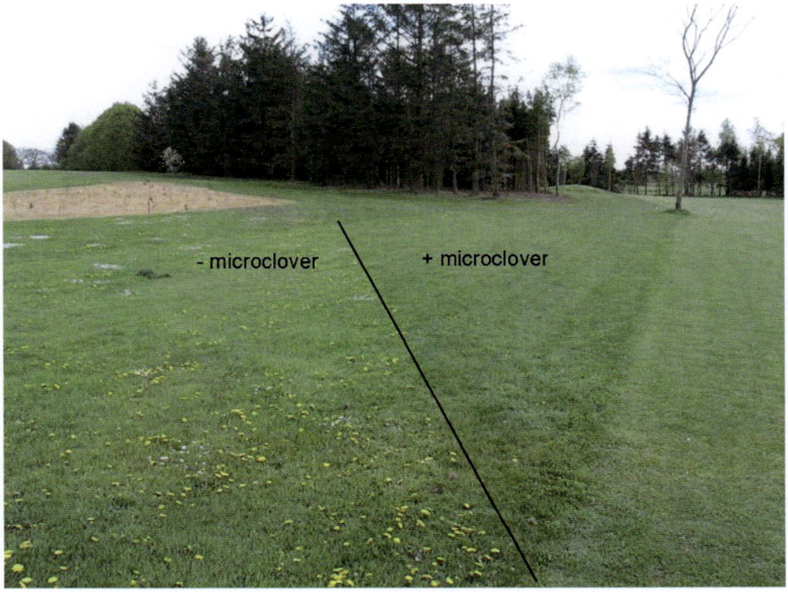

Fig. 36.3 The effect of micro clover on fairway and semi rough on a golfcourse

as a path crossing the fairway and the semi rough. The grass-clover mixture appears much greener and more vigorous, at the same time showing a substantial reduction in the amount of weeds compared to the parts without clover.

However the use of microclover does not solve all problems faced under low input conditions.

Improvements have to be made in respect of water and nitrogen use efficiency and in respect to disease resistance. For these important traits, where progress with conventional breeding is often very slow and difficult, genetic markers clearly offer a great help to reach a breakthrough. The traditional QTL approach for developing marker assisted selection (MAS) is however entirely dependent on the phenotyping quality of the plants in the mapping population. Usually, phenotyping is done on single spaced plants of the mapping population, although, for many traits, there is little or no correlation to what is observed when the grass is growing in swards, making a useful MAS system very difficult to develop. One possible solution can be the production of half-sib offspring of the mapping population by a top cross, enabling harvest of sufficient seed for establishment of plots. However, when using this approach half of the variation will be lost. Even in plots, diseases like red thread (*Laetisarium fuciformis*) or snow mould (*Fusarium nivale*) are difficult to score which will make the phenotyping even more difficult. Furthermore, for many of these traits, there is a large GxE interaction, requiring for the phenotyping to be done at many locations over several years in order to obtain sufficient precision. On the marker side, the amount of available SSR markers is often not enough to ensure a reliable QTL mapping with high enough map density.

Recent years have seen rapid development of SNP-technologies by companies such as Illumina (www.illumina.com), and these technologies are already being used for animal breeding and human disease diagnostics. How far these developments for use in humans have proceeded is illustrated on the web site www.23andme.com. The company offers investigations of private persons DNA profile by running the DNA on an Illumina 550 K SNP chip monitoring the alleles of 550 000 SNP markers which will be used for risk assessments of 29 mapped diseases and conditions like diabetes 2, Parkinson's disease and Alzheimer, and with even another 83 diseases and conditions to come in the near future. The price for this service is only $399 clearly illustrating the power of this new generation of marker techniques.

The vast amounts of marker data generated by these techniques open up for association or linkage disequilibrium mapping approaches in grasses. Compared to the traditional mapping approach, these have the outstanding advantage that also plot based agronomic data generated by the breeders can be used for the development of MAS systems thereby solving problems arising from missing or little correlation between the single spaced plants and plots. However, statistical methods handling allele frequencies in populations compared to single genotypes like humans yet need to be developed.

In conclusion, with these new technologies in hand once more a close interaction is needed between grass breeders, molecular marker scientists and statisticians to combine data from careful trait phenotyping, large amounts of marker information and newly developed statistical models into breeding tools that are capable to respond to the demands of the future.

Genetic Gain of Forage Grass

Compared to many other species, genetic gain for quantitative traits in grasses is little documented and in a number of available studies the gains resulting from genetic improvement is confounded with improved agronomic agricultural practice and changing climatic conditions.

The one-dimensional approach that generally is used such as treated yield that is adopted from other species as maize and wheat is not a good indicator to measure genetic gain in grasses and legumes. This approach is not taken in account the progress that has been made by increasing several traits simultaneously, taken in account that many of these traits are negatively correlated or even pleiotropic. Any experiment to measure genetic gains should be designed specifically for the purpose and it is important to find a balance in measuring the traits of interest and describe the relationship between the traits to come to scientific conclusions.

Wilkins and Humphreys (2003) described a number of components of genetic improvement:

- Dry matter yield and nitrogen recovery
- Persistency

- Nutritional value for ruminants or in vitro dry matter digestibility (DMD or DOMD).
- Resistance to invertebrate pests and fungal, bacterial and viral diseases.
- Tolerance to environmental stress (freezing temperatures, prolonged snow cover, low light intensities during winter, heat, drought, high light intensities leading to oxidative damages ...)
- Seed yield.

However, this list is not exhaustive as can be seen from the following table that is describing the current objectives within temperate pasture plant breeding programmes in New Zealand example.

Breeding objective	Forage legumes	Forage grasses
Herbage yield	↑ Annual DM yield Improved seasonal distribution of DM yield ↑ Persistence ↑ Grazing tolerance ↑ Stolon density	↑ Annual DM yield Improved seasonal distribution of DM yield ↑ Persistence ↑ Grazing tolerance
Herbage quality	↑ Metabolisable energy Improved protein/energy balance ↑ Condensed tannins ↑ Rumen undegradable protein	↑ Metabolisable energy ↑ Digestibility ↓ Leaf sheer strength Delayed heading date ↓ Aftermath heading
Pest and disease resistance	↑ Clover root weevil resistance ↑ Nematode resistance (stem, root-knot and clover cyst nematodes) ↑ Clover flea resistance ↑ Virus resistance (WCIMV,AIMV,CYVV) ↑ Resistance to *Sclerotinia*	↑ Crown rust resistance ↑ Endophyte-mediated resistances (e.g. Argentine stem weevil, mealy bug, porina, black beetle) ↑ Scald resistance
Edaphic stress tolerance	↑ Water use efficiency Improved root and stolon morphology ↑ Aluminium tolerance	↑ Water use efficiency ↑ Root strength ↑ Aluminium tolerance
Animal health, welfare and fertility	↓ Formononetin content ↑ Anthelmintic properties	↓ Ryegrass staggers ↓ Fescue toxicosis
Environmental	↓ Greenhouse gas emissions	↓ Greenhouse gas emissions
Symbiotic associations	↑ Nitrogen fixation at low temperatures	↑ Non-toxic endophytes
Product value	Improve flavour and texture characteristics ↑ Lipid composition	Improve flavour and texture characteristics ↑ Lipid composition

Fig. 36.4 Objectives of a temperate pasture breeding programme in New Zealand (Woodfield and Easton, 2004). DM = dry matter; WCIMV = white clover mosaic virus; AIMV = alfalfa mosaic virus; CYVV = clover yellow vein virus (Woodfield and Easton, 2004)

Table 36.2 Genetic gains in harvested annual dry matter yield (DMY) of temperate grasses during 1950–2008

Grass species	Country	Genetic gain (%/decade)	Reference
Perennial ryegrass	UK	6	Aldrich (1987)
	UK	4	Wilkins and Humphreys (2003)
	Belgium	5	Van Wijk and Reheul (1991)
	France	2.5	Allerit (1986)
	Italy	2.5	Veronesi (1991)
	Germany	0	Laidig et al. (2008)
	EU	3.5	Humphrey (1999)
Italian ryegrass	Netherlands	2	Van Wijk and Reheul (1991)
	France	4.5	Allerit (1986)
Tall fescue	France	1	Veronesi (1991)
	Italy	5.5	
	Spain	3.5	
Cocksfoot/orchard grass	France	1	Veronesi (1991)
	Italy	5.5	Veronesi (1991)
	Germany	–0.2	Laidig et al. (2008)
	USA (hay)	0 – 1	Casler et al. (2000)
	USA (grazing)	0 – 1	Casler et al. (2001)
Smooth brome grass	USA (hay)	0	Casler et al. (2000)

Adapted from Wilkins and Humphreys (2003)

In Table 36.2, an overview of genetic gain in Europe and USA is given.

These data are supported by the gains for early heading perennial ryegrass in the UK of 0.5% per year, however for late heading perennial ryegrass, the progress was only half (Camlin,1997).

From unpublished sources, the following gains were reported 0.1% for *Lolium perenne* in Germany (Feuerstein pers. comm.) and 0.3% per year in Belgium (Chavez, Eucarpia, 2009, France).

The genetic gains for New Zealand are very well documented in several papers by various authors:

Wilkins and Humphreys (2003) wrote that very few routine variety trials, or experiments specifically designed to test gains from breeding in *persistency*, have been extended beyond 3 full harvest years. But differences among ryegrass varieties in annual DMY are usually greater in the second and third year harvests than in the first harvest year, indicating that improved persistency has contributed to the gains.

Wilkins and Humphreys (2003) stated that breeding and variety assessment for DM digestibility (DMD) was more recent than breeding for DMY and much fewer data are available. Mean DMD of the five perennial ryegrass varieties first listed on or before 1989 was 694 g/kg while the mean of the 9 varieties listed on or since 1999 was 710 g/kg, i.e. 10 g/kg per decade. In smooth brome grass, the genetic gain is reported to be 2 g/kg per decade.

Table 36.3 Genetic gains in dry matter production for forage grasses in New Zealand (Easton et al. 2002)

Grass species	Time frame (years)	Annual genetic gain (%)	Reference
Perennial ryegrass	?	0.25–0.73	Woodfield (1999)
	25 – 30 60	0.4 (0.7 in autumn and 0.1 in spring)	Easton (1999); Easton et al. (2001)
		0.25–1.5%	Woodfield and Easton (2004)
White clover	40	0.6	Woodfield and Caradus (1994)
	?	1.21–1.49	Woodfield (1999)
	?	0.2–4.0	Woodfield and Easton (2004)
Italian ryegrass	25–30	1.5	Easton et al. (2001)

Table 36.4 Genetic gains in resistance of forage grasses

	Easton et al. (2002)	Wilkins and Humphreys (2003)
Perennial ryegrass	Crown and stem rusts (Easton, 1999)	Crown rust (Charles 1973)
Cocksfoot	Rust (Rumball, 1982)	
White clover	Leaf disease and nematode tolerance (Cooper and Chapman, 1993) (Taylor, 2008)	
Smooth bromegrass		Brown leaf spot Casler et al. (2000)

Quesenberry and Casler (2001) reviewed also the gain for forage quality traits. The development of an *in vitro* procedure for rapid, repeatable and relative inexpensive evaluation tool for forage digestibility by Tilley and Terry (1963) is seen as the single most important event in the evolution of cool-season forage grass selection criteria. However, despite this breakthrough, only limited information is available on cool-season forages as alfalfa, orchard grass, perennial ryegrass, smooth brome grass and timothy (Casler and Vogel (1999) in Quesenberry and Casler (2001)). The rate of genetic progress in these species ranged from 8 to 45 kg per cycle (1.3 to12.1% per cycle or 0.7 to 2.5% per year).

The following authors referred to genetic gains in resistances to several diseases:

In a more recent study by Sampoux et al. (2010, in press), in which performance of 18 forage varieties of perennial ryegrass (*Lolium perenne*) listed in France in the period 1975–2005 was compared.

A summary of a part of the results are given in Table 36.5.

The yield components except for spring growth, were also significant as well the yields in the individual harvest years. The significantly increased yield confirms the conclusion by Wilkins and Humphreys (2003) that new varieties have better persistence, but to have a final proof the trials should be extended for one or two more years.

Despite the need for higher seed yielding varieties, the tested varieties did not show a significantly increased seed yield. It seems that the evaluation of agronomic traits and the resulting listing of varieties on official lists are hampering progress on this very important aspect of the economics of the grass seed business.

Table 36.5 Increase in major forage traits in the period 1975–2005

Trait	Improvement of trait in 10 years	Significance level
Persistency	0.32 points	**
Reheading	–0.36 points	*
Crown rust	0.66 points	***
Average annual dry matter yield	0.29 ton/ha	***
Water soluble Carbohydrates	0.67%	**
ADL	–0.05%	***
NDF	–0.23%	*
Digestibility of NDF	+0.79%	**
DOMD	+0.51%	**

*** = significant effect at 1% level – Ns no significant effect

Discussion

Compared to other crops as wheat and maize where genetic gains are reported ranging from 1 to 2.5% per year the genetic gain in forage grasses are reported to be considerable lower. This is to be expected on theoretical considerations.

The formula for selection response is:

$R = cis_g{}^2/y\sigma_p$

c = pollen factor (1/2 after pollination, 1 before pollination, 2 recombine selfed progenies)

y = numbers of years per cycle.

i = selection differential expressed as number of σ_p

$s_g{}^2$ = genetic variance of type of selection practised

σ_p = phenotypic standard deviation for the progenies evaluated.

(Empig et al., 1972 in Quantitative Genetics, Genomics and Plant Breeding (Kang, 2002))

In perennial grasses the selection cycle is in general at least two to three years compared to maximum one year for wheat, barley or maize. Besides that the selection differential is in much smaller due to the fact that breeders are not focussing on one crop target but are spreading their resources over many species and within species have to work in different maturity classes and ploidy levels. Moreover, the breeding objectives are more difficult to define and measuring of the traits via direct methods in animals is impractical or even impossible. Further are many important traits negatively correlated and can influence negatively the economic feasibility of a variety due to the negative effect on seed yield (Wilkins and Humphreys, 2003).

Measuring genetic gains in grasses is for a number of reasons more complex than for other species because many traits play an equally important role, there are a large amount of good candidate varieties offered each year to official testing authorities and the impact of GXE interaction in grasses makes it difficult to estimate significantly yield differences (Camlin, 1997; Conaghan et al., 2008).

It could also be questioned if the protocol of the current official variety testing systems in the different countries in Europe is the best way to estimate genetic gain, because these protocols serve a different purpose, namely the listing of varieties on national lists. For measuring genetic gains specific designed trials are necessary but due to limited resources and the large number of varieties from the different species, maturity classes and ploidy levels makes the testing process is laborious and in general only limited resources are available for this purpose.

In the chapter on genetic gain of turf grasses the use of molecular markers is discussed. However, the development of new tools as for example marker assisted breeding is rather slow. Many QTLs are mapped but information on successful use in commercial breeding programmes in field crops and forages is rather scarce. The implementation in grasses is further slowed down due to the genetic complexity and the fact that in grasses we have to deal with quantitative traits that are difficult to phenotype. Although there are positive results reported by Barrett et al. (2001), Faville et al. (2003), Humphreys et al. (2003), Koelliker et al. (2005), and Skot et al. (2005), there is no example of successful implementation in commercial breeding programmes. However, often contrasting information about the results is published that makes it difficult to develop a general breeding strategy because it is unclear on beforehand if the selection response is positive or negative. A good example is the use of genetic distance to predict the performance of the synthetics or hybrids (Koelikker et al., 2005; Bertan et al., 2003). The generally poor implementation of marker assisted selection in reviewed by Heffner et al. (2009) and as an alternative is suggested the use genomic selection. Genomic selection is developed for human and animal genetics and is now successful implemented in commercial dairy companies in Europe and New Zealand.

The technology looks very promising but an important difference between forage and turf species and animals is that with animals individual phenotyping is possible but in forage and turf species the correlated response of the selection of the individual is very poorly correlated with the performance in a sward. The successful genomic selection in plants will very much depend on the development of efficient phenotyping tools and less on the design of high end 60 k DNA chips or other new technologies.

Estimating genetic gains in forage and turf species is due to the complexity of the species, its perennial status and the multi-dimensional approach. Despite of these difficulties, positive results can be reported. To make comparisons to other crops it would be useful to develop an index approach for each of these crops that also take in account the economic value.

Conclusion

There is only limited evidence available that suggests that progress in improving DMY and DMD of forage grasses has been very variable (Quesenberry and Casler,

2001; Wilkins and Humphreys, 2003; Woodfield and Easton, 2004; Sampoux et al., in press):

- For perennial ryegrass in North western Europe and New Zealand, there has seen substantial genetic gains in DMY and DMD; as for DMY in other species in France, Italy and Spain. The situation is different in Germany (Laidig et al., 2008) where according to the official trials a gain of 0% was estimated for perennial ryegrass and orchard grass. However, U. Feuerstein (pers. comm.) estimated in private trails a gain of 0.1%
- In USA, the development of endophyte-free tall fescue has been a major development but there is no published evidence of genetic gains in DMY or DMD of tall fescue.
- There is no reported overall improvement in DMY and DMD of cocksfoot in USA in the last 50 years

Pasture plant breeding in perennial ryegrass in (New Zealand) has achieved significant incremental gains in herbage yield potential, disease resistance and in range of maturity (Easton et al., 2001)

The contribution made to UK grassland within a relatively short timescale, in particular by the tetraploid perennial types, is worthy of special mention in any review of achievements of plant breeding (Camlin, 1997).

Acknowledgement We thank Dr Radha Ranganathan from the International Seed Federation and the members of the section Forage and Turf for providing information and synthesis on genetic gains for forage grasses.

References

Abberton, M.T., MacDuff, J.H., Marshall, A.H., Humphreys, M.W. 2007. The genetic improvement of forage grasses and legumes to reduce greenhouse gas emissions. FAO-report.

Aldrich, D.T.A. 1987. Developments and procedures in the assessment of grass varieties at NIAB 1950–1987. J. Natl. Inst. Agric. Bot. 17:313–327.

Allerit, R. 1986. Espèces fourragères pérennes: progrès réalises depuis 25 ans, apprécies a travers 11 expérimentation officielle. Fourragères 107:17–33.

Barrett, B., Griffiths, A., Mercer, C., Ellison, N., Faville, M., Easton, S., Woodfield, D.R. 2001. Marker-assisted selection to accelerate forage improvement. Proceedings of the New Zealand Grassland Association. 63:241–245.

Betran, F.J., Ribaut, J.M., Beck, D., Gonzalez de Leon. 2003. Genetic diversity, specific combining ability and heterosis in tropical maize under stress and nonstress environments. Crop Sci. 43:797–806.

Camlin, M S. 1997. Plant Breeding – Achievements and Prospects: Grasses. Proceedings of the British Grasslands Symposium 31:2–13.

Casler, M.D., Fales, S.L, Undersander, D.J., McElroy, A.R. 2001. Genetic progress from 40 years of orchardgrass breeding in North America measured under management intensive rotational grazing. Can. J. Plant Sci. 81:713–721.

Casler, M.D, Vogel, K.P. 1999. Accomplishments and impact from breeding for increased forgae nutritional value. Crop Sci. 39:12–20.

Casler, M.D., Vogel, K.P., Balasko, J.A., Berdahl, J.D. Miller, D.A. Hansen, J.L., Fritz, J.O. 2000. Genetic progress from 50 years of Smooth Bromegrass Breeding. Crop Sci. 40:13–22.

Charles, A.H. 1973. A comparison of ryegrass populations from intensively managed permanent pastures and leys. J. Agric. Sci.-Cambridge 81:99–106.

Conaghan, P., Casler, M.D., McGilloway, D.A., O'Kiely, P., L.J. Dowley 2008. Fenotype x environment interactions for herbage yield of perennial ryegrass sward plots in Ireland. Grass and Forage Science, 63:107–120.

Cooper, B.M., Chapman, D.F. 1993. Grasslands Presige (G39), a white clover cultivar originating from northern New Zealand. Proceedings of the New Zealand XVII International Grassland Congress (pp. 458–459).

Easton, H. S., Amyes, J. M., Cameron, N. E., Green, R. B., Kerr, G. A., Norriss, M. G., A. V. Stewart. 2002. Pasture plant breeding in New Zealand: Where to from here? Proceedings of the New Zealand Grassland Association 64:173–179.

Easton, H.S. 1999. Endophyte in New Zealand ryegrass pastures, an overview. Grassland Res. Pract. series. 7:1–9.

Easton, H.S., Christensen, M.J., Eerens, J.P.J., Fletcher, L.R., Hume, D.E., Keogh, R.G. Lane, G.A., Latch, G.C.M., Pennell, C.G.L., Popay, A.J., Rolston, M.P., Sutherland, B.L., Tapper, B.A. 2001. Ryegrass endophyte: A New Zealand Grassland success story. Proceedings of the New Zealdn Grassland Association. 63:37–46.

Empig, L.T, Gradner, C.O., Compton, W.A. 1972. Theoretical gains of different population improvement procedures. Univ. Nebraska Agric. Exp. Stn. Bull. Misc. 26 (revised).

Faville, M., Barret, B., Griffiths, A., Schreiber, M., Mercer, C., Baird, I., Ellison, N., Bryan, G., Woodfield, D., Forster, J., Ong, B., Sawbridge, T., Spangenberg, G., Easton, H.S. 2003. Implementation of molecular marker technology in forage improvement. Proceedings of the New Zealand Grassland Association 65:229–238.

Heffner, E.L., Sorrels, M.E., JL Jannink. 2009. Genomic selection for crop improvement. Crop Sci. 49:1–12.

Humphreys, M. 1999. The contribution of conventional plant breeding of forage crops. In: Porc. XVIII Intl. Grassl. Congr. 1: 4-71-4-78. 8-19 June 1997. Winnipeg, Manitoba and Saskatoon, Saskatchewan, Canada.

Humphreys, M., Turner, L., Skot, L., Humpheys, M., King, I., Armstead, I., Wilkins, P. 2003. The use of genetic markers in grass breeding. Czech J. Genet. Plant Breed. 39:112–119.

Kang, M.S. (ed.) 2002. Quantiative genetics, genomics and plant breeding. CABI Publishing.

Koelliker, R., Boller, B., Widmer, F. 2005. Marker assisted polycross breeding to increase diversity and yield in perennial ryegrass (Lolium perenne L.). Euphytica. 146:55–65.

Laidig, F., Drobek, T., Meyer, U. 2008. Genotypic and environmental variability of yield for cultivars from 30 different crops in German official variety trials. Plant Breed. 127:541–547.

Lancashire, J. 2005. The importance of exotic forage germplasm in feeding New Zealand's livestock. Proceedings of the 4th International symposium on the molecular breeding of forage and turf, a satellite workshop of the XXth International Grassland Congress, July 2005, Aberystwyth, Wales.

Piano, E. (ed.). 2004a. Inerbimenti e Tappeti Erbosi. Vols 1–4.

Piano E (ed.). 2004b. Proceedings of the convention on Inerbimenti e tappeti erbosi per l'agricultora, l'ambiente e la societa. Vol 1 Comunicazioni Poster.

Quesenberry, K.H., Casler, M.D. 2001. Achievements and perspectives in the breeding of temperate grasses and legumes (pp. 517–524). In: A. Gomide, J.A., Mattos, W.R.S., da Silva, S.C. (eds.), Proc. Int. Grasslands Congress, XIX, Sao Pedro, Sao Paulo, Brazil. 11–21 Feb. 2001. FEALQ, Piricicaba, Sao Paulo, Brazil.

Rumball, W. 1982. "Grasslands Wana" cocksfoot (Dactlyis Glomerata L.). N.Z. J. Exp. Agric. 10:51–52

Skot, L., Humphreys, M.O., Armstead, I., Heywood, S., Skot, K.P., Sanderson, R., Thomas, I.D., Chorlton, K.H., Hamilton, N.R.S. 2005. An association mapping approach to identify flowering time genes in natural populations of Lolium perenne (L.) Molecular Breeding. 15:233–245.

Soussana, J.F., Luscher, A. 2007. Temperate grasslands and global atmospheric change: a review. Grass and Forage Science 62:127–134.

Taylor, L.T. 2008.. A century of clover breeding developments in the United States. Crop Sci. 48:1–13.

Tilley, J.M.A, Terry, R.A. 1963. A two-stage technique for the in vitro digestion of forage crops. J. Br. Grassl. Soc. 18:104–111.

Van Wijk, A.J.P, Reheul, D. 1991. Achievements in fodder crop breeding in maritime Europe. In fodder crops breeding: achievements, novel strategies and biotechnology. In: de Nijs, A.P.M., Elgersma, A. (eds.), Proceedings of the Eucarpia Fodder Crops and Amenity Grass Section for 1990, Wageningen, The Netherlands (pp. 13–18). Pudoc, Wageningen.

Veronesi, F. 1991. Achievements in fodder crop breeding in Mediterranean Europe. In fodder crops breeding: achievements, novel strategies and biotechnology. In de Nijs, A.P.M., Elgersma, A. (eds.), Proceedings of the Eucarpia Fodder Crops and Amenity Grass Section for 1990, Wageningen, The Netherlands (pp. 25–30). Pudoc, Wageningen.

Wilkins, P W., Humphreys, M.O. 2003. Progress in breeding perennial forage grasses for temperate agriculture. J. Agr. Sci. 140:129–150.

Woodfield, D.R, Caradus, J.R.. 1994. Genetic improvement in white clover representing six decades of palnt breeding. Crop Sci. 34:1205–1213.

Woodfield, D.R., Easton, H.S. 2004. Advances in pasture plant breeding for animal productivity and health. NZ. Veter. J. 52(6):300–310.

Woodfield, D.R. 1999.. Genetic improvements in New Zealand Forage cultivars. Proceedings of the New Zealand Grassland Association 61:3–7.

Chapter 37
Review of the Protocols Used for Assessment of DUS and VCU in Europe – Perspectives

Trevor J. Gilliland and Vincent Gensollen

Abstract The multipurpose function of turf and forage grasses and legumes makes determining Distinctness, Uniformity and Stability (DUS) extremely complex and makes providing definitive assessments of Value for Cultivation and Use (VCU) a multifaceted process. DUS testing of grasses and legumes is internationally harmonised and the principles, logistics and effectiveness of this system is reviewed. The outline procedures and characteristics involved in the VCU testing of both turf and forage varieties is explained and evidence of their effectiveness is evaluated against how well they meet the end-users requirements. The future challenges and prospects for all three systems are discussed, with particular reference to three issues varieties; the DUS challenge of providing protection for ever rising numbers of existing; the proposed pan-EU harmonised turf VCU testing scheme; the acute need for forage VCU to address serious issues concerning more specific evidence of ruminant production benefits from 'improved' varieties.

Keywords Variety evaluation · Turf · Forage · VCU · DUS

Introduction

Up until the 1940's, agricultural production across Europe operated within a framework of nationally imposed tariffs and protectionist measures to free trade (Perren, 1995). In contrast, plant breeding had minimal legislative protection to ensure fair return from new varieties. While there were a patchwork of national voluntary levy and protection schemes, these were largely unsatisfactory as they often only protected the variety name and provided little effective guard against others releasing the same variety under a different name (Laclaviere, 1965). The

T.J. Gilliland (✉)
Agri-Food Biosciences Institute (AFBI), Plant Testing Station, Belfast BT6 9SH, UK
e-mail: trevor.gilliland@afbini.gov.uk

C. Huyghe (ed.), *Sustainable Use of Genetic Diversity in Forage and Turf Breeding*,
DOI 10.1007/978-90-481-8706-5_37,

driving force for change in agriculture was that security of food supply became a political priority across Europe, given the experiences of the preceding world wars. From the late 1940's multilateral trade negotiations began dismantling national trade barriers and tariffs (Laird and Yeats, 1990) and in parallel came international guidelines on the protection of varieties. These were established by inter-governmental agreement under the auspices of the Union for the Protection of Varieties (UPOV, www.upov.int, May 2009). This led to national plant protection legislation being implemented in most European countries (eg. UK Plant Varieties and Seeds Act 1964). These national regulations prohibited ownership of a variety until it was officially proven to be novel and also implemented the requirement for improved value for cultivation and use before rights to commercialise the new entity would be granted. This legislation was implemented primarily on agricultural species involved in the food chain, including forage grasses and legumes, but in some countries was also extended to non-food species, such as for turf grasses in France. Testing of fitness for use gave independent quantification of the value of superior new varieties and excluded those that were inferior from the market. This created consumer confidence and ensured the best varieties gained greatest market share and prices (Culleton and Cullen, 1992, Gilliland et al., 2007). The net effect was to drive competing breeders to achieve even higher performances from their new releases, which benefited growers and end users.

An additional benefit for the breeders was that the regulated national systems facilitated international seed trade and ensured the flow of funds back to the breeders. Of specific importance within Europe was that on the 27 May 1952 the European Defence Community (EDC) treaty was signed by Belgium, France, Germany, Italy, Luxembourg and Netherlands. This created the progenitor of the European Union and the current 27 EU countries now maintain individual national lists of varieties, compiled using the DUS and VCU testing principles. Once listed in any member state a new variety is automatically listed on the EU Common Catalogue of varieties of agricultural plant species, giving rights to commercialise across the EU. The accumulative effect of these regulatory controls have transformed plant breeding into an industry with a more secure income stream than when operating in an unregulated market. The questions that now arise are whether they continue to fulfil their purpose and whether they are equipped to meet the challenges of future breeding innovation.

In grasses and legumes the market value of the seed is much less than for the major arable crops. This is partly because annual resowing is not obligatory for these perennial grasses and legumes, but also because herbage is not an end-product, unlike grain. This has been particularly true for the forage variety market, due to the difficulty in demonstrating increased ruminant product profit to offset reseeding costs. Therefore, herbage seed prices have not kept pace with world market prices nor stayed instep with the major arable crops. The turf seed market is more diverse with better margins than for forage in the high quality sector, but falling to a price cutting commodity market at the lower end of the scale. So the challenge for the testing authorities is to protect existing elite varieties and adequately describe their value for use. Thereby permitting breeders to gain fair remuneration for their achievements and encouraging them to develop further genetic improvements within a competitive seed market, to the benefit of end-users across Europe.

Discussion

The regulatory systems across Europe all comply with the basic model of novelty and value testing to determine whether national listing is appropriate and this then leads to automatic listing on the EU common catalogue. In some countries this can be followed by varying extents of advisory level testing to determine a more detailed or regionalised recommendation.

Despite this integration on the administrative level, the DUS plus Turf and Forage VCU systems each function independently at the technical level and so need to be considered separately.

Distinctness, Uniformity and Stability – Testing Procedures

The primary objective of this testing system is to provide protection for the intellectual property rights (IPR) residing in existing novel varieties. It is entirely detached from any assessment of end-use value. To protect the rights of the existing varieties the DUS test requires that any newly registered variety is *D*istinct from all other varieties in 'common knowledge', *U*niform in the expression of its distinguishing characteristics (consistent with its reproductive mode) and *S*table, such that it can be reproduced true to type in multiple cycles of seed production. 'Common Knowledge' is intended to protect all existing varieties regardless of the scale or location of commercialisation. Within the EU, the Community Plant Varieties Office (CPVO, Angers, France) limits this, where appropriate, to the EU common catalogue of varieties for entrustment of DUS test centres across the EU.

The testing scheme is controlled by legislation in all EU member states and is conducted in compliance with the international guidelines compiled by UPOV. It is, therefore, a highly harmonised testing system. UPOV specifies the characters that should be measured and also the methodology. These specifications define whether characters are visually assessed or measured (some by either method) on single plants or groups of plants (Table 37.1).

While the UPOV guidelines establish approved characters and agreed procedures and standards of test it is not completely prescriptive and countries will use those characters that are likely to achieve variety discriminations under their climatic conditions. Table 37.2 lists the UPOV approved character use in Germany, Netherlands and the United Kingdom for perennial ryegrass. In addition, Germany

Table 37.1 Example numbers of UPOV approved DUS characters (2009)

	Visual	Measured	Both	Total
Grass Species				
Cocksfoot (*Dactylis glomerata*)	3	6	1	10
Perennial Ryegrass (*Lolium perenne*)	7	12	2	21
Red fescue (*Festuca rubra*)	9	8	0	17
Smooth Stalked Meadow Grass (*Poa pratensis*)	7	6	1	14
Legume Species				
Alfalfa/Lucerne (*Medicago sativa*)	10	11	1	22
Red Clover (*Trifolium pratense*)	10	8	0	18
White Clover (*Trifolium repens*)	6	13	1	20

Table 37.2 Example of UPOV character assessment in perennial ryegrass in three EU states

UPOV No.	Character description	DE	NL	UK
1	Ploidy	√	√	√
2	Plant: growth habit in autumn of year of sowing	√	√	√
3	Plant: tendency to form inflorescences in year of sowing	√	√	–
5	Leaf: colour in autumn of year of sowing	√	√	–
6	Plant: growth habit in spring	√	–	√
7	Plant: natural height in spring	√	√	–
8	Time of inflorescence emergence in 2nd year	√	√	√
9	Plant: natural height at inflorescence emergence	√	√	√
10	Flag leaf: length at inflorescence emergence	√	√	√
11	Flag leaf: width	√	√	√
12	Stem: Length of longest stem 30 days after MDEE	√	√	√
13	Inflorescence: length	√	√	√
14	Inflorescence: number of spikelets	–	√	√

assesses a further 12 plant characteristics and the Netherlands assesses an extra five. The UK has 15 extra and also produces computer-derived characteristics from these for use as special tests for indistinguishable pairs. In all cases, however, the UPOV guidelines for DUS standards are adhered to.

In the allogamous grasses and legumes assessments are normally made on 60 plants sown in 3 or 6 replications and on rows of plants. Consistency in discriminating difference has to be demonstrated by reoccurrence in two separate environments from different representative samples of seed, which normally means testing in two separate trial years. Therefore, a typical testing programme will last a minimum of 3 years from application, though routinely a third cycle of testing is normally performed where distinctness from all existing varieties has not been established after the first two test years (Table 37.3). Thereafter, special tests can be performed to examine for differences in a phenotypic characteristic that is not routinely measured. These specific examinations are conducted to test the claim of a breeder that they will demonstrate the novelty of the candidate from a registered variety, despite being indistinguishable using the normal test characters. The same standards of precision and repeatability are applied to assessing these extra characters as is performed in the standardised testing programme.

Table 37.3 Example DUS testing scheme

Year 1	Test Year (2)	Test Year (3)	Test Year (4)	Year 5
Sow Trial I	Test Trial I			
↑	Sow Trial II	Test Trial II	↑	↑
		Sow Trial III	Test Trial III	

Application Award/Refuse PBR

Distinctness, Uniformity and Stability – Achievements and Challenges

Evidence of whether this testing system is operating as intended can be derived from the level of past/fail decisions that are achieved. Essential to this consideration is the fact that 100% distinctness is not the primary objective of the DUS system as it is in essence an IPR protection scheme. In doing so it should not, however, be a restrictive barrier to the release of new novel varieties. The example pass/fail ratios from the UK (Table 37.4) show that between 74 and 87% of candidates are passed as DUS, with some failing due to lack of uniformity and around 12% failing to express a distinguishing morphological difference from an existing protected variety.

It cannot be definitively determined whether the variation in annual discrimination levels reflects differing novelty of candidates between years or the impact of annual growing conditions, but as the fail decisions are taken on a 3-year data set, the impact of individual annual growing conditions is reduced. So variety uniqueness rather than conditions are probably the dominant factor. Nonetheless, growing conditions can have very significant effects and so not all DUS centres may achieve similarly high and consistent discrimination levels. Nonetheless, this evidence suggests that while protection is being afforded to the existing varieties, this high number of passes is evidence that novel creations can still get registered despite large reference collections of protected varieties.

The primary challenge for DUS testing is to maintain the necessary protection for all existing varieties while coping with the increasing workloads and costs caused by the ever expanding reference collection. Table 37.5 shows the increasing number of candidate and control varieties of ryegrass and clover in DUS trials from 1987 to 2008, at the UK test centre. Underlining this is the fact that the current EU common catalogue comprises 965 ryegrass, 333 red fescue, 195 tall fescue, 380 alfalfa, 202 red clover and 134 white clover varieties.

Table 37.4 Comparison of pass/fail ratios (%) for candidate varieties in trial years 2000–2008 for *Lolium spp.* and *Trifolium repens* (UK)

Sowing year	2000	2001	2002	2003	2004	2005	2006	2007	2008
Pass	86%	74%	87%	82%	81%	74%	83%	74%	75%
Fail distinctness	11%	15%	3%	12%	7%	13%	10%	16%	19%
Fail uniformity	3%	11%	8%	5%	8%	14%	6%	9%	5%
Fail both	0	0	2%	2%	3%	0	1%	1%	1%
Number of accessions	63	82	63	60	59	72	94	80	104

Table 37.5 Total number of control and candidate varieties in trial years 1987–2008 for *Lolium spp.* and *Trifolium repens* (UK)

Sowing Year	1987	1990	1993	1996	1999	2002	2005	2008
Number of accessions	626	829	953	1101	1192	1297	1337	1354

Unlike many arable, vegetable and ornamental species there are very few qualitative characteristics in grasses and legumes that can be used to definitively separate the reference collections into distinct groups (except ploidy). This would allow reduced testing of the reference collection to only those varieties from the group that had the identical qualitative character expression as the candidate variety. This is partly because of the virtual absence of fixed state morphological features in grasses and legumes and also because their allogamous nature means the varieties are populations of genetically differing plants that express a range within each characteristic. One novel approach has been to create virtual groups based on statistical techniques (Camlin et al., 2001). This cyclic planting scheme capitalises on the 3 year testing cycle by dividing the reference collection into three groups and rotationally sowing only two each year (Table 37.6).

Table 37.6 Cyclic planting scheme for managing DUS reference collections of grasses and legumes

Control Variety Groups				Candidate Test Years		
Year	2004	2005	2006	2007	2008	2009
Group A	Past	Past		Test	Test	**X**
Group B		Past	Past	**X**	Test	Test
Group C	Past		Past	Test	**X**	Test

Past = previous trials on controls, Test = controls + candidates, **X** = unsown group in that year

The missing comparisons each year are computed by the measured relationship between the control varieties in the groups during the preceding 3 years. This substantially reduces the effective size of the reference collection, but as is evident from Table 37.5, this 33% reduction only returns workloads to that around 1992–1993. As reference collections continue to rise this may prove to only be an interim solution.

Distinctness, Uniformity and Stability – Future Developments

The DUS system has a very specific objective and the evidence presented above indicates that it is achieving the necessary levels of protection without being restrictive of novel breeding. Given, therefore, that it currently meets legislation and customer requirements, future developments will most likely focus on new technologies to improve the workload and cost implications of the current system. Molecular genetics is a powerful tool for both generating genetic novelty and as a taxonomic tool. The GM debate within the EU is currently focused on environmental and food safety issues, but even if these were resolved, they are not expected to present the DUS system with unique challenges. This is because UPOV intends to treat GM and conventional varieties equally, by still requiring the phenotypic novelty of standard DUS tests to maintain current degrees of genetic distance between registered varieties. The sensitive issues of patents and right to release into the environment and human food chains will remain a national responsibility.

The highly discriminating taxonomic potential of molecular techniques is unquestioned and there are many publications reporting differences between grass varieties including in the current issue (Jensen et al., 2009). Such molecular markers are normally not causal or indicative of a phenotypic state and the molecular identity of individual plants differ within grass and legume varieties due to their allogamous nature. This unfortunately creates two impediments to adopting these highly successful discriminating methodologies for DUS testing. Firstly the primary function of DUS testing is to protect the existing varieties and with molecular markers it is possible to selectively multiply a sub sample of plants from a registered variety to change the overall molecular identity without changing the phenotypic identity (or end-use value) of the new plagiarised entity. Secondly, is the risk of eroding genetic distances between varieties and the undesirable implications of multiple examples of varieties that appear indistinguishable to the end-user. It may be possible to address the latter problem by imposing minimum distances to molecular discriminations, but the problem of plagiaristic exploitation has yet to be resolved. The key role for molecular markers in the near future will be for measuring the genetic distance between varieties and assessing the probability that a new variety has been derived out of an existing one. Such 'Essentially Derived' varieties cannot be marketed without agreement with the registered owner of the original source variety and the International Seed Federation (ISF www.worldseed.org, May 2009) has previously approved an AFLP methodology for ryegrass (Roldán-Ruiz et al., 2000) and is currently considering a newer system based on SSR markers.

The continuing problem of excess workloads and costs is, therefore, expected to be addressed in the near future by further statistical innovations. Currently, the German, Dutch and UK testing authorities are investigating a novel 'Sun – Satellite concept'. In this, one centre is envisaged as being the Sun for a species and will sow the reference collection plus its candidates. The Satellite centres will only sow their candidates plus a reduced number of linking controls, to allow a sharing of the reference collection identities. The proposal is still in the early stages of investigation, but if successful, has considerable potential for resolving the current workload and cost challenges.

Turf Value for Cultivation and Use – Testing Procedures

Unlike the DUS testing scheme, assessing the Value for Cultivation and Use (VCU) of turf grasses is not subject to any internationally agreed guidelines and within the EU is not conducted under legislative control, except in France. Therefore, there are no harmonised testing protocols and the work is conducted to meet commercial market requirements rather than government legislative controls.

Despite the absence of any coordinated guidelines, there is much similarity in the various testing procedures conducted in EU countries. This is to a considerable extent due to the clearly defined and highly specific attributes that a successful turf grass must possess to meet end-user needs. Although the uses to which turf grasses can be employed are in themselves diverse, it is widely accepted that certain common characteristics determine the VCU. Therefore, amenity grass testing programmes typically involve three replicate randomised plot (2 m^2) trials, with and without wear treatments and frequently mowed to around 25–30 mm. Virtually all assessments are made by visual scores on a 9 or 10 points scale, which can be done very rapidly by an expert trials scientist. Assessments typically include

- Establishment capability
  Scored as ground cover after 2 and 6 months
- Sward structure
  Turf ground cover after wear treatment; Turf ground cover without wear treatment;
  Persistency at the end of the trial period
- General aesthetic quality
  Leaf colour (summer/winter); Fineness of leaf; Cleanness of cut; Presence of weeds; Reactions to climate (Frost, drought, waterlogging etc); Shoot growth (less being superior)
- Disease tolerance:
  Rusts (*Uromyces, Puccinia* spp); Drechslera diseases (*Helminthosporium* spp); Red Thread (*Corticium fuciforme*); Fusarium Patch (*Microdochium nivale, F. nivale, F. roseum*); Dollar Spot (*Sclerotinia homoeocarpa*); Anthracnose (*Colletotrichum graminicola*); Leaf Spot (*Drechslera* spp, *Bipolaris* spp., *Curvularia* spp); Myxomycose.

The involvement of wear treatment depends on the intended use (eg sport or ornament) and responses will differ depending on species, soil type and climatic conditions. Either entire plots, but more frequently parts of each plot are subjected to the wear treatment. This is imposed by using specialist machines with drums that rotate at different speeds or artificially simulate the wear effects that occur on a sports ground. These can assess resistance to sport use damage especially in winter or the response to compaction in summer.

The requirements for establishment ability, sward structure and aesthetic qualities are similar across the EU, with the major factors changing the VCU of a variety being its reaction to the disease pressures and climatic/edaphic factors specific to the test ecozone. Turf grasses still have to satisfy the IPR requirements by undergoing DUS testing as described earlier, but in most EU countries, if passed, they are placed on their national list for 'Non Forage' use and then progress automatically to the EU Common Catalogue. In France, the official system first requires a VCU assessment from which a variety can gain a listing on the French national list for

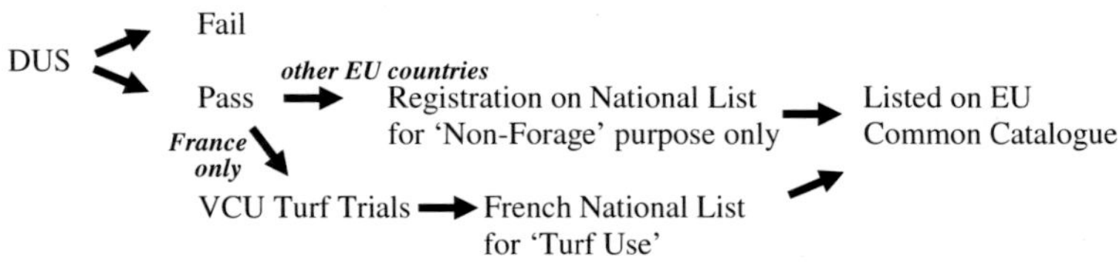

'Turf Use' and this permits their incorporation in turf mixes produced and marketed in France.

The trials are typically run over two test years, though with variations to account for specific uses and the number of test sites involved. Therefore, the official French testing system involves multiple sites with each trial involving a sowing year followed by two test years and National Listing by the 4th year after the application. As the Sports Turf Research Institute, Bingley, England, has only a single site the trials are resown in the 3rd year. So after an initial listing in year 4 this is confirmed by the further trial results by year 6 of testing. Winter sports trials are also sown in September, with assessments made in the following March and July to give an initial listing in year 3, with a confirming second sowing completed by year 5. Despite such timescale differences between systems and testing organisations across the EU, the basic methodologies and recordings remain similar, with the same turf characteristics required but with regional variation in varieties driven by disease pressure, climatic and edaphic factors.

Turf Value for Cultivation and Use – Achievements and Challenges

The best evidence of whether turf testing systems are operating as intended is available from the French official testing statistics. Figure 37.1a shows the progressive

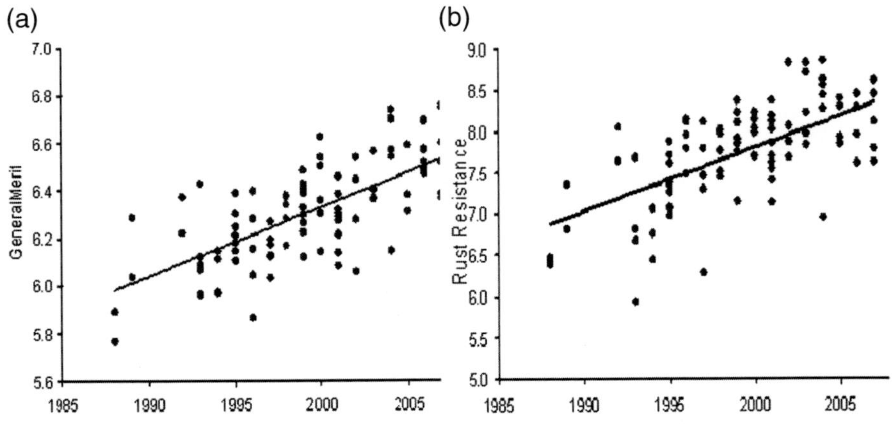

Fig. 37.1 Improvement of turf varieties in French statutory VCU trials, (**a**) increasing general lawn quality, (**b**) increasing rust resistance

rise in general lawn quality, presented as an overall index since the late 1980's, with a similar rise in rust resistance (Fig. 37.1b). Among the biggest achievements have been the increased density and fineness of leaf in ryegrass over the past 10–15 years, plus the increased use of tall fescue for its drought tolerance, which has become possible due to new varieties with much finer leaves in the last decade.

In addition to these performance increases, is the fact that the French turf seed sales volume has been increasing annually and is now at 20,000 t (second only to Germany – 23,000 t and much larger than the next biggest EU market, UK at 11,000 t). Furthermore the mean turf seed prices in France are higher than the EU mean. The French seed market is quality driven due to high consumer expectations with a dynamic supply chain comprising numerous SME (Small and Medium Enterprises), a vigorous breeding sector and a continuing rise in candidate varieties to the official French VCU system.

Turf Value for Cultivation and Use – Future Developments

It is clearly evident from the preceding section that turf grass testing is highly focused on consumer requirements. It is achieving progressive improvement in the key attributes that are essential to the performance of the turf in practical use and it is unlikely that substantial changes to testing procedures will be required in the near future. While it is possible to devise specific tests for small niches uses, fee incomes from such minor sectors could not support independent testing programmes and as the current testing procedures are delivering progressive changes in characters likely to contribute to improved niche use, such expansions are unlikely.

A change that is envisaged is an initiative by the European Seed Association (www.euroseeds.org, May 2009). This new network may merge with the GEVES network. The data from these networks could be used by other independent testing schemes across the EU and provide turf VCU recommendations on an ecozone basis across EU member states. This is expected to achieve an even greater degree of harmonised testing procedures as that achieved in DUS testing by the UPOV guidelines.

Forage Value for Cultivation and Use – Testing Procedure

Assessing Value for Cultivation and Use (VCU) of forage grasses and legumes is subject to national statutory testing within all EU member states, but there are no internationally agreed guidelines in the EU. Therefore, similar to turf VCU, there are no harmonised testing protocols between, in this case, official testing authorities. In addition, there are various complexities of secondary testing schemes run in some countries after national listing is completed. These seek to provide more detailed, precise and/or regionally specific recommendations, to identify which varieties have the superior VCU potential. Assessment of forage varieties is also more laborious and costly than assessing turf varieties as visual scores alone are insufficient and analysis of the dry matter output and nutritive quality of the herbage produced requires physical measurement.

In April 2009, GEVES conducted a survey of the perennial ryegrass VCU testing procedures in eight EU member states. A summary of the findings (Table 37.7) appears to show a degree of commonality between the countries. All countries assess dry matter yield, most assess an aspect of sward persistence/density and an aspect of the seasonal distribution of yield and winter hardiness.

Rust (*Puccinia* spp.) is monitored by all but one country with other diseases included depending on importance in each country. Surprisingly, however, testing for nutritional value is very limited despite the end product of grass and legume production being animal fodder.

This summary, however, obscures considerable differences in exactly what measurements are made and how they are presented to the end-user. For example, total yield assessment can involve either or both simulated grazing and conservation managements in separate trials or a combination of both managements within the one trial, or even actual animal grazing. Similarly, assessments of nutritive value, when assessed, involve digestibility by either DMD or DOMD methods and can be presented as the absolute values, seasonal averages or even combined with yield to give an estimate of digestible yields. Furthermore, assessment of pass/fail standards and presentation of percentage data can be against a declared standard yield, as the percentage of a control group of varieties, of a threshold level within the existing range of a national list, of the best variety on a list, or even of the worst variety on a list. This diversity of assessment approaches leads to a plethora of data sets, all scientifically sound measures of forage value, but creating different standards for success in different jurisdictions.

Typically testing involves three test years after sowing for persistent species such as cocksfoot, fescues, perennial ryegrass and timothy, and one or two years for short term grasses such as Italian or hybrid ryegrasses. Table 37.8 shows the current testing scheme in England with the UK wide statutory national list programme being followed by advisory regional testing.

Forage Value for Cultivation and Use – Achievements and Challenges

To determine whether VCU testing of forage grasses and clovers is successful, reviews typically consider what increases in total DM output have been achieved in previous years. Change in variety use is also often used to highlight how previous elite performers have all been surpassed by newer superior material (eg in perennial ryegrass, Aber S23 bred *c.*74 years ago, Vigor bred *c.*46 years ago, Talbot bred *c.*36 years ago, Condesa bred *c.*26 years ago). There is a history of such analyses since at least the late 1980's (Aldrich, 1987), to recent (Wilkins and Humphreys, 2003) and present (Chapter 41) reviews. While these often excellent analyses have shown annual yield improvements of up to 0.5%, they cannot assess whether the end-user

Table 37.7 Key characteristics assessed in VCU testing of perennial ryegrass in eight EU countries

	UK	Germany	France	Netherlands	Italy	Spain	Denmark	Hungary
Candidates	30–35	30	25	25	few	few	few	few
Duration	3 yr/trial	3 yr/trial	3 yr/trial	3 yr/trial	3 yr/trial	3 yr/trial	3 yr/trial	3 yr/trial
No Trials	6	10 (+9obs)	8	4	3	6–8	4 (+1)	4
Yield	Total DM	Total DM	Total DM	Total DM	Total DM	Total DM	Total DM	Total DM
Growth Rhythm	Seasonal	1st cut yield stability	Seasonal	1st cut yield	2nd cut seasonal	Annual output	Seasonal	Seasonal
Disease	Rust Leaf Spot Mildew	Rust	Rust	Rust	Rust Leaf Spot		Rust Leaf Spot	Rust Leaf Spot Mildew
Behaviour	Win. hard Gr. cover	Win. hard Persist. Gr. cover Lodging Re-heading	Win. hard Re-heading	Win. hard Persist. Durability	Win. hard Persist.		Persist.	Win. hard Persist. Lodging
Quality	Digestibility		Digestibility (in progress)				Digestibility	Total crude protein

Gr. cover = Ground cover; Win. hard = Winter hardiness; Persist. = Persistence; Seasonal = Seasonal yields

Table 37.8 Example National list and regional advisory testing timescales (England & Wales)

2009	2010	2011	2012	2013	2014	2015	2016	2017	2018
Sow I	test	test	test						
	Sow II	test	test	test					
				NL Listing					
				Additional regional testing scheme in England	RL'P' Listing				
					Sow III	Test	Test	Test	
								RL Full Listing	

NL = national listing decision; RL 'P' = regional recommended list provisional listing

benefits have been equally improved as official testing systems do not directly measure this. Controversially, a recent survey of around 800 ruminant sector farmers' at the Royal Agricultural Show in England revealed that total yield production was less important to their enterprise than persistence, quality and seasonality (Billings, British Seed Houses 2008, pers. commun.). Furthermore, on-farm assessments in the 1940's showed that grass contributed around 65% of total animal feed supply, but by 1980–1990's this was down to 60% of dairy cow and only 79% of beef cattle diets. Recent advisory estimates in Northern Ireland, a primary grass growing region of the EU, indicates that even top grassland farms operate at around 73% feed energy from forage. This contrasts with current grass dependence in New Zealand where dairy bulls & steers are finished for market at 18 months on a 100% grass/clover diet (Department of Agriculture and Rural Development for N. Ireland internal report 2009). In addition, forage seed markets have been falling across the EU for many years (eg. Gilliland et al., 2007). A major factor in this is that grassland farmers are not convinced that 100 kg of herbage from the newest varieties is a substantially better feed for their animals than that from a variety bred 30+ years ago. Unlike arable farmers, therefore, new varieties are not valued as a key financial benefit to the farm enterprise and so reseeding is often avoided for as long as possible. The overall effect is that although the best varieties are still the most successful in the market place, reseeding is too infrequent, seed prices are heavily depressed with poor profit margins leading to an ailing plant breeding sector. This is evident from the large number of breeding programmes that have been terminated, merged or taken over in the past two decades to achieve sustainable economy of scales. Forage VCU testing unquestionably fulfils its statutory obligation by promoting better yielding varieties, but unlike the DUS or turf VCU schemes, it does not appear to adequately service either its breeder or farming customers needs.

Forage Value for Cultivation and Use – Future Developments

From the preceding section, it is clear that the primary challenge for VCU testing is to better assess the nutrient value of forage grass and legume varieties within the testing system, to promote varieties with not just better yielding and persistent characteristics but also better nutritional benefits to the ruminant animal. Currently a major tool for doing this is the development of Near Infra-Red Spectrometry (NIRS). This technology offers, for the first time, rapid and cost effective means of assessing a multitude of different nutritional parameters such as crude protein, organic matter digestibility, neutral detergent fibre, acid detergent fibre, water soluble carbohydrates, buffering capacity, fatty acids (as conjugated lactic acids) and their precursors (α-linoleic acid and α-linolenic acid). This brings both the advantage of specific nutritional assessment of varieties, but also the considerable risk that without some harmonisation of approach, the potential multitude and complexity of differing test strategies will be counterproductive. Unlike the turf VCU systems, there has yet to be established a consensus of how grass value should be assessed for the ruminant animal. Each variant testing procedure across Europe is another assessment standard that breeders have to account for in their selection programmes. At best these can represent another compromise in the selection criteria and at worst actually impede progress if too many differing test standards evolve. It is also confusing for farmers if the measures for elite feeding potential in varieties are not standardised. As many remain to be convinced of the full value of new varieties, this is an important consideration.

Conclusions

The allogamous reproductive mode of most grasses and legumes make their varieties complex genetic entities that present unique challenges for variety testers. The multipurpose function of turf and forage grasses and legumes adds further to the difficulty of quantifying elite performing varieties. Based on the presented evidence, DUS testing appears to function as intended by successfully protecting existing varieties without impeding breeding for novelty. Similarly, the available evidence shows that VCU testing of turf grasses has successfully identified elite varieties that continue to fulfil the diverse needs of an exacting but profitable market sector. Forage VCU testing has also successfully promoted breeding improvement in yield and persistence, but has not demonstrated sufficient benefits in ruminant nutrition from new varieties for farmers to justify the frequent reseeding investment capable of maintaining a vibrant plant breeding service. As always, new scientific innovations will present new challenges to the testing systems, but within the foreseeable future the DUS testing programmes will need to continue seeking means of managing ever expanding reference collections. The envisaged harmonised turf VCU testing in the EU can be expected to stimulate greater consumer demand for elite performing new varieties throughout the EU, as achieved in France by their official testing scheme. Forage VCU must urgently quantify 'forage value' to farmers in terms of direct ruminant benefit. The official testing authorities will be challenged to achieve this within capped or contracting public funding and with the need to satisfy EU policies on environmental issues such as reduced carbon footprint of ruminant farming. For these reasons VCU testing authorities and breeders in Europe need to act collectively and redress this knowledge void, by developing and agreeing harmonised procedures as currently achieved for the DUS and turf VCU systems.

References

Aldrich, D.T.A. 1987. Developments and procedures in the assessment of grass varieties at NIAB 1950–1987. J. Nat. Inst. Agric. Bot. 17:313–327.

Camlin, M.S., Watson, S., Waters, B.G., Weatherup, S.T.C. 2001. The potential for management of reference collections in herbage variety registration trials using a cyclic planting system for reference varieties. Plant Variet. Seed. 14:1–14.

Culleton, N., Cullen, T. 1992. Trends in herbage seed use in Ireland. Plant Variet. Seed. 5:63–69.

Gilliland, T.J., Johnston, J., Connolly, C.P. 2007. A review of forage grass and clover seed use in Northern Ireland. GrassForage Sci.62:1–8.

Jensen, L.B., Roulund, N., Nielsen, K.K., Deneken, G., Lubberstedt, T. 2010. Application of molecular markers for variety protection in ryegrass (*Lolium perenne*) This volume.

Laclaviere, B. 1965. The convention of Paris of December 2, 1961, for the protection of new varieties of plants and the international union for the protection of new varieties of plants. Indus. Prop. 10:224–28.

Laird, S., Yeats, A. 1990. Trends in nontariff barriers of developed countries, 1966–1986. Rev. World Econ. 126:299–325.

Perren, R. 1995. Agriculture in Depression, 1870–1940. Economic History Society, Cambridge University Press.

Roldán-Ruiz, I., Calsyn, E., Gilliland, T.J., Coll, R., Van Eijk, M.J.T., De Loose, M. 2000. Estimating genetic conformity between related ryegrass (*Lolium*) varieties, II. AFLP characterisation. Mol. Breeding. 6:593–602.

Wilkins, P.W., Humphreys, M. O. 2003. Progress in breeding perennial forage grasses for temperate agriculture J. Agric. Sci., 140:129–150.

Chapter 38
Modelling Adaptive Responses Across Agricultural Environments as a Prerequisite for Identifying Adaptive Traits and Plant Ideotypes

Paolo Annicchiarico and Luciano Pecetti

Abstract Modelling cultivar yield responses across agricultural environments by additive main effects and multiplicative interaction (AMMI) or factorial regression is useful for understanding genotype × environment interaction patterns, simplifying and improving the targeting of cultivars, and helping breeding programs in defining selection strategies, adaptive traits and plant ideotypes by itself and by acting as a benchmark for physiological studies. Examples are provided from results of joint field testing across several Mediterranean countries and further physiological work performed within the EU-funded project PERMED (Improvement of native perennial forage plants for sustainability of Mediterranean farming systems). Cultivar adaptive responses across Mediterranean sites depended mainly on the site drought-stress level for cocksfoot and sulla and on the level of drought, mowing frequency and soil salinity for lucerne. Specific adaptation of cocksfoot material was largely affected by its summer dormancy. Lucerne cultivars which showed contrasting adaptation pattern were evaluated for shoot and root traits in metal containers (55 cm × 12 cm × 75 cm deep) under different drought-stress levels. These artificial environments were able to reproduce the cultivar adaptive responses across agricultural environments (unlike another experiment using 30 cm-deep pots), revealing a conservative water strategy in a drought-tolerant Italian landrace.

Keywords Drought tolerance · Root development · *Dactylis glomerata* · *Medicago sativa*

Introduction

Forage breeding for European or north-African Mediterranean environments is constrained by the small market of certified variety seed and the consequent limited funding for private programmes, the budget constraints imposed on public

P. Annicchiarico (✉)
CRA – Centro di Ricerca per le Produzioni Foraggere e Lattiero-Casearie, viale Piacenza 29, Lodi 26900, Italy
e-mail: paolo.annicchiarico@entecra.it

C. Huyghe (ed.), *Sustainable Use of Genetic Diversity in Forage and Turf Breeding*,
DOI 10.1007/978-90-481-8706-5_38,

programmes, and the heterogeneity of its possible target environments. On the other hand, forage cropping is increasingly recognized as fundamental for sustainable agriculture and as a hindrance to the expansion of desertification due to aridity and overgrazing which is expected in the region as a consequence of climate change (Fischer et al., 2002). However, forage breeding programmes of Mediterranean countries that share a similar diversity of target environments could co-operate to also share resources and become, on the whole, more cost-efficient. A major aim of the EU-funded project PERMED (Improvement of native perennial forage plants for sustainability of Mediterranean farming systems) is supporting these programmes by defining selection strategies and adapted plant types for lucerne, sulla, tall fescue and cocksfoot and verifying, also on this ground, the opportunities for international co-operation. The emphasis on perennials is justified by their ability relative to annuals to extend the feeding season by earlier and faster growth at the onset of autumn rains and more prolonged growth in spring.

Additive main effects and multiplicative interaction (AMMI) and factorial regression outstand from other modelling techniques for their usefulness in understanding genotype × environment (GE) interaction patterns, simplifying and improving the predictive ability of cultivar recommendation, and helping breeding programs in defining adaptation strategies, selection environments, adaptive traits and plant ideotypes (Annicchiarico, 2002; Gauch et al., 2008). Modelling cultivar adaptive responses is also a prerequisite for reliable physiological studies, as it defines material possessing specific traits (e.g., drought tolerance) and allows to verify the ability of artificial environments used in physiological experiments to reproduce the cultivar adaptive responses as observed across agricultural environments.

This study aims to briefly summarize some preliminary indications issued by PERMED with respect to: (i) GE interaction patterns, similarity of target environments, adaptive traits and plant ideotypes for some forage species as depicted by modelling cultivar yield responses across Mediterranean sites; (ii) plant survival and root development of lucerne cultivars grown in large containers, and (iii) the ability of artificial environments used for physiological studies to reproduce the cultivar adaptive responses as observed across agricultural environments.

Modelled Cultivar Yield Responses Across Mediterranean Sites

Cocksfoot

Seven cocksfoot cultivars were grown at seven rainfed sites of Algeria, France, Italy, Morocco, Portugal and Tunisia. Figure 38.1 reports AMMI-modelled cultivar forage yields recorded over the second cropping year (i.e., between the second and the third summer of growth) as a function of the first GE interaction principal component (PC 1). The graph exploits one useful feature of AMMI analysis (Gauch et al., 2008), namely the clear display of one-dimensional GE interaction patterns as nominal yields obtained by subtracting the site main effect (irrelevant for entry ranking)

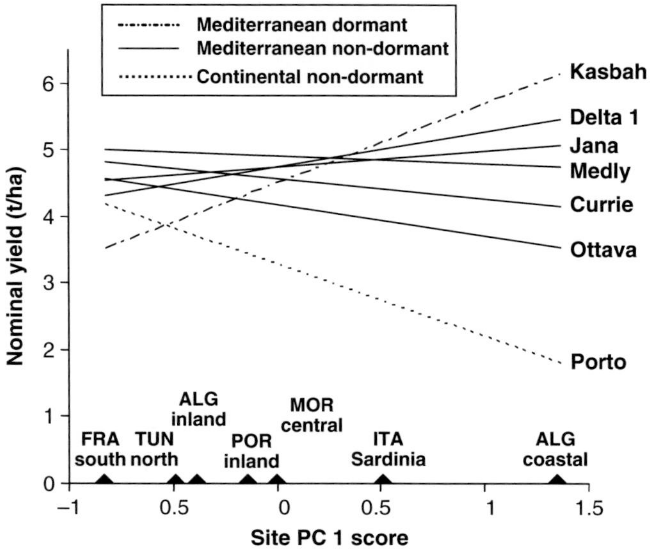

Fig. 38.1 Nominal second-year dry-matter yields of seven cocksfoot varieties as a function of the first cultivar × location interaction principal component (PC 1) of seven Mediterranean sites (Source: Annicchiarico, 2008; AMMI analysis of data by A. Abdelguerfi, S. Ben Taamallah, H. Bouzerzour, R. Kallida, F. Lelièvre, C. Porqueddu, N. Simoes, F. Volaire)

from the modelled data. Site ordination on PC 1 was a negative indicator ($r = -0.68$, $P < 0.10$) of spring and summer water (April through August) over the first two cropping years. Mediterranean dormant material (which is dormant in summer also in the presence of available water according to Volaire and Norton, 2006) proved specifically adapted to severely drought-stressed sites, while Mediterranean non-dormant entries were specifically adapted to moderate-stress environments. These results set a limit to the selection of Mediterranean non-dormant, widely-adapted material able to persist in very dry summers and to respond to water possibly available in late spring and summer (Piano et al., 2004), supporting the selection of dormant varieties in severely drought-stressed areas.

Factorial regression modelling of 3-year forage yields confirmed drought-stress level of the site as the main determinant of cultivar adaptation patterns (unpublished data).

Sulla

Results for sulla have already been reported by Annicchiarico et al. (2008). Cultivar yield responses showed cross-over GE interaction which depended only on the site water available over the crop cycle in a factorial regression model, suggesting to breed either for rainfed cropping in semi-arid or nearly-so environments, or for definitely sub-humid or irrigated environments. Material selected in rainfed environments of southern Italy was more drought-tolerant than germplasm selected or evolved in more favourable areas of Italy or Tunisia.

Lucerne

Fourteen lucerne varieties or landraces originated in northern Africa, Europe, Australia or USA were evaluated in ten rainfed or irrigated environments of Algeria, Tunisia, Morocco and Italy. GE interaction patterns were not as ecologically simple as those in the other species, as they depended on cultivar variation for response to mowing frequency, spring-summer water available and soil electrical conductivity in a factorial regression model (unpublished data). Specific adaptation to frequent mowing featured a landraces from northern Africa, i.e. 'Demnat', which evolved under favourable, oasis conditions. Distinct drought-tolerance was showed by the Sardinian landrace 'Mamuntanas', which evolved under rainfed cropping in a semi-arid area.

Plant Survival and Root Development of Lucerne Cultivars in Artificial Environments

Some cultivars with contrasting adaptation were studied in metal containers (55 cm × 12 cm × 75 cm deep) under two levels of imposed drought stress. Each container included 21 plants at 2.5 cm spacing on a single row. Stress levels were applied soon after the second harvest, after irrigating at field capacity. Mild stress implied 20 days, and severe stress 45 days, of suspended watering from the appearance of wilting signs on the most susceptible entry. Crown and root biomass and content of nitrogen and carbohydrate reserves were observed on three sets of containers opened prior to stress application, at the end of mild stress and at the end of severe stress, respectively. Plant survival at the end of mild and of severe stress was assessed on two additional sets of containers, after rewatering the plants. Soil moisture at the end of each stressing period was assessed by the gravimetric method. Stress treatments were on main plots, and cultivars on subplots, of a split-plot experiment with four replications.

Results on plant survival, root biomass prior to stress application and water used between mild and severe stress are presented in Table 38.1 for four cultivars, i.e.: (i) 'Mamuntanas' (drought-tolerant); (ii) 'Demnat' (adapted to frequent mowing); (iii) 'Sardi 10' (featuring wide adaptation in the multi-site evaluation network); (iv) 'Prosementi' (adapted to moderately favourable conditions). They suggested that drought tolerance of 'Mamuntanas' depends on a drought-avoidance, water-conservation strategy based on limited root development, which implies more water available in late, severe stress periods. Limited root biomass already emerged as a feature of material specifically adapted to water-limited Italian environments

Table 38.1 Plant survival under contrasting drought stress levels, root biomass, and water used between end of mild and of severe stress, for lucerne cultivars evaluated in metal containers

Cultivar	Origin	Plant survival after mild stress (%)	Plant survival after severe stress (%)	Root dry weight (2-68 cm depth; g/plot)	Soil water available (%)
Mamuntanas	Sardinia	93.4	67.0	8.9	1.5
SARDI 10	Australia	92.1	54.6	14.9	1.2
Prosementi	North Italy	93.4	46.3	12.0	0.7
Demnat 203	Morocco	90.5	45.2	16.4	0.5
LSD ($P = 0.05$)		NS	14.0	3.4	0.6

(Annicchiarico, 2007), probably because the large investment of photosynthates required for growth, function and maintenance of an extensive root does not repay the additional water uptake in rainfed Mediterranean environments (where spring and summer rainfalls hardly reach deep soil layers). Large root development featured 'Demnat', in agreement with previous indications on the importance of this trait in favorable cropping environments (Annicchiarico, 2007) and/or under frequent mowing. In these conditions, greater storage of nitrogen reserves associated with the larger root contributes to better persistence and ability to withstand the severe intra-specific competition for light and nutrients (Avice et al., 1997).

The cultivar ranking for plant survival after severe stress in the artificial environment was consistent with that for 3-year forage yield observed in a rainfed environment of inland Algeria which was the most drought-stressed among those in the network of evaluation trials (unpublished data). On the contrary, a preliminary study with two stress levels performed within the project in 30-cm deep pots failed to reproduce the cultivar adaptive responses in the stressful site (unpublished data), suggesting caution in choosing artificial environments to mimic field conditions in physiological studies.

Acknowledgements This work was carried out within the EU-funded project 'Improvement of native perennial forage plants for sustainability of Mediterranean farming systems' (PERMED).

References

Annicchiarico, P. 2002. Genotype × environment interactions: Challenges and opportunities for plant breeding and cultivar recommendations. FAO plant production and protection paper no. 174. FAO, Rome, Italy.

Annicchiarico, P. 2007. Lucerne shoot and root traits associated with adaptation to favourable or drought-stress environments and to contrasting soil types. Field Crops Res. 102:51–59.

Annicchiarico, P. 2008. Ecological approaches in breeding of forage crops (pp. 88–94). In: Kobiljsky, B. (ed.), Conventional and molecular breeding of field and vegetable crops. Institute of Field and Vegetable Crops, Novi Sad, Serbia.

Annicchiarico, P., Abdelguerfi, A., Ben Younes, M., Bouzerzour, H., Carroni, A.M., Pecetti, L., Tibaoui, G. 2008. Adaptation of sulla cultivars to contrasting Mediterranean environments. Aust. J. Agric. Res. 59:702–706.

Avice, J.C., Ourry, A., Lemaire, G., Volenec, J.J., Boucaud, J. 1997. Root protein and vegetative storage protein are key organic nutrients for alfalfa shoot regrowth. Crop Sci. 37:1187–1193.

Fischer, G., Shah, M., van Velthuizen, H. 2002. Climate change and agricultural vulnerability. IIASA, Vienna, Austria.

Gauch, H.G., Piepho, H.-P., Annicchiarico, P. 2008. Statistical analysis of yield trials by AMMI and GGE: further considerations. Crop Sci. 48:866–889.

Piano, E., Pecetti, L., Annicchiarico, P., Carroni, A.M., Fornasier, F., Romani, M. 2004. Combining drought tolerance and responsiveness to moisture availability in cocksfoot (*Dactylis glomerata* L.) germplasm grown in Mediterranean environments. Aust. J. Agric. Res. 55:1197–1204.

Volaire, F., Norton, M. 2006. Summer dormancy in perennial temperate grasses. Ann. Bot. 98:927–933.

Chapter 39
Evaluation of Drought Tolerance Variability in Mediterranean Alfalfa Cultivars in the Field Under Moroccan Conditions

Abdelaziz Bouizgaren, Rajae Kallida, and Chaouki Al Faiz

Abstract Sixteen alfalfa cultivars originating from the Mediterranean basin were tested in an experimental station in Morocco, located in the semi arid bioclimatic area. This research was conducted in PERMED project during 2006–2008 and aimed to evaluate the adaptation of cultivars to drought stress. The trial was conducted under two irrigation treatments. The first treatment was normally irrigated by providing an amount of water corresponding to the potential evapotranspiration of the crop and in the second treatment with water deficit which was applied by stopping the irrigation during 9 weeks in summer. Results showed that water stress during summer significantly reduced aerial biomass of all cultivars. This reduction varied between 25 and 41% according to cultivars in comparison with a normal irrigated treatment. The difference between cultivars for biomass production was significant only in stress treatment ($P< 0.001$ with 15 df). Some cultivars showed high forage yield potential, mainly Ameristand, ABT 805, Sardi10, Siriver, Gabes-2355, Rich2, and Erfoud1 even in presence of stress. In the end of August (before restart of irrigation for treatment with suspended summer irrigation), the mean rate of leaf senescence of cultivars was 84% with no significant difference between cultivars. The row cover estimated in the end of summer of the third year varied between 12 and 40% according to the cultivars. Cultivars Gabes-2355, Ameristand and ABT805 showed a smaller number of dead plants under water stress. Therefore, those cultivars could be used by local farmers in this region.

Keywords Alfalfa · Drought · Irrigation · Biomass · Perenniality

Introduction

Alfalfa (*Medicago sativa L.*) is the largest cultivated forage crop in Morocco with 100,000 ha (22% of cultivated area of forage crops) and its cultivation is conditioned by irrigation. As a reason of its distribution in oasis ecosystems and in irrigated

A. Bouizgaren (✉)
Institut National de la Recherche Agronomique, INRA, Marrakech BP: 533.
Av Mohamed VI., Marrakesh, Morracco
e-mail: bouizgaren_abdelaziz@yahoo.fr

C. Huyghe (ed.), *Sustainable Use of Genetic Diversity in Forage and Turf Breeding*,
DOI 10.1007/978-90-481-8706-5_39, © Springer Science+Business Media B.V. 2010

areas, alfalfa is considered as a crop which is able to survive drought but it also uses a large amount of water.

The availability of water to agriculture in Mediterranean region in general and in Morocco in particular is an issue of growing concern because of increasing population pressure and great incidence of drought determined by climatic changes (Fisher et al., 2002). Therefore, the efficiency of water use for forage production must be maximized. The evaluation of alfalfa Mediterranean cultivars under Moroccan conditions is important in order to select materials which persist under drought conditions especially in summer when the availability of water is very critical because of the high demand for irrigation imposed by the competition with more profitable irrigated crops.

The present study aimed to evaluate the adaptation of cultivars of diversified geographic origin to drought summer.

Material and Methods

The trial was carried out in an experimental Station located in north-east of Marrakech and characterized by a loamy clay soils. This experiment has been conducted for 3 years and included three French varieties, one variety and two ecotypes from Italy, four varieties from Australia or USA, and six ecotypes from Morocco, Algeria and Tunisia. The trial was conducted under two irrigation treatments. The first treatment was irrigated by providing an amount of water corresponding to the potential evapotranspiration of the crop, and in the second treatment water deficit was applied by stopping the irrigation during 9 weeks in summers.

Forage dry matter yield was assessed on a plot area of 5 m^2 and was estimated for both normally irrigated and stressed trials. Cultivar persistence was provided by percent of alfalfa row cover. This character was assessed by scoring linear density rate of living plants visually (scale 0–100) at the end of the second and third summers. Leaf summer senescence was visually scored by rating the green-leafiness of plants (rate from 0 to 100%) in each plot: before drought onset, 2 weeks after stress, and at end of August (before restart of irrigation for treatment with suspended summer irrigation).

An ANOVA was performed for each trait, separating the top- and bottom-ranking cultivar means according to the LSD at $P< 0.05$. A t-test was performed to compare the two treatments.

Climatic Conditions

The three evaluation seasons were rather stressful with a total of rainfall of 388 mm, 144 mm and 227 mm respectively for the first, second and third season (Table 39.1). The spring and summer seasons in our experimental site are normally characterized by a severe drought stress associated with high temperatures.

Table 39.1 Mean climatic variables in the evaluation seasons

Climatic variable	2005–06	2006–07	2007–08
Rainfall September–March (mm)	323	91	182
Rainfall April–June (mm)	37	42	28
Rainfall July–August (mm)	28	11	17
Mean of daily min. temp. January–February (°C)	9.5	5.7	6.7
Absolute min. temp. in winter (°)	–0.5	–2	1
No. of frost days	1	3	0
Mean of daily max. temp. May–June (°C)	30.4	29.1	29.8
Mean of daily max. temp. July–August (°C)	30.6	36.7	36.6

Results and Discussion

The comparison between the mean yields of dry matter obtained in both treatments (normal and stressed) showed significant differences between cultivars ($t = 28.4$ with $P<0.001$). This reduction varied between 25 and 41% according to cultivars in comparison with a normal irrigated treatment (Prosementi, Africaine, Silicano, and Demnat203 are most sensible). Those cultivars were selected under rainfed environments, such as Italian cultivars (Pecetti et al., 2008), or under irrigation like Demnat203. For this reason, their productions in stressed conditions are moderate. This difference was due to the number of cuts and also to the number of plants that died under the drought stress when little irrigation was provided generating a lack of persistency. The highest mean cultivar yield over the 3 years in normal irrigated treatment is obtained by the ecotype Demnat203. This ecotype originates from the mountain ecosystem of the Moroccan Atlas Mountain where the conditions of development of alfalfa are not limited and was selected for high forage yield and absence of fall dormancy (Birouk, 1997). Tamantit and Coussouls were the lowest yielding cultivars. However, in the presence of summer stress, the top mean yielding set of cultivars included Ameristand, Sardi10 and 2355-Gabes. The lowest forage yield was obtained by Africaine and Coussouls. These varieties are not well adapted in Moroccan conditions. Indeed, Coussouls was selected for grazing tolerance, and Africaine was sensible to diseases and insects especially (*Hypera variabilis*) and has a low perenniality. For these reasons it was abandoned by farmers (Bouizgaren et al., 2004).

The persistency of cultivars at the end of the third summer reflected largely the response of cultivars for forage yield. ABT 805 and Ameristand are the most persistent cultivars (Table 39.2). Among the north-African cultivars, the high rate of persistence was showed by Gabès-2355 and did not differ from other north-African populations ($P<0.001$).

Before stress, visual leaf senescence of all cultivars was around 23% with significant differences between cultivars ($P = 0.002$ with 15 df). In the end of August the leaf senescence rate was very high with a mean of 84% with no significant difference

Table 39.2 Three-year forage dry matter yield, persistence, plant high and leaf senescence of the evaluated alfalfa cultivars

Cultivar	Origin	Forage yield (t/ha)		Persistence %	Plant high (cm)		Leaf senescence (%)		
		Normally irrigated treatment	Summer stressed treatment		Normally irrigated treatment	Summer stressed treatment	before stress	2 week after stress	End of August
Tamantit	Algeria	$41,4^{a}$	$30,9^{abc}$	17^{ab}	$72,5^{bc}$	60^{abcd}	30	30	84
Siriver	Australia	$51,3^{ab}$	$34,4^{abc}$	$15,6^{ab}$	$72,5^{bc}$	$62,5^{cd}$	22	29	87
Sardi10	Autralia	$51,4^{ab}$	$35,9^{bc}$	$13,2^{a}$	$71,2^{bc}$	$67,5^{d}$	25	32	85
coussouls	France	$42,4^{a}$	$27,8^{ab}$	$13,7^{a}$	$66,2^{ab}$	51^{8a}	22	40	84
Magali	France	$46,7^{ab}$	$30,2^{abc}$	17^{ab}	$65,6^{ab}$	$56,2^{abc}$	22	36	86
Melissa	France	$47,9^{ab}$	$31,3^{abc}$	$14,5^{ab}$	$81,8^{d}$	$66,2^{d}$	21	41	86
Prosementi	Italy	$45,8^{ab}$	$27,5^{ab}$	$13,1^{a}$	$59,3^{a}$	$53,1^{ab}$	17	34	82
Mamuntanas	Italy	$47,4^{ab}$	$33,3^{abc}$	17^{ab}	$66,8^{ab}$	$60a^{bcd}$	22	35	86
Silicano	Italy	$48,5^{ab}$	$29,7^{abc}$	$14,3^{ab}$	$66,8^{ab}$	$54,3^{abc}$	25	37	82
Rich2	Morocco	$46,5^{ab}$	$31,5^{abc}$	$14,9^{ab}$	$78,7^{cd}$	$63,1^{cd}$	21	39	85
Erfoud1	Morocco	$49,2^{ab}$	$32,9^{ahc}$	$16,2^{ab}$	$76,8^{cd}$	$61,2^{bcd}$	25	42	86
Demnat203	Morocco	$53,1^{b}$	$34,1^{abc}$	$16,8^{ab}$	$84,3^{d}$	$66,8^{d}$	25	36	85
Africaine	Tunisia	$45,0^{ab}$	$26,4^{a}$	12^{a}	$78,1^{cd}$	$65,6^{d}$	25	41	87
Gabes-2355	Tunisia	$48,1^{ab}$	$33,1^{abc}$	$22,1^{ab}$	70^{bc}	$56,2^{abc}$	17	44	82
ABT 805	USA	$48,4^{ab}$	$31,1^{abc}$	$30,7^{b}$	65^{ab}	$55,6^{abc}$	22	34	81
Ameristand	USA	$49,6^{ab}$	$37,1^{c}$	$30,4^{b}$	$71,8^{bc}$	$63,7^{cd}$	31	37	84

Values with the same letter do not differ ($P \leq 0.05$)

between cultivars ($P = 0.372$ with 15 df). The reason of these differences could be different root systems allowing different water absorption under low water potential.

Acknowledgements This study was funded by EU within the project "Improvement of native perennial forage plants for sustainability of Mediterranean farming systems" (PERMED). We thank farmers who have made available their materials for this study. We are grateful to PERMED coordination for their help.

References

Birouk, A. 1997. Les ressources génétiques de luzerne (*Medicago sativa L.*) au Maroc: Potentiel agronomique et diversité génétique. Dans: Ressources phytogénétiques et développement durable. Edité par A. Birouk et M. Rjdali. Actes éditions

Bouizgaren, A., Thami Alami, I., Abbad Andaloussi, F., Birouk, A., Julier, B., Al Faiz, C. 2004. Développement de la production fourragère dans les régions subsahariennes et les oasis. Cas des vallées de Ziz et Draâ. In Séminaire International sur: «Développement des cultures fourragères: une nécessité pour améliorer les productions fourragères et atténuer la dégradation des ressources naturelles. Rabat, 8 et 9 mars.

Fisher, G., Shah, M., and van Velthuisen, H. 2002. Climate change and agricultural vulnerability. IIASA, Vienna.

Pecetti, L., Carroni, A.M., Annicchiarico, P., Manunza P., Longu A., and Congiu G. 2008. Adaptation, summer survival and autumn dormancy of Lucerne cultivars in south European Mediterranean region (Sardinia). In Option Mediterranéennes "Sustainable mediterranean grasslands and their multi-functions". N A-79. CIHEAN editions.

Chapter 40
Field Resistance of *Festuca Rubra* Varieties to Red Thread (*Laetisaria Fuciformis*)

Bohumír Cagaš, Magdalena Ševčíková, and Radek Macháč

Abstract To assess the extent of progress in breeding for resistance to some typical turfgrass diseases, a field collection of 6 varieties of sheep fescue (*Festuca ovina*) and 35 varieties of red fescue (*Festuca rubra* s.l.) was used to assess the incidence of red thread (*Laetisaria fuciformis*) in the years 2004–2007. Whereas in the first crop years (2004–2005) disease incidence was very low and did not provide enough information about intervarietal differences, in the years 2006–2007 the levels of disease in the collection of red fescue varieties were very high. As sheep fescue varieties generally had high resistance red thread incidence was not recorded. In the collection of *F. trichophylla* varieties, Barpearl, Baroyal and Viktorka had high resistance (area damaged did not exceed 5%). They differed significantly from other varieties. Among the varieties of *F. rubra* ssp. *rubra* there was none with higher resistance (damage was 95%). In the collection of *F. nigrescens* the varieties Barborka and Citera had high resistance (damage was less than 10%) and differed significantly from other varieties of this species. There was no connection observed between disease incidence and experimental treatments (wear, cutting frequency). A hypothesis was confirmed that with low levels of fertilizer application there is considerable expression of this disease, especially in older stands and the selection of donors of resistance is optimal.

Keywords Red thread · *Laetisaria fuciformis* · Field resistance · *Festuca rubra* · *Festuca ovina*

Introduction

Red thread, a disease of turfgrasses, caused by the fungus *Laetisaria fuciformis* (McAlp.) Burdsall 1979 (anamorph *Isaria fuciformis* Berk.) is one of the most common and widespread diseases of turfgrasses on the European continent. The agent

B. Cagaš (✉)
OSEVA PRO Ltd., Grassland Research Station Rožnov-Zubří, Hamerská 698, 756 64 Zubří, Czech Republic
e-mail: cagas@oseva.cz

C. Huyghe (ed.), *Sustainable Use of Genetic Diversity in Forage and Turf Breeding*,
DOI 10.1007/978-90-481-8706-5_40,

produces circular patches in the lawns in the growing season and the symptoms of the disease are withering of leaf tips and later entire plants. Red thread most commonly affects turfgrass varieties of perennial ryegrass, Kentucky bluegrass and red fescue. With respect to irregular occurrence of this disease, which affects mostly older stands, there is not enough information about varietal resistance or factors that might influence it. The aim of the study was to find out if there were differences in resistance to red thread in sheep fescue and red fescue varieties and what factors influenced its expression. This communication should also indicate whether field trials carried out for several years are still a suitable method and tool of screening for plant material with higher resistance.

Material and Methods

Red thread occurrence was studied in a field collection of genetic resources of turfgrass varieties which was established with the aim of assessing lawn qualities (including health) at Zubří (345 m above sea level, annual rainfall 865 mm, annual average temperature of 7.5°C) in April 2003. The collection contained 6 varieties of sheep fescue (*Festuca ovina* L.) and 35 varieties of red fescue (*Festuca rubra* s.l.). The plot size of 4.5 m^2 per variety was divided into three sub-plots with different type of management: (a) intensive cutting, (b) intensive cutting and wear treatment in the second crop year, (c) extensive cutting. In all three treatments the cut biomass was removed. In the first 3 years (2004–2006) intensive treatments were fertilized four times a year with a rate of 30 gm^{-2} fertilizer with a long lasting form of N, the first two rates of Compo Rasen Floranid (20% N, 5% P_2O_5, 8% K_2O, 2% MgO) and the following two rates of Compo Floranid NK (14% N, 19% K_2O, 3 % MgO). The plots were cut first with a rotary lawn mower and later with a reel mower to a height of 30 mm. The number of cuttings in these years ranged from 29 to 40 (a period from 15 April to 15 November) and the number of wearing treatment was 21 (a period from 2 May to 22 October 2005). In the year 2007 after termination of a 3-year assessment of the lawn collection, the plots were used for demonstration purposes and the intensity of their treatment decreased (only 13 cuttings per growing season and the first three rates of N of the above mentioned system of fertilizer application). The extensive treatment was cut only twice per growing season and grown without any fertilizer application.

The occurrence and the intensity of infection of red thread in grasses were assessed annually starting from the first crop year (2004) till the year 2007 using the following scale:

Table 40.1 Relationship between score of resistance and damages on the plots

Degree of resistance	Damaged area of the plot (%)
1	more than 95
3	41–95
5	16–40
7	6–15
9	less than 5

The statistical processing of the differences in intensity of infection was based on the program STATISTICA 8.0. Meteorological data were provided by the station located in the field experimental area.

Results and Discussion

In the population of 6 varieties of sheep fescue the occurrence of red thread was not recorded in the course of the study. However, there were significant differences in resistance among 12 varieties of slender creeping red fescue (*Festuca trichophylla*) (Table 40.2). The highest levels of resistance were found in the varieties Barpearl, Baroyal and Viktorka. The difference between these three varieties and the rest of the population was significant.

Similar differences in field resistance to red thread were found in a population of Chewings fescue (*Festuca rubra* subsp. *commutata*, syn. *F. nigrescens*). Varieties Citera and Barborka were significantly more resistant, compared to some other varieties (Table 40.3).

Even among strong creeping red fescue (*Festuca rubra* subsp. *rubra*) varieties there were differences in resistance which, however, became quite invisible at the end of the observation period (2007) (Table 40.4).

The effects of other factors on the disease symptoms (besides the genetic basis of the variety) – stand age, fertilizer application, intensity of management (number of cuttings and mechanical impact),or climatic effects (air temperature and precipitation) were assessed?

Stand age. In the year of establishment and in the first two crop years (2004, 2005) the occurrence of red thread in all varieties studied was sporadic. Sporadic areas of infection in the year 2005 were observed only in varieties Gentil and Diego (degree 7). A sharp increase in the intensity of occurrence was observed only in

Table 40.2 Field resistance of a population of slender creeping red fescue (*Festuca trichophylla*) to red thread

	Date of assessment/Degree of resistance			
Variety	31 July 2006	7 August 2006	24 August 2007	Mean
Barpearl	9	9	9	9.0a
Baroyal	9	7	9	8.3a
Viktorka	9	7	9	8.3a
Barlotte	9	5	5	6.3b
Samanta	5	5	5	5.0bc
Smirna	5	5	3	4.3 cd
Suzette	5	5	3	4.3 cd
Gentil	5	5	3	4.3 cd
Napoli	5	5	3	4.3 cd
Kami	5	5	3	4.3 cd
Estica	5	3	3	3.7 cd
Lifine	3	3	3	3.0d

Table 40.3 Field resistance of varieties of chewing fescue (*Festuca rubra* subsp. *commutata*, syn. *F. nigrescens*) to red thread

	Date of assessment/Degree of resistance			
Varieties	31 July 2006	7 August 2006	24 August 2007	Mean
Citera	7	7	9	7.7a
Barborka	9	7	7	7.7a
Juliska	9	7	5	7.0ab
Enjoy	9	7	3	6.3abc
PF 27/95	7	7	5	6.3abc
Bardiva	5	7	5	5.7abcd
Liroyal	9	3	3	5.0bcde
Darwin	5	5	5	5,0bcde
Makyta	5	3	5	4.3cde
Medina	5	3	3	3.7de
Lirubin	5	3	3	3.7de
Raymond	3	3	3	3.0e
Simone	3	3	3	3.0e
Bareagle	3	3	3	3.0e

the third and the fourth crop year. Although Berestetski and Kastirr (2001) did not show any relationship between the age of perennial ryegrass and the intensity of infection under artificial infection conditions, a severe infection was reported under field conditions just in the last year of study (in the oldest stands).

Fertilizer application. Fertilizer application was the lowest in the last year of the trial (2007). Only then there were striking differences observed in resistance and in the intensity of the disease among the varieties. Snijders and Winkelhorst (1994) regarded a lack of nitrogen in the soil as a factor stimulating the incidence of this disease in lawns. Tredway et al. (2001) also reported the 'protective' action of nitrogen (predominantly in the fast release form).

Table 40.4 Field resistance of varieties of strong creeping red fescue (*Festuca rubra* subsp. *rubra*) to red thread

	Date of assessment/Degree of resistance			
Variety	31 July 2006	7 August 2006	24 August 2007	Mean
Diego	7	7	3	6.3a
Aniset	7	7	3	5.7ab
Engina	7	7	3	5.7ab
Cindy	7	7	3	5.7ab
Laxton	7	7	3	5.7ab
Camilla	5	5	3	4.3ab
Barcorsa	3	3	3	3.0b
Barustic	3	3	3	3.0b
Felix	3	3	3	3.0b

Cutting and turf impact. In the first 2 years when the number of cuttings was the largest disease occurrence in the intensive treatment was negligible. Only in the third year and even more in the following year there was a marked expression of the disease. The number of cuttings or the wearing treatment is unlikely to be the primary factor influencing the occurrence of red thread. However, the motion of the lawnmower and the operator may have some effect on the spread of the disease. The compaction and wearing had no effect on the occurrence of turfgrass disease.

Climatic changes. The highest intensity of red thread infection was observed in the last decade of August 2007, which was characterized by drought. Extreme weather conditions, no matter how short, may initiate a number of lawn diseases (Licht, 2008).

The intensity of red thread in red fescue varieties was affected predominantly by varietal resistance (most probably genetically controlled). The intensity of the disease was markedly influenced by the age of stand. Marginal effects were also exerted by drought and a smaller supply of nitrogen in the soil. It was proved that long-term field trials may be useful in searching for valuable sources of resistance necessary for further breeding.

Acknowledgements The financial support of the Ministry of Agriculture of the Czech Republic (National Programme on Conservation and Utilization of Plant Genetic Resources and Agro-biodiversity) and of the Ministry of Education, Youth and Sport of the Czech Republic (Research Project no. 2B06034) is acknowledged.

References

Berestetski, A., Kastirr, U. 2001. Factors affecting red thread disease of turf grasses. New aspects of resistance research on cultivated plants: fungal diseases. Proceedings of the 8th Aschersleben Symposium, Aschersleben, Germany, 16–17 November 2000. Beiträge zur Züchtungsforschung Bundesanstalt für Züchtungsforschung an Kulturpflanzen. 2001, 7: 3–6.

Licht, B. 2008. Wetterextreme – Einfluss auf das Auftreten von Rasenkrankheiten. 49. Fachtagung des DLG-Ausschusses 'Gräser, Klee und Zwischenfrüchte'. Züchtungsperspektiven und Saatgutproduktion bei Gräsern, Klee und Zwischenfrüchten. Vorträge der Fachtagung vom 4. November in Bonn. pp. 69–71.

Snijders, C.N.A., Winkelhorst, G.D. 1994. Red thread disease (*L. fuciformis*) in *Lolium perenne* and *Festuca rubra* by artificial inoculation. In: Reheul, D.,Chesquiere, A. (eds.), Breeding for quality. Proceedings of the 19th Fodder Crops Section Meeting held in Brugge, Belgium, 5–8 October 1994, pp.73–76.

Tredway, L.P., Soika, M.D., Clarke, D.D. 2001. Red thread development in perennial ryegrass in response to nitrogen, phosphorus, and potassium fertilizer applications. Intern. Turfgrass Soc. Res. J. 9:715–722.

Chapter 41
Change in Agronomic Performance of *Lolium perenne* and *Lolium multiflorum* Varieties in the Past 40 Years Based on Data from Belgian VCU Trials

Barbara Chaves, Alex De Vliegher, Johan Van Waes, Lucien Carlier, and Bram Marynissen

Abstract By using data of Belgian trials for Value of Cultivation and Use (1963–2007) the change in agronomic performance of ryegrass varieties was quantified. Since the genetics of 'Vigor' and 'Lemtal' have remained identical to those in 1963, these were used as constant standards to measure improvements of new varieties. Dry matter yield (DMY) of 'Vigor' and 'Lemtal' varied annually but did not show a progressive change, indicating that cultural changes in VCU trials of ryegrasses were small. By expressing DMY, persistency and rust resistance of the candidate varieties relative to Vigor and 'Lemtal', the change in agronomic performance due to breeding was determined. DMY increased with 0.3% of 'Vigor' and 'Lemtal' annually. Persistency showed an annual increase around 0.5% of the standard varieties. Before 1990, rust resistance of ryegrass varieties varied around 100, and only after 1990, an annual increase of 3.6% relative to 'Vigor' and 'Lemtal' was found.

Keywords Dry matter yield · Persistency · Rust resistance · Ryegrasses

Introduction

Since 1963, official variety trials for Value of Cultivation and Use (VCU) with forage grasses have been organised in Belgium with a view to admission or prolongation of varieties on the Belgian National Catalogue. Based on the results of the Belgian VCU trials, the improvement in agronomic performance of candidate varieties to the Belgian testing scheme over the past 40 years can be evaluated. 'Vigor' a diploid, late *L. perenne* variety and 'Lemtal' a diploid *L. multiflorum* variety are two varieties which have always been included in the official VCU variety trials in Belgium as a collection or standard variety, respectively. During the past 40 years,

A. De Vliegher (✉)
Plant, Crop Husbandry and Environment, Institute for Agricultural and Fisheries Research (ILVO), Burg. Van Gansberghelaan, Merelbeke 109, 9820, Belgium
e-mail: alex.devliegher@ilvo.vlaanderen.be

C. Huyghe (ed.), *Sustainable Use of Genetic Diversity in Forage and Turf Breeding*,
DOI 10.1007/978-90-481-8706-5_41,

the management of grassland and the protocol of VCU trials have changed: different harvesting machinery, different levels of fertilization, different methods of observation, etc., but the genetics of 'Vigor' and 'Lemtal' are still the same. Hence, data from 'Vigor' and 'Lemtal' allow us to quantify the effect of changes in cultural practices in Belgian VCU trials. Furthermore, by expressing the performances of the candidate varieties relative to 'Vigor' and 'Lemtal', management changes are captured within the relative performances in function of 'Vigor' and 'Lemtal' and this makes it possible to attribute all gains such as crop yield, disease resistance and persistency to genetic breeding. The objective of this study was to quantify the change in agronomic performance of ryegrasses due to (1) changes in cultural practices, and (2) forage grass breeding.

Materials and Methods

Data from 144 *L. perenne* varieties and 69 *L. multiflorum* varieties were available. For data entry on the graphs, the year of inscription on the National Catalogue was used. In all years, VCU trials for ryegrasses were organized in randomized complete block designs with 4 replicates, plot size of at least 8 m^2, sowing density of 1400 viable seeds m^{-2}, and on at least five locations in the most important agricultural regions in Belgium. The schematic presentation of the field trials is shown in Fig. 41.1. In all years, the plots were only cut: before 1992, with a hand-pushed cutter bar and a cutting frequency between 3 and 4 per year, and after 1992, with a Haldrup plot harvester, a cutting frequency of 4 to 6 per year, and with an earlier

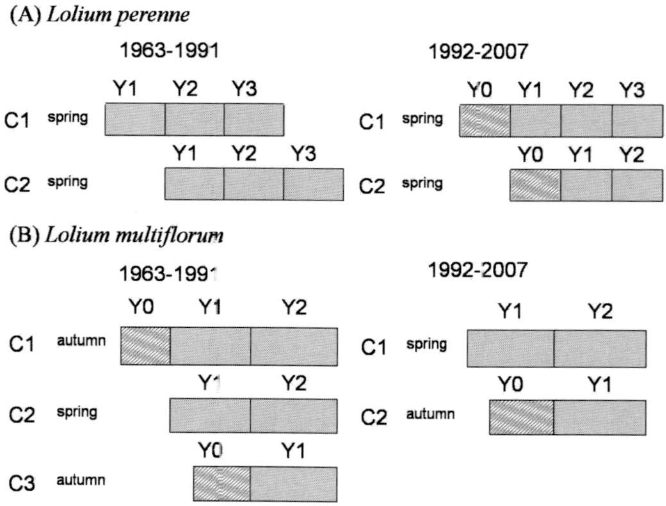

Fig. 41.1 Schematic presentation of the field trials for *L. perenne* (**a**) and *L. multiflorum* (**b**); sowing season: spring or autumn; Y0: harvest year with no observations; Y1, Y2, Y3: harvest year with observations; C1, C2, C3: growing cycle 1, 2 or 3

first cut. From 1963 until 1980, N fertilization was generally around 250 kg N ha^{-1} for *L. perenne* and around 300 kg N ha^{-1} for *L. multiflorum*. From 1981 until 2000, N fertilization was between 350 and 450 kg N ha^{-1}. Since 2000, N fertilization was decreased to 350 kg N ha^{-1}.

In a cycle, *dry matter yield (DMY)* of a variety was the mean of all DMY (in t.ha^{-1}) of all locations (per location: DMY was the sum of all cuts) and all years. For each variety, DMY was the mean DMY from all growing cycles. *Persistency* was visually rated by observing the percentage of the sown grass variety still present in the sown lines (i.e. lack of gaps) some days after the last cut in the final year (rating scale from 1 to 9, with 1: very low persistency, 9: very good persistency). Each year, crown rust resistance (*Puccinia coronata f. sp. lolii)* was visually rated in the field trials after natural infection (rating scale from 1 to 9, with 1: very high rust infection, 9: no rust infection). In a cycle, the rust resistance was the mean value of all years.

Results and Discussion

Cultural Practices of the VCU Trials

DMY of 'Vigor' increased between 1966 (on average 9 t.ha^{-1}) and 1981 (on average 12 t.ha^{-1}) (Fig. 41.2). After 1981, DMY of 'Vigor' varied around 12 t.ha^{-1}, and

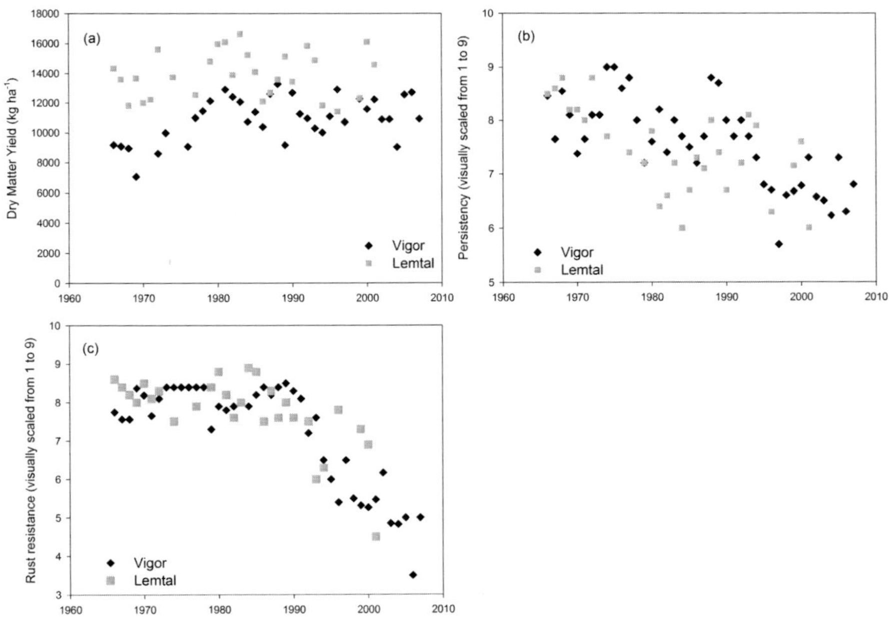

Fig. 41.2 Change in dry matter yield (**a**), persistency (**b**) and rust resistance (**c**) of Vigor (=*L. perenne*) and Lemtal (=*L. multiflorum*) from 1966 until 2007

DMY of 'Lemtal' varied around 14 t.ha^{-1}, but there was no progressive rise. The rather constant DMY of 'Vigor' and 'Lemtal' during the past 40 years was probably due to the small changes in cultural practices in VCU trials with ryegrasses (no need for weed and pest control, no increase in plant density). The only cultural practices that did change in the VCU trials of ryegrasses were (1) a higher N fertilization in the *L. perenne* trials after 1980, and (2) the changed trial setup in 1992 (DMY from year of sowing was no longer included; higher cutting frequency, earlier first cut). The score for persistency and rust resistance of 'Vigor' and 'Lemtal' decreased during the 40 years. This was due to (1) yearly independent scores, and (2) a new calculation of the mean annual value per variety in 1992 when only trials where actual differences in persistency and rust resistance between varieties, were included. This actually lowered the values, as there was no difference in trials where all varieties behaved very well.

Agronomic Performance of Candidate Varieties Relative to Vigor and Lemtal

During the past 40 years, the relative DMY of ryegrasses significantly increased by 0.31% per year (Fig. 41.3a). This rather small improvement in DMY was due

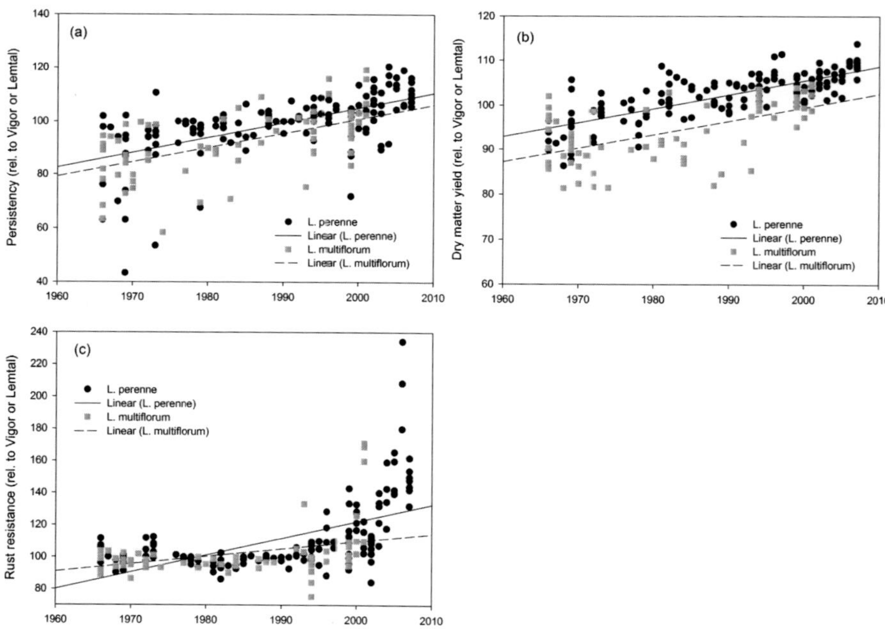

Fig. 41.3 Change in dry matter yield (**a**), persistency (**b**) and rust resistance (**c**) of candidate ryegrass varieties relative to Vigor (=100 for *L. perenne*) and Lemtal (=100 for *L. multiflorum*) from 1966 until 2007

to a low scope for altering the harvest index of grasses, costly and complicated breeding work, no reliable and economical system to hybridize inbreds, and the lower financial investment due to a low sale of grass seed (Wilkins and Humphreys, 2003). DMY of early *L. perenne* varieties did not increase. The annual improvement was 0.40% per year for late and 0.30% per year for intermediate varieties. For both ryegrasses, the improvement in DMY over the past 40 years was similar for diploid and tetraploid varieties.

During the past 40 years, persistency of ryegrasses significantly increased by 0.55% per year (Fig. 41.3b). Persistency of early *L. perenne* varieties did not increase. For intermediate and late varieties the annual improvement in relative persistency was on average 0.57% of 'Vigor' per year. The annual improvement in persistency was significantly lower for diploid *L. perenne* varieties (0.46% of 'Vigor') than for tetraploid *L. perenne* varieties (0.73% of 'Vigor'). Diploid *L. multiflorum* varieties showed a significant increase in persistency (+0.73% of 'Lemtal' per year), while no significant regression could be fitted for tetraploid *L. multiflorum* varieties. Good persistency is highly desirable because full cultivation and reseeding of permanent pasture is relatively expensive. Presently, differences in persistency between *L. perenne* varieties registered on the Belgian National Catalogue are relatively small but still important.

The main increase in rust resistance occurred after the 1990s, when it significantly increased with 3.5% per year for *L. perenne* candidate varieties and with 3.8% per year for the *L. multiflorum* varieties (Fig. 41.3c). This was due to (1) more attention for rust resistance in the 1990s since limitations to N fertilization had increased problems of rust infection and (2) a more efficient screening of rust resistance (e.g. controlled artificial infections; Reheul and Ghesquière, 1996). The annual improvement in rust resistance was significantly larger for intermediate (1.26% per year) and late (0.92% per year) *L. perenne* varieties than for early *L. perenne* varieties (0.43% per year). No significant difference could be found in the annual improvement in rust resistance between diploid and tetraploid varieties.

The agronomic performance of early *L. perenne* varieties improved less compared to late and intermediate varieties since the latter were more interesting for the farmers (used for both grazing and cutting), and hence they received more attention in breeding.

Conclusions

In Belgium, during the last 40 years, the agronomic performance of ryegrass varieties improved: an increase in DMY, persistency and rust resistance was observed. The improvement was mainly due to plant breeding since changes in management of VCU trials were rather small and did not have a univocal effect on the agronomic performance of the grasses.

Forage producers are accustomed to occasionally adopting new and improved cultivars. This study, however, showed that given the progressive year-on-year improvements in DMY, persistency and rust resistance, it is important for farmers to use the most recent varieties.

Acknowledgements The authors would like to thank M. Van Hyfte for collecting all the data, and all the members of the Variety Testing team during 1963 and 2007 for their outstanding work in the Variety Research in Belgium.

References

Reheul, D., Ghesquière, A. 1996. Breeding perennial ryegrass with better crown rust resistance. Plant Breed. 115:465–469.

Wilkins, P.W., Humphreys, M.O. 2003. Progress in breeding perennial forage grasses for temperate agriculture. J. Agric. Sci. 140:129–150.

Chapter 42
Impact of Four Decades of Breeding on Molecular Differentiation Between Forage and Turf Cultivars of *Lolium Perenne*

Marc Ghesquière, Philippe Barre, Gilles Boutet, Isabelle Cameleyre, Sandrine Flajoulot, Jean-Baptiste Pierre, Charles Poncet, Michel Romestant, Kirsten Vangsgaard, and Jean-Paul Sampoux

Abstract How much differentiated are forage and turf type cultivars within *L. perenne*? To estimate this, we used 10 SSR/STS markers for genotyping a collection of 7 natural populations, 50 forage and turf cultivars and 4 old cultivars of dual usage registered since 1965–2004. We showed that differentiation between usage types has steadily increased since the opening of a turf national list in France and that it has mostly involved 3 markers, among which 2 were mapped onto linkage group 1 in *L. perenne*. Relative to natural populations, assumed to sample genetic diversity in perennial ryegrass when breeding started, turf cultivars were found to be more distantly related than forage cultivars, especially those which were recently registered. However, genetic differentiation remained primarily between cultivars whatsoever type they were. Differentiation between cultivars has increased to be about twice higher on average than between natural populations, even of quite distant geographical origin. Loss of genetic variability after 40 years of breeding was found to be very low. All alleles present in natural populations were sampled again in the collection of cultivars we investigated. The results are briefly discussed in conclusion as respect to phenotypic differentiation and efficiency of breeding methods in the grasses.

Keywords Molecular markers · *F-statistics* · Selection · Genetic drift · Effective size · Metapopulation

Introduction

Life-history traits make genetic differentiation in grass species relatively low; using isozyme loci Charmet et al. (1993) estimated F_{ST} to be only 0.054 within a collection of 550 French ecotypes of *Lolium perenne*. However, breeding may deeply change

M. Ghesquière (✉)
UR4 (Unité de Recherche Pluridisciplinaire Prairies et Plantes Fourragères), INRA, Centre Poitou-Charentes, Lusignan, F-86600, France
e-mail: Marc.Ghesquiere@lusignan.inra.fr

C. Huyghe (ed.), *Sustainable Use of Genetic Diversity in Forage and Turf Breeding*, DOI 10.1007/978-90-481-8706-5_42,

genetic structure in crops following genetic drift and selection and considerably decrease genetic variability initially available in ecotypes. From dominant molecular polymorphism (*RAPDs*), Bolaric et al. (2005) found that the between-component of genetic distance variance slightly increased from 29% among ecotypes to 33% among cultivars. Herein, we aimed to evaluate the impact of 40 years of breeding perennial ryegrass with special emphasis on divergence between forage and turf cultivars and by using co-dominant markers (*SSRs*).

Material and Methods

The genotype sampling was identical to that used by Chapter 46 for assessing genetic progress among 54 diploid ryegrass cultivars registered on European national lists from 1965 to 2004, i.e. 32 turf and 18 forage cultivars, 4 undifferentiated old cultivars and 7 ecotypes originating from various climatic areas in Europe. 10 markers among 197 public *SSR* or *STS* markers were selected for maximum rate of amplification, high number of alleles, transferability from Li-Cor *IR2* to ABI Prism 3100 sequencer and covering each of the 7 linkage groups in *L. perenne* with at least one marker (Table 42.1). A total of 2300 individuals were genotyped using *GeneMapper* software and exploited for differentiation analysis. Assuming that populations and cultivars were close to equilibrium (i.e. $F_{IS} \sim 0$), *F-statistics*

Table 42.1 *F-statistics* between and within forage and turf cultivars of *L. perenne* at 10 markers

Marker	Linkage group	Allele number	Overall F_{ST} Turf vs Forage	F_{ST} between forage cv.	F_{ST} between turf cv.
DLF27[a]	LG1	7	0.0426	0.1696	0.1244
NFFA059[b]	LG5	10	0.0384	0.1143	0.1708
NDPK[c]	LG1	4	0.0153	0.1349	0.1178
NFFA015[b]	LG6	7	0.0075	0.0749	0.1372[f]
DLF20[a]	LG7	14	0.0064	0.1015	0.0938
NFFA023[b]	LG2	5	0.0028	0.0867	0.0733
Syn20738[d]	LG5	7	0.0019	0.0707	0.1156
NFFA113[b]	LG4	8	0.0017	0.0388	0.1102[f]
S7F7[e]	LG6	21	0.0010	0.1175	0.0942
UNI001[a]	LG3	24	0.0010	0.1059	0.1155
		Mean :	0.0118		
			Multilocus F_{ST} estimate by usage type :	0.1029	0.1143
			Confidence interval 0.95 :	0.0843–0.1208	0.1000–0.1297
			Overall multilocus F_{ST} estimate	0.1192	
			Confidence interval 0.95 :	0.1023–0.1385	

[a]Jensen et al. (2005); [b]Saha et al. (2004); [c]Lem and Lallemand (2003); [d]Barre et al. (2009); [e]courtesy of M. Fujimori, NILGS/NARO, Nishinasuno, Japan (unpublished); [f]Significantly higher between turf cv than between forage cv

were computed (*Genetix* software, release 4.05, Belkhir et al., 2004) by emphasizing only the estimation of the average genetic differentiation between usage type and between cultivars within each usage type (i.e F_{ST}). An ascending hierarchical classification with clustering according to Ward criterion was then performed on all cultivars and ecotypes using the procedure *proc cluster* (SAS/STAT, release 8.1 for SunOS, SAS Institute Inc., Cary, NC, USA) on frequency profile of 107 alleles at the 10 markers. Consistency of the clustering was estimated by the *a posteriori* probabilities of classification given by a canonical discriminant analysis (*proc discrim* procedure in SAS/STAT).

Results

F-statistics showed that differentiation between turf and forage as a whole accounted for only 0.0118 while between cultivars within either usage type, F_{ST} accounted in a non-significant way for 0.1029 in turf and for 0.1143 in forage (Table 42.1). Among markers, NNFA015 and NNFA113 were found to discriminate significantly better among turf cultivars than among forage cultivars.

When cumulated over year of registration, F_{ST} between forage and turf increased at a 0.0003 rate each year on average (Fig. 42.1). It is noteworthy that not all markers contributed in the same way to increase differentiation; apparently, drift or possibly selection primarily affected linkage group 1 in *L. perenne* in the course of breeding forage vs turf (Fig. 42.1). The classification consistently produced 6 groups among which, variance accounted for 37% of total variance (Fig. 42.2). Only 2 cultivars remained unclustered because allele frequency profile was either almost exactly intermediate between turf and forage (e.g. cv *Idole*) or it was of exceptionally high homozygosity (e.g. cv *Aberavon*).

Group I and III typically gathered forage cultivars with either old cultivars of dual usage (cv *Perma* and *Belida*, 1974) or with very first turf cultivars (cv *Bianca*,

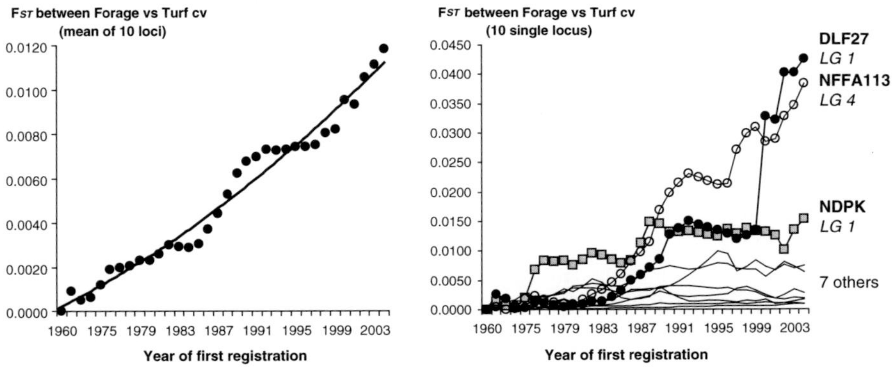

Fig. 42.1 Rate of evolution of F_{ST} between forage and turf cultivars over year of registration: on average (*left*), marker by marker (*right*)

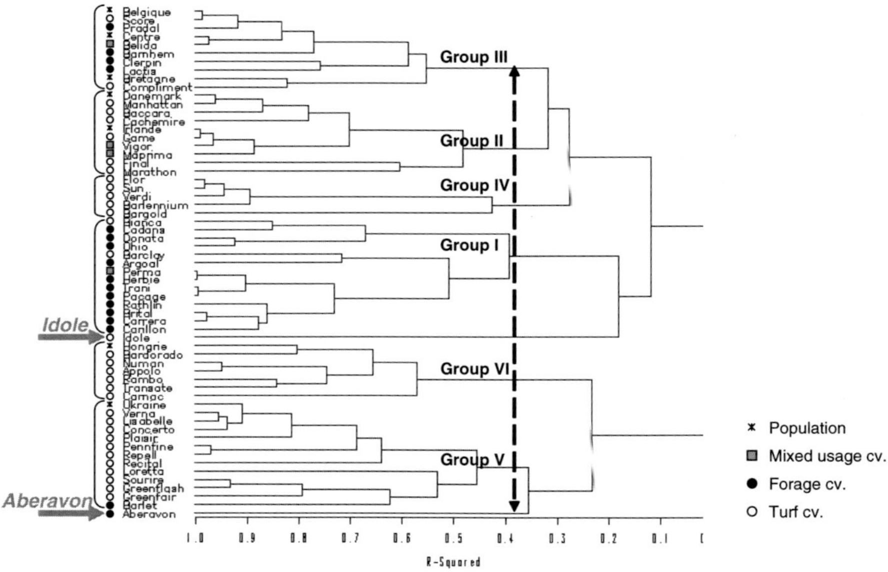

Fig. 42.2 Ascending hierarchical classification of 61 ecotypes, forage and turf cultivars of *L. perenne* using frequency profile of 107 alleles at 10 SSR/STS markers

1979; *Score*, 1980; *Barclay*, 1982). Close to Group III, Group II was found to be composed of only turf cultivars including the two other old cultivars of undifferentiated usage (cv *Vigor*, 1971 and *Maprima*, 1980). At last, Group IV, V and VI were the most homogenous with only turf cultivars (except cv *Barlet*, 1980, as a forage in Group V) as well as the most differentiated from the other groups.

The canonical discriminant analysis validated the clustering into 6 main groups with only 2 populations, Ireland and Belgium, which reciprocally permuted from groups II to III. When plotted onto the first two canonical axes (not shown), populations from Ukraine and Hungary clearly stand out from the rest of ecotypes while populations of Ireland, Belgium, French Brittany and central part of France, embedded in Group I, II and III, look as they have been a major source of plant material for starting breeding in *L perenne*.

Discussion

Differentiation from molecular markers suggests that breeding forage and turf started with an initial plant material of quite large effective size which genetic variability has been remarkably preserved likely in continuously incorporating genetic variability from natural populations and/or from previously registered cultivars of either usage type. However, divergence of turf *vs* forage populations of cultivars seems to be occurring when molecular differentiation is plotted over years, possibly in relationship with the recent registration of turf cultivars bred in US. The contribution of markers mapped on particular linkage groups to the overall differentiation remains to be confirmed as does the nature of the differentiation – genetic drift or selection, but this suggests that turf and forage cultivars in *L. perenne* could evolve in the future as separate gene pools practically not exchanging genetic variability anymore. High phenotype differentiation and response to selection found by Chapter 46 in the same collection of cultivars can be explained only by the large genetic base of plant material combined with an intense mass selection for highly heritable traits as those controlling leaf morphogenesis (Ghesquière et al., 1994; Hazard et al., 1996). If turf and forage have to evolve separately with consequently a pronounced decrease of genetic variability, one can wonder whether genetic progress could be maintained unless breeding develops more effective selection methods (*e.g.* progeny tests). Alternatively, this renews the interest of structuring more closely genetic variability within *L. perenne*, *e.g.* in building heterotic (at least complementary) gene pools with possible development of population hybrids (e.g. Chapter 75; Chapter 85).

Acknowledgements This work was financially supported by a grant (2005–2008) from the French Ministry of Agriculture and Fisheries.

References

Barre, P., Moreau, L., Mi, F., Turner, L., Gastal, F., Julier, B., Ghesquière, M. 2009. Leaf length QTLs in perennial ryegrass. Grass and Forage Sci. 64:310–321.

Belkhir, K., Borsa, P., Chikhi, L., Raufaste, N., Bonhomme, F. 2004. GENETIX 4.05, laboratoire génome, populations, interactions, CNRS UMR 5171, Université de Montpellier II, Montpellier, France.

Bolaric, S., Barth, S., Melchinger, E., Posselt, U.K. 2005. Molecular genetic diversity within and among German ecotypes in comparison to European perennial ryegrass cultivars. Plant Breed. 124:257–262.

Charmet, G., Balfourier, F., Ravel, C. 1993. Isozyme polymorphism and geographic differentiation in a collection of French perennial ryegrass populations. Genet. Resour. Crop Evol. 40:77–89.

Ghesquière, M., Hazard, L., Bétin, M. 1994. Breeding for management adaptation in perennial ryegrass (*Lolium perenne* L.). II. Genetic variability and heritability of leaf morphogenesis components. Agronomie 14:267–272.

Hazard, L., Ghesquière, M., Barraux, C. 1996. Genetic variability for leaf development in perennial ryegrass populations. Can. J. Plant Sci. 76:113–118.

Jensen, L.B., Muylle, H., Arens, P., Andersen, C.H., Holm, P.B., Ghesquière, M., Julier, B., Lübberstedt, T., Nielsen, K.K., De Riek, J., Roldan-Ruiz, I., Roulund, N., Taylor, C., Vosman, B., Barre, P. 2005. Development and mapping of a public reference set of SSR markers in *Lolium perenne* L. Mol. Ecol. Notes 5:951–957.

Lem, P., Lallemand, J. 2003. Grass consensus STS markers: An efficient approach for detecting polymorphism in *Lolium*. Theor. Applied Genet. 107:1113–1122.

Saha, M.C., Rouf, Mian M.A., Eujayl, I., Zwonitzer, J.C., Wang, L., May, G.D. 2004. Tall fescue EST-SSR markers with transferability across several grass species. Theor. Applied Genet. 109:783–791.

Chapter 43
Chemical Composition of the First Cut of Forage Ryegrass (*Lolium*) Species

Vilma Kemešytė and Nijolė Lemežienė

Abstract Four ryegrass species: annual (*Lolium multiflorum* var. *westerwoldicum*), Italian (*Lolium multiflorum*), hybrid (*Lolium boucheanum*) and perennial (*Lolium perenne*) were studied at the Lithuanian Institute of Agriculture during 2007–2008. Crude protein (CP), water soluble carbohydrates (WSC), crude fibre (CF) and dry matter digestibility (DMD) were evaluated.

The Italian and hybrid ryegrass had significantly highest DMD (86.28% and 85.56%) and WSC (37.37% and 35.98%). These species also had the lowest CF percentage (18.8% and 19.33%).

Annual ryegrass had significantly lowest DMD (62.21%) and highest CF (26.15%). The annual ryegrass had the best proportion of CP and WSC – 0.83. This species also had the significantly highest CP and lowest WSC among all studied species. The CP variation was average in annual, Italian and hybrid ryegrass, and low– in perennial. The variation WSC was average in annual ryegrass and low in all other studied species. The CF and digestibility variation was low for all species. Ryegrass varieties were distinguished by the best chemical composition: annual– 'Elunaria', 'Weldra' and 'Energa'; Italian – 'Talvike', 'Delecta' and 'Corbes'; hybrid – 'Abereve', 'Agata' and 'Lorry'; perennial – 'Žvilgė' and 'Raminta'.

Keywords Chemical composition · Ryegrass species

Introduction

Herbage quality is becoming increasingly important in milk and meat production from ruminants. Forage yield and quality are determined by various factors: genotype, cropping technology, weather conditions, cutting time and stage of plant growth (Butkute and Paplauskiene, 2004; Harkot, 2005). The ryegrasses are

V. Kemešytė (✉)
The Lithuanian Institute of Agriculture, Instituto al. 1, Akademija, Kaunas, Lithuania
e-mail: vilma@lzi.lt

C. Huyghe (ed.), *Sustainable Use of Genetic Diversity in Forage and Turf Breeding*,
DOI 10.1007/978-90-481-8706-5_43, © Springer Science+Business Media B.V. 2010

considered to be high quality forages and their high digestibility makes them suitable for all type ruminants (Aganga and Omphile, 2004). Ryegrass accumulates more water soluble carbohydrates, other digestible compounds and less crude fibre than timothy, orchard grass or other forage grasses. There is a great interest in combining ryegrass ability to produce WSC with effective use of nitrogen (Wilkins et al., 2003).

Chemical ryegrass composition has been investigated in various aspects: different ryegrass species and varieties, various growth stage in a pure crop and in mixtures with red clover (Szyszkowska and Sowiński, 2001; Wilkins et al., 2003; Aganga and Omphile, 2004; Marley et al., 2007). The chemical composition of perennial ryegrass has been compared to other grasses (Butkutė and Paulauskienė, 2006). There is a lack of data on different ryegrass species compared during the same growing season.

The aim of the current study was to analyse the average quality data of different ryegrass species ant to identify the best varieties.

Material and Methods

Four ryegrass species, annual (*Lolium multiflorum* var. *westerwoldicum*), Italian (*Lolium multiflorum*), hybrid (*Lolium x boucheanum*) and perennial (*Lolium perenne*) were studied at the Lithuanian Institute of Agriculture during 2007–2008. Ten varieties of each species were investigated. The standard annual ryegrass variety was Lithuanian 'Varpė', Italian – Lithuanian 'Ugnė', perennial – Lithuanian 'Veja' and hybrid – Latvian 'Saikava'.

Plants per each genotype were planted at 50 × 50 cm distances, using randomized complete block design with three replications. Samples were collected when ryegrass was at heading growth stage. Contents in crude proteins (CP), water soluble carbohydrates (WSC), crude fibre (CF) and dry matter digestibility (DMD) were evaluated using near-infrared spectrometer NIRS-6500 (Perstorp Analytical, Silver Spring, Maryland, USA) (Butkute et al., 2003).

The winters during the experimental years were mild. Spring of year 2007 was cold, there were frequent frosts. Early spring of year 2008 was warm and humid, so the conditions for rapid plant growth were favourable, but later growth was reduced due to drought in May.

The statistical analysis computer software STAT developed by *P.* Tarakanovas in the Visual Basic of Application as macro program to run in the EXCEL was used (Tarakanovas and Raudonius, 2003).

Results and Discussion

CP is the only forage component that contains nitrogen and it is the source of amino acids, essential for growth and lactation, reproductive functions (Butkutė and Paulauskienė, 2006). Among all studied species, the annual ryegrass had the

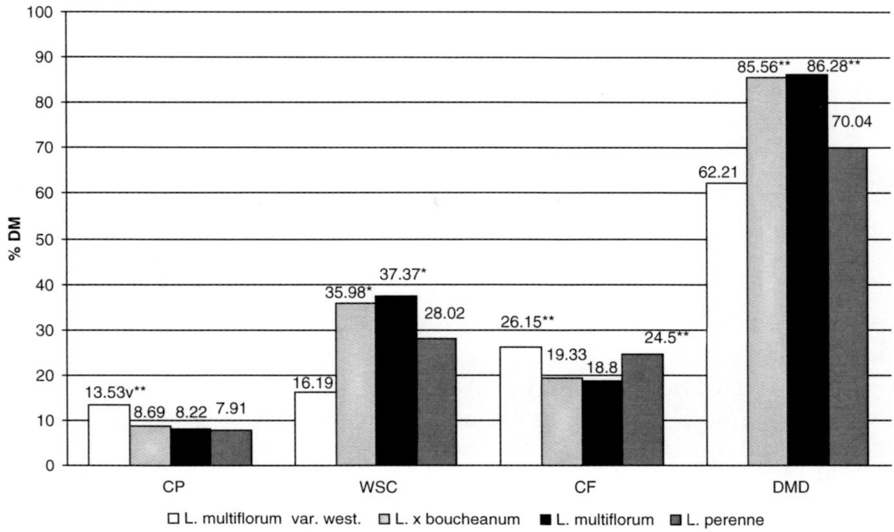

Fig. 43.1 Chemical composition of the first cut of ryegrass species, $*P < 0.05$, $**P < 0.01$

significantly ($P < 0.01$) highest (13.53%) and perennial ryegrass – the lowest CP (7.91%) (Fig. 43.1).

Cattle can fully and rapidly assimilate starch and WSC, WSC and starch content in forage is therefore important for cattle weight gain. WSC are also important for ryegrass, white and red clover mixture silage. Clover has a high content of protein, but its carbohydrate fermentation is low (Wilkins et al., 2003; Butkutė and Pauplauskienė, 2006). Most researchers state that ryegrass WSC range from 12 to 27% (Wilkins et al., 2003; Marley et al., 2007), but there are also experiments stating that the range of WSC variation can be from 9.4 to 47.4% (Tabacco et al., 2004). Our results confirm the latter opinion. The Italian and hybrid ryegrass had significantly ($P < 0.05$) highest WSC (37.37% and 35.98%) (Fig. 43.1). Proportion of crude protein and water soluble carbohydrates is a very important ryegrass characteristic. Annual ryegrass had the best proportion of CP and WSC – 0.83. The perennial ryegrass proportion of CP and WSC was 0.28, hybrid- 0.24 and Italian – 0.22. The annual ryegrass also had the significantly ($P < 0.05$) highest CP and lowest WSC among all studied species (Fig. 43.1).

Fibre is the main compound of plant cell wall. It consists of cellulose, hemi-cellulose and lignin (Jones and Moseley, 1993). The annual ryegrass had the significantly ($P < 0.01$) highest CF (26.15%) compared to other species. The Italian and hybrid ryegrass had lower CF percent (18.8% and 19.33%) (Fig. 43.1).

The Italian and hybrid ryegrass had the significantly ($P < 0.01$) highest DMD (86.28% and 85.56%). The annual ryegrass had the significantly ($P < 0.01$) lowest DMD (62.21%) (Fig. 43.1).

Table 43.1 Variation of chemical composition parameters of forage ryegrass species (CV)

Parameters / Species	Crude Protein	Water soluble carbohydrate	Crude fibre	Dry matter digestibility
L. multiflorum var. west.	10.49	11.15	6.95	8.54
L. multiflorum	13.27	6.13	5.04	3.42
L. x boucheanum	12.67	7.2	6.95	3.85
L. perenne	7.94	8.1	7.35	6.08

Within all investigated species, the variation of chemical composition was moderate or low (Table 43.1). Chemical composition parameters varied most in annual ryegrass. Average CP variation was assessed in annual (10.49%), Italian (13.27%) and hybrid (12.67%) ryegrass, and low CP variation (7.94%) – in perennial. Average WSC variation was estimated in annual ryegrass (11.15%) and low in all other studied species. The crude fibre and digestibility variation was low in all species (Table 43.1).

Annual ryegrass variety 'Elunaria' had the best proportion of CP and WSC – 0.65 (Table 43.2).

Varieties 'Weldra' and 'Energa' had a high CP percent. Variety 'Weldra' distinguished for the higher DMD (75.3%). Italian ryegrass varieties 'Talvike', 'Delecta' and 'Corbes' and hybrid – 'Abereve', 'Agata' and 'Lorry' had a high DMD and WSC and a low CF percent. The Lithuanian perennial ryegrass varieties 'Žvilgė' and 'Raminta' distinguished for the highest DMD (73.9% and 75.4%) (Table 43.2).

The choice of the ryegrass species should be based on the purpose it is intended for –grazing, hay or silage production and on whether it will be grown in mixture or pure swards.

Table 43.2 Ryegrass varieties with best chemical composition

Parameters / Varieties	Crude protein, %	Water soluble carbohydrate, %	Crude fibre, %	Dry matter digestibility, %
Lolium multiflorum var. westerwoldicum (n = 10)				
'Weldra'	15.5	15.1	21.9	75.3
'Elunaria'	12.2	18.8	25.6	64.2
'Energa'	15.8	12.6	25.6	64.2
Lolium multiflorum (n = 10)				
'Talvike'	9.8	38.0	18.3	91.1
'Delecta'	9.1	36.7	18.0	88.6
'Corbes'	9.1	39.1	18.6	87.2
Lolium x boucheanum (n = 10)				
'Abereve'	9.0	36.8	17.6	90.0
'Agata'	7.6	41.5	18.6	89.4
'Lorry'	8.4	37.9	18.5	88.5
Lolium perenne (n = 10)				
'Raminta'	8.9	30.0	22.0	75.4
'Žvilgė'	8.7	30.2	22.2	73.9

References

Aganga, A.A., Omphile, U.J. 2004. Chemical composition of ryegrass (*Lolium multiflorum*) at different stages of growth and ryegrass silages with additives. J. Biol. Sci. 4(5):645–649.

Butkutė, B.; Mašauskienė, A., Paplauskienė, V. 2003. Duomenu bazes sudarymas ir lygčiu sukurimas varpiniu žoliu kokybes analizei spektrometru NIRS-6500. Zemdirbyste/Agriculture. 82:157–168 (In Lithuanian).

Butkutė, B., Paplauskienė, V. 2004. Changes in the quality of some Lithuanian grasses as affected by cutting time in spring. Grassland Sci. Europe 9:909–911.

Butkutė, B., Paplauskienė, V. 2006. Daugiamečių varpinių žolių pašarinės vertės potencialas. Zemdirbyste/Agriculture.93(3):172–187 (In Lithuanian).

Harkot, W. 2005. Differences in the phonologic development of forage on mineral and organic soil. Grassland Sci. Europe 10:251–254.

Jones, D.I.H., Moseley, G. 1993. Laboratory methods for estimating nutritive quality. In: Davies, A., Baker, R.D., Grant, S.A., Laidlaw A.S. (eds.), Sward Measurement Handbook. (2nd ed., pp. 265–283). British Grassland Soc., Reading, Berks.

Marley, C.L., Fraser, M.D., Fisher, W.J. et al. 2007. Effects of continuous or rotational grazing of two perennial ryegrass varieties on the chemical composition of the herbage and the performance of finishing lambs. Grass Forage Sci. 62:255–264.

Szyszkowska, A., Sowiński, J. 2001. Botanical composition and nutritional value of two-component mixtures containing red clover and different grass species. Electronic journal of Polish Agricultural Universities. 4. http://www.ejpau.media.pl/volume4/issue2/animal/art-07.html

Tabacco, E., Borreani, G., Valente, M.E., Peiretti, P. G. 2004. Dry matter and water-soluble carbohydrate contents of Italian ryegrass at cutting as affected by environmental factors. Ital. J. Agron. 8(1):63–74.

Tarakanovas, P., Raudonius, S. 2003. Agronominių tyrimų duomenų statistinė analizė taikant kompiuterines programas *ANOVA, STAT, SPLIT-PLOT* iš paketo *SELEKCIJA* ir *IRRISTAT.*: 56 (In Lithuanian).

Wilkins, P.W., Lovatt, J.A., Jones, M.L. 2003. Improving annual yield of sugars and crude protein by recurrent selection within diploid ryegrass breeding populations, followed by chromosome doubling and hybridization. Czech J. Genet. Plant Breed. 39:95–99.

Chapter 44
Variability and Correlative Relations of Important Traits for Red Clover (*Trifolium pratense* L.) Half Sib progenies

Lugic Zoran, Radovic Jasmina, Sokolovic Dejan, Jevtic Goran, and Milenkovic Jasmina

Abstract The level of variability and correlative relationships of breeding material for the most important agronomic traits are very important in red clover breeding programs. Two diploid red clover populations, which showed good or satisfactory results in agro-ecological conditions in Serbia, were used in experiment. 60 plants from each of 45 half-sib progenies from both red clover populations, obtained by a cross-pollinated plants procedure suggested by Comstock and Robinson (1949), were researched during 2 years in NC I design. Obtained results mostly depended on genotype and the year of research. The population 1 had higher average values for observed traits in both research years. In both research years higher values of genetic and phenotypic variation coefficients for all the researched traits, except for total sugar content, were found in population 2. Values of genetic and phenotypic correlation coefficients in population 2 were lower than in population 1. The highest values of positive genetic correlation coefficient for both populations, were recorded for a dry matter yield and plant height (0,726** and 0,443**) and for crude cellulose content (0,754** and 0,351*). Average, strong and significant correlations between the number of tillers per plant and total sugar content, and crude cellulose content were found in both observed populations. Negative values of genetic and phenotypic correlation coefficient for both researched populations were found between dry matter yield per plant and crude protein content, and also between crude protein and crude cellulose content.

Keywords Correlation coefficient · Population · Red clover · Variability

Introduction

Red clover is the second most important perennial forage legume in Serbia. It is cultivated in pure stands on about 130,000 ha and in grass-legume mixtures on about 100,000 ha. Its positive traits are high dry matter yield potential, high protein

---

L. Zoran (✉)
Institute for Forage Crops Krusevac, Trg Kosturnice 50, Krusevac 37000, Serbia
e-mail: zoran.lugic@ikbks.com

C. Huyghe (ed.), *Sustainable Use of Genetic Diversity in Forage and Turf Breeding*,
DOI 10.1007/978-90-481-8706-5_44,

content, micro and macro elements, as well as good amino-acid balance. These traits give it a high potential for ruminants' feeding (Ignjatovic, 2002). The most frequent use of red clover in Serbia is for hay production. Other harvest methods, which have been used in the last 10 years, are silage and haylage (Dinic, 1990). Smith (2000) concluded that, because of its high seedling vigor, excellent forage quality, ease of establishment, competitiveness with grass and importance to soil conservation, red clover may play a major role in world agriculture. Due to cross pollination and the synthetic structure of the varieties, red clover has a high genetic variability for most agronomic traits. The aim of red clover breeding is to create genotypes with high values for key agronomic traits (dry matter yield, high quality, resistance to diseases and abiotic stress). Locally grown populations may exhibit adaptability and resistance to most important diseases and pests. They can represent the first source of variability in breeding (Taylor and Quesenberry, 1996). The existence of genetic variation is the first condition for successful selection. The correlations among the most important traits are also very important to the breeders. Baring in mind that the red clover is a perennial legume, the understanding of these correlations may thus accelerate the breeding process (Muntean and Savatti, 2003).

Material and Methods

Two diploid breeding populations were used: 1. domestic population K-17 and 2. introduced grown population Reinsberger. The progenies were obtained by hand pollination of randomly chosen plants from both populations in accordance to a procedure for cross-polinated plants suggested by Comstock and Robinson in 1948. During 2 years, 45 half-sib progenies from both populations, separated in three sets (5 male parents crossed with 3 females parents per set) were studied. There were 60 plants within each progeny. During the 1st year, the following traits were measured: dry matter yield per plant (DMY), plant height, number of stems per plant, number of internodes per stem (average of the three stems), protein content (PC, $g.kg^{-1}$), cellulose content (CC, $g.kg^{-1}$) and total sugar content (TSC, $g.kg^{-1}$). During the 2nd year, the following traits were researched: DMY, plant height, number of tillers and number of internodes per stem. The data were calculated using ANOVA for NC Design I. The interval of variation (IV) is difference between minimal and maximal values. The calculation of genetic correlations was done by analysis of covariance. The significance of correlation coefficient was tested by t-test.

Results and Discussion

Due to good conditions during the 1st year, high levels of dry matter yield were obtained in both populations (110,28–130,55 g/plant). In both years, the domestic population (no 1) showed higher values for dry matter yield per plant and plant height (Table 44.1). In the 1st year, the domestic population (no 1) had more stems

Table 44.1 Averages and variation coefficient for examined traits

Trait	Population	First year Average	IV	CV_G (%)	CV_F (%)	Second year Average	IV	CV_G (%)	CV_F (%)
DMY ($gplant^{-1}$)	1	130.55	115.10	16.25	17.78	146.37	153.00	17.93	22.27
	2	110.28	91.00	22.09	22.42	141.13	143.00	21.42	24.89
Plant height (cm)	1	59.23	15.50	5.10	6.38	56.64	27.00	4.31	5.06
	2	56.64	10.60	6.86	7.99	54.68	22.00	7.21	10.02
Number of stems	1	9.56	15.70	14.10	20.47	43.54	70.10	26.79	29.07
	2	9.46	10.60	12.32	21.21	46.89	65.60	16.15	17.79
Number of internodes	1	5.87	1.70	2.58	5.44	5.42	3.40	5.56	6.63
	2	6.04	2.50	10.84	12.67	4.94	2.30	13.81	14.89
PC (gkg^{-1})	1	175.70	4.70	9.15	9.19	–		–	–
	2	164.50	5.10	12.36	12.41	–		–	–
CC (gkg^{-1})	1	281.90	8.30	12.49	12.65	–		–	–
	2	295.30	8.50	12.47	12.64	–		–	–
TSC (gkg^{-1})	1	96.10	4.58	16.92	16.97	–		–	–
	2	94.50	4.91	14.52	14.54	–		–	–

per plant, whereas in the 2nd year, the foreign population (no 2) exceeded the domestic, without effect on dry matter yield per plant. Moisa (1998) wrote about high variability level of different red clover genotypes concerning yield and plant height. The domestic population also achieved better results for quality parameters, proteins content and cellulose content.

The variation among individual plants was very high for each observed trait, except the quality parameters. This is indicated by a wide variation interval. The highest genetic and phenotypic variability coefficients were obtained for dry matter yield per plant and the number of stems per plant. The lowest results were obtained for the plant height and the numbers of internodes per stem in the 2nd year. Ledda et al. (2000) stated wide interval of biomass yield per plant of red clover. Compared to phenotypic coefficient, larger genetic correlation coefficients were obtained for all observed traits in both researched years.

Correlation coefficients were different according to year and population investigated. Within both populations the strongest correlations were observed in the dry matter yield per plant, plant height and cellulose content (Table 44.2). Medium strong correlation were observed in dry matter yield per plant and the number of stems per plant, while the protein content, cellulose content, plant height and dry matter yield per plant had the low positive or negative correlation. In the 2nd year highest correlation coefficients were calculated for DMY per plant and plant height and DMY per plant and number of stems per plant (Table 44.3). Similar correlations for the dry matter yield per plant and height were found by Lugic et al. (1994), whereas Vasiljevic et al. (1998) stated slightly larger correlation values for dry matter yield per plant, plant height and the number of stems per plant among 17 red clover genotypes. According to Pederson et al. (1980) strong correlations were found between dry matter yield per plant, root diameter and regrowth after cutting.

Table 44.2 Genetic (above) and phenotypic (below a diagonal line) correlation in the 1st year

Population	Trait	DMY	Plant height	Number of stems	PC	CC	TSC
1	DMY		0.726**	0.325	0.125	0.754	0.016
2			0.443**	0.311	−0.110	0.351*	−0.075
1	Plant height	0.664**		−0.153	−0.404*	0.319	−0.161
2		0.421*		−0.248	−0.187	0.314	−0.043
1	Number of stems	0.321	−0.075		0.167	0.327	0.438*
2		0.212	−0.150		0.317	−0.294	0.146
1	PC	0.114	−0.366	0.117		−0.832**	−0.057
2		−0.100	−0.187	0.143		−0.691	−0.187
1	CC	0.692**	0.308	0.284	−0.749**		−0.143
2		0.341	0.314	0.285	−0.681		−0.228
1	TSC	0.014	−0.152	0.329*	−0.058	−0.112	
2		−0.073	−0.043	0.473*	−0.184	−0.223	

*Significant at 0.05 probability level; **Significant at 0.01 probability level

Table 44.3 Genetic (above) and phenotypic (below a diagonal line) correlation in the 2nd year

Population	Trait	DMY	Plant height	Number of stems	Number of internodes
1	DMY		0.674*	0.747*	0.148
2			0.712	0.623	0.093
1	Plant height	0.643*		−0.06	0.186
2		0.613*		−0.163	0.125
1	Number of stems	0.633**	0.04		0.148
2		0.611**	−0.149		0.120
1	Number of internodes	0.121	0.159	0.073	
2		0.088	0.076	0.100	

*Significant at 0.05 probability level; **Significant at 0.01 probability level

Conclusion

Domestic population (no 1) achieved better performance in yield and yield components, and to a lesser extent in quality parameters.

The obtained results in both populations pointed out the high variability available for all observed traits, particularly for the dry matter yield per plant and the number of stems per plant.

The largest correlations were obtained for the dry matter yield per plant, plant height and the number of stems per plant, while the protein content exhibited low correlations with almost all observed traits.

Acknowledgements This research has been financed by Ministry of science and technology of Republic of Serbia (Project TR20048).

References

Comstock, R.E., Robinson, H.F. 1949. The components of genetic variance in population of biparental progenies and their use in estimating the average degree of dominance. Biometrics 4:254.

Dinic, B. 1990. Uticaj provenjavanja silo-krme cevene dateline i konzervansa na kvalitet silaže. Arhiv za poljoprivredne nauke 51:235–244.

Ignjatovic, S. 2000: Biochemical-genetic and Biochemical characteristic of new domestic cultivars of Red Clover (*Trifolium pretense* L.) K-9 and white clover (Trifolium repens L.) K-33 and their nutrition biological values in different phenophases (pp. 1–1007). University of Belgrade Faculty of chemistry.

Ledda, L., Roggero, P.P., Veronesi, F. 2000. Comparisons among different plant breeding approaches applied to red clover. CIHEAM-Options Mediterraneennes, 63–67.

Lugic, Z., Konstantinov, K., Ivanovic, M., Marinkovic, D, Milosevic, G. 1994.: Variability of red clover (*Trifolium pratense* L.) for the process of biological nitrogen fixation. Genetika 26(3):141–146.

Moisa, F. 1998. The Variability and Heredity of some Characteristics in Red Clover. Proceeding of 2nd Balcan Symposium on Field Crops. 1:453–455.

Muntean, L., Savatti, M. 2003. Phenotypic correlations between productivity elements of red clover (*Trifulium pratense* L.). J. Cent. Europ. Agri.:185–190.

Pederson, G.A., Hill R.R., Leath, K.T. 1980. Host-Pathogen Variability for Fusarium-Caused Root Rot in Red Clover. Crop Science, 20(6):787–789.

Smith, R.R. 2000. Red clover in the 'Twenty-first' Century. www.uwex.edu/ces/forage/wfc/proceedings2000/smith.htm

Taylor, N.L., Quesenberru K.H. 1996. Red Clover Science. Kluwer Academy Publishers, Dordrecht

Vasiljevic, S., Surlan-Momirovic, G., Lukic, D., Katic, S., Zivanovic, T. 1998. Interdependence of seed yield components in varieties and population of red clover (*Trifolium pratense* L.). Proceeding of 2nd Balcan Symposium on Field Crops. 1:449–452.

Chapter 45
Genetic Diversity Within and Among Alfalfa Varieties for Some Traits

Jasmina Radovic, Dejan Sokolovic, Zoran Lugic, Snežana Anđelković, and Ratibor Štrbanović

Abstract Increasing variability in selection material could be achieved by introducing distance alfalfa varieties, as new source of diversity. The aim of this investigation was to determine the productivity and morphological traits of different varieties from USA in comparation with domestic varieties in order to find the genotypes with good agronomical traits, suitable for improvement of domestic varieties. Investigation was carried out in eight cuts in 2nd and 3rd year of utillization. Results showed a significant differences between varieties in almost all investigated traits. The highest variability among and within varieties was obtained for green and dry matter yield per plants and number of steam in all cuts. The lower coefficient of variation was noticed for other traits. Cluster analysis for investigated traits, calculated by the Ward method, using the Euclid distance, showed diversity among alfalfa varieties. Wide genetic variability of agronomic important traits in alfalfa cultivars provides a good basis for the improvement of domestic varieties and creation of new cultivars with great potential for high and quality forage yield.

Keywords Alfalfa · Forage yield · Morphological traits · Genetic diversity

Introduction

The nature of alfalfa, allogamy and autotetraploidy, contribute to large genetic variability among and within alfalfa populations or varieties. High alfalfa variability for many morphological and agronomical important traits was registered (Mikic et al., 2005, Radović et al., 2006). Besides variability among varieties variability of individual plants within varieties were much higher for numerous traits (Julier et al.,

J. Radovic (✉)
Institute for forage crops, Trg Kosturnice 50, Kruševac 37 000, Serbia
e-mail: jasmina.radovic@ikbks.com

C. Huyghe (ed.), *Sustainable Use of Genetic Diversity in Forage and Turf Breeding*,
DOI 10.1007/978-90-481-8706-5_45,

2000, Annicchiarico, 2006, Salvia et al., 2006). Variability among and within population were confirmed using molecular analysis (Zaccardeli et al., 2003). Forage yield variability of alfalfa is frequently used in breeding program for developing cultivars with high forage production and quality. Increasing variability in selection material could be achieved by introducing distinct alfalfa varieties, as new source of diversity. The aim of this investigation was to determine the productivity and morphological traits of different varieties from USA in compare with domestic varieties and their diversity in order to find the distinct genotypes with good agronomical traits, suitable for improvement of domestic varieties.

Material and Methods

The investigation was carried out in experimental field of Institute for forage crops in Krusevac, Serbia. Eleven varieties from USA and four domestic (K-22 and K-28 created in Institute for forage crops in Krusevac and ZA 83 and Krajina from Center for Agriculture in Zajecar) and one local population were investigated. The varieties were sown as space plants with 60 plants per variety. In all cuts in 2nd and 3rd year of utillization green and dry matter yield per plant, plant height and number of steam per plant were examined, while in 2nd year of utillization steam thickness, life size and number of internodes are mesaured. The results were analysed by ANOVA and the differences between genotypes were tested by the LSD test. The variability of the plant material for the chosen characteristics was expressed by among-cultivar variance ($\delta^2 c$) and within cultivar variance ($\delta^2 w$). Cluster analysis based on all examined traits, was calculated by the Ward method, using the Euclid distance.

Results and Discussion

Results showed significant differences between cultivars in both year of investigation (Table 45.1). Varieties Integrity, Archer and Magna showed higher green and dry matter yield in both years, but not significant in compare with domestic cultivars. Besides those varieties, Tru Test and Durango achieved high green and dry matter yield in 3rd year of vegetation. Highest number of stem per plant was observed in Ameri stand 201, but varieties Integrity, Archer and Magna had a high value for this trait, too. Lowest values for forage yield and number of stems per plant were observed at Pointer and WL 530 HQ in both years. Differences between varieties for plant height were less expressed.

Higher value for forage yield and number of stems per plant for almost all varieties was achieved in 3rd year. It can explain with favorable environmental conditions for alfalfa grown in 3rd year of vegetation, but with stage of plant grow, too. The number of stems per plant is important forage yield component which typically increased with plant age, and strongly affect on plant yield, especially on space plant

Table 45.1 Average values, among-cultivar variance (δ^2c) and within cultivar variance (δ^2w) in 2nd (2006) and 3rd year of utilization (2007) for forage yield, plant height and number of stems

		2006				2007			
	Varieties	Plant height (cm)	No. of steam	GMY (g per plant)	DMY (g per plant)	Plant height (cm)	No. of steam	GMY (g per plant)	DMY (g per plant)
1	Pointer	69,69	45,79	1281,1	281.85	71.43	56.23	1625.2	401.55
2	WL 530 HQ	73,84	40,06	1137,5	252.53	80.95	56.39	1511.6	335.59
3	Dakota	80,30	51,22	1591,3	346.91	81.38	74.33	1768.9	385.64
4	Tru Test	75,87	55,11	1432,5	314.01	83.13	74.15	2010.5	440.71
5	Durango	79,59	49,68	1523,1	335.09	80.52	72.72	2012.5	442.75
6	K 22	79,74	54,89	1790,9	370.32	76.22	66.85	1894.0	392.89
7	K 28	78,87	57,03	1842,1	383.27	77.20	75.04	1994.1	442.70
8	Local population	75,66	54,05	1611,4	357.75	77.66	71.40	1782.6	395.74
9	Ameri stand 201	71,24	70,30	1671,9	367.48	66.35	82.90	1694.9	372.54
10	Affinity	74,82	67,10	1771,0	389.44	77.32	80.18	1855.5	408.04
11	Ladak	79,87	62,05	1899,0	417.79	73.46	75.01	1858.7	408.92
12	Integrity	78,51	67,51	1838,1	404.02	74.04	79.31	2094.9	460.47
13	Archer	81,73	57,22	1873,5	414.06	77.54	85.50	2263.0	500.14
14	Magna	79,39	62,10	1914,4	421.18	78.10	82.17	2382.1	524.07
15	ZA 83	80,18	41,09	2086,9	457.05	81.49	59.35	1810.5	396.51
16	Krajina	78,78	58,46	1783,8	392.46	82.37	60.00	1852.3	407.51
	Mean	77,81	55,85	1690.5	369.08	77.45	71.97	1913.2	419.74
	LSD 0.05	6.25	5.52	186	40.95	5.52	6.75	228.6	42.38
	$\delta^2 c$	4.22	28.56	247.42	53.70	4.36	8.21	216.96	47.30
	$\delta^2 w$	11.37	71.70	640.84	140.50	14.74	30.60	949.16	204.07

Table 45.2 Average values, among-cultivar variance (δ^2c) and within cultivar variance (δ^2w) in 2nd (2006) year of utilization for number of internodes, leaf size and stem thickness

	Varieties	No. of internodes	Leaf width (cm)	Leaf length (cm)	Stem thickness (cm)
1	Pointer	13,58	0,90	2,67	0,34
2	WL 530 HQ	13,29	0,97	2,52	0,35
3	Dakota	14,56	0,89	2,67	0,34
4	Tru Test	12,88	0,91	2,55	0,34
5	Durango	12,44	0,84	2,46	0,35
6	K 22	13,58	0,92	2,58	0,38
7	K 28	13,75	0,88	2,64	0,37
8	Local population	13,77	0,88	2,64	0,33
9	Ameri stand 201	12,86	0,90	2,63	0,33
10	Affinity	14,40	0,83	2,50	0,31
11	Ladak	13,82	0,96	2,62	0,34
12	Integrity	14,32	0,89	2,60	0,32
13	Archer	14,14	0,97	2,61	0,36
14	Magna	13,59	0,99	2,59	0,31
15	ZA 83	13,43	0,80	2,40	0,30
16	Krajina	13,64	0,91	2,57	0,34
	Mean	13,62	0,91	2,57	0,34
	LSD 0.05	0.97	0.30	0.92	0.23
	δ^2 c	0.59	0.08	0.08	0.02
	δ^2 w	0.73	0.23	0.34	0.07

in nursery. High variability among alfalfa populations for forage yield and number of stems was reported in numerous articles (Mikic et al., 2005, Radovic et al., 2004).

Considering both years of investigation, cultivars Archer, Integrity and Magna achieved excellent forage yield and high number of stems per plant. This showed that those varieties had good adaptability on environmental conditions in Serbia. Domestic varieties K-22, K-28 and ZA-83 achieved lower production than those varieties, but differences were not significant. Varieties Pointer and WL 530 HQ showed significantly lowest value for forage yield and forage yield components.

Differences among varieties for other traits were lower (Table 45.2). Varieties Dakota, Affinity and Integrity had higher number internodes per steam then other, but differences were not significant for most varieties. For stem thickness and leaf size observed differences were not significant.

Variability within population was much higher than variability among populations, for all investigation traits. For the forage yield and number of stems within cultivar variation was greater than for other traits. The large within-cultivar variation confirming the importance of selection within population for exploiting the genetic resources of alfalfa and should be used in breeding programs to maximize genetic progress.

Julier et al. (2000) found that for the morphological and yield traits, within cultivar variation was greater than for quality traits. Wide variability within the alfalfa ecotype Ampurdan for morphological and other traits of agronomic interest was reported by Salvia et al. (2006). Similar result showed Annicchiarico (2006) for Italian alfalfa landraces. Beside high alfalfa variability for morphological traits, Zaccardeli et al. (2003) were found large within population variation for molecular markers.

High yielding cultivars are those that perform well in all environments and across all harvests. Large within-cultivar variation for yield and morphological traits may be needed to achieve high yield under various environmental conditions. This may relate to the need to avoid inbreeding depression for growth and yield by incorporating a broad genetic base in cultivars (Julier et al., 2000). Cluster analysis for investigated traits showed diversity among investigated alfalfa varieties (Fig. 45.1). The resulting dendrogram showed three clusters. Varieties Archer and Magna formed one cluster. Varieties Pointer, WL 530 HQ, Tru Test and Durango formed a separate cluster from the other investigated varieties, while all other varieties formed the third cluster.

Cluster analyses showed that domestic alfalfa varieties have similarity with some USA cultivars. Cultivar Ladak was introduced in our breeding program years ago. It could be the reason why cluster analyses showed low distinctness between these varieties.

For further work varieties Archer, Magna and Integrity, which achieved higher forage yield and numerous stems per plant in both year then domestic cultivars, are especially interesting, as varieties Tru Test and Durango which achieved high forage yield in the 3rd year.

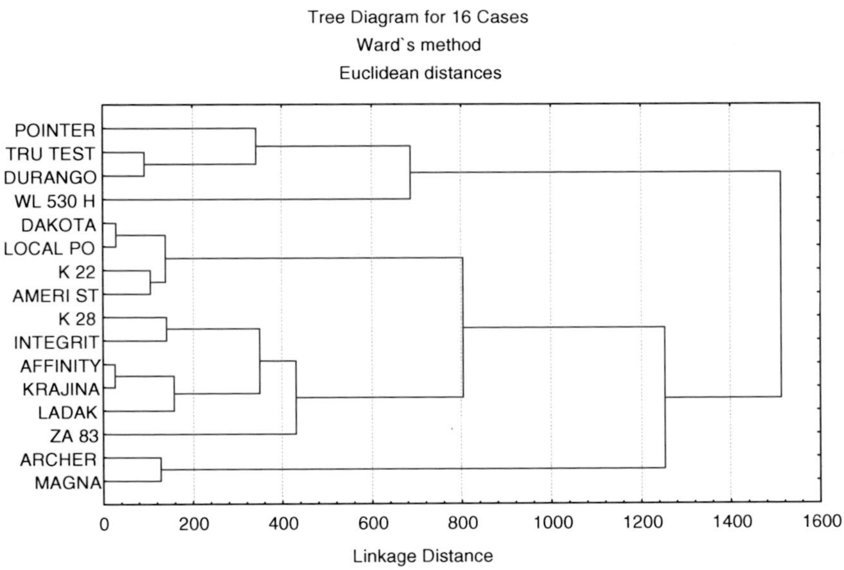

Fig. 45.1 Cluster diagram of 16 alfalfa varieties based on investigated traits

Wide genetic variability of agronomic traits among and within investigated varieties provides a good basis for the improvement of domestic varieties and creation of new cultivar with great potential for high and quality forage yield. Investigation will be carried on to observe field persistence of those varieties in pure stands.

Acknowledgements This research has been financed by Ministry of science of Republic of Serbia (Project 20048).

References

Annicchiarico, P. 2006. Diversity, genetic structure, distinctness and agronomical value of Italian lucerne (*Medicago sativa* L.) landraces. Euphytica 148:269–282.

Julier, B., Huyghe, C., Ecalle, C. 2000. Within and among-cultivar genetic variation in alfalfa: Forage quality, morphology and yield. Crop Science 40:365–369.

Mikic, V., Radovic, J., Mrfat-Vukelic, S., Lugic, Z., Lazarevic, D. 2005. Variability of agronomic characteristic in eight lucerne genotypes. Grassland Sci. in Europe 9:565–568.

Radovic, J., Lugic, Z., Ignjatovic, S., Delic, D. 2004. Yield and quality of different alfalfa varieties. Acta Agric. Serbica IX(17):109–114.

Radovic, J., Lugic, Z., Sokolovic, D., Delic, D., Stanisavljevic, R. 2006. Genetic variability for seed yield and seed yield component in alfalfa. Proceedings of XXVI Meeting of the EUCARPIA Fodder Crops and amenity Grasses Section, Perugia, Italy, pp. 121–123.

Salvia, J., Serra, J., Delgado, I., Caparros, G., Loveras, J. 2006. Morphology characterization of the alfalfa ecotype Ampurdan. Proceedings of XXVI Meeting of the EUCARPIA Fodder Crops and amenity Grasses Section, Perugia, Italy, pp. 169–171.

Zaccardeli, M., Gnocchi, S., Carelli, M., Scotti, C. 2003. Variation among and within Italian alfalfa ecotypes by means of bio-agronomical characters and amplified fragment length polymorphism analyses. Plant Breed. 122:61–65.

Chapter 46
Genetic Improvement in Ryegrass (*Lolium perenne*) from Turf and Forage Breeding Over the Four Past Decades

Jean-Paul Sampoux, Rémi Métral, Marc Ghesquière, Pierre Baudouin, Bernard Bayle, Vincent Béguier, Pierre Bourdon, Jean-François Chosson, Koos de Bruijn, François Deneufbourg, Christophe Galbrun, Wieslaw Pietraszek, Bernard Tharel, and Anne Viguié

Abstract In order to assess the efficiency of breeding in ryegrass (*Lolium perenne*), we tested the agronomic performances of a set of diploid cultivars including turf and forage cultivars registered on European national lists from 1965 to 2004. Seven ecotypes, originating from the main European climatic areas were also tested. Linear regressions fitting agronomic traits to the year of registration enabled to test the genetic improvement in turf and forage breeding through time. The genetic improvement in turf breeding was highly significant (p value of regression slope < 0.001) for most target traits (aesthetic aspect, disease resistance, wear tolerance, summer aspect, persistency). In forage breeding, the genetic improvement was highly significant (p value < 0.001) for autumn dry matter yield, rust resistance, and persistency, and was significant (p value < 0.05) for summer dry matter yield and reduction of aftermath heading. The rate of improvement of annual dry matter yield reached 0.29 ton per 10 years. Forage breeding was additionally associated with a highly significant decrease of lignin and crude protein content, and with a highly significant increase of soluble carbohydrate content. No significant change in grain production was noted either in forage or turf breeding, but conversely a large range of seed production was pointed out in both ancient and recent cultivars.

Keywords Forage breeding · Genetic improvement · *Lolium perenne* · Turf breeding

J.-P. Sampoux (✉)
INRA, Centre Poitou-Charentes, UR4 (Unité de Recherche Pluridisciplinaire Prairies et Plantes Fourragères), Lusignan F-86600, France
e-mail: jpsampoux@lusignan.inra.fr

C. Huyghe (ed.), *Sustainable Use of Genetic Diversity in Forage and Turf Breeding*,
DOI 10.1007/978-90-481-8706-5_46,

Introduction

The most direct way to assess the efficiency of plant breeding in a species is to compare within a common experiment the performances of cultivars released over a time span covering several decades. In order to assess the efficiency of breeding in perennial ryegrass (*Lolium perenne*), we tested the agronomic performances of a set of diploid cultivars including turf and forage cultivars registered on European national lists from 1965 to 2004.

Material and Methods

The set of cultivars included 32 turf cultivars, 18 forage cultivars, and 4 dual use (forage and turf) cultivars (Table 46.1). Seven ecotypes, originating from the main European climatic areas were also tested. Turf and forage trials were performed at six and four locations, respectively. All turf and forage cultivars and ecotypes were furthermore assessed at two locations in spaced plant trials and in seed production trials.

Results and Discussion

A principal component analysis based on traits recorded in spaced plant trials highlighted two main independent dimensions of variation among cultivars and ecotypes (Fig. 46.1): a first dimension relating to morphology and vegetative biomass, and a second dimension relating to disease resistance.

Linear regressions fitting agronomic traits to the year of registration enabled to test the genetic improvement in turf and forage breeding through time. The genetic improvement in turf breeding was highly significant (p value of regression slope < 0.001) for most target traits: aesthetic aspect, disease resistance, wear tolerance, summer aspect, persistency (Table 46.2 and Fig. 46.2). The rate of improvement of the aesthetic aspect equalled 56% of the standard deviation of the cultivar set range per 10 years.

In forage breeding, the genetic improvement was highly significant (p value < 0.001) for autumn dry matter yield, rust resistance and persistency, and was significant (p value < 0.05) for summer dry matter yield and aftermath heading (Tables 46.3 and 46.4, and Fig. 46.3). The autumn dry matter yield was positively correlated with rust resistance (0.82) and negatively with aftermath heading (−0.48). Forage breeding significantly increased rust resistance and reduced aftermath heading; improvements obtained for these two traits might contribute to the noted increase in autumn dry matter yield. On the other hand, spring dry matter yield was mainly correlated with morphological traits, particularly leaf blade

Table 46.1 Panel of ryegrass cultivars and ecotypes used to assess the genetic improvement brought by forage and turf breeding

Cultivar	Breeder	Registration year	Cultivar	Breeder	Registration year
Turf cultivars			**Forage cultivars**		
Apollo	Seed Research of Oregon	1992	*Aberavon*	IGER	2000
Baccara	Carneau	1995	*Argoal*	R2N	2004
Barclay	Barenbrug	1982	*Barlet*	Barenbrug	1985
Bardorado	Barenbrug	2003	*Barnhem*	Barenbrug	1999
Bargold	Barenbrug	2000	*Brital*	R2N	2000
Barlennium	Barenbrug	2001	*Cadans*	Cebeco	1996
Bianca	Van der Have	1979	*Carillon*	Carneau	2002
Cachemire	Carneau	2002	*Carrera*	Carneau	2000
Carnac	Carneau	2000	*Clerpin*	INRA	1996
Concerto	R2N	1997	*Compliment*	Van der Have	1997
Final	Clause	1986	*Donata*	Van der Have	1977
Flor		1990	*Herbie*	Van der Have	1990
Game	Joordens	1974	*Lactis*	Limagrain - DLF	2003
Greenfair	Limagrain - DLF	1999	*Ohio*	Green Genetics - DLF	1990
Greenflash	Limagrain - DLF	2003	*Pacage*	Carneau	1988
Idole	Vilmorin	1979	*Pradal*	R2N	2001
Lisabelle	DSV	1983	*Rathlin*	PBSA (N. Ir.)	1982
Loretta	Steinach	1975	*Trani*	DLF	1977
Manhattan	Rutgers State Univ.	1974	**Dual use (forage and turf) cultivars**		
Marathon	Vilmorin	1989	*Perma*	Cebeco	1974
Numan	Pure Seed	1986	*Vigor*	Rikjsstation Belgium	1971
Pennfine	Pennsylvania Ag.	1976	*Maprima*	Malchow	1980
Plaisir	Semunion - REGA	1995	*Belida*	DLF	1974
Rambo	Vilmorin - DLF	1989	**Ecotypes**		
Recital	R2N	2003	*Belgium*		
Repell	Loft's Seed	1988	*France (Brittany)*		
Score	Zwaan & De Wiljes	1980	*France (Centre)*		
Sourire	Semunion - REGA	1998	*Denmark*		
Sun		1992	*Hungary*		
Transate	REGA	2004	*Ireland*		
Verdi	Loft's Seed	1995	*Ukraine*		
Verna	DLF	1965			

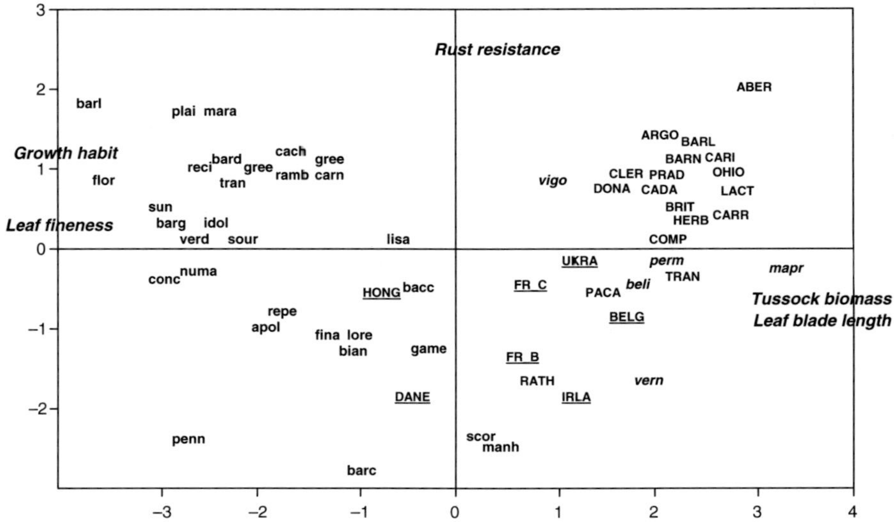

Fig. 46.1 Projection of perennial ryegrass cultivars and ecotypes on the first two principal axes yielded by a principal component analysis of spaced plant trial traits. *Capital letters*: forage cultivars, *small letters*: turf cultivars, *italics*: dual use (forage and turf) cultivars, *underlined*: ecotypes

Table 46.2 Genetic gains per 10 years for ordinal traits in ryegrass turf breeding expressed as a percentage of trait standard deviation for the turf cultivar set, and *p* values of the slope of trait regression on cultivar registration year computed with turf cultivars only

Trait	Genetic gain per 10 years	p value
Aesthetic aspect	+ 56.4	<0.0001
Summer stay green	+ 23.1	< 0.0001
Winter stay green	+ 12.3	0.03
Turf density	+ 60.7	<0.0001
Fineness	+ 62.8	<0.0001
Persistency	+ 32.5	0.0004
Wear tolerance	+ 50.0	<0.0001
Rust resistance	+ 39.0	0.0002
Red thread	null	0.06

length (0.52). However, forage breeding did not change leaf blade length. It could be hypothesized that a direct breeding effort to increase leaf blade length would favour a more substantial improvement of spring dry matter yield, which is the largest contributor to the annual dry matter yield. The rate of improvement of annual dry matter yield reached 0.29 ton per 10 years, a value similar to that estimated by Camlin (1997) and Tabel and Allerit (2005). Forage breeding additionally led to significant

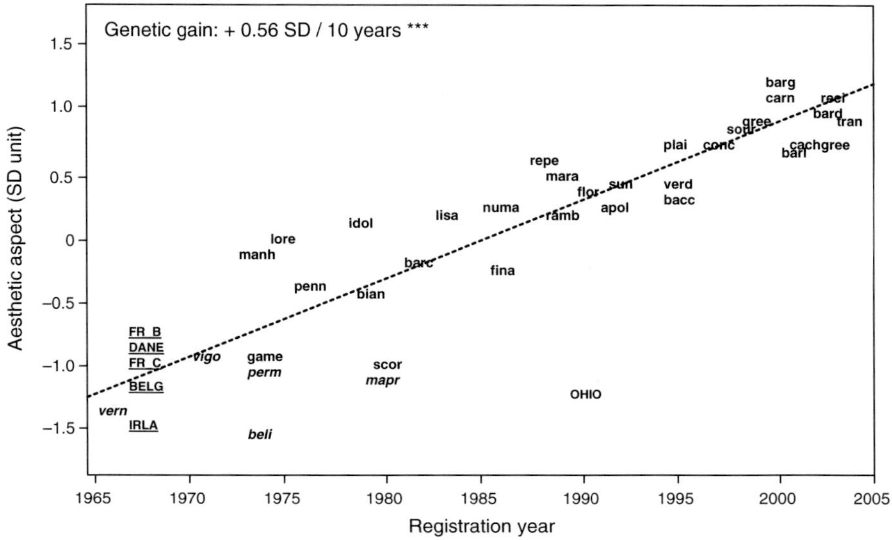

Fig. 46.2 Plot of the aesthetic aspect of cultivars assessed in turf trials against registration year. The aesthetic aspect is expressed in standard deviation unit with zero mean. The dotted line is the regression line fitting the aesthetic aspect to the registration year computed with turf cultivar data only. *Capital letters*: forage cultivars, *small letters*: turf cultivars, *italics*: dual use (forage and turf) cultivars, *underlined*: ecotypes

Table 46.3 Genetic gains per 10 years for ordinal traits in ryegrass forage breeding expressed as percentage of trait standard deviation for the forage cultivar set, and *p* values of the slope of trait regression on cultivar registration year computed with forage cultivar data only

Trait	Genetic gain per 10 years	p value
Aftermath heading	– 21.8	0.03
Rust resistance	+ 41.5	0.0004
Leaf blade length	null	0.99
Growth habit	null	0.16
Tillering	null	0.12

correlative changes in chemical composition: a highly significant decrease in lignin and crude protein contents, and a highly significant increase in soluble carbohydrate content (Table 46.4).

No significant change in seed yield production was noted either in forage or turf breeding, but conversely a large range of seed yield was pointed out in both ancient and recent cultivars.

Table 46.4 Genetic gains per 10 years for quantitative traits in forage breeding, *p* values of the slope of trait regression on cultivar registration year and trait mean values computed with forage cultivar data only. Genetic gains and mean values are expressed in trait units (tons/ha, % dry matter or % NDF)

Trait	Genetic gain per 10 years	p value	Mean
Annual dry matter yield (tons/ha)	+0.29	0.0004	9.7
Spring dry matter yield (tons/ha)	+0.12	0.16	6.1
Summer dry matter yield (tons/ha)	+0.09	0.04	4.3
Autumn dry matter yield (tons/ha)	+0.15	<0.0001	2.2
Soluble carbohydrate content (% dry matter)	+0.67	0.003	14.2
Lignin (ADL) content (% dry matter)	−0.05	0.0001	1.8
NDF content (% dry matter)	−0.23	0.03	44.7
Crude protein content (% dry matter)	−0.27	0.006	15.2
NDF digestibility (% NDF)	+0.40	0.14	77.5

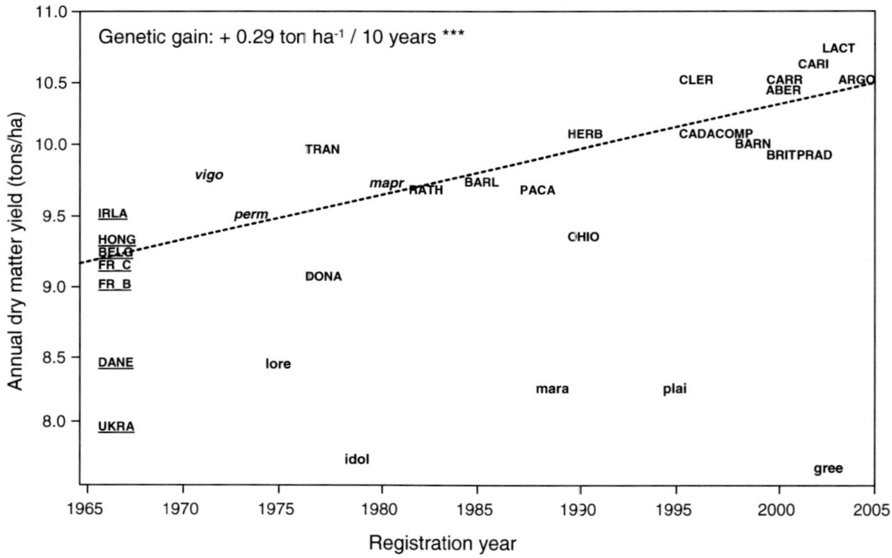

Fig. 46.3 Plot of the annual dry matter yield of cultivars assessed in forage trials against registration year. The *dotted line* is the regression line fitting the annual dry matter yield to the registration year computed with forage cultivar data only. *Capital letters*: forage cultivars, *small letters*: turf cultivars, *italics*: dual use (forage and turf) cultivars, *underlined*: ecotypes

References

Tabel, C., Allerit, R. 2005. Bilan du progrès génétique obtenu pour différents caractères et différentes espèces. Fourrages 183:365–376.

Camlin, M.S. 1997. Grasses, Seeds of progress. Proceedings of the BGS/BSPB/NIAB/SAC Conference, Nottingham, UK, 18–19 February 1997, pp. 2–14.

Chapter 47
The EUCARPIA Multi-Site Rust Evaluation – Results 2007

Franz Xaver Schubiger, Joost Baert, Bohumir Cagas, Vladimir Cernoch, Jean François Chosson, Elzbieta Czembor, Fred Eickmeyer, Ulf Feuerstein, Stefan Hartmann, Hana Jakesova, Bernhard Krautzer, Hennietha Leenheer, Hans Lellbach, Laurence Poinsard, Ulrich Posselt, Massimo Romani, Luigi Russi, Sabine Schulze, Marie Claire Tardin, François VanHee, Evelin Willner, Lukas Wolters, and Beat Boller

Abstract The EUCARPIA rust evaluation trial was established in 2007 for the third time. The same ryegrass cultivars (33 perennial, 15 Italian and 3 hybrid ryegrass) were tested as in the first and second trial of 2001 and 2004, respectively (Boller et al., 2003; Schubiger et al., 2007). In addition, the Italian ryegrass cultivar Crema, the hybrid ryegrass cultivar Gosia and the perennial ryegrass cultivar Maja were included in 2007. The trials were sown at 27 sites in 12 countries of Europe. Twenty one sites were the same as in the first and second trial. Crown rust (*Puccinia coronata* f. sp. *lolii*) was again the most frequently observed rust on both ryegrass species. Variation in resistance to crown rust among cultivars was significant at 22 sites for Italian and 16 sites for perennial ryegrass. Stem rust (*Puccinia graminis* f. sp. *graminicola*) occurred mainly on perennial ryegrass. There was a significant difference in mean stem rust scores among cultivars at 12 sites. The new cultivar Gosia and Tarandus showed the highest level of resistance of all the Italian and hybrid ryegrass cultivars tested. Bocage, Gwendal and Lacerta were the most crown rust resistant perennial ryegrass cultivars. The ranking of the mean crown rust susceptibility of the cultivars was highly correlated with the corresponding ranking of cultivars in the 2001 and 2004 trials, respectively. This was true for both Italian and perennial ryegrass.

Keywords Crown rust · Stem rust · Italian ryegrass · Perennial ryegrass · Hybrid ryegrass

---

F.X. Schubiger (✉)
Agroscope Reckenholz-Tänikon, Research Station ART, Zurich CH-8046, Switzerland
e-mail: franz.schubiger@art.admin.ch

C. Huyghe (ed.), *Sustainable Use of Genetic Diversity in Forage and Turf Breeding*,
DOI 10.1007/978-90-481-8706-5_47,

Introduction

The EUCARPIA multisite rust evaluation trial was initiated in 2000 by the EUCARPIA Fodder crops and Amenity grasses section at the meeting on the Azores. The aim of the trial is to assess the rust susceptibility of different Italian (*Lolium multiflorum*), hybrid (*L. boucheanum*) and perennial (*L. perenne*) ryegrass cultivars in widely distributed regions across Europe. In repeating this trial every 3 years, starting in 2001, we wanted to assess if rust resistance of individual cultivars breaks down and is overcome by the rust pathogen over the course of time. The detailed results of the 2001 and 2004 trials were published in the proceedings of the "Eucarpia Fodder crops and Amenity grasses" meetings in Braunschweig 2002 and Perugia 2006, respectively (Boller et al., 2003; Schubiger et al. 2007). The present paper outlines the results of the experiments established in 2007.

Material and Methods

In 2007 the trial was sown at 27 sites in 12 European countries. Twenty one sites were identical as in the first (2001) and second trial (2004). The same 33 perennial, 15 Italian and 3 Hybrid ryegrass cultivars were tested as in 2001 and 2004, respectively (Boller et al., 2003; Schubiger et al., 2007). In addition, the Italian ryegrass cultivars Crema (already included in 2004), the hybrid ryegrass cultivar Gosia and the perennial ryegrass cultivar Maja were included. Eleven Italian, two hybrid and fifteen perennial ryegrass cultivars were tetraploid, the other cultivars were diploid.

The experimental design was the same as described in the two previous publications. The trials were scored for rust incidence between July and the beginning of November one to four times during different growth cycles and periods of abundant rust development. Participants were asked to score the cultivars for each rust species occurring in the field, separately. A scale of 1 to 9 was used; with 1= no rust disease, 2 = trace of rust, 3 = 5%, 4 = 10%, 5 = 25%, 6 = 40%, 7 = 60%, 8 = 75% and 9 = more than 75% of the foliage covered with rust. The rating values represented a relative estimate of leaf area occupied by rust pustules, and not reaction type. Scoring data with an average score of at least 2 at a particular site and date were included in the analysis. If there were sites with more than one valid scoring per year, means of the scorings (per row) were calculated and used in further analysis.

Results and Discussion

A relevant incidence (average score of at least 2) of crown rust (*Puccinia coronata*) on Italian ryegrass was observed at 22 out of 26 sites (Table 47.1) and on perennial ryegrass at 16 out of 27 sites (Table 47.2), respectively. Stem rust (*P. graminis*) occurred less frequently as it was observed on Italian ryegrass in 3 sites (Table 47.3) and on perennial ryegrass in 12 sites (Table 47.4), respectively. This finding confirms previous observations that stem rust is present almost exclusively on perennial ryegrass and more in the eastern and southern part of Europe.

Table 47.1 Crown rust (*Puccinia coronata*) disease scores of 16 Italian and 4 hybrid ryegrass cultivars at 22 sites (means of 1–4 scoring per site). Cultivars are ranked according to the mean of all sites

cultivar	Asendorf (D)	Bornhof (D)	Druelle (F)	Gumpenstein (A)	Hohenheim (D)	Lelystad C (NL)	Lodi (I)	G. Luesewitz (D)	Malchow (D)	Merelbeke (B)	Montours (F)	Orchies (F)	Ottersum (NL)	Perugia (I)	Pulling (D)	Radzikow (P)	Rennes (F)	Les Rosiers (F)	Roznov Zubri (CZ)	Spitalhof (D)	Steinach (D)	Zurich (CH)	mean of all sites
Gosia (4n)^2	2.0	2.0	2.4	1.5	2.9	3.3	4.4	4.3	1.8	1.8	1.3	3.4	2.8	3.0	2.0	3.4	2.3	2.3	1.0	1.3	4.0	1.3	2.5
Tarandus (4n)	2.4	1.8	2.3	1.5	3.5	3.5	1.4	6.3	2.3	2.3	1.9	3.3	3.3	2.5	1.6	3.0	3.4	3.3	1.3	1.3	2.5	1.6	2.5
Zorro (4n)	2.8	2.5	2.9	1.5	2.9	3.4	4.5	5.6	2.3	2.3	1.5	3.0	3.5	3.5	2.0	3.1	2.9	2.8	1.3	1.5	3.0	1.5	2.7
Caballo (4n)	2.4	2.5	2.9	1.3	4.1	3.9	4.1	5.0	2.5	2.5	2.0	2.6	3.5	3.0	2.0	3.1	3.3	3.0	2.3	1.8	3.3	1.6	2.8
Domino (4n)	2.4	2.1	2.8	1.3	4.1	4.4	4.8	4.9	2.3	3.5	2.0	3.5	4.3	3.5	1.8	2.5	2.8	3.2	1.0	1.8	2.5	1.8	2.9
Tonyl (4n)	2.6	2.9	2.9	1.8	3.8	5.1	4.1	5.3	2.3	3.0	1.8	3.6	3.5	2.8	2.1	2.5	3.3	4.4	2.5	2.0	3.3	1.7	3.0
Bolero (4n)	3.1	2.6	3.3	2.0	3.4	4.4	3.3	5.4	3.3	2.8	2.0	3.9	4.3	3.3	1.8	3.5	4.0	4.4	1.3	2.0	2.5	1.9	3.1
Barprisma	2.4	1.6	2.6	1.8	3.9	4.9	2.0	6.4	3.5	4.8	2.9	4.7	3.8	3.8	1.6	3.1	4.3	3.7	1.5	1.8	4.3	1.8	3.2
Fastyl	3.4	2.0	3.1	2.0	4.5	5.0	2.4	6.5	3.3	4.3	3.1	4.7	4.0	3.5	2.1	3.0	4.6	4.1	1.3	2.0	3.8	2.2	3.4
Aberexcel (4n)^2	3.3	2.4	3.1	1.8	3.5	5.3	5.5	5.9	2.3	6.5	1.5	5.1	4.0	3.8	2.5	3.3	4.3	4.7	1.5	2.5	3.8	1.4	3.5
Ellire (4n)	3.6	2.8	4.0	2.0	5.0	5.6	6.9	5.8	2.8	4.0	3.3	4.5	5.0	3.8	2.3	3.3	4.1	4.4	3.8	1.8	4.0	3.3	3.9
Crema	5.3	3.0	2.8	2.5	4.1	5.5	1.3	7.8	5.3	5.3	5.0	6.7	5.8	4.5	2.5	4.3	5.8	5.6	3.0	3.0	4.8	3.3	4.4
Pirol^2	5.0	2.4	3.8	2.5	5.3	5.9	3.8	7.8	4.0	7.0	5.0	5.9	6.0	4.3	2.6	5.3	6.3	4.9	1.3	2.3	4.5	3.1	4.5
Meryl	5.1	2.9	4.1	3.0	5.3	6.6	3.3	8.0	5.5	7.5	5.3	5.8	6.5	4.0	2.6	4.4	5.3	5.2	2.0	2.3	6.3	3.8	4.8
Danergo (4n)	6.5	4.5	4.6	2.3	5.8	8.3	7.9	7.6	6.0	8.3	3.0	6.7	8.0	5.3	3.1	5.1	6.5	5.8	1.5	2.5	6.8	3.8	5.4
Lolita (4n)	5.8	6.0	4.8	2.3	5.8	7.5	7.2	7.6	6.8	7.3	4.1	6.9	8.0	4.8	3.4	5.0	6.3	5.8	5.5	2.8	7.5	4.3	5.7
Lema	7.3	4.6	5.3	3.3	5.9	7.6	7.1	8.4	6.3	9.0	6.6	7.5	8.3	4.8	3.6	6.0	7.5	7.3	1.8	3.5	7.8	5.3	6.1
Ligrande	7.5	4.9	5.4	3.8	6.5	8.0	6.9	8.8	6.5	8.8	7.4	7.5	8.0	4.8	3.4	6.0	7.4	7.0	2.0	3.3	8.0	5.6	6.2
Gordo	7.4	5.3	6.3	4.5	6.5	7.9	6.3	8.8	6.8	9.0	6.9	7.5	8.5	5.3	3.6	6.1	8.0	6.7	1.3	4.0	8.0	5.3	6.3
Gumpensteiner^2	7.3	4.0	5.9	3.3	6.8	8.1	8.4	8.5	6.8	9.0	7.8	7.7	8.3	4.8	3.6	6.0	8.4	6.2	6.8	3.8	8.0	5.8	6.6

Table 47.1 (continued)

cultivar	Asendorf (D)	Bornhof (D)	Druelle (F)	Gumpenstein (A)	Hohenheim (D)	Lelystad C (NL)	Lodi (I)	G. Luesewitz (D)	Malchow (D)	Merelbeke (B)	Montours (F)	Orchies (F)	Ottersum (NL)	Perugia (I)	Pulling (D)	Radzikow (P)	Rennes (F)	Les Rosiers (F)	Roznov Zubri (CZ)	Spitalhof (D)	Steinach (D)	Zurich (CH)	mean of all sites
mean	4.4	3.1	3.7	2.3	4.7	5.7	4.8	6.7	4.1	5.4	3.7	5.2	5.5	3.9	2.5	4.1	5.0	4.7	2.2	2.3	4.9	3.0	4.2
LSD (P=0.05)	0.7	1.3	0.6	0.9	1.0	1.0	2.3	1.0	1.2	1.3	0.9	1.0	1.3	0.9	0.5	0.6	1.0	1.2	0.7	0.8	1.5	0.8	0.5
correlation [1]	0.94	0.74	0.94	0.93	0.91	0.95	0.67	0.87	0.90	0.96	0.88	0.98	0.94	0.92	0.93	0.89	0.95	0.94	0.42#	0.93	0.95	0.91	

[1]Spearman rank order correlation with mean of all sites (all values are significant at $P<0.05$ except # = not significant),[2] hybrid ryegrass

Table 47.2 Crown rust (*Puccinia coronata*) disease scores of 34 perennial ryegrass cultivars at 16 sites (means of 1–4 scoring per site). Cultivars are ranked according to the mean of all sites

cultivar	Asendorf (D)	Bornhof (D)	Druelle (F)	Lelystad C (NL)	Lodi (I)	G. Luesewitz (D)	Malchow (D)	Merelbeke (B)	Montours (F)	Orchies (F)	Ottersum (NL)	Pulling (D)	Rennes (F)	Les Rosiers (F)	Steinach (D)	Zurich (CH)	mean of all sites
Bocage (4n)	3.4	2.5	1.9	3.1	1.0	6.3	1.8	2.8	2.0	2.1	2.0	2.0	2.8	2.9	2.3	1.8	2.5
Gwendal (4n)	3.3	2.6	1.8	3.4	1.0	5.3	3.0	2.0	2.0	1.5	2.8	2.0	5.0	3.1	2.0	1.4	2.6
Lacerta (4n)	3.5	2.5	2.4	2.8	1.0	5.5	2.0	2.8	2.3	2.3	3.5	2.3	2.6	2.7	2.3	2.2	2.6
Pastoral (4n)	3.3	3.3	2.5	3.0	1.8	6.6	2.3	2.3	2.0	1.7	3.0	2.0	3.9	3.1	1.8	2.3	2.8
Carrera	2.8	1.8	1.9	2.1	1.0	8.0	3.3	4.0	3.0	2.7	2.3	2.0	3.1	2.8	3.0	1.8	2.8
Option	2.9	2.6	3.0	2.8	1.0	6.5	3.8	5.0	2.3	3.5	2.3	2.3	3.1	3.1	3.5	2.1	3.1
Vincent	3.3	2.9	2.6	2.8	1.0	7.1	3.0	5.5	2.0	3.6	2.5	2.8	2.9	3.3	4.0	1.5	3.2
Aubisque (4n)	3.9	2.9	2.9	3.8	1.8	6.5	2.5	3.3	2.5	2.4	3.0	2.0	5.0	4.2	2.0	2.8	3.2
Heraut	3.4	2.4	2.4	2.8	1.0	6.5	4.0	5.8	2.5	3.4	2.5	2.8	3.5	3.4	4.0	2.6	3.3
Elgon (4n)	4.6	3.8	2.3	3.3	1.8	6.4	2.5	3.5	5.3	2.3	3.3	2.3	5.5	4.4	2.0	3.3	3.5
Barnhem	5.6	2.4	2.5	3.3	1.1	7.9	3.8	5.3	3.5	3.3	2.5	2.0	4.4	3.9	4.0	2.6	3.6
Fennema	4.5	2.5	3.1	3.1	1.0	7.0	3.5	5.5	4.5	3.6	3.0	3.3	4.0	3.6	3.8	3.0	3.7
Roy (4n)	4.4	3.6	3.0	4.4	3.0	6.5	2.5	4.5	3.8	2.6	3.5	2.5	5.0	3.8	3.8	3.3	3.7
Terry (4n)	5.0	4.6	2.8	4.3	1.0	7.4	2.5	4.3	4.0	2.3	4.3	2.3	4.9	4.4	3.3	3.4	3.8
Kells	4.9	2.5	2.4	2.9	1.0	7.8	4.8	5.0	5.0	3.7	4.0	2.8	3.9	3.9	3.5	3.1	3.8
Weigra	4.1	4.0	2.6	2.9	1.0	7.5	4.0	5.3	5.0	3.3	3.0	2.5	4.0	4.1	4.3	3.8	3.8
Arabella	5.0	3.9	2.8	3.6	1.0	7.4	3.5	5.8	4.0	3.6	2.3	3.0	4.4	4.4	4.3	2.7	3.8
Kentaur (4n)	5.1	3.9	2.8	4.9	3.5	6.8	2.0	4.3	3.3	2.7	3.5	2.8	4.8	3.9	4.0	4.4	3.9
Sponsor	5.3	2.5	2.6	3.4	1.0	8.0	4.0	5.3	5.5	3.7	2.8	2.3	4.6	4.3	3.3	4.2	3.9
Orval	3.0	3.0	2.3	2.6	5.0	7.6	3.0	5.0	5.8	4.5	3.3	2.0	7.1	4.6	2.0	2.3	3.9
Maja (4n)	5.1	3.5	3.0	3.8	5.3	7.0	2.0	3.5	5.0	2.9	4.3	2.5	5.8	4.2	3.5	3.5	4.0
Corbet	5.5	3.6	2.8	2.5	1.0	7.8	4.8	4.8	7.0	3.8	3.3	2.8	5.0	4.3	4.0	2.1	4.0

Table 47.2 (continued)

cultivar	Asendorf (D)	Bornhof (D)	Druelle (F)	Lelystad C (NL)	Lodi (I)	G. Luesewitz (D)	Malchow (D)	Merelbeke (B)	Montours (F)	Orchies (F)	Ottersum (NL)	Pulling (D)	Rennes (F)	Les Rosiers (F)	Steinach (D)	Zurich (CH)	mean of all sites
Aberdart	5.3	2.4	2.9	3.4	1.0	8.1	4.8	6.3	5.5	3.0	3.0	2.5	6.5	4.1	3.5	5.0	4.2
Aristo	6.0	3.4	2.8	5.5	1.0	8.4	3.8	6.5	5.0	3.5	5.0	2.5	6.0	4.8	4.0	4.0	4.5
Litempo (4n)	5.1	5.3	3.3	4.0	5.8	7.0	3.0	5.0	5.0	2.7	3.8	3.3	5.8	4.6	4.8	4.2	4.5
Foxtrot	6.3	3.4	3.4	5.0	1.3	8.1	3.3	5.0	5.5	3.3	5.0	2.5	6.3	4.9	4.5	4.8	4.5
Gladio	5.8	3.6	2.9	4.0	4.5	8.1	3.8	5.0	5.0	4.0	4.3	3.0	5.9	4.4	4.0	4.8	4.6
Tivoli (4n)	4.4	6.3	3.9	5.5	3.3	7.1	3.3	5.5	6.5	2.7	4.8	2.5	5.5	5.1	2.8	4.4	4.6
Sirocco (4n)	5.9	5.4	3.5	5.0	6.5	7.4	3.3	5.0	6.8	3.3	4.3	3.5	6.3	5.6	4.5	5.0	5.1
Helmer (4n)	6.5	5.9	3.6	5.8	6.0	7.4	3.5	5.3	4.5	2.8	5.3	3.5	5.9	5.3	6.3	4.5	5.1
Condesa (4n)	5.5	5.4	3.8	5.4	3.8	7.6	3.8	6.5	6.8	3.0	5.3	3.0	7.0	5.7	4.5	5.8	5.2
Guru	6.3	2.5	3.9	2.8	1.0	7.8	5.3	8.5	6.3	4.6	7.3	3.5	7.6	6.3	5.3	4.6	5.2
Lipresso	6.4	3.9	3.8	4.1	2.0	8.0	5.0	7.8	6.0	4.5	5.3	4.3	6.6	5.1	6.8	5.4	5.3
Aurora	6.6	4.8	5.0	6.4	2.0	8.3	7.3	9.0	8.3	4.0	6.0	6.0	8.4	6.6	8.0	8.0	6.5
mean	4.8	3.5	2.9	3.8	2.2	7.2	3.5	5.0	4.5	3.1	3.7	2.7	5.1	4.3	3.8	3.5	4.0
LSD (P=0.05)	0.7	1.5	0.8	1.1	2.3	0.9	1.2	1.3	0.9	0.6	1.3	0.6	0.9	0.6	1.4	1.2	0.56
correlation [1]	0.83	0.58	0.78	0.58	0.52	0.65	0.56	0.67	0.86	0.55	0.80	0.74	0.84	0.92	0.73	0.87	

[1] Spearman rank order correlation with mean of all sites (all values are significant at $P<0.05$)

Table 47.3 Stem rust (*Puccinia graminis*) disease scores of 16 Italian and 4 hybrid ryegrass cultivars at 3 sites. Cultivars are ranked according to the mean of all sites

cultivar	H. Zivotice (CZ)	Lodi (I)	Malchow (D)	Mean of all sites
Caballo (4n)	3.1	1.3	2.3	2.2
Zorro (4n)	3.3	1.5	2.0	2.3
Domino (4n)	3.3	1.3	2.5	2.3
Gosia (4n)[2]	3.4	1.5	2.3	2.4
Tonyl (4n)	3.3	1.5	2.8	2.5
Bolero (4n)	3.7	1.0	3.3	2.6
Tarandus (4n)	3.7	1.8	2.5	2.6
Aberexcel (4n)[2]	3.5	1.8	2.8	2.7
Ellire (4n)	3.5	1.9	3.5	3.0
Lolita (4n)	3.6	1.4	4.0	3.0
Danergo (4n)	3.5	1.0	5.5	3.3
Barprisma	4.2	3.0	3.8	3.6
Fastyl	4.2	4.0	3.0	3.7
Meryl	3.6	3.5	4.3	3.8
Pirol[2]	4.0	3.9	4.8	4.2
Gordo	4.0	2.5	6.3	4.3
Ligrande	4.2	1.9	6.8	4.3
Gumpensteiner[2]	4.2	2.5	7.5	4.7
Lema	4.6	5.3	6.0	5.3
Crema	5.2	5.3	6.5	5.6
mean	3.8	2.4	4.1	3.4
LSD (P=0.05)	0.6	1.7	1.2	1.1
correlation[1]	0.88	0.77	0.93	

[1] Spearman rank order correlation with mean of all sites (all values are significant at $P<0.05$),
[2] hybrid ryegrass

There was a highly significant difference ($P<0.001$) in mean crown rust scores among Italian ryegrass cultivars at each site and over all sites (Table 47.5). The cultivars Gosia and Tarandus showed the highest level of crown rust resistance of all the Italian / hybrid ryegrass cultivars tested (Table 47.1). Despite the occurrence of significant interactions of cultivars with sites, the Spearman rank order correlation between the data of a particular site and the mean of all sites was in all but one case (Roznov Zubri) significant.

Mean crown and stem rust susceptibility scores of perennial ryegrass cultivars at each site (with an average score of at least 2) are presented in Tables 47.2 and 47.4. Analysis of variance revealed highly significant differences in crown and stem rust susceptibility at each site and over all sites, respectively (Table 47.5). The ranking of the cultivars in respect to crown or stem rust susceptibility was very consistent. The Spearman rank order correlations between the data of each site and the mean of all sites were significant for both rust species. Bocage and Gwendal were the most crown and stem rust resistant perennial ryegrass cultivars. However, the rank order correlation between the mean disease scores of the cultivars for the two rust pathogens was low ($r_s = 0.37$, $P < 0.05$).

Table 47.4 Stem rust (*Puccinia graminis*) disease scores of 34 perennial ryegrass cultivars at 12 sites (means of 1–4 scoring per site). Cultivars are ranked according to the mean of all sites

cultivar	Gumpenstein (A)	H. Zivotice (CZ)	Hohenheim (D)	Jevicko (CZ)	Lodi (I)	Malchow (D)	Montours (F)	Perugia (I)	Radzikow (P)	Roznov Zubri (CZ)	Spitalhof (D)	Zurich (CH)	mean of all sites
Gwendal (4n)	2.5	3.4	2.7	2.0	1.1	2.6	1.0	1.3	2.8	1.0	1.8	1.3	1.9
Bocage (4n)	2.0	3.2	2.7	2.0	1.6	1.9	1.5	2.5	2.4	1.3	1.5	1.1	2.0
Pastoral (4n)	2.3	3.1	3.5	2.3	1.6	1.8	1.3	1.8	2.9	1.0	1.0	1.6	2.0
Maja (4n)	2.0	3.1	3.2	2.0	1.5	2.0	2.0	1.8	3.6	1.0	1.8	1.0	2.1
Roy (4n)	2.3	3.3	3.3	2.0	1.6	1.9	1.0	2.0	3.9	2.0	1.5	1.1	2.2
Orval	3.0	3.0	2.1	2.0	1.6	2.8	1.5	2.3	4.3	1.0	3.0	1.8	2.4
Lacerta (4n)	2.0	4.6	2.8	2.0	3.3	2.3	1.8	2.5	3.1	1.5	2.0	1.3	2.4
Tivoli (4n)	2.8	3.9	3.7	2.0	1.8	3.2	1.8	1.8	3.6	1.0	2.0	1.9	2.4
Elgon (4n)	2.5	3.5	3.9	2.0	1.8	2.5	2.5	1.8	4.0	1.8	2.0	1.1	2.4
Terry (4n)	2.0	3.5	2.7	2.3	2.4	2.3	3.8	2.3	3.8	1.0	1.8	2.0	2.5
Aubisque (4n)	2.0	4.0	3.2	2.3	1.6	2.2	2.0	2.5	3.4	3.3	2.0	1.9	2.5
Kentaur (4n)	2.3	4.1	3.3	2.0	2.1	2.3	1.8	2.5	4.1	2.8	2.0	1.8	2.6
Carrera	2.3	3.4	3.7	2.3	2.5	2.9	1.5	2.5	4.1	1.5	2.8	1.5	2.6
Litempo (4n)	2.3	4.8	3.3	2.0	3.0	2.5	3.3	2.8	3.6	1.0	2.0	1.4	2.7
Condesa (4n)	2.3	3.8	4.7	2.0	1.6	2.7	1.0	2.8	5.0	2.0	2.0	2.4	2.7
Helmer (4n)	2.8	4.0	3.4	2.0	2.0	2.2	2.0	3.3	4.1	3.8	1.8	1.9	2.8
Sirocco (4n)	2.5	4.7	4.5	2.0	2.8	3.2	2.5	2.8	4.1	3.3	1.5	2.1	3.0
Aberdart	3.3	4.1	4.2	2.8	1.8	3.5	2.0	2.5	5.1	3.0	2.8	2.3	3.1
Aristo	3.0	4.4	4.0	2.5	2.5	3.5	2.3	2.5	4.3	3.3	2.8	2.5	3.1
Foxtrot	3.0	4.6	4.3	3.0	2.4	3.0	1.8	3.3	5.3	2.3	2.5	2.4	3.1
Gladio	3.0	4.8	3.1	2.5	3.3	3.1	1.0	2.8	4.9	4.8	2.5	2.4	3.2
Option	2.8	4.9	3.5	2.5	3.5	3.3	1.3	4.0	4.3	4.3	2.5	3.3	3.3

Table 47.4 (continued)

cultivar	Gumpenstein (A)	H. Zivotice (CZ)	Hohenheim (D)	Jevicko (CZ)	Lodi (I)	Malchow (D)	Montours (F)	Perugia (I)	Radzikow (P)	Roznov Zubri (CZ)	Spitalhof (D)	Zurich (CH)	mean of all sites
Guru	4.0	5.1	3.8	3.0	2.8	3.9	1.0	3.3	5.4	2.3	3.0	2.5	3.3
Heraut	3.3	4.9	3.7	3.8	2.1	3.5	1.3	3.5	5.0	4.3	2.8	2.9	3.4
Fennema	3.0	5.4	4.2	3.5	2.5	3.7	1.3	3.0	5.4	3.0	3.0	3.4	3.4
Kells	3.0	5.3	4.6	2.8	3.1	3.9	1.8	3.5	4.9	2.8	2.8	3.3	3.5
Sponsor	3.3	5.4	4.0	2.8	2.9	3.3	3.3	3.5	4.9	3.8	3.0	3.1	3.6
Vincent	3.0	4.8	3.8	3.5	2.9	3.1	3.5	3.5	5.0	4.0	3.0	3.4	3.6
Weigra	2.8	5.0	4.7	2.8	4.8	3.5	2.3	3.5	5.1	3.5	3.3	2.9	3.7
Barnhem	3.0	4.9	3.6	3.3	4.0	3.6	4.5	3.3	4.5	3.8	4.0	3.0	3.8
Corbet	2.5	5.5	4.9	3.5	3.4	4.4	3.3	3.3	4.9	3.8	3.0	3.8	3.8
Arabella	2.8	6.1	4.9	3.0	4.4	3.5	3.0	3.8	5.0	3.8	3.3	3.5	3.9
Lipresso	3.5	5.6	4.6	4.3	5.5	4.5	2.3	3.3	6.1	2.5	4.0	3.3	4.1
Aurora	5.0	6.9	6.2	4.8	4.0	5.7	2.0	3.5	6.9	5.5	3.8	5.3	4.9
mean	2.8	4.4	3.8	2.6	2.6	3.1	2.0	2.8	4.4	2.7	2.5	2.3	3.0
LSD (P=0.05)	0.6	0.8	1.3	0.6	1.3	0.7	0.8	1.1	0.6	0.7	0.6	0.9	0.43
correlation [1]	0.68	0.92	0.74	0.86	0.81	0.85	0.40	0.87	0.83	0.76	0.83	0.93	

[1] Spearman rank order correlation with mean of all sites (all values are significant at $P<0.05$)

Table 47.5 Analysis of variance of mean rust disease scores. All F-values are significant at the *P*<0.001 level

Ryegrass species	*L. multiflorum/boucheanum*		*L. perenne*	
Rust species	*P. coronata*	*P. graminis*	*P. coronata*	*P. graminis*
No. of cultivars	20	20	34	34
No. of sites	22	3	16	12
F-value for cultivars	303.2	17.7	75.6	60.6
F-value for sites	206.5	92.7	281.6	211.5
F-value for cultivar x site interaction	4.8	5.1	4.4	3.3

The results of the Eucarpia multisite rust evaluation trial 2007 revealed that several Italian and perennial ryegrass cultivars are resistant to crown and stem rust in a wide range of sites across Europe. The European rust populations were not virulent to the majority of plants of the most resistant cultivars. Moreover, the ranking of the cultivars in terms of crown and stem rust resistance was quite consistent with the ranking of the trials in 2001 and 2004, respectively. However, some diploid perennial ryegrass cultivars, such as Orval, were ranked considerably poorer in crown rust resistance than in 2001, suggesting a breakdown of resistance against certain rust populations.

References

Boller, B., Schubiger, F.X., and Streckeisen, P. 2003. The Eucarpia multisite rust evaluation – results 2001. Vortr. Pflanzenzüchtung 59:198–207.

Schubiger, F.X., Streckeisen, P., and Boller, B. 2007. The EUCARPIA Multisite Rust Evaluation – Results of the trials 2004. In Rosellini, D. and Veronesi, F. (eds.), Breeding and seed production for conventional and organic agriculture. Proceedings of the Eucarpia meeting Perugia, pp. 154–158.

Chapter 48
Dry Matter Production and Nutritive Value of Perennial Ryegrass Cultivars Collection

Dejan Soković, Snežana Babić, Jordan Marković, Jasmina Radović, Zoran Lugić, Bogovid Živković, and Aleksandar Simić

Abstract Perennial ryegrass (*Lolium perenne* L.) is one of the most important perennial cool-season forage grasses. It is highly productive, with the highest nutritive value among forage grasses, adapted for frequent defoliation and grazing. Market demands focus interest of perennial ryegrass breeders to production of cultivars with high, stable yield and good dry matter quality, tolerant to drought, frost and other stressful environmental conditions, with different maturity. According to that, initial breeding material must be heterogeneous, with a range of different genotypes, either cultivars or wild populations. In this article the diversity in a collection of perennial ryegrass cultivars has been investigated at the beginning of the breeding process. The collection consisted of 21 genotypes which originated from Europe and USA. Within a 2-year period time of tillering, crop height in first and regeneration in second cut, annual yield and nutritive value of dry matter were investigated. The data were analysed by ANOVA on the basis of 2-year mean values and all traits showed great level of variability. Highest annual dry matter yield of about 14tha^{-1} was achieved by cultivars Mara and K-11, both being intermediate diploid varieties.

Keywords Breeding · Cultivar diversity · Dry matter yield · Forage quality · Perennial ryegrass

Introduction

Perennial ryegrass (*Lolium perenne* L.) is the dominant forage grass in temperate regions of the world where intensive grassland management and grazing are basic methods of forage utilization. Perennial ryegrass has a number of positive attributes

D. Soković (✉)
Institute for forage crops, Trg Kosturnice, Kruševac, Serbia
e-mail: dejan.sokolovic@ikbks.com

C. Huyghe (ed.), *Sustainable Use of Genetic Diversity in Forage and Turf Breeding*,
DOI 10.1007/978-90-481-8706-5_48,

including high seedling vigor, leafiness, high palatability, rapid recovery after harvest and high DM yield and quality (Peeters, 2004). Although it has limited heat and drought tolerance, marginal winter hardiness and rust susceptibility, it shows a significant presence in Serbian grasslands. Considering the specific Serbian climate, interest of grass breeders is focused on the production of perennial ryegrass cultivars with high, stabile yield and good quality of dry matter, which differ in maturity and tolerance to drought, high temperature and frost.

According to that, initial breeding material must be genetically heterogeneous, derived from different cultivars (Gilliland et al., 2000), or wild populations (Charmet and Balfourier, 1991). Often, autochthonous material exhibits a desirable level of adaptability but with lack of yielding capacity. On the other hand, imported cultivars are frequently not adapted to local agroecological conditions and show weak vigor in subsequent years of grassland utilisation. Both, hopefully, will bring part of desired characteristics to new genotypes.

High dry matter yield and quality associated with resistance to drought and rust are the most important criterion in perennial ryegrass breeding. It is known that quality and digestibility of forage is strongly related to cell wall content (NDF) and its lignification (Van Soest, 1994). With an increased regrowth period, the cell-wall contents (neutral detergent fiber (NDF), acid detergent fiber (ADF) and acid detergent lignin (ADL)) increase and the composition of the cell wall changes (Wilman and Altimimi, 1982). Therefore, forage nutritive value is dominantly influenced by the growing season and maturity of the sward, but with genotype also (Buxton and Casler, 1993; Sokolovic et al., 2002). In this study genetic diversity of production, morpho-phenological traits and chemical composition of dry matter of 21 perennial ryegrass genotypes were studied in order to evaluate the heterogeneity among cultivars and to find possibilities for future cultivar improvement.

Material and Methods

A collection of 20 perennial ryegrass genotypes (cultivars and breeding populations), both diploid and tetraploid, widely differing in maturity and origin, and one hybrid (Table 48.1) was investigated during 2 years (2005 and 2006). Genotypes were sown in the experimental field of the Institute for forage crops in completely randomized block design in three repetitions with $2m^2$ basic plots.

Within a 2-year period time of heading (days after May 1st), crop height in first cut (cm), regeneration (30 days after first cut; cm), average annual dry mater yield (DMY; tha^{-1}) and nutritive value of dry matter (crude protein (CP), neutral detergent fiber (NDF), acid detergent fiber (ADF), hemicelluloses (HEM) and acid detergent lignin (ADL) (Goering and Van Soest, 1970) were investigated. Annual DMY was obtained as sum of three cats. First cut was done immediately after heading, second in first decade of July and third in second decade of September. The data were analysed by ANOVA on the basis of 2-year mean values and cluster analysis was made by the Ward method, using the Euclidean distances.

Table 48.1 Time of heading, crop height in first cut (cm), regeneration after first cut (cm) and dry matter yield (t.ha^{-1}) of 21 perennial ryegrass cultivars or breeding populations (values are averaged for 2 years)

Genotypes	Genotype description (type, origin, ploidy, maturity)	Time of heading	Crop height	Regeneration	DMY
Arion	Cv (cultivar), Switzerland, 2n, early	14.5	84.0	34.0	8.19
Cavia	Cv, Switzerland, 2n, early	14.0	91.7	26.0	11.43
Talpa	Cv, Switzerland, 2n	14.5	95.7	39.7	11.56
Respect	Cv, West Europe, 2n	20.5	87.7	27.7	12.98
Mara	Cv, East Europe, 2n, medium-late	17.0	71.3	44.7	14.28
MSP 3366	Breeding population, USA, 2n	14.5	75.7	24.7	9.98
MSP 3372	Breeding population, USA, 2n	15.5	66.7	34.0	8.12
K-11	Cv, Serbia, 2n, intermediate	17.5	77.3	40.3	13.93
Shandon	Cv, Ireland, 2n, intermediate	15.0	83.3	32.0	9.95
Cashel	Cv, Ireland, 2n, intermediate	14.5	84.3	37.0	9.15
BG 34	Cultivars Mixture, 2n, East Europe	23.5	60.7	46.3	10.96
Grand Daddy	Cv, USA, 4n, early intermediate	14.5	85.0	25.3	9.23
Calibra	Cv, West Europe, 4n, intermediate	19.0	82.7	41.7	9.75
Magician	Cv, Ireland, 4n, intermediate	14.0	93.7	47.3	11.72
Glenstal	Cv, Ireland, 4n, intermediate	15.0	90.0	40.7	11.04
Greengold	Cv, Ireland, 4n, intermediate	24.0	76.0	37.7	8.37
Meillenium	Cv, Ireland, 4n, late	29.5	65.3	37.0	9.80
Sarsfield	Cv, Ireland, 4n, late	29.0	61.0	36.0	7.55
Glencar	Cv, Ireland, 4n, late	29.0	75.3	25.3	11.09
Maja	Cv, East Europe, 4n	20.5	88.0	36.7	9.90
Spring green	Cv, Festulolium hybrid, lolium type, USA	15.0	98.0	42.0	10.40
Lsd 0.05		0.78	4.71	2.74	1.59
0.01		1.05	6.30	3.66	2.13

Results and Discussion

A great level of variability was obtained for all investigated traits, as indicated by Table 48.1.

Time of heading ranged between 14 and 29.5 days after May 1st. The three cultivars from Switzerland, Irish cultivars Magician and Cashel and two genotypes originated from USA showed early heading in Serbian ecological conditions. The latest cultivars were three tetraploid cultivars from Ireland. Crop height was between 60 and 98 cm in first cut and from 25 to 47 cm in regeneration after first cut. Generally the tallest cultivars in first cut were the highest in regrowth (Talpa, Magician and Spring green). The diploid cultivars Mara (14,28 tha^{-1}), K-11 (13,93 tha^{-1}) and Respect (12,98 tha^{-1}) showed the highest annual DMY. Mara is winter hardy and summer tolerant cultivar, while K-11 is a new cultivar bred from an autochthonous population, released in Serbia 2006 (Sokolovic et al., 2007). DMY of tetraploid cultivars was lower, with a maximum value of 11,7 tha^{-1}, obviously because they were not adapted to the dry cropping conditions in Serbia.

The first two cuts were analysed for chemical parameters because yield in first cut and regrowth in second represent a large part of annual yield. Chemical parameters are presented as overall value of first and second cut (Table 48.2) weighted by the dry matter yield in each cut. Variability among genotypes for nutritive value

Table 48.2 Dry matter chemical composition ($g.kg^{-1}$) of 21 perennial ryegrass cultivars or breeding populations (overall value of first and second cut)

Genotype (abbreviation)	CP	NDF	ADF	HEM	Lignin
Arion (Ar)	109,50	618,78	334,64	284,14	44,11
Cavia (Cv)	107,10	619,67	338,99	280,69	36,45
Talpa (Ta)	114,56	637,22	336,83	300,40	41,09
Respect (Rs)	105,36	583,04	319,16	263,89	32,50
Mara (Mr)	111,44	573,67	307,45	266,30	36,59
MSP 3366 (M66)	113,36	572,28	301.84	270,35	54,67
MSP 3372 (M72)	114,91	548,11	292.91	255,19	29,06
K-11 (K)	112,37	551,28	299,85	251,44	47,67
Shandon (Sh)	117,40	596,43	317,04	279,40	51,74
Cashel (Cs)	117,73	583,46	306,02	277,45	35,58
BG 34 (Bg)	118,94	594,08	315,90	278,27	34,11
Grand Daddy (Gd)	106,60	604,44	317,03	287,41	26,30
Calibra (Cb)	118,60	568,04	305,45	262,60	50,20
Magician (Mg)	121.34	569,51	289,22	280,37	40,19
Glenstal (Gls)	125.47	617,63	316,59	301,04	30,08
Greengold (Gr)	146,89	716,64	365,50	351,14	59,52
Meillenium (Ml)	127,71	590,95	300,04	300,48	43,65
Sarsfield (Sr)	127,81	548,81	276,71	272,11	52,17
Glencar (Glk)	126,48	570,37	283,15	287,24	37,54
Maja (Mj)	127,30	599,17	303,22	295,95	27,29
Spring green (Sg)	107,45	575,45	303,39	272,06	33,11
Lsd 0.05	8.30	43.65	24.38	23.12	2.07
0.01	11.34	58.76	33.99	30.87	2.96

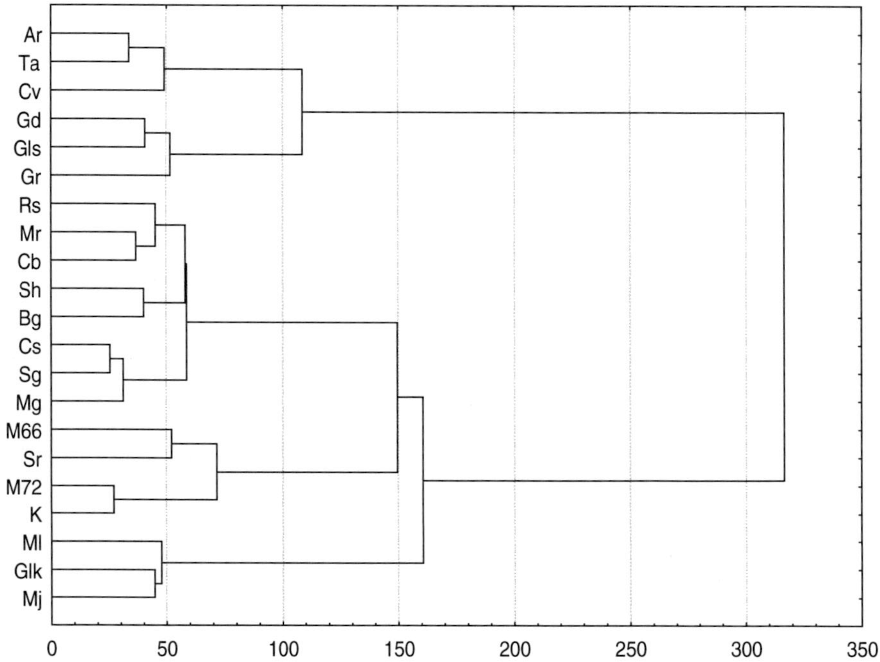

Fig. 48.1 Cluster diagram of 21 perennial ryegrass cultivars

of forage was lower than for productive traits, but differences were statistically significant (ANOVA).

CP content and especially cell wall components were slightly different from values reported in literature for perennial ryegrass (Jensen et al., 2003; Smit et al., 2006). This was expected according to the low number of harvests in this trial and the difference in maturity stage.

Cluster analysis divided cultivars in two major groups composed of several subgroups and showed genetic distances and similarities based on productive traits and chemical composition of DM (Fig. 48.1). Except Swiss cultivars, grouping was not based on geographic origin. Obtained results show that this collection of perennial ryegrass displays considerable traits diversity and indicates that these materials may be suitable for breeding of cultivars with superior DM yield and quality under difficult cropping condition (especially water deficit and high summer temperatures).

Acknowledgements This research has been financed by Ministry of science and technology of Republic of Serbia (Project TR20048).

References

Buxton, D.R., Casler, M.D. 1993. Environmental and genetic effect on cell wall composition and degradability (pp. 685–714). In Jung, H.G. et al. (ed.), Forage Cell Wall Structure and Digestibility. ASA, CSSA and SSSA, Madison, WI.

Charmet, G., Balfourier, F. 1991. Further evaluation of exotic germplasm of perennial ryegrass for use in French breeding programmes. Euphytica 57(1): 67–76.

Gilliland, T.J., Coll, R., Calsyn, E., De Loose, M., van Eijk, M.J.T., Roldan-Ruiz, I. 2000. Estimating genetic conformity between related ryegrass (*Lolium*) varieties. 1. Morphology and biochemical characterisation. Mol. Breed. 6:569–580.

Goering, H.K., Van Soest, P.J. 1970. Forage fiber analyses: Apparatus, reagents, procedures and some applications. USDA-ARS. Agric. Handbook 379. US Gov. Print. Off.

Jensen, K., Waldron, B., Asay, K., Johnson, D., Monaco, T. 2003. Forage nutritional characteristics of orchardgrass and p. ryegrass at five irrigation levels. Agron, J. 95:668–675.

Peeters, A. 2004. Wild and Sown Grasses (pp. 198–207). FAO and Blackwell publishing, Oxford.

Smit, H., Tamminga S., Elgersma A. 2006. Dairy cattle grazing preference among six cultivars of perennial ryegrass. Agron. J. 98:1213–1220.

Soković, D., Tomić Z., Ignjatović S., Šurlan-Momirović, G., Živanović, T. 2002. Genetic variability of perennial ryegrass (*Lolium perenne* L.) autochthonous populations. II. Dry mater yield and chemical composition. Grasslands science in Europe, 7:92–93.

Soković, D., Lugić, Z., Radović, J., Tomić, Z., Babić, S., Vučković, M. 2007. Agronomic traits of new perennial ryegrass cultivar Kruševački 11 (K-11). Proceeding of XI Symposium on forage crops of Republic of Serbia, Novi Sad, No. I, pp. 169–175.

Van Soest, P.J. 1994. Nutritional ecology of the ruminant. Comstock Public Ass., Ithaca, NY.

Wilman, D., Altimimi, K. 1982. The digestibility and chemical composition of plant parts in Italian and perennial ryegrass during primary growth. J Sci. Food Agric. 33(7):595–602.

Chapter 49
Variability and Correlation Between the Seed Yield, Seed Yield Components and Quality of Alfalfa Seed

Rade Stanisavljevic, Jasmina Milenkovic, Jasmina Radovic, Dragoslav Djokic, Dragan Terzic, Goran Jevtic, and Dejan Dodig

Abstract Seed yield and seed yield components (stem height, number of stems per m^2, panicle branches) and quality of alfalfa seed (germination and mean seed weight), were studied for 3 years of research. Three cultivars of alfalfa from Serbia (NS-Slavija, NS-Medijana and Zajecarska 83) and French cultivar Europe were observed. In the sowing year, the highest variability was determined for seed yield (CV=56.6%) and percent of dormant seed (CV=19%), while other traits showed low variability. The lowest variability was determined for seed germination (CV=0.2%). In the 2nd and 3rd year, the highest variability was determined for percent of dormant seed (B_1 CV=13.6% and B_2 CV=13.7%) and for seed yield (B_1 CV=7.4% and B_2 CV=17.5%). Other investigated traits expressed low variability. The lowest variability was expressed in seed germination (B_1 CV=1.0% and B_2 CV=0.6%).

Seed germination is in very strong positive correlation with yield (r=0.90). Number of secondary shoots in correlation with yield showed almost complete negative correlation (r=−0.94).

Keywords Alfalfa · Seed yield components · Correlation · Variability

Introduction

Alfalfa is the most cultivated forage legume in Serbia in the average area of 200,000 ha. (Đukić, 2005). Production of alfalfa seed in Serbia is mostly done during the second growth cycle, while the first cut is used for livestock feed production. Many factors affect alfalfa seed yield in this region, but choice of cultivar with

R. Stanisavljevic (✉)
Institut for forage crops, Trg Kosturnica, Krusevac 37000, Serbia
e-mail: radestanisavljevic@ikbks.com

C. Huyghe (ed.), *Sustainable Use of Genetic Diversity in Forage and Turf Breeding*,
DOI 10.1007/978-90-481-8706-5_49,

adequate traits has a direct impact on seed yield. The aim of this study was to investigate variability of seed yield and seed yield components in four alfalfa cultivars during a 3 years period.

Material and Methods

Sowing was done on April the 16th, 2002 with usual agronomic practices, using 18 $kgha^{-1}$ of seed and row distance was 20 cm. Three alfalfa cultivars from Serbia (NS-Slavija, NS-Medijana and Zajecarska 83) and French cultivar Europe (factor A) were observed, during 3 years period, with different climatic conditions (Table 49.1). In the 1st year (2002-B_0), the first growth was used for the study of alfalfa seed yield and seed quality. In the second and 3rd years (2003-B_1 – 2004, B_2 – factor B), the second growth was used.

Stem height (cm) has been assessed by measuring of 30 stems from each repetition. The number of stems per m^2 was determined by counting stems in 1m of row length in four repetitions, and than calculated to stem number per m^2. A panicle branch per stem (average total number) total number of branches was assessed by sampling 30 stems from every repetition. Seed yield ($kgha^{-1}$) was measured after threshing.

Seed quality was examined using the ISTA handbook for seed quality. Statistic analysis of results was done using methods of variance analyses (ANOVA). Significance between mean values was determined by LSD test. For traits expressed in percent, transformation data was done. Also variation coefficient (CV%) and correlation coefficients (*r*) were determined.

Table 49.1 Temperatures T (°C) and amount of precipitation P (l m^{-2}), 2002, 2003, 2004

	Year					
	2002-B_0		2003-B_1		2004-B_2	
Months	T (°C)	P. l/m^{-2}	T (°C)	P. l/m^{-2}	T (°C)	P. l/m^{-2}
January	0.4	15.7	−0.3	53.1	−2.1	90.6
February	6.1	1.9	−3.3	19.4	2.2	41.8
March	8.8	15.3	4.2	7.8	6.6	46.6
April	1.3	50.3	10.2	89.0	11.9	46.4
May	18.6	45.2	22.9	60.5	14.9	27.6
June	22.6	43.5	22.5	43.3	19.5	81.3
July	23.8	117.3	22.3	55.6	21.9	49.0
August	20.8	118.0	24.3	1.3	20.5	62.1
September	15.7	73.1	15.6	67.6	15.9	35.6
Oktober	10.6	53.6	9.1	149.0	12.2	45.9
November	6.6	30.3	6.5	27.2	6.6	48.6
December	−2.2	63.3	−0.5	35.7	1.9	34.8
Aver. or total	11.8	627.5	11.1	609.5	11.0	641.2

Results and Discussion

Stem height indicates the general crop condition and in that way it affects seed yield. In the sowing year, alfalfa cultivars had average stem height of 46.4 cm (Table 49.2). The cultivar Zajecarska 83 had the highest stem of 47.0 cm while Novosadjanka H-11 had the shortest stems (45.7 cm). Average stem height in second and 3rd year was 83.2 cm, the tallest being Zajecarska 83 (84.8 cm), while the shortest was Novosadjanka H-11 (80.9 cm). For this trait, narrow variability was determined (B_0 CV=1.2% in the sowing year, B_1 CV=3.9% and B_2 CV=5.2% in 2nd and 3rd year). There was a strong positive correlation ($r = 0.74$) between stem height and seed yield. These results are relevant with Bolanos – Aguilar et al. (2002).

On the contrary, Karimov (1987) pointed that stem height and seed yield were strongly negatively correlated ($r = -0.67$) mostly in years with high level of summer precipitation. For a good initial development of alfalfa, well distributed rainfalls are needed. This was the case in year of sowing, and affected large number of stems (average 777 per m^2), from 785 per m^2 (NS Slavia) to 719 per m^2 (Europe). In the 2nd and 3rd years, average stem number was similar to the value in the 1st year, with significant cultivar influence (Table 49.3). Variability for this trait in the sowing year was CV=5.8%, in 2nd year was CV=6.3%, and in 3rd year was CV=2.5%.

Stem number and seed yield were almost complet negatively correlated ($r = -0.94$) (Tab. 4). In correlation with these results, Karagić (2004) pointed a negative relationship ($r = -0.44$) in agroecological conditions of Vojvodina, while a strong positive relationship ($r = 0.74$) was only met between the number of fertile stems and seed yield.

Studied alfalfa cultivars had average 2.0 branches per stem in the year of sowing. Number of branches per stem were largest in cultivars of Europe, Zajecarska and

Table 49.2 Seed yield, seed yield components and seed quality of alfalfa cultivars in the sowing year, B_0

	Seed yield components				Seed quality component		
Cultivar	Plant height (cm)	Stem number per m^2	No. of panicle branches per stem	Seed yield ($kgha^{-1}$)	Germ. (%)	Hard seeds (%)	1000 - seed (g)
Novosađanka H-11	45.7	776	1.9	51.2	76.7	8.3	2.2
NS-Slavija	46.5	785	2.0	102.3	76.9	9.6	2.1
Zaječarska-83	47.0	829	2.0	24.5	76.5	6.3	2.1
Europe	46.5	719	2.0	51.6	76.6	7.0	2.0
Mean	46.4	777	2.0	57.4	76.7	7.8	2.1
CV (%)	1.2	5.8	2.1	56.6	0.2	19.0	2.6
LSD 0.05	3.32	32.38**	0.14	4.99**	0.87	1.20**	0.13
LSD 0.01	4.77	46.52	0.21	7.17	1.25	1.72	0.19

significant; * : $p \geq 0.05$ ** : $p \geq 0.01$.

Table 49.3 Seed yield, seed yield components and seed quality of alfalfa cultivars in the 2nd and 3rd years, B_1-B_2

		Seed yield components				Seed quality component		
Cultivar	Year	Plant Height (cm)	Stem number per m²	No. of panicle branches per stem	Seed yield (kgha^{-1})	Germination (%)	Hard seeds (%)	1000 – seed (g)
Novosadja nka H-11	B_1	81.2	659	3.7	607	92	12.3	1.8
	B_2	80.5	864	2.8	300	78	17.5	1.9
	Mean	80.9	753	3.3	454	85	14.8	1.9
NS-Slavija	B_1	87.1	687	3.8	708	90	13.8	1.8
	B_2	79.6	848	2.9	413	78	20.1	1.8
	Mean	83.4	768	3.4	561	84	16.9	1.8
Zajecarska −83	B_1	86.0	618	3.8	638	92	9.9	1.6
	B_2	83.5	841	2.7	326	78	14.4	1.8
	Mean	84.8	730	3.3	482	85	12.3	1.7
Europe	B_1	89.1	598	3.8	702	91	10.2	1.8
	B_2	79.2	815	2.9	431	77	17.4	1.9
	Mean	84.2	707	3.4	558	84	13.6	1.9
CV (%)	B_1	3.9	6.3	1.3	7.4	1.0	15.9	5.7
	B_2	5.2	2.5	3.4	17.5	0.6	13.7	3.1
LSD A	0.05	6.15	76.91	0.14	40.766**	3.99	3.16	0.17
	0.01	8.54	103.19	0.19	56.582	5.54	4.39	0.24
B	0.05	4.35*	54,10**	0.10**	28.826**	2.82**	2.24**	0.12
	0.01	6.04	75.09	0.14	40.009	3.92	3.10	0.17
AxB	0.05	10.77	128,63*	0.29	0.69**	7,08*	6.09*	0,34
	0.01	12.06	142.36	0.36	0.84	8.29	7.63	0.48

*A-cultivar; B-year. significant; * : $p \geq 0.05$, **: $p \geq 0.01$.

Slavia (2.0 branches per stem) and the lowest for cultivar of Novosadjanka H-11 (1.9 branches per stem). In the 2nd and 3rd years, favorable growing conditions positively affected branching (on average 3.4 branches per stem). The cultivar did not significantly affect this trait in the sowing year and in 2nd and 3rd year of life, that showed low variability (CV=2.1% in the sowing year, CV=1.3% in second and CV=3.4% in 3rd year). Very strong positive dependence (r=0.79) was determined between number of branches per stem and seed yield (Table 49.4).

In the year of sowing examined cultivars achieved very low seed yield (57.4 kgha^{-1}), and cultivar had a significant effect (Table 49.2). Average yield in 2nd year was 663.8 kgha^{-1}, with cultivar variability CV=7.4%. Average yield in 3rd year was 367.5 kgha^{-1}, with cultivar variability CV=17.5%. Numerous researchers pointed that the choice of cultivar can significantly affect seed yield (Bolanos-Aguilar at *al.*, 2002; Pelikan, 2003; Bekovic, 2005; Radovic, 2006). In 1st year,

Table 49.4 Correlation coefficient of seed yield and seed yield components of alfalfa cultivars during, B_0, B_1, B_2

	II	III	IV	V	VI	VII
Seed yield (I)	0.74**	−0.94**	0.79**	0.90**	−0.56**	−0.40*
Plant height (II)	-	−0.76**	0.72**	0.71**	0.58**	0.32
Stem number (III)		-	−0.54**	−0.81**	0.58**	0.32
No. of branches per stem (IV)			-	0.90**	0.34*	−0.88**
Germination seed (V)				-	−0.06	−0.72**
Hard seeds (VI)					-	−0,45*
Masa of 1000 seed (VII)						-

significant; * : $p \geq 0.05$, ** : $p \geq 0.01$. n=36

average germination was 76.7%, and impact of cultivars was not significant (Tab. 4). Cultivar as factor, in the 2nd and 3rd years of study, was not significant for germination. There is very strong positive correlation dependence between seed germination and seed yield ($r = 0.90$). According to Berngardt (1988), water ratio is significant factor in the stages of seed formation and grain filling. Cultivar impact on hard seed percent in the sowing year, was significantly higher than in the following years (B_0 CV=19%, versus B_1 CV=15.9% and B_2 CV=13.7%) (Tab. 3). In the sowing year, the examined cultivars had average weight of 1000 seeds of 2.1 g, with small cultivar impact (CV=2.6%), while average value for this trait for the 2nd and 3rd years was 1.8 g with small cultivar impact too (Tab. 3).

During 2nd and 3rd year of investigation ecological conditions (factor B) had high significant impact ($p \geq 0.01$) on all traits except plant height (significant impact $p \geq 0.05$). On 1000 seed mass there is no environment impact in this investigation. At the same period interaction of cultivar and environment had high significant impact on seed yield ($p \geq 0.01$) and significant on hard seed percent, stem number and germination ($p \geq 0.05$). There is not any significant impact on other examined traits. Between weight of 1000 seeds and seed yield exist weak negative correlation dependence ($r = -0.40$).

References

Beković, D. 2005. Influence of ecological conditions and sowing method, on seed yield and forage of alfalfa. PhD thesis, Faculty of Agriculture, Priština-Lešak, pp. 153.

Berngardt, I.I. 1988. Osobenosti formirovanija semjan i ih urožaj pri raznih sposobah poseva i orošenii. Nekatorie aspekti semenovodstva ljucerni. Selekcija i semenovodstvo, br, 45–47

Bolanos-Aguilar, E.D., Huyghe, C., Ecalle, C. 2001. Genetic control of alfalfa seed yield and its components. Plant Breed. 120, 1, 67–72.

Bolanos-Aguilar, E.D., Huyghe, C., Ecalle, C., Hacquet, J., Julier, B. 2002. Effect of Cultivar and Environment on Seed Yield in Alfalfa. Crop Sci. 42, 45–50.

Đukić, D. 2005. Lucerne state and perspectives in Europe and Serbia and Montenegro. Periodical of scientific research on field and vegetable crops (Vol. 41, pp. 155–169), Institute of Field and vegetable Crops, Novi Sad.

Hadživuković, S. 1991. Statistic methods (pp. 634). Faculty of agriculture, University of Novi Sad, Novi Sad.

International Rules for Seed Testing (ISTA), 2008. Chapter 5: Germination, 5-31

Karagić, Đ. 2004. Yield components, yield and quality of alfalfa seed depending on cutting schedule. PhD thesis, Faculty of Agriculture Novi Sad, pp. 174.

Karimov, H.Z. 1989. Urožaj semjan lucerni pri raznih sposobah poseva. Selekcija i semenovodstvo, br. 3: 43–45.

Pelikan, J., Gottwaldova, P., Vymyslicky, T. 2003. Evaluation of the Alfalfa Assortment under the Conditions of the Czech Republic. Czech J. Genetich Plant Breed.. 39: 228–231.

Radović, J., Lugić, Z., Sokolović, D., Jevtić, G., Stanisavljević, R. 2006. Genetic variability for seed yield and seed yield components of alfalfa (*Medicago sativa* L.). Procceding of 26th EUCARPIA Fodder Crops and Amenity Grasses Section Meeting and 16th EUCARPIA Medicago spp. Group Meeting. Perugia, Italy, pp. 121–123.

Chapter 50
Influence of Forage Species, Cultivar and Cut on Lipid Metabolism During the Ensiling Process

Gijs Van Ranst, Veerle Fievez, Joost Baert, Muriel Vandewalle, Jan De Riek, and Erik Van Bockstaele

Abstract Lipid metabolism in silage, mainly lipolysis, of different species (perennial ryegrass, red clover and white clover) and three cultivars of white and red clover at three cutting dates in one growing season (April, July and October) were studied. Lipolysis in silage was influenced by cutting date, species and to some extent by cultivar. Furthermore, in some cuts silages of red and white clover displayed a lower lipolysis than silage of perennial ryegrass. On average, over the three cutting dates, respectively 90.3%, 86.4% and 85.7% of the membrane lipids in perennial ryegrass, red clover and white clover were hydrolysed during ensiling. In red clover this could be due to the lipid-protecting properties of polyphenol oxidase (PPO) activity. This was not observed in perennial ryegrass or white clover. Nevertheless, differences in lipolysis in silage between cultivars of red clover were not correlated with PPO activity.

Keywords Fatty acid metabolism · Silage · Clover · Lipolysis · *Trifolium*

Introduction

For human health concerns, there is increasing interest in the fatty acid (FA) composition of animal products and more in particular ruminant products as these are known to contain high amounts of saturated FA mainly due to an extensive hydrogenation of unsaturated FA in the rumen. Saturated FA are generally said to be detrimental for human health. However, through feeding the FA composition of these products can be significantly enhanced by reducing the proportion of saturated FA and increasing the proportion of polyunsaturated FA. This study focuses

G. Van Ranst (✉)
Institute for Agricultural and Fisheries Research (ILVO), Plant Science Unit, 21 Caritasstraat, 9090 Melle, Belgium
e-mail: Gijs.vanranst@ugent.be

C. Huyghe (ed.), *Sustainable Use of Genetic Diversity in Forage and Turf Breeding*,
DOI 10.1007/978-90-481-8706-5_50, © Springer Science+Business Media B.V. 2010

on forages and in particular ensiled ryegrass, red and white clover. During ensiling an extensive lipid metabolism occurs, leading to an accumulation of free FA in the silage. Due to this pre-ruminal lipolysis, a more extensive hydrogenation can occur in the rumen. However, there are indications that in red and white clover silages lipolysis is inhibited by specific plant compounds. For red clover this is most probably polyphenol oxidase, a diphenol oxidising enzyme; for white clover this is less clear, although saponines have been suggested. However, the content and/or activity of these specific plant compounds that reduce lipolysis can vary between cultivars and cuts. Therefore, the goal of this experiment was to study the lipid metabolism in the silage of one ryegrass cultivar, three white and three red clover cultivars and in three cuts.

Material and Methods

The plant material used, included red clover (*Trifolium pratense* L. cultivars Merula, Lemmon and Milvus), white clover (*Trifolium repens* L. cultivars. Barbian, Barblanca and Riesling) and perennial ryegrass (*Lolium perenne* L. cv. Aberdart). All species were grown in the same location: 50° 59′ N, 3° 46′ E. All cultivars were sown in separate fields in three replicates. Herbage was cut on five occasions in 2007: perennial ryegrass on 23 April, 4 June, 16 July, 27 August and 8 October; red and white clover on 8 May, 11 June, 17 July, 27 August and 8 October. For this study only the herbage from the first, third and fifth cuts were used. The harvested herbage was stored overnight at 4°C. The following morning, the herbage was crushed using a metal roller of approximately 50 kg. Next, the herbage was wilted in an oven at 35°C for approximately 8 h until a DM content of approximately 350 g DM kg^{-1} had been reached. After wilting, herbages were ensiled by means of vacuum-packing in thick (200 μm) polyethylene bags. No additive was used. Every silage treatment was made in three replicates. Silages were opened after 9 weeks.

FA were extracted with chloroform/methanol (2/1, v/v) according to Lourenço et al. (2007).

Lipid fractions were separated using solid-phase extraction (SPE) sequentially over two different kinds of SPE columns. After methylation (Raes et al., 2001) FA content and composition was determined by gas chromatography. Lipolysis in silage was calculated as the difference in proportion of total FA that occurred in the membrane lipids (ML) between the wilted herbage and the ensiled forages divided by proportion of total FA in the ML of the wilted herbage. For PPO activity measurement a protein extract from fresh red clover was made. This purified protein extract was used to measure PPO activity according to Robert et al. (1995) using 4-methyl-catechol (Sigma, Bornem, Belgium). Absorbance (A) was measured at 400 nm every 10 s for 1 min. PPO activity is expressed as ΔA min^{-1} mg^{-1} protein. Protein content was analysed using Folin-Ciocalteau reagent (Sigma, Bornem,

Belgium) according to Winters and Minchin (2005). Absorption was measured at 650 nm. Standards of 0 to 0.8 mg mL^{-1} casein were assayed simultaneously.

Results and Discussion

In Fig. 50.1, the lipolysis of total FA from the ML fraction is presented. An extensive lipolysis of ML (proportionately >0.80) occurred in all silages. Nevertheless, at the first cutting date lipolysis was lower in red clover compared to perennial ryegrass silages, white clover was intermediate and did not differ from the other species. At the third cutting date there was no difference between species and at the last cutting date, lipolysis in white clover silages was lower than in red clover and perennial ryegrass silages. At the first cutting date cultivar Barbian showed the lowest and cultivar Riesling the highest lipolysis within white clover. At the third cutting date cultivar Milvus displayed a lower lipolysis than the two other red clover cultivars. At other cutting dates, no effect of cultivar occurred. Lipolysis at the fifth cutting date was higher than at the two other cutting dates.

Due to plant lipases, which are activated by plant stress and damage, and microbial lipases, ML are hydrolysed resulting in an accumulation of free FA in the silage.

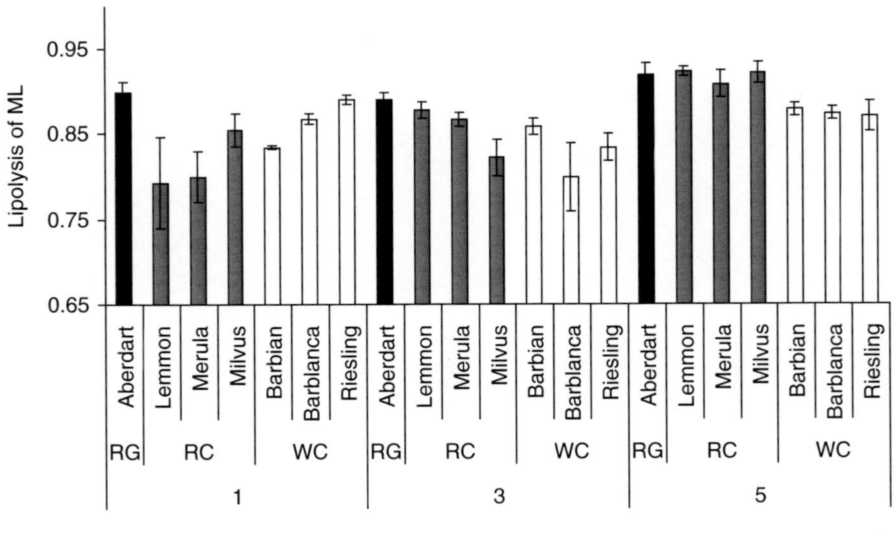

Fig. 50.1 Lipolysis of membrane lipids (ML), calculated as (proportion of total fatty acids (FA) in ML in wilted herbage – proportion of total FA in ML in ensiled forage)/ proportion of total FA in ML in wilted herbage, in silages of cultivars of three species (perennial ryegrass (RG), red clover (RC) and white clover (WC)). Herbages at three cutting dates (1, 23 April; 3, 16 July; and 5, 8 October) were ensiled. *Error bars* indicate 2 × the standard error of mean (n=3)

Lipolysis in silages is a very extensive process. It has already been reported in several studies (Steele and Noble, 1984; Van Ranst et al., 2009). In the present study, effects of species, cultivars and cutting dates were found. The effect of cutting date, however, could have been confounded with differences in DM content at ensiling. Van Ranst et al. (2009) reported that a higher DM content can result in a lower lipolysis in perennial ryegrass. Nevertheless, in the current study, the cutting date with the highest DM content (fifth cutting date) resulted in the highest lipolysis. It has also been suggested that lipolysis could be influenced by fermentation (Van Ranst et al., 2009). The lower lactic acid concentration at the third cutting date, and the higher ammonia and acetic acid concentrations at the fifth cutting date of the red clover (data not shown), could indicate an extended or less efficient fermentation, explaining the higher lipolysis in these silages of red clover. Nevertheless, the difference between white clover and perennial ryegrass, and between red clover and perennial ryegrass at the first cutting date, could suggest that there was a lower lipolysis in red and white clover silages than in perennial ryegrass silages.

Differences in lipolysis, between species as well as between cultivars, could be explained by the presence of lipase-inhibiting compounds in the herbage as suggested by Lourenço et al. (2005). However, the saponins, which were suggested to be responsible for lipase inhibition in white clover, were not studied in this experiment.

Furthermore, variations between species and/or cultivars in activities of enzymes, which are important in lipid metabolism (e.g. lipoxygenase, hydrolase and digalactosylacyl transferase), could also be expected and thus explain at least part of the observed differences. Variations in these enzyme activities, however, were not measured. Nevertheless, the PPO activity, which can influence lipolysis in silages (Lee et al., 2008), was measured in cultivars of red clover (Fig. 50.2).

Yet, no relationship between measured PPO activity and lipolysis in silages with cultivars of red clover was observed. This could have been due to the wilting

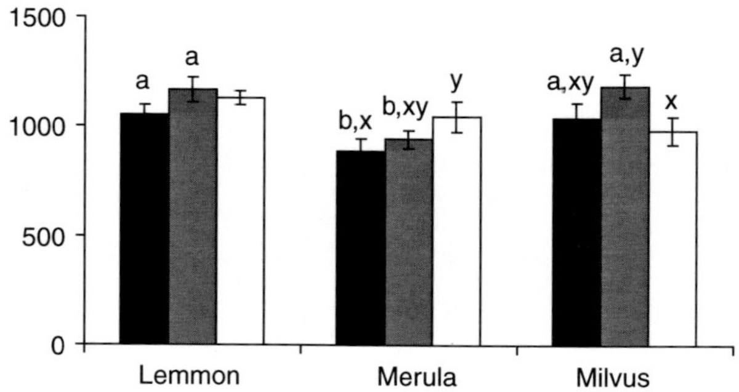

Fig. 50.2 Polyphenol oxidase (PPO) activity, expressed as change in absorbance (A) min^{-1} mg^{-1} protein in a protein extract of three red clover cultivars sampled at three cutting dates (■, 23 April; ▩, 16 July; and □, 8 October). Error bars indicate 2 × the standard error of mean (n=3). a and b, and x and y, indicate significant (P< 0.05) differences between cultivars within cuts and between cuts within cultivars, respectively

time (8 h) since PPO needs an activation period after being damaged, to generate quinones (Lee et al., 2009). These quinones have been suggested to inhibit lipolysis in two ways: firstly by directly binding to the enzymes and secondly by binding to the ML, and protecting them against the lipases. A longer wilting time could have resulted in more quinones binding to lipases and thus in a higher inhibition, as PPO needs oxygen to oxidize phenols to quinones, no further production of quinones can be expected in the silage. Thus a longer wilting time could have reduced lipolysis in silages of red clover even more than silages of perennial ryegrass and make differences between cultivars of red clover with different PPO activities more distinct.

Acknowledgements PhD. Grant of Gijs Van Ranst is funded by the Institute for Agricultural and Fisheries Research (ILVO) – Belgium. The research was partially funded by IWT (Institute for the Promotion of Innovation by Science and Technology). The technical assistance of Charlotte Melis and Sjarai Deschildre was greatly appreciated. Results were also presented in a paper which was accepted for publication in Grass and Forage Science.

References

Lee, M.R.F., Tweed, J.K.S., Minchin, F.R., and Winters, A.L. 2009. Red clover polyphenol oxidase: activation, activity and efficacy under grazing. Anim. Feed Sci. Technol. 149(3–4):250–264.

Lourenço, M., Van Ranst, G., De Smet, S., Raes, K., and Fievez, V. 2007. Effect of grazing pastures with different botanical composition by lambs on rumen fatty acid metabolism and fatty acid pattern of *longissimus* muscle and subcultaneous fat. Animal 1(4):537–545.

Lourenço, M., Van Ranst, G., and Fievez, V. 2005. Difference in extent of lipolysis in red or white clover and ryegrass silages in relation to polyphenol oxidase activity. Comm. Agric. Appl. Biol. Sci. 70(2):169–172.

Raes, K., De Smet, S., and Demeyer, D. 2001. Effect of double-muscling in Belgian Blue young bulls on the intramuscular fatty acid composition with emphasis on conjugated linoleic acid and polyunsaturated fatty acids. Anim. Sci. 73(2):253–260.

Robert, C.M., Cadet, F.R., Rouch, C.C., Pabion, M., and Richardforget, F. 1995. Kinetic-Study of the Irreversible Thermal Deactivation of Palmito (*Acanthophoenix rubra*) Polyphenol Oxidase and Effect of Ph. J. Agric. Food Chem. 43S(5):1143–1150.

Steele, W. and Noble, R.C. 1984. Changes in Lipid-Composition of Grass during Ensiling with or without Added Fat or Oil. Proc. Nutr. Soc. 43(2):A51–A51.

Van Ranst, G., Fievez, V., De Riek, J., and Van Bockstaele, E. 2009. Influence of ensiling forages at different dry matters and silage additives on lipid metabolism and fatty acid composition. Anim. Feed Sci. Technol. 150(1–2):62–74.

Winters, A.L. and Minchin, F.R. 2005. Modification of the Lowry assay to measure proteins and phenols in covalently bound complexes. Anal. Biochem. 346(1):43–48.

Chapter 51
Variability of the Rumen Escape Protein and Fatty Acid Composition of Grass and Clover Species and Cultivars

Muriel Vandewalle, Gijs Van Ranst, Veerle Fievez, Johan De Boever, Chris Van Waes, Jan De Riek, and Joost Baert

Abstract Increase of rumen escape protein (REP) and higher content of polyunsaturated fatty acids (PUFA) in forage are getting risen interest. In this work, fast, reliable and cheap methods were developed to predict REP and C18:3 content using dried, ground and stored samples of grasses and clovers. Prediction of REP was possible using a regression curve and robust NIRS calibrations for in vitro digestibility and ADF, as well as the determination of moisture. Evaluation of C18:3 content was also possible by NIRS. Samples from plots (3 replicates, 5 cuts, 2 N fertilizer levels) of perennial and Italian ryegrass, timothy, orchard grass, meadow fescue, tall fescue, red clover and white clover were analysed using these methods. Variation for REP and C18:3 content was present between and within most species. Orchard grass presented the highest average REP while perennial ryegrass had the lowest one. For C18:3, timothy showed on average the highest content while Italian ryegrass had the lowest one. There were no significant differences in C18:3 content and REP within the fescue species. The white clover presented significantly lower REP and higher C18:3 content than the red clover. The red clover presented no significant differences within species for REP. In general, more variation for REP was present in the grasses while the clovers presented more variation for C18:3 content.

Keywords Clover · Fatty acid · Grass · Rumen escape protein · Variability

Introduction

Rumen escape protein (REP), sugar content and fatty acid composition are key factors for forage quality. In general, proteins of grasses and clovers are extensively degraded in the rumen, resulting in a low nitrogen use efficiency by cattle and a

---

M. Vandewalle (✉)
Institute for Agricultural and Fisheries Research (ILVO), Plant Sciences Unit, 21 Caritasstraat, 9090 Melle, Belgium
e-mail: muriel.vandewalle@ilvo.vlaanderen.be

C. Huyghe (ed.), *Sustainable Use of Genetic Diversity in Forage and Turf Breeding*,
DOI 10.1007/978-90-481-8706-5_51,

high burden on the environment. Poly-unsaturated fatty acids (PUFA) are known to have many positive effects on animal and human health, but are also highly degraded in the rumen. Breeding for improved REP and higher PUFA content is getting risen interest. However there is a lack of simple techniques for the screening of a high number of plants as well as a lack of insight in the variation present between and within species.

Our project aims to acquire the knowledge and techniques to allow the breeding of grass and clover for farm production of high quality forage. Fast, reliable and cheap methods were developed to predict the REP and C18:3 content of high number of samples. The variation of REP and C18:3 content between and within species has been studied on plots of grasses and clovers.

Material and Methods

Development of Prediction Methods

Fast, reliable methods were developed to predict REP and C18:3 content using dried, ground and stored samples of grasses and clovers. The prediction methods were developed from analysis of samples from ryegrasses (*Lolium perenne* and *L. multiflorum*), timothy (*Phleum pratense*), orchard grass (*Dactylis glomerata*), meadow fescue (*Festuca pratensis*), tall fescue (*F. arundinacea*), red clover (*Trifolium pratense*) and white clover (*T. repens*).

The method for the evaluation of REP was developed by analyzing fresh grass and clover samples through in sacco incubations in the rumen of fistulated cows according to the reference CVB protocol (CVB, 2003). Corresponding dried and ground samples were chemically analyzed (moisture, crude protein (CP), ash, neutral detergent fiber (NDF), acid detergent fiber (ADF), acid detergent lignin (ADL), water soluble carbohydrates (WSC) and in vitro digestibility of the dry matter with cellulase (Dc)). The Near Infrared Reflectance Spectroscopy (NIRS) spectra of all fresh and dried samples were taken. No robust NIRS calibration was obtained for the direct evaluation of REP, which was not surprising since the complexity of the trait and the measurements on fresh samples. By means of multiple regression analysis an equation based on Dc, dry matter content (DM) and ADF as independent variables could explain 84% of the variation in REP with a residual error of 3.1%-units: **REP = 33.83 – 0.334 Dc + 0.0542 DM + 0.066 ADF**. NIRS could be used for the determination of the individual parameters included in the regression equation. The prediction of REP was thus possible using a regression curve and robust NIRS calibrations for Dc and ADF, as well as the determination of moisture.

Gas chromatography (GC) and NIRS were applied on fresh, dried-ground and dried-ground-stored (6 months) samples of grasses and clovers for fatty acid determination. A significant effect of drying the samples was observed in the GC results: the total PUFA and C18:3 content decreased but good correlations were found between the results of the fresh and dried samples ($r = 0.83$ for the grasses and $r = 0.65$ for the clovers). No significant difference was observed between the dried sample analysed directly and that analysed after storage. No robust NIRS calibration for C18:3 content could be obtained from the fresh material. Calibrations on dried (stored or not) samples gave good results and evaluation of C18:3 content by NIRS analysis was made possible.

Both developed methods allow the use of dried samples and NIRS which facilitate their application in common breeding programs.

Variability Analysis

The variability of REP and C18:3 content between and within species has been studied on plots of grasses and clovers using the software SPSS 11.1.5 (Norusis, 2002). The plot experiment included 3 replicates / object, five cuts in the year after sowing (around every 6 weeks starting 23/04/07 for the grasses and 05/05/07 for the clovers) and two nitrogen fertilization levels for the grasses. Twelve *L. perenne*, 4 *L. multiflorum*, 2 *P. pratense*, 1 *D. glomerata*, 2 *F. pratensis*, 3 *F. arundinaceae*, 4 *T. pratense* and 3 *T. repens* varieties representing the different types and ploidy levels of each species were present in the plot trial. All samples were dried 48 h at 75°C and ground.

Results and Discussion

Results for REP

The REP of the grasses in plots ranged form 30.4 to 56.2% with an average of 41.2%. Significant interaction was found between factors species, cut and fertilization level. There was a significant effect of the cut. For all species, the first cut presented the lowest REP followed by the fifth cut. The cut with the highest REP varied for the different species but was in most cases the second cut. There was no significant effect of fertilization level. There was a significant effect of the species (Table 51.1): *D. glomerata* had the highest REP and *L. perenne* presented the lowest REP. Within species, there was significant differences between varieties of *L. perenne*, *L. multiflorum*, *P. pratense,* but there was no significant difference within the *Festuca* species (n.b. the number of varieties / species varied and there was only one *D. glomerata* in trial and two *P. pratense*) (Fig. 51.1). There was a significant difference between diploids and tetraploids of both ryegrasses (diploid showing the highest values). The early heading varieties of *L. perenne* had significantly higher REP than the intermediate types which had also significantly higher REP than the late types. The REP was negatively correlated with Dc, CP content and WSC content. It was highly positively correlated with ADF and NDF.

For the clovers in plots the average REP was 34.3% (*min*= 27.7%, *max*= 39.8%). There was interaction between the factors cut and species. There was a significant effect of the cut with the fifth cut having the highest REP and the third cut the lowest. On average red clover varieties had higher REP than the white clover ones.

Table 51.1 Average rumen escape protein (REP) percentage and C18:3 content of the different grasses and clovers species for the 3 replicates, 5 cuts and for the grasses 2 fertilization levels; Lp: *Lolium perenne*, Lm: *L. multiflorum*, Pp: *Phleum pratense*, Dg: *Dactylis glomerata*, Fp: *Festuca pratensis*, Fa: *F. arundinacea*, Tp: *Trifolium pratense, Tr: T. repens*; for each trait averages with a same letter are not significantly different (*P*>0.05); SD = standard deviation

	Lp	Lm	Pp	Dg	Fp	Fa	Mean +/– SD grasses	Tp	Tr	Means +/– SD clovers
REP (%)	40.0 d	42.8 b	41.7 c	44.1 a	41.7 c	42.3 bc	41.2 +/– 5.0	36.5 a	31.7 b	34.3 +/– 3.2
C18:3 content (g/kg DM)	9.0 bc	7.5 d	10.2 a	8.7 c	9.1 bc	9.4 B	8.9 +/– 2.6	13.0 b	14.8 a	13.8 +/– 4.4

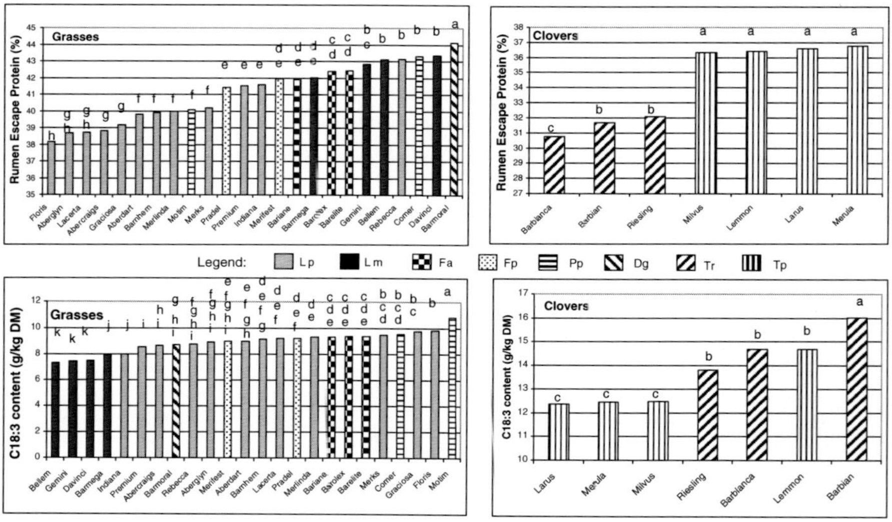

Fig. 51.1 Average rumen escape protein percentage and C18:3 content of the different varieties from the grass and clover species studied; Lp: *Lolium perenne*, Lm: *L. multiflorum*, Pp: *Phleum pratense*, Dg: *Dactylis glomerata*, Fp: *Festuca pratensis*, Fa: *F. arundinacea*, Tp: *Trifolium pratense*, Tr: *T. repens*; within each graph, varieties with a same letter are not significantly different (*P*>0.05)

Differences within species were found only for the white clovers. Correlations had the same trends as for the grasses except that no correlation with WSC and higher positive correlation with yield were observed.

Results for C18:3 Content

The results obtained for C18:3 content are illustrated in Table 51.1 and Fig. 51.1. The C18:3 content of the grasses in plots was on average 8.9% (*min*=1.7%, *max*=15.5%) (Table 51.1). Significant interactions between species, cut and fertilization level were found as for the REP. There were significant effects of the cut, fertilization level and species. The second and third cut had the lowest mean C18:3 content and the fifth cut the highest content. A higher fertilization led to a higher C18:3 content. The C18:3 content of *P. pratensis* was significantly higher and the C18:3 content of *L. multiflorum* significantly lower than the other species. Within species, there were significant differences between varieties of *L. perenne*, *L. multiflorum*, *P. pratensis,* but no significant difference within the studied *Festucas* species. For both ryegrass species, tetraploid varieties had a higher C18:3 content than diploid ones. The early heading *L. perenne* presented the lowest C18:3 content while the late types had the highest C18:3 content.

The C18:3 content of the clovers in plots ranged from 6.9 to 24.3% with an average of 13.8%. Interaction was found between cut and species. There was a significant effect of the cut with the fifth cut presenting the highest C18:3 content. The cut with the lowest content was different for the two clover species (cut 2 for the white clover and cuts 1 and 3 for the red clover). There were significant differences between the two clover species, the white clover having higher content than the red clover (Table 51.1). There were significant differences between varieties of both clover species.

For the grasses and the clovers, the C18:3 content was highly positively correlated with the CP content and to a low extent with digestibility. A moderate positive correlation with WSC was found for the white clovers.

A significant negative correlation between REP and C18:3 content was found for all grasses and clovers in plots for the first harvest year.

Conclusions

Significant differences between and within grass and clovers species for REP and C18:3 content were found in plots. The two traits studied often presented opposite trends. More variation for REP was present in the grasses while the clovers showed more variation for C18:3 content. In the plots, *Dactylis glomerata* (orchard grass) presented the highest mean REP content while *Lolium perenne* (perennial ryegrass) had the lowest one. For C18:3 content, *Phleum pratense* (timothy) showed on average the highest content while *L. multiflorum* (Italian ryegrass) had the lowest one. There were significant differences between the two clover species for both traits. There was no significant difference in REP and C18:3 content within the fescue species, neither differences in REP within the red clover cultivars. The variation identified for both traits should allow the breeding for higher REP and C18:3 content. Results of the second harvest year of the plots still need to be analysed. In order to evaluate the developed methods and the heritability of the studied traits, grass and clover individual plants were analyzed and used in divergent selections (results expected in 2010).

Acknowledgments This project is supported by the IWT-Flanders – contractual research in agriculture.

References

Norusis N.J. 2002 SPSS for Windows release 11.1.5, http://www.spss.com, Chicago, Illinois.

CVB, 2003. Protocol voor in situ pensincubatie: bepaling van afbraaksnelheid en uitwasbare fracties van eiwit, zetmeel, celwanden en organische restfractie (p. 14). Centraal Veevoeder Bureau, Lelystad (Nederland).

Chapter 52
Comparison of Different Low-Input Lignocellulosic Crops as Feedstock for Bio-ethanol Production

Steven Van Hulle, Isabel Roldán-Ruiz, Erik Van Bockstaele, and Hilde Muylle

Abstract Lignocellulosic biomass is a renewable carbon source and can significantly contribute to a reduction of the use of fossil resources for the production of energy, chemicals and materials. Lignocellulose serves as feedstock a.o. for the production of energy (by direct combustion or by fermentation), building materials (wood, thatch, straw, particle board), paper and cardboard. In this study several low-input energy crops are being compared for their potentiality as feedstock for bio-ethanol production. In May 2007, a yield trial was installed with the following crops: two *Miscanthus* species, two varieties of *Phalaris arundinacea* (reed canary grass), two varieties of *Panicum virgatum* (switchgrass), one accession of *Phragmites australis* (common reed) and one willow (*Salix* spp.) cultivar as short rotation coppice reference. The trial is being conducted under low input conditions: no fertilizer is applied and the biomass is harvested once a year in late winter – early spring. In 2008 and 2009, dry matter yield was determined as well as the cell wall composition (ADL, ADF and NDF). Two-year experience and results are discussed.

Keywords Bio-ethanol · Biomass production · Lignocellulosic biomass

Introduction

The increasing worldwide energy consumption, the concerns about environment protection, and the decrease of the world reserves of fossil energy have increased the demand for alternative energy sources. Bio-ethanol is one of the possible alternatives and is already commercially available in several countries all over the world as a first generation bio-fuel. First generation bio-ethanol is produced by fermentation of sugar-rich food crops such as wheat or sugar beet. However, the cultivation of

S. Van Hulle (✉)
Plant Sciences Unit – Growth and Development, Institute for Agricultural and Fisheries Research (ILVO), Caritasstraat 21, Melle B-9090, Belgium
e-mail: Steven.VanHulle@ilvo.vlaanderen.be

C. Huyghe (ed.), *Sustainable Use of Genetic Diversity in Forage and Turf Breeding*,
DOI 10.1007/978-90-481-8706-5_52, © Springer Science+Business Media B.V. 2010

these 'traditional' crops is energetically demanding and first-generation transformation processes exploit only a small part of the biomass produced. Second generation bio-ethanol is generated from lignocellulosic biomass (Somerville, 2007). This biomass can be obtained from agricultural or forestry waste but also from dedicated energy crops. Important characteristics of an energy crop are high yield, low input requirements and perenniality. In addition, quality of biomass will be an important determinant of the potential of a crop as raw material for bio-ethanol production. Good-quality biomass, when focusing on bio-ethanol production, should contain a large amount of polysaccharides which can be released and hydrolyzed into monosaccharides. The large amounts of cellulose and hemicelluloses contained in plant cell walls are the sugar resources for bio-ethanol production, while lignin constitutes a limiting factor in the conversion process (Keating et al., 2006).

Limited information is available on the yield potential and the advantages and disadvantages of dedicated energy crops under Belgian conditions. A field trial was therefore set-up with the main purpose to compare the biomass yield and biomass quality of *Panicum virgatum* (SwitchGrass), *Phalaris arundinacea* (Red Canary Grass), *Miscanthus spp.* (Elephant grass), *Phragmites australis* (common reed) and *Salix spp* (willow). Here we report the results of the first 2 years after installation.

Material and Methods

Yield Trial

In May 2007, a field trial was installed including two cultivars of *Panicum virgatum* (Cave in Rock and Kanlow), two cultivars of *Phalaris arundinacea* (Bamse and Palaton), two *Miscanthus spp.* clones (*Miscanthus x giganteus* – Nl and *Miscanthus sinensis* – Goliath), one *Salix spp.* clone and one *Phragmites australis* variety. The trial was arranged in a randomized block design with three replicates. Each crop was sown or planted at the density recommended for large scale plantations in plots of 28 m^2. *P. virgatum, P. arundinacea* and *P. australis* were sown at a density of 1000 seeds/m^2. *Miscanthus* rhizomes were planted at a density of 2/m^2. For *Salix,* cuttings were planted following a typical short rotation coppice design. During installation (April–May 2007) weeds were controlled with herbicides. No fertilization was applied. The different crops were harvested once a year in February–March except, for willow which will be harvested once in 3 years. Fresh and dry matter yields were determined.

Cell Wall Composition

Cell wall composition of the harvested material was determined according to the NDF, ADF and ADL method described by Van Soest et al. (1991).

Tabl2 52.1 Dry matter yield of 7 objects in spring 2008 en spring 2009 and cell wall composition of harvested material in spring 2009

	DMY2008 (ton DM/ha)		DMY2009 (ton DM/ha)		NDF (g/100 g DM)		ADF (g/100 g DM)		ADL (g/100 g DM)	
Object	Mean	SD	Mean	SD	Mean	SD	Mean	SD	Mean	SD
Reed Canary Grass cv. 'Palaton'	4,00[c]	0,62	8,59[b]	1,25	87,87[b]	0,83	53,47[b]	0,79	8,68[b]	0,41
Reed Canary Grass cv. 'Bamse'	5,64[d]	1,10	9,47[b]	1,42	85,32[a]	1,69	50,99[a]	1,27	8,99[b]	0,38
Switchgrass cv. 'CIR'	2,52[b]	0,63	12,69[c]	1,10	94,08[d]	0,49	59,84[c]	0,10	11,48[d]	0,13
Switchgrass cv. 'Kanlow'	2,58[b]	0,53	12,90[c]	2,58	91,90[c]	0,16	53,15[b]	1,04	7,78[a]	0,65
Miscanthus x *giganteus*	3,31[b,c]	0,86	15,44[d]	0,68	96,57[e]	0,21	68,26[d]	0,16	13,04[e]	0,51
Miscanthus sinensis cv. 'Goliath'	0,49[a]	0,18	4,09[a]	0,55	95,55[d,e]	0,71	58,50[c]	0,37	7,64[a]	0,36
Phragmites australis	0,92[a]	0,25	3,72[a]	0,62	92,11[c]	0,31	58,62[c]	0,58	10,30[c]	0,11

[a, b, c, d, e] homogeneous subsets by use of Duncan's multiple range test ($p<0.05$)

Results and Discussion

Biomass was harvested in March 2008 and March 2009 for all crops included in the trial except for willow, which will only be harvested every 3 years. In Table 52.1, the dry matter yield (DMY) for 2008 and 2009 are represented. In 2008, the mean yield was 2.78 t DM/ha and in 2009 9.56 t DM/ha. The first year after installation, reed canary grass performed the best (5.64 t DM/ha). In the second year, *Miscanthus x giganteus* performed the best with 15.44 t DM/ha followed by switchgrass cv. 'Kanlow' (12,90 ton DM/ha). The yield obtained is comparable to yield reports found in literature (Möller et al., 2007).

Table 52.1 shows cell wall composition of the biomass harvested in 2009. It is clear that most grass species (and cultivars) included in the trial are similar with regards to cell wall composition, except for *Miscanthus x giganteus* which shows the highest ADL (13.04 g/100 g DM). Taking into consideration that lignin is a limiting factor in the bio-ethanol production process, the grass energy crops should be a better resource for ethanol than the short rotation coppice (ADL values of 31%).

Conclusion

These preliminary results indicate that *Miscanthus x giganteus* might be the most productive lignocellulosic crop, also under Belgian conditions. Reed canary grass was the most productive crop in the first year after installation, but was not able to reach the *M. x giganteus* yield in the second year.

This trial will be assessed for at least three more years for yield and cell wall composition. A saccharification test is currently being optimized and will be used to determine the potential of the different crops for bio-ethanol production.

References

Keating, J.D., Panganiban, C., Mansfield, S.D. 2006. Tolerance and adaptation of ethanologenic yeasts to lignocellulosic inhibitory compounds. Biotechnol. Bioeng. 93:1196–1206.

Möller, R., Toonen, M., van Beilen, J., Salentijn, E., Clayton, D. 2007. *Crop platforms for cell wall biorefining: Lignocellulose feedstocks* (p. 176). In: Outputs from the EPOBIO Project, Newbury Berks, UK, CPL press.

Somerville, C. 2007. Biofuels. Curr. Biol.17:R115–R119.

Van Soest, P.J., Robertson, J.D., Lewis, B.A. 1991. Methods for dietaryfiber, neutral detergent fiber and non-starch polysaccharide in relation to animal nutrition. J. Dairy Sci. 74:3583–3597.

Chapter 53
Occurrence of *Colletotrichum trifolii* (Bain *et* Essary), the Inducer of Alfalfa Anthracnose in Serbia

Vasić Tanja, Radović Jasmina, Lugić Zoran, Marković Jordan, Jevtić Goran, and Gajić Sanja

Abstract Southern anthracnose or crown rot, caused by *Colletotrichum trifolii* (Bain *et* Essary), is a disease which has been detected on alfalfa in the 9th decade of last century in Serbia. During a 3 year period, especially in summer and autumn, plants with anthracnose symptoms were studied in alfalfa field. Stem infection resulted in wilting and death of the upper portion of the steam, giving rise to the characteristic "shepherd's crook" symptom. Strains of *C. trifolii* were -isolated from diseased alfalfa stems collected from different locations in Serbia. According to preliminary pathogenicity examination four strains were chosen for further investigations. For determining strains pathogenicity two methods of inoculation were used. Infected plants showed symptoms of "shepherd's crook" in both cases. Phytopathological investigations of seed of four commercial alfalfa cultivars (K-28, NS Mediana, Affinity+Z and Alfagraze) were conducted also, by the agar-plate methods. Investigated cultivars had different reactions to *C. trifolii* isolates. Conidia were hyaline, rounded at the ends, nonsepted with average size 7.85 × 3.87 μm. Average sizes of appressoria were 7.5–16.5 × 5.5–8.9 μm. Molecular analyses, in comparison with referent strains CBS 158.83, confirmed that the isolates Luc-7, Luc-17, Luc-27 and Luc-33 were *Colletotrichum trifolii.*

Keywords *Colletotrichum trifolii* · Morphology · Pathology · Alfalfa

Introduction

Colletotrichum trifolii Bain *et* Essary, inducer of alfalfa anthracnose, is widespread in many areas in Serbia. Considering that and the damage it can cause, anthracnose is an economically very important disease in Serbia (Robotic et al., 1983).

V. Tanja (✉)
Institute for forage crops, Trg Kosturnice 50, Kruševac 37000, Serbia
e-mail: tanja.vasic@ikbks.com

C. Huyghe (ed.), *Sustainable Use of Genetic Diversity in Forage and Turf Breeding*,
DOI 10.1007/978-90-481-8706-5_53,

The damage caused by this pathogen agent can be seen through decreased quantity and quality of green mass from 10 to 30%, depending of the alfalfa cultivar, pathogen species and edaphic factors (Stuteville and Erwin, 1990).

Anthracnose decreases the vigour of the individual plants and thins out the plant population. During the summer and autumn, diseased plants start to appear in the alfalfa fields. These diseased plants have typical appearance. Several stems of the plant have yellow to silverish colour and start to show signs of wilting. The apical part of the plants is curved down forming the so-called "shepard's crook".

The aim of this study was to collect the strains of *C. trifolii* in Serbia and to study their pathogenicity and morphological characteristics, to understand the pathologic processes and possibilities for breeding the resistant alfalfa cultivars.

Material and Methods

Using standard mycological methods, several fungi strains were obtained. For the further study, four of these strains and CBS 158.83 strains which was imported from the Netherlands were selected. Molecular analyses performed according to Chen and Dickman (2002), in comparison with referent strains CBS 158.83, confirmed that the strains Luc-7, Luc-17, Luc-27 and Luc-33 were *Colletotrichum trifolii*. Pathogenicity assessment of these five selected *Colletotrichum trifolii* strains (Luc-7, Luc-17, Luc-27, Luc-33 and CBS 158.83) were further tested by using two different methods. The plants were sampled from fields and transplanted into vessels. The first method (A) is based on inoculation of the damaged alfalfa plants by applying fragments of *Colletotrichum trifolii* colonies of studied strains. Before inoculating, all plantes were damaged in the stem bases. Using the sterile lancete, the surface layer of stem is removed in the form of square "windows". The second method (B) is based on inoculation of the undamaged alfalfa plants by applying colonies fragments. In both methods the control group of plants were inoculated with the substrate that did not contain mycelium. For plant scoring, scores in range from 1 to 4 were used (Barnes et al., 1969).

Susceptibility of different alfalfa cultivars to *Colletotrichum trifolii* was investigated on four cultivars. Krusevacka-28 (K-28), NS Mediana are domestic cultivars mostly spread in our agroecological conditions and there are not information about their resistance to *Colletotrichum trifolii* and American cultivars, Affinity + Z, highly resistant (HR) and Alfagraze, moderate resistant (MR) to *Colletotrichum trifolii*. Phytopathological studies of seeds of the four commercial alfalfa cultivars were conducted using the agar-plate methods according to Graham et al. (1976). Plants were categorised based on the anthracnose reaction using the 1–5 scale (1 = no stem lesions or only few small water-soaked or black spots; 2 = stems with elongated black lesions but without acervuli; 3 = stems with long, wide, but non-gridling lesions, with acervuli present; 4 = large, coalescing and sporulating lesions which gridle and kill upper part of seedling; 5= seedling dead) by Ostazeski et al. (1969), where 1 and 2 are assigned to resistant plants. Plants were scored individually, 2 weeks after inoculation. Disease intensity or severity of infected plants was calculated using the severity index. The morphological traits of the five selected

C. trifolii isolates (Luc-7, Luc-17, Luc-27, Luc-33 and CBS158.83) were studied on the nutrient medium using the method of Baxter et al. (1983). The morphological traits of appressoria of the studied isolates were determined using the modified method of Hawksworth and Graham (1974). The presence or absence of sets in the culture was determined using the method by Smith and Black (1990). Also, the possibility of forming the teleomorphic state in the isolates was studied according to Baxter et al. (1983).

Results and Discussion

Pathogenicity Assessment

All studied isolates caused the symptoms of alfalfa anthracnose on the plants that were treated using both methods (Table 53.1).

Table 53.1 The pathogenicity of the five *Colletotrichum trifolii* isolates on the alfalfa, assessed using two different inoculation methods

	Method A		Method B	
Isolates	After 15 days	After 30 days	After 15 days	After 30 days
Luc-7	+	+ +	+	+ +
Luc-17	+ +	+ + +	+	+ + +
Luc-27	+ +	+ + +	+	+ + +
Luc-33	+ +	+ + +	+	+ +
CBS158.83	+ +	+ + +	+ +	+ + +
Control	–	–	–	–

LEGEND: – no lesions (class 1), + lesions are present on the stem, but there are no withered stems (class 2), + + numerous lesions on the stem, stems starts to wilt (class 3), + + + stems are withered and dry (class 4)

However, the intensity of the symptoms was different between the used methods after 15 days. The intensity of the symptoms in the first method (A) was somewhat stronger than the intensity of the symptoms in the method B. This can be explained with the fact that conidia need some time to start germinating, to form the appressoria, to penetrate cuticle and to infect the plant. Thirty days after the infection, there was no significant difference in expression of the symptoms between the inoculation methods.

Study of the Susceptibility of Different Alfalfa Cultivars to C. trifolii

Investigated cultivars showed various resistances to different *C. trifolii* isolates (Table 53.2).

Table 53.2 The resistance of different alfalfa cultivars to *C. trifolii*

		Plants categorised by Ostrazeski scale (%)						
Isolates	Alfalfa cultivar	1	2	3	4	5	Severity index (%)	Resistant plants (1+2%)
Luc-7	K-28	0	0	45,90	29,56	24,59	3,79	0
	NS M	0	0	29,78	34,04	36,17	4,06	0
	Aff.+Z	0	9,16	13,46	26,92	50,00	4,17	9,16
	Alfag.	26,15	32,31	30,77	10,77	0	2,26	58,46
Luc-17	K-28	5,77	26,92	28,85	23,07	15,38	3,15	32,69
	NS M	0	0	27,27	31,81	40,90	4,14	0
	Affi.+Z	25,00	19,23	28,85	15,38	11,54	2,69	44,23
	Alfag.	24,56	31,79	22,81	12,28	8,77	2,49	56,35
Luc-27	K-28	0	24,00	22,00	34,00	20,00	3,50	24
	NS M	17,95	38,46	23,08	20,51	0	2,64	56,41
	Affi.+Z	0	4,55	22,73	43,18	29,55	3,98	4,55
	Alfag.	31,25	33,33	16,67	12,50	6,25	2,29	64,58
Luc-33	K-28	0	12	20	30	38	3,94	12
	NS M	0	0	20	50	30	4,10	0
	Affi.+Z	6,81	11,36	20,45	38,64	22,73	3,59	18,17
	Alfag.	28,26	30,43	19,65	15,17	6,25	2,41	58,69
CBS 158.83	K-28	21,05	31,58	18,42	28,95	0	2,55	52,63
	NS M	7,69	15,38	12,82	43,59	20,51	3,54	23,07
	Affi.+Z	25,40	26,98	20,63	19,48	7,94	2,57	52,38
	Alfag.	28,33	35,00	25,00	8,33	3,33	2,23	63,33

LEGEND: 1 – no stem lesions or only few small water-soaked or black spots, 2 – stems with elongated black lesions but without acervuli; 3 – stems with long, wide, but non-gridling lesions, with acervuli present ; 4 – large, coalescing and sporulating lesions which gridle and kill upper part of seedling; 5 – seedling dead

Cultivar Alfagraze had the highest proportion of healthy plants, that ranged of 64.5% for strain Luc-27–56.3% for Luc-17. Variety NS Mediana showed lowest resistance for four isolates (severity index 3.54–4.14), but it was very tolerant to Luc-27 (56.41% resistant plants). Cultivars resistance to strain CBS 158.83 were highest. Until now, three pathogenic races of *C. trifolii* have been identified worldwide. Irwin et al. (2006) showed that alfalfa clones had different resistance on various races of *C. trifolii*. It is possible that strains used in our investigation belong to different *Colletotrichum trifolii* races. It can explaned different response of investigated cultivars on applied *Colletotrichum trifolii* strains.

Despite the large number of studies, which indicated that resistance to anthracnose is controlled by major dominant or recessive genes and other resistance mechanisms, resistance to *C. trifolii* is not clear enough. The possibility of combining non-specific resistance with major gene resistance for anthracnose may, therefore, exist (Melotto et al., 2000).

Morphological Characteristics

The colonies developed uniformly. They gained olive green to gray coloration on the entire surface. They also produced acervulae and pink spore mass. Acervulae were small in size, 100–250 μm in diameter. Baxter et al. (1983) already pointed out that when cultivated on KDA medium *Colletotrichum trifolii,* formed acervulae.

It was also determined that the isolates Luc-17 and CBS 158.83 formed sets in the conidiomates. Sets were septed with 1–3 septs, and 45.5–65.45 × 3.2–5 μm in size. Graham et al. (1976) indicated that phytopathogen fungus *C. trifolii* rarely formed sets in the culture.

The average colony size was 7.85 × 3.87 μm. Consistently to Baxter et al. (1983), conidia were cylindrical, rounded at the both ends. Formed appressoria were light brown or hyalic at the first. Average dimensions of the appressoria were 7.5–16.5 × 5.5–8.9 μm. Baxter et al. (1983) pointed that phytopatogen fungus of the genus *Colletotrichum* had the ability to form appressoria when infectious hyphae touched solid surface.

All studied *Colletotrichum trifolii* isolates obtained from Serbia, as well as the control isolate CBS 158.83 obtained from Netherlands did not form perithecia. Baxter et al. (1983) reported that teleomorphic state was never found in association with anamorphic state in the culture.

After the study of pathogenicity of the *C. trifolii* isolates obtained from Serbia, it was determined that inoculated plants showed symptoms of anthracnose.

References

Barnes, D.K., Ostazeski, S.A., Schillinger, J.A., Hanson, C.H. 1969. Effect of anthracnose (*Colletotrichum trifolii*) infection on yield, stand and vigor of alfalfa. Crop Sci. 9:344–346.

Baxter, A.P., Westhuizen, G.C.A., van der; Eicker, A. 1983. Morphology and taxonomy of South African isolates of *Colletotrichum*. South African J. Bot. 2:259–289.

Chen, C., Dickman, M.B. 2002. *Colletotrichum trifolii* TB3 Kinase, a COT1 homolog, is light inducible and becomes localized in the nucleus during hyphal elongation. Eukaryotic Cell 1(4):626–633.

Graham, J.H., Devine, T.E., Hanson, C.H. 1976. Occurrence and interaction of three species of *Colletotrichum* on alfalfa in the Mid-Atlantic United States. Phytopath. 66:538–541.

Hawksworth, D.L., Graham, S.O. 1974. Mycologists handbook. Commonwealth mycological Institute, Kew, England.

Irwin, J.A.G., Aitken K.S., Mackie, J.M., Musial, J.M. 2006. Genetic improvement of lucerne for anthracnose (*Colletotrichum trifolii*) resistance. Aust. Plant Pathol. 35(6):573–579.

Melotto, M., Balardini, R.S., Kelly, J.D. 2000. Host-pathogen interaction and variability of *Colletotrichum lindemuthianum* (pp. 346–361.). In: Prusky. D., Freeman, S., Dickman, M.B. (eds.), "*Colletotrichum* host specificity, pathology, and host-pathogen interaction". The American phytopatological society St. Paul, Minnesota.

Ostazeski, S.A., Barnes, D.K., Hanson, C.H. 1969. Laboratory selection of alfalfa for resistance to Anthracnose, *Colletotrichum trifolii*. Crop Sci. 9:351–354.

Robotic, V., Klokocar-Smit, Z., Djukic, D. 1983. Reakcija nekih sorti lucerke prema *Colletotrichum trifolii* Bain et Essary. Zbornik naučnih radova sa IV Jugoslovenskog simpozijuma o krmnom bilju, Novi Sad 378–386.

Smith, B.J., Black, L.L. 1990. Morphological, cultural and pathogenic variation among *Colletotrichum* species isolated from strawberry. Plant Dis. 74:69–76.

Stuteville, D.L., Erwin, D.C. 1990. Compendium of alfalfa diseases (2nd ed.). The American Phytopathological Society St. Paul, Minnesota.

Chapter 54
Resistance of Red Clover to Broad Spectrum of *Sclerotinia trifoliorum*

Tim Vleugels, Joost Baert, Kurt Heungens, Marianne Malengier, Gerda Cnops, and Erik van Bockstaele

Abstract In this research a diversity study on different European isolates of the pathogenic fungus *S. trifoliorum* will be performed using mycelial compatibility grouping and AFLP. The next step is the development of a bio-test to screen red clover plants for their resistance level against clover rot. The third step in this research is evaluating different European strains for their virulence and evaluating a broad spectrum of red clover varieties for their resistance against *S. trifoliorum*. Over 100 varieties will be evaluated, including cultivars, landraces and wild varieties. Finally the inheritage of clover rot resistance in red clover will be evaluated by a QTL study.

Sclerotinia isolates have been collected from clover fields among different European countries. Mycelial compatibility has shown a large variability within fields. The DNA extraction has been optimized. Sequencing of the ITS-region will be used to determine the exact species of every isolate. Primer combinations are currently being tested for the AFLP study. Different culture media were tested for their capacity to induce the production of multiple big sclerotia. The most optimal medium is being used to produce sclerotia from every isolate. Sclerotia are induced to apothecia formation and formed ascospores will be used to construct the bio-test.

Keywords AFLP · Bio-test · Diversity study · *Sclerotinia trifoliorum* · *Trifolium pratense*

Introduction

Since the 16th century, red clover has been grown to a large extent in Europe. After the Second World War, the area of red clover has been severely reduced. Nowadays there is a growing interest in red clover again, although some setbacks

T. Vleugels (✉)
ILVO (Institute for Agricultural and Fisheries Research), Unit Plant, Caritasstraat 21, Melle 9090, Belgium
e-mail: tim.vleugels@ilvo.vlaanderen.be

C. Huyghe (ed.), *Sustainable Use of Genetic Diversity in Forage and Turf Breeding*, DOI 10.1007/978-90-481-8706-5_54,

still remain. Low persistence and susceptibility to diseases are two important ones. One very important disease in Europe is clover cancer or clover rot, caused by the necrotrophic ascomycete fungus *Sclerotinia trifoliorum*.

S. trifoliorum survives the summer as sclerotia, buried in the soil. In the fall, temperatures will drop and small mushrooms (apothecia) will appear on the sclerotia. These mushrooms produce ascospores that can infect red clover leaves. During the winter the fungus grows systemically through the plant and at the end of the winter, the plant finally dies. In this dead plant material, new sclerotia are formed to provide inoculum for the next fall. The development of clover cancer is highly dependent on the weather conditions: a humid fall, necessary for the germination of the ascospores, and a warm winter with short periods of frost are favourable. Cold winters slow mycelium growth down too much and thereby prevent the disease from spreading (Raynal et al., 1991).

Clover cancer is difficult to control. Fungicides are uneconomical and in some European countries not admitted to be used in fodder crops (Raynal et al., 1991). Resistant cultivars have yet to be developed and the breeding for resistant material is being slowed down by the lack of a bio-test usable in breeding programs. Because of the great annual variation in disease severity, natural infection is not an effective means to screen for resistant material.

The first aim of this project is to investigate the diversity of *S. trifoliorum* among European isolates. Diversity can be assessed by vegetative compatibility testing and molecular methods (AFLP). The ITS-region will be sequenced to determine the exact species of every isolate.

Vegetative compatibility tests can be used to investigate the diversity of *S. trifoliorum* genotypes within a field (Öhberg, 2008). Vegetative compatibility is controlled by at least 3 loci in *S. trifoliorum*. There are probably many alleles for each locus and these alleles are very sensitive to mutation and recombination, so that many different allelic combinations exist. Mycelia with exactly the same allelic combination belong to the same compatibility group (MCG) and their hyphae can fuse. Each compatibility group is composed of highly related, if not identical individuals. Mycelia belonging to different MCGs show a reaction line at their interaction point, which can be seen clearly on some culture media (Durman et al., 2003; Glass and Kaneko, 2003).

An AFLP study has been used to determine the diversity among isolates of *Sclerotinia homeocarpa* (DeVries, 2006). The used enzymes and primer combinations are a point of start for *S. trifoliorum*.

The second aim is to construct a bio-test that is easy to conduct and applicable on a large sale. Different infection assays have already been developed in the past, with some of them infecting through ascospores (Marum et al., 1994; Öhberg, 2008; Vaverka et al., 2003). Although it has not been shown that ascospore inoculation reveals disease resistance better than the customary infection with mycelium fragments, our preference goes to a bio-test infecting through ascospores, since it is the natural inoculum. We also prefer young, cold-treated plants because it has been suggested that cold treatments are required for the expression of resistance against *S. trifoliorum* (Öhberg, 2008). Ascospores are produced by apothecia that appear on buried sclerotia in the fall. Sclerotia and ascospores can also be produced in vitro. One gram of dry sclerotia can produce up to 150 million ascospores in its lifetime (Delclos and Raynal, 1995). Hence many sclerotia will have to be produced. Different culture media referred to by the literature will be evaluated in this research.

The third step in this research is the evaluation of the virulence of fungal isolates by screening them on 6 cultivars with different resistance levels, using the new bio-test. Next, a broad spectrum of red clover varieties will be evaluated for their resistance against the most virulent *S. trifoliorum* isolates. We plan to evaluate 122 red clover varieties (cultivars and wild varieties), including 85 varieties from the USDA red clover gene bank core collection. Resistant genotypes could lead to the development a new resistant cultivar.

The final step is to investigate the inheritance of clover rot resistance in red clover. Inheritance will probably be controlled by more than one gene, since complete resistance has not yet been found although there is a certain degree of variation in resistance level between different varieties. In this step, resistant plants will be crossed with susceptible ones and a QTL-analysis will be started.

Material and Methods

Fungal Isolates

Fungal isolates have been collected and received from Belgium (4 isolates), Switzerland (5 isolates), the UK (1 isolate), Poland (3 isolates), Czech Republic (1 isolate), Lithuania (2 isolates), Sweden (2 isolates) and Finland (1 isolate). We also used 3 control isolates, purchased from the BCCM-MUCL genebank for fungi. More isolates will be added later on. Isolates are stored as mycelium on potato dextrose agar at 4°C.

Diversity Study

Different media have been evaluated for their ability to show vegetative incompatibility. The most suitable medium consists of 20 g/l malt extract, 20 g/l glucose, 1 g/l peptone and 15 g/l agar. A clear reaction line between incompatible genotypes is seen on this medium. Mycelium compatibility grouping was used to assess the amount of distinct genotypes in an infected red clover seed lot from Sweden (15 sclerotia) and in two Belgian clover fields in the same region (25 sclerotia).

For DNA extraction, mycelium plugs cut with a sterile cork borer were inoculated in 100 ml Erlenmeyer flasks containing 20 ml potato dextrose broth (PDB) or other culture media. The flasks were incubated in the dark for 3 days at 22°C under continuous shaking at 200 rpm. The most suitable method for DNA extraction has been experimentally assessed and proved to be the PureGene® kit. This method has been used for the next experiments.

To determine the species of every isolate, we sequenced the ITS-region (internal transcribed spacer). A PCR with ITS1 and ITS4 primers (35 cycles) was performed to amplify the ITS-fragment. Gel analysis showed the amplified fragments just below 500 bp length. The bands were cut out of the stained gel and the ITS-fragments were purified using the Nucleospin® extract II gel purification kit. After setting the volume to 20 μl, the fragments were sequenced.

For the AFLP-study, DNA extraction was performed as described above. Multiple primer combinations are currently being tested.

Bio-test Development

For production of ascospores, many sclerotia are needed. 5 different media were tested for their ability to induce sclerotia production. From each medium, 5 jars with equal amount were inoculated with the UK-strain. After 25 days incubation at 22°C in the dark, the sclerotia were collected and weighed.

Harvested sclerotia were incubated at 30°C for 4 weeks to break their dormancy. Afterwards they were placed in plastic containers filled with wet vermiculite and incubated at 15°C under light. Apothecia appeared after 3 weeks and produced ascospores for 2–3 weeks. Spores were collected using a 5 μm membrane filter connected to a vacuum pump and stored at –20°C in a tube with a drying agent (silica gel).

Results and Discussion

Diversity Study

The Swedish isolate (15 sclerotia) consisted of 7 MCGs, while the Belgian isolates housed 10 MCGs. Because the Belgian isolates were collected in two small fields (less than 1 hectare), it is surprising that there are so many MCGs present. Hence, there is a large variability of *Sclerotinia* genotypes in Belgian red clover fields. Öhberg (2008) proved that there was a large variability of *S. trifoliorum* genotypes in red clover fields in Sweden.

The DNA extraction from fungal tissue has been optimized for *S. trifoliorum*. We originally extracted DNA from mycelium cultures grown in liquid culture medium using the CTAB protocol, generally the most used protocol for DNA extractions. If applied on *S. trifoliorum* though, the DNA yield was low and there were problems with complex carbohydrates as shown by gel analysis and Nanodrop® measurements. Carbohydrate contamination was less when 3-days-old or even younger cultures were used. Different liquid culture media were tested, but the medium proved to have little influence on the quality of the extracted DNA, so PDB was used in the following experiments. Other DNA extraction kits have been tested on 3-days-old mycelium cultures for their ability to extract DNA in a high concentration with limited contamination. The PureGene® kit proved to be the best kit for *S. trifoliorum* fungal tissue and has been used for the following experiments.

First experiments with ascospores suggested that some isolates were not *S. trifoliorum*. The shape of the apothecia and the size of the ascospores in the asci suggested that some isolates were *S. sclerotiorum* instead of *S. trifoliorum*. This is why the exact species is determined for every isolate using the ITS-region sequence. ITS sequencing has been performed on some of the isolates, but it revealed little variation in sequence between the genotypes. BLAST search indicated that the sequences of all strains had over 99% homology with known *S. trifoliorum* ITS sequences, but also with *S. sclerotiorum* sequences. This result suggests that the ITS region might not be ideal to discriminate between *S. sclerotiorum* and *S. trifoliorum*, so we are now looking to other genes, like the Elongation factor α.

Bio-test Development

From the 5 tested culture media, oatmeal with a red clover infusion induced the highest production of sclerotia (Fig. 54.1). This medium has been used in the following experiments.

Sclerotia are being produced from each fungal isolate and the induction of ascospore production is ongoing. Ascospores are being stored –20°C and will be used to construct the bio-test.

Jars were filled with an equal amount of medium, inoculated and incubated for 25 days at 22°C in the dark. Statistical analysis was performed with ANOVA ($\alpha = 0.05$)

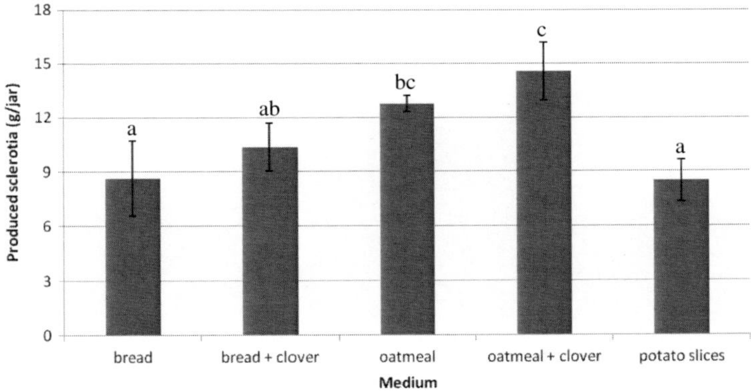

Fig. 54.1 Sclerotia production on 5 different media

Conclusions

There is a large variability of *Sclerotinia* genotypes in Belgian red clover fields. A suitable gene for species identification is being looked for.

A suitable medium for the production of sclerotia has been found, the method to produce ascospores has been optimized and ascospores are being produced.

Acknowledgements This research is completely financed by ILVO. Special thanks to all the people across Europe who kindly collected *Sclerotinia* samples in their red clover fields and donated them to ILVO for the benefit of this research.

References

Delclos, B., Raynal, G. 1995. Comparison of techniques for the production of *Sclerotinia trifoliorum* ascospores in the laboratory for forage legume resistance tests. J. Phytopathol. 143:345–3480.

DeVries, R.E. 2006. Genetic diversity of *Sclerotinia homoeocarpa* delineated by AFLP and vegetative compatibility. Master of Science Degree Thesis, University of Tennessee, Knoxville, USA.

Durman, S.B., Menéndez A.B., Godeas, A.M. 2003. Mycelial compatibility groups in Buenos Aires field populations of *Sclerotinia sclerotiorum* (Sclerotiniaceae). Aust. J. Botany 51: 421–427.

Glass, N.L., Kaneko, I. 2003. Fatal attraction: nonself recognition and heterokaryon incompatibility in filamantous fungi. Euk. Cell 2:2–8.

Marum, P., Smith R.R., Grau, C.R. 1994. Development of procedures to identify red clover resistant to *Sclerotinia trifoliorum*. Euphytica. 77:257–261.

Öhberg, H. 2008. Studies of the persistence of red clover cultivars in Sweden, with particular reference to *Sclerotinia trifoliorum*. Doctoral Thesis.

Raynal, G., Gayraud, P., Mousset-Declas, C., Serpeille, A. 1991. Possibilités de la lutte contre la sclérotinoise de trèfle violet. Fourrages 127:335–344.

Vaverka, M., Vaverka, S., Vichova, J. 2003. Resistance of the Czech assortment of red clover *Trifolium pratense* L. to the stem and crown rot *Sclerotinia trifoliorum* Erikss. Czech. J. Genet. Plant. Breed. 39:326–329.

Part V
Molecular Biology and Biotechnologies for an Appropriate Management and Creation of Genetic Diversity

Chapter 55
Marker-assisted Selection in Forage Crops and Turf: A Review

Isabel Roldán-Ruiz and Roland Kölliker

Abstract Researchers and breeders have long been aware that the combination of conventional breeding approaches and molecular tools would benefit forage and turf cultivar development. In spite of this, the forage and turf cultivars currently released are still conventionally bred. This contrasts with the increasing number of research reports about DNA-marker assisted characterization of germplasm resources and QTL mapping for a variety of traits in different species. In this paper, we describe the few attempts undertaken thus far to make use of marker-assisted selection (MAS) in forage species. We also present an overview of the most-researched crops and traits of the past decennia, for which substantial genetic and genomic knowledge is currently available. The expected impact of association mapping approaches and cost-effective high-throughput genotyping technologies, as well as the wealth of information available from model organisms, is discussed. Achievements in breeding species such as maize demonstrate the effectiveness of international, multidisciplinary collaborations. Taking into consideration the level of individual industrial/academic investments currently typical in forage and turf breeding, we emphasize the relevance of international collaborative efforts and suggest ways to share resources (e.g., at the level of association mapping populations) for effective implementation of MAS programs in a typical forage or turf species.

Keywords SNP · SSR · QTL-mapping · LD-mapping · Genome-wide Selection

Introduction

DNA-marker technologies have proven valuable in a range of applications, including Marker-assisted Selection (MAS). There is great interest in MAS, as illustrated by more than 1,000 published articles in 2003 containing the term "marker-assisted

I. Roldán-Ruiz (✉)
Plant Sciences Unit – Growth and Development, Institute for Agricultural and Fisheries Research (ILVO), Caritasstraat 21, Melle B-9090, Belgium
e-mail: isabel.roldan-ruiz@ilvo.vlaanderen.be

C. Huyghe (ed.), *Sustainable Use of Genetic Diversity in Forage and Turf Breeding*,
DOI 10.1007/978-90-481-8706-5_55,

selection" (Xu and Crouch, 2008). However, most of these articles discuss the potential use of research results in MAS, rather than actual applications. This is certainly the case for forage and turf species. This congress reflects this trend, as revealed by the content of the presentations and posters offered for the section "Molecular biology and biotechnologies for an appropriate management and creation of genetic diversity". According to the information provided in the abstracts, the following topics are covered: DNA-marker development (2), enabling research (6), QTL mapping (10), association mapping (2), candidate gene identification and/or metabolite profiling (6). Only Ratan et al. describe a research program that combines genetic mapping and marker-assisted selection of traits.

Genetic improvement in breeding of crops such as maize and soybean has been accelerated thanks to MAS (Eathington et al., 2007). In contrast, forage and turf products lag behind in marker-assisted breeding. Similarly to what is speculated for other crop species (Xu and Crouch, 2008), it is possible that the lag in seeing products from marker assisted breeding is related to the time required for the development of efficient genotyping platforms, sufficient gene-based markers and QTL information for traits of interest, and to the long product development cycle in plant breeding. Further, large genome sizes, often in combination with outbreeding, and the population-based selection schemes typical of current breeding practice, complicate the development and application of MAS tools in forage and turf species.

Past Achievements

In the past decade, researchers have developed molecular genetic tools and discovered genomic regions affecting monogenic and polygenic traits, namely in *Lolium, Festuca, Trifolium* and *Medicago* species. After an initial lack of sequence-specific markers, a substantial number of genomic and EST-derived SSR markers is now available (Hirata et al., 2006; Sato et al., 2005; Studer et al., 2008). This enabled the first consensus linkage map in a forage species to be constructed (Isobe et al., 2009). QTL mapping approaches have been used to investigate the genetic control of vernalization response, heading date or lodging resistance in ryegrasses; winter hardiness and growth characteristics in alfalfa (reviewed in Inoue et al., 2007); and persistence and seed yield parameters in red clover (Herrmann et al, 2006. 2008). Disease-resistance traits in forage and turf grasses have garnered particular attention, owing to their economic impact. QTL for resistance to diseases such as crown rust, bacterial wilt or leaf dollar spot have been identified (Bonos et al., 2006; Muylle et al., 2005; Studer et al., 2006, 2007). In addition, QTL analysis has been used to elucidate the role of fructan in growth and drought response and the genetic control of self-incompatibility in perennial ryegrass (Turner et al., 2008; Yang et al., 2008).

While few, reports of research on MAS approaches in forage and analogous systems do indicate several possibilities to add value to forage breeding programs using DNA marker information (Brummer and Casler, 2009). Stendal et al. (2006) demonstrated the use of MAS to screen out deleterious recessive alleles in smooth bromegrass. Barrett and colleagues (2009) demonstrated the potential of MAS for improving seed yield in white clover. Using SSR markers linked to a seed yield QTL of moderate resolution, they analyzed marker-trait associations in twelve breeding pools. Diagnostic SSR markers explained an average of 38% of the difference in seed yield, which illustrates the value of MAS in white clover breeding. MAS may also support parental selection in forage crop breeding programs. A study on perennial ryegrass showed that selection of genetically diverse parents may lead to better agronomic performance in first and second generation progenies (Kölliker et al., 2005). However, studies in other crops failed to show a direct correlation between molecular marker diversity and heterosis (Cerna et al., 1997; Riday et al., 2003).

To our knowledge, there has been no systematic intention to exploit the disease-resistance QTL described above for MAS. However, it would be possible to use molecular markers to introgress several resistance genes or QTL into single genotypes or cultivars. This has been proven to be particularly successful for bacterial blight resistance in rice (Ashikari and Matsuoka, 2006; Singh et al., 2001). The main limiting factors seem to be the limited availability of sequence specific markers closely linked to genes or QTL, as well as the lack of efficient crossing and selection schemes for the introgression of specific chromosome segments into cultivars commercialized as populations.

Current Trends

Past work regarding identification of marker-trait associations was based on the analysis of biparental populations. In recent years, this has shifted towards the use of association mapping strategies (Skøt et al., 2005, 2007; Smith et al., 2009). This eliminates the need to develop dedicated mapping populations and allows the identification of QTL within breeding germplasm. Association mapping strategies allow for the estimation of QTL-allele values that can be directly used in MAS without the need for an extra validation step. These strategies work best for populations with a low level of structure. As demonstrated by Auzanneau et al. (2007), synthetic cultivars with a broad genetic basis (e.g., cv. Herbie) may be used for this purpose. The growing interest in association mapping approaches is related to the increasing availability of SNP markers in genes of species such as perennial ryegrass or white clover (Cogan et al., 2006, 2007). Nevertheless, the example of *L. perenne*, one of the most intensively studied species, shows that the number of current publicly-available SNP markers is clearly insufficient to carry out a genome-wide association mapping study. In addition, the extent of linkage disequilibrium (LD) in perennial ryegrass seems to be extremely low, as expected for an obligate outbreeder. Smith et al. (2009) studied the extent of LD in four genes in a diverse collection of germplasm, two ecotypes and two cultivars. This study revealed that LD decays to levels below $r^2 = 0.2$ over 500–3,000 bp. These two factors (reduced availability of SNP markers and short LD distances) have led to the proposal of applying candidate-gene based strategies (Skøt et al., 2007; Smith et al., 2009). This approach is based on the assumption that genes with a proven or predicted function in a "model" species (functional candidate genes) or genes that are co-localized with a trait-locus (positional candidate genes) could control a similar function or trait in a crop of interest (Salentijn et al., 2007). This "translational" approach would allow crop breeders to exploit the enormous amount of information that is still being generated in model organisms.

A valuable alternative, as yet unexplored, would be to use the *Lolium/Festuca* DArT array that is currently being developed at the Laboratory of Molecular Cytogenetics and Cytometry (Olomouc, Czech Republic) for DNA-marker screening, instead of SNP markers. DArT technology should provide sufficient genome coverage for genome-wide association mapping (see Kopecký et al., the current congress). Most of the clones spotted on the array are probably not associated to expressed regions, but this could be compensated by the large number of markers screened. It could be argued that AFLPs could do the same (Skøt et al., 2005). However, one clear advantage of DArT markers is that they are based on cloned DNA-fragments, which greatly facilitates sequence determination of interesting markers.

It remains to be seen whether the application of association mapping approaches will significantly accelerate the assimilation of MAS by commercial breeding of forage and turf. The quality of marker-phenotype associations identified by association mapping depends on several factors, similar to what is known for QTL-mapping. Sufficient marker density is required, along with the use of appropriate population structures, and robust phenotypic data (in most cases multi-year, multi-location data). Given the level of individual industrial and academic investments typical for forage and turf breeding, the large population sizes required for association mapping might soon become prohibitive. International collaborative efforts and resource sharing (e.g., at the level of association mapping populations) might be an appropriate strategy in this case.

The power of detecting useful associations between candidate genes and traits can be increased by comparing populations that have been subjected to divergent selection for specific traits. For example, Castonguay et al. (2009) used bulk segregant analysis to identify genetic polymorphisms closely associated to adaptation to freezing among populations derived from the alfalfa cultivar Apica that had been recurrently selected for tolerance to freezing. Polymorphisms were uncovered in candidate genes including homologs of galactinol synthase (Castonguay et al., 2006) and genes involved in central metabolism or known to be responsive to environmental changes. MAS-derived progenies were generated, but to our knowledge, the comparison of these progenies with the initial unselected and selected populations to assess the contribution of these four genomic regions to the gains in freezing tolerance achieved has not yet been reported. The advantage of this approach is that the selected progenies could be advanced for the production of improved commercial cultivars.

An Optimistic View of the Future

Brummer and Casler (2009) asked "Given that we can identify markers associated with traits of interest, either in biparental mapping populations or by association mapping, how do we effectively integrate them into a recurrent selection scheme to improve the rate of genetic gain?" Some breeding programs based on inbred line development, F_2 enrichment, and marker-assisted recurrent selection (MARS) have already been used to increase the frequency of favourable QTL alleles at multiple loci (Bernardo, 2008). But this is not the case for fodder and turf species: there is a lack of theoretical and modeling research about the design of MAS programs tailored to these species. A theoretical framework must be developed to explore the feasibility and appropriateness of alternative strategies. Bernardo and Yu (2007) propose taking this exercise one step further. Instead of viewing markers as an add-on to the breeding process, one can start fresh from the question "How can molecular markers best be used to achieve breeding progress?" without conditioning the answer on how breeding has traditionally been done. The following ideas deserve particular attention and may provide the basis for the development of successful MAS strategies in forage and turf species.

QTL-mapping has two limitations: one, biparental mapping populations used for QTL studies are not often readily translated to breeding applications; and two, the statistical methods used to identify target loci and implement MAS are inadequate for improving polygenic traits controlled by many loci of small effect (Heffner et al., 2009). These limitations are exacerbated in highly heterogeneous and heterozygous crops with breeding processes based on population improvement.

Forage and turf scientists are working to make QTL studies more useful for breeding applications. Their research is shifting away from the use of biparental crosses towards very diverse LD mapping populations. The analytical tools used for LD mapping come from human and animal genetics. However, the range of analytical techniques that has been developed by human and animal geneticists to accommodate complex familiar interrelationships and intermediate levels of relatedness, structure and stratification is much broader than this (Sneller et al., 2009). Plant geneticists have only recently started to explore the possibilities offered by these other analytical tools when analyzing plant populations with complicated pedigrees and structure (e.g., Stich et al., 2006). These approaches deserve consideration; they may prove extremely useful for the development of new conceptual models in forage and turf breeding.

Alternatively we can question the identification of candidate genes or QTL for the traits of interest is really necessary when developing efficient selection tools. In this context, MAS approaches which are not based on the use of QTL information have been proposed. These approaches are gaining importance. For example, Genomic Selection, or GS, (originally proposed by Meuwissen et al., 2001) predicts the breeding values of lines in a population by analyzing their phenotypes and high-density marker scores. The availability of increasing numbers of DNA-markers spread over crop genomes and the decrease in the cost per marker-data-point makes this possible. GS incorporates all marker information in a prediction model, thereby avoiding biased marker effect estimates and capturing more of the variation due to small-effect QTL (Heffner et al., 2009). Bernardo and Yu (2007) demonstrated that genome-wide selection, as a brute-force and black-box procedure that exploits cheap and abundant molecular markers, is superior to MARS in maize breeding. Simulation studies have been used to demonstrate that the Bayesian multiple marker approach introduced by Meuwissen et al. (2001) might be suitable for predicting genetic values, even with reduced density datasets where large numbers of markers are not yet available (Cleveland and Deeb, 2009).

The enormous technical developments in the area of DNA sequencing and SNP genotyping will also accelerate the development and efficient deployment of molecular genetic tools in plant breeding. The sequencing of large cereal genomes is within reach, thanks to second-generation sequencing techniques such as high-throughput pyrosequencing (Stein, 2007). Further developments may lower sequencing costs to $1,000 per genome. This will not only enable the sequencing of genomes of numerous species, but will also allow for the re-sequencing of entire genomes or genomic subsets of a number of individuals within species or populations (Shendure and Ji, 2008). In humans, over 3.1 million SNPs were genotyped in 270 individuals from four geographically diverse populations. This provides an invaluable resource for understanding the structure of genetic variation and its link to particular phenotypes (The International HapMap Consortium, 2007). It is very likely that SNPs will be identified in a large number of genes in crop plants within a few years. Ultimately, this will enable efficient management of genetic diversity on a whole-genome level (Ganal et al., 2009). Before these strategies can be applied to population-based forage and turf breeding, one must consider the issues noted above, as well as the range of genetic diversity relevant for breeding.

MAS in forage and turf species has yet to live up to the expectations of plant breeders and the scientific community, but it is still in its infancy. Recent technical developments, and lessons learned from other crop species, are moving the research forward. Future progress will depend largely on conceptual models for applying MAS in forage and turf species. These conceptual models are the key to maximal exploitation of existing and arising resources.

References

Ashikari, M., Matsuoka, M. 2006. Identification, isolation and pyramiding of quantitative trait loci for rice breeding. Trends Plant Sci. 11:344–350.

Auzanneau, J., Huyghe, C., Julier, B., Barre, P. 2007. Linkage disequilibrium in synthetic varieties of perennial ryegrass. Theor. Appl. Genet. 115:837–847.

Barret, B., Baird, I., Woodfield, D. 2009. White clover seed yield: A case study in marker-assisted selection (pp. 241–250). In: Yamada, T., Spangenberg, G. (eds.), Molecular breeding of forage and turf. Springer Science + Business Media, New York, USA.

Bernardo, R. 2008. Molecular markers and selection for complex traits in plants: learning from the last 20 years. Crop Sci. 48:1649–1664.

Bernardo, R., Yu, J. 2007. Prospect for genomewide selection for quantitative traits in maize. Crop Sci. 47:1082–1090.

Bonos, S.A., Clarke, B.B., Meyer, W.A. 2006. Breeding for disease resistance in the major cool-season turfgrasses. Annu. Rev. Phytopathol. 44:213–234.

Brummer, E.C., Casler, M.D. 2009. Improving selection in forage, turf, and biomass crops using molecular markers (pp. 193–210). In: Yamada, T. Spangenberg, G. (eds.), Molecular breeding of forage and turf. Springer Science + Business Media.

Castonguay, Y., Cloutier, J., Laberge, S., Bertrand, A., Michaud, R. 2006. A bulk segregant approach to identify genetic polymorphisms associated with cold tolerance in alfafa (pp. 88–102). In: Chen, T.H.H., Uemura, M., Fijikawa, S. (eds.), Cold hardiness in plants: molecular genetics, cell biology and physiology. CAB International, Wallingford, UK.

Castonguay, Y., Cloutier, J., Michaud, R., Bertrand, A., Laberge, S. 2009. Development of marker-assisted selection for the improvement of freezing tolerance in alfalfa (pp. 221–228). In: Yamada, T., Spangenberg, G (eds.), Molecular breeding of forage and turf. Springer Science + Business Media.

Cleveland, M.A., Deeb, N. 2009. Evaluation of a genome-wide approach to multiple marker association considering different marker densities. BMC Proceedings, 3 (suppl. I): S5.

Cerna, F.J., Cianzio, S.R., Rafalski, A., Tingey, S., Dyer, D. 1997. Relationship between seed yield heterosis and molecular marker heterozygosity in soybean. Theor. Appl. Genet. 95:460–467.

Cogan, N.O.I., Drayton, M.C., Ponting, R.C., Vecchies, A.C., Bannan, N.R., Sawbridge, T.I., Smith, K.F., Spangenberg, G.C., Forster, J.W. 2007. Validation of in silico-predicted genic SNPs in white clover (*Trifolium repens* L.), an outbreeding allopolyploid species. Mol. Genet. Genomics 277:413–425.

Cogan, N.O.I., Ponting, R.C., Vecchies, A.C., Drayton, M.C., George, J., Dracatos, P.M., Dobrowolski, M.P., Sawbridge, T.I., Smith, K.F., Spangenberg, G.C., Forster, J.W. 2006. Gene-associated single nucleotide polymorphism discovery in perennial ryegrass (*Lolium perenne* L.). Mol. Genet. Genomics 276:101–112.

Eathington, S.R., Crosbie, T.M., Edwards, M.D., Reiter, R.S., Bull J.K. 2007. Molecular markers in a commercial breeding program. Crop Sci. 47(S3):S154–S163.

Ganal, M.W., Altmann, T., Roder, M.S. 2009. SNP identification in crop plants. Curr. Opin. Plant Biol. 12:211–217.

Heffner, E.L., Sorrells, M.E., Jannik, J.-L. 2009. Genomic selection for crop improvement. Crop Sci. 49:1–12.

Herrmann, D., Boller, B., Studer, B., Widmer, F., Kölliker, R. 2006. QTL analysis of seed yield components in red clover (*Trifolium pratense* L.). TAG Theor. Appl. Genet. 112:536.

Herrmann, D., Boller, B., Studer, B., Widmer, F., Kölliker, R. 2008. Improving persistence in red clover – insight from QTL analysis and comparative phenotypic evaluation. Crop Sci. 48:269–277.

Hirata, M., Cai, H., Inoue, M., Yuyama, N., Miura, Y., Komatsu, T., Takamizo, T., Fujimori, M. 2006. Development of simple sequence repeat (SSR) markers and construction of an SSR-based linkage map in Italian ryegrass (*Lolium multiflorum* Lam.). Theor. Appl. Genet. 113:270–279.

Inoue, M., Fujimori, M., Cai, H. 2007. Forage crops (pp. 51–75). In: Kole, C. (ed.), Genome mapping and molecular breeding in plants, vol. 6, Technical Crops. Springer-Verlag, Berlin Heidelberg.

Isobe, S., Kölliker, R., Hisano, H., Sasamoto, S., Wada, T., Klimenko, I., Okumura, K., Tabata, S. 2009. Construction of a consensus linkage map and genome-wide polymorphism analysis of red clover. BMC Plant Biol. 9:57.

Kölliker, R., Boller, B., Widmer, F. 2005. Marker assisted polycross breeding to increase diversity and yield in perennial ryegrass (*Lolium perenne* L.). Euphytica 146:55–65.

Meuwissen, T.H.E., Hayes, B.J., Goddard, M.E. 2001. Prediction of total genetic value using genome-wide dense marker maps. Genetics 157:1819–1829.

Muylle, H., Baert, J., Van Bockstaele, E., Pertijs, J., Roldán-Ruiz, I. 2005. Four QTLs determine crown rust (*Puccinia coronata*f. sp. *lolii*) resistance in a perennial ryegrass (*Lolium perenne*) population. Heredity 95:348–357.

Riday, H., Brummer, E.C., Austin Campbell, T., Luth, D., Cazcarro, P.M. 2003. Comparison of genetic and morphological distance with heterosis between *Medicago sativa* subsp. *sativa* and subsp. *falcata*. Euphytica 131:37–45.

Salentijn, E.M.J., Pereira, A., Angenent, G.C., van der Linden, D.G., Krens, F., Smulders, M.J.M., Vosman, B. 2007. Plant translational genomics: From model species to crops. Mol. Breed. 20:1–13.

Sato, S., Isobe, S., Asamizu, E., Ohmido, N., Kataoka, R., Nakamura, Y., Kaneko, T., Sakurai, N., Okumura, K., Klimenko, I., Sasamoto, S., Wada, T., Watanabe, A., Kohara, M., Fujishiro, T., Tabata, S. 2005. Comprehensive structural analysis of the genome of red clover (*Trifolium pratense* L.). DNA Res. 12:301–364.

Shendure, J., Ji, H.L. 2008. Next-generation DNA sequencing. Nat. Biotechnol. 26:1135–1145.

Singh, S., Sidhu, J.S., Huang, N., Vikal, Y., Li, Z., Brar, D.S., Dhaliwal, H.S., Khush, G.S. 2001. Pyramiding three bacterial blight resistance genes (Xa5, Xa13 and Xa21) using marker-assisted selection Into indica rice cultivar Pr106. Theor. Appl. Genet. 102:1011–1015.

Skøt, L., Humphreys, M.O., Armstead, I., Heywood, S., Skøt, K.P., Sanderson, R., Thomas, I.D., Chorlton, K.H., Sackville Hamilton, R. 2005. An association mapping approach to identify flowering time genes in natural populations of *Lolium perenne* (L.). Mol. Breed.15:233–245.

Skøt, L., Humphreys, J., Humphreys, M.O., Thorogood, D., Gallagher, J., Sanderson, R., Armstead, I.P., Thomas, I.D. 2007. Association of candidate genes with flowering time and water-soluble carbohydrate content in *Lolium perenne* (L.). Genet. 177:535–547.

Smith, K.F., Dobrowolski, M.P., Cogan, N.O.I., Spangenberg, G.C., Forster, J.W. 2009. Utilizing linkage disequilibrium and association mapping to implement candidate gene based markers in perennial ryegrass breeding (pp. 259–274). In: Yamada, T., Spangenberg, G. (eds.), Molecular breeding of forage and turf. Springer Science + Business Media.

Sneller, C.H., Mather, D.E., Crepieux, S. 2009. Analytical approaches and population types for finding and utilizing QTL in complex plant populations. Crop Sci. 49:363–380.

Stein, N. 2007. Triticeae genomics: advances in sequence analysis of large genome cereal crops. Chromosome Res. 15:21–31.

Stendal, C., Casler, M.D., Jung, G. 2006. Marker-assisted selection for neutral detergent fiber in smooth bromegrass. Crop Sci. 46:303–311.

Stich, B., Melchinger, A.E., Piepho, H.-P., Heckenberger, M., Mauer, H.P., Reif, J.C. 2006. A new test for family-based association mapping with inbred lines from plant breeding. Theoretical and Appl. Genet. 113:1121–1130.

Studer, B., Boller, B., Bauer, E., Posselt, U.K., Widmer, F., Kölliker, R. 2007. Consistent detection of QTLs for crown rust resistance in Italian ryegrass (*Lolium multiflorum* Lam.) across environments and phenotyping methods. Theor. Appl. Genet. 115:9–17.

Studer, B., Boller, B., Herrmann, D., Bauer, E., Posselt, U.K., Widmer, F., Kölliker, R. 2006. Genetic mapping reveals a single major QTL for bacterial wilt resistance in Italian ryegrass (*Lolium multiflorum* Lam.). Theor. Appl. Genet. 113:661–671.

Studer, B., Asp, T., Frei, U., Hentrup, S., Meally, H., Guillard, A., Barth, S., Muylle, H., Roldán-Ruiz, I., Barre, P., Koning-Boucoiran, C., Uenk-Stunnenberg, G., Dolstra, O., Skøt, L., Skøt, K.P., Turner, L.B., Humphreys, M.O., Kolliker, R., Roulund, N., Nielsen, K.K., Lubberstedt, T. 2008. Expressed sequence tag-derived microsatellite markers of perennial ryegrass (*Lolium perenne* L.). Mol. Breed. 21:533–548.

The International HapMap Consortium. 2007. A second generation human haplotype map of over 3.1 million SNPs. Nature 449:851–861.

Turner, L.B., Cairns, A.J., Armstead, I.P., Thomas, H., Humphreys, M.W., Humphreys, M.O. 2008. Does fructan have a functional role in physiological traits? Investigation by quantitative trait locus mapping. New. Phytol. 179:765–775.

Xu, Y., Crouch, J.H. 2008. Marker-assisted selection in plant breeding: From publications to practice. Crop Sci. 48:391–407.

Yang, B., Thorogood, D., Armstead, I. 2008. How far are we from unravelling self-incompatibility in grasses? New Phytol. 178:740–753.

Chapter 56
Allelic Diversity for Candidate Genes and Association Studies: Methods and Results

Toshihiko Yamada and Leif Skøt

Abstract The increasing ease with which molecular markers can be generated makes it possible for plant geneticists to use these genomic technologies for better exploitation of the available genetic variation in breeding populations. Identifying markers based on conventional bi-parental mapping populations is most likely not the best way to implement a marker assisted selection (MAS) program, although this approach is useful for introgression of alleles from wild germplasm. Instead, association mapping may be used in a more practical approach, by measuring both phenotypes and markers directly on the plants in the breeding nursery. Conventional quantitative trait loci (QTL) mapping enables one to identify chromosomal regions of 5–20 cM containing genes underlying the trait of interest. However, that still leaves several hundred potential candidate genes. Association mapping enables the exploitation of the wider genetic diversity and incorporate a larger number of recombinations. Synthetic populations used for genetic improvement of self-incompatible crops including many forage and turf species, present a useful tool for incorporating association mapping and genotype building using molecular markers. This is particularly true for traits that have not previously been selected for, since linkage disequilibrium (LD) is less likely to have been built up. We show some preliminary data from a experiment to illustrate population structure, LD and associations with candidate genes in synthetic populations not previously selected for this trait. Some recent research on association analysis in perennial ryegrass and clovers are also reviewed. We also briefly describe genomic selection (GS) that can predict the breeding values of lines in a population by analyzing phenotypes and high-density marker scores as a way to incorporate MAS into the breeding process.

Keywords Association mapping · Genomic selection · Linkage disequilibrium · Quantitative trait loci

---

T. Yamada (✉)
Field Center for Northern Biosphere, Hokkaido University, Kita 11, Nisha 10, Kita-ku, Sapporo 060-0811, Japan
e-mail: yamada@fsc.hokudai.ac.jp

C. Huyghe (ed.), *Sustainable Use of Genetic Diversity in Forage and Turf Breeding*,
DOI 10.1007/978-90-481-8706-5_56,

Introduction

Many forage crops have been continuously improved through the introduction of natural occurring beneficial alleles by spontaneous and/or artificial crossing, and by recurrent selection. In the last century, systematic plant breeding and many forage crop cultivars developed through breeding have significantly improved forage and livestock systems. However, a limitation in breeding programs is that even though we can control the resultant phenotypes to a certain extent, mechanisms governing the genetic control of quantitative traits such as yield and forage quality is still poorly understood and under-exploited. In the last decade, genetic dissection of 'quantitative trait loci' referred to as QTL has become a common approach, and created a new paradigm in plant breeding and genetics. Many QTL have been identified using linkage maps constructed from DNA markers such as simple sequence repeat (SSR) in forage crops (Yamada and Spangenberg, 2009).

Conventional QTL analysis based on bi-parental mapping population is not always the best way to implement a marker assisted selection (MAS) program in obligate outbreeding forage crops, although this approach is useful for introgression of alleles from wild germplasm. Because of their out-breeding habit forage crops are likely to have limited linkage disequilibrium (LD). LD is the non-independence of alleles, i.e. non-random allelic association at different loci, and is inversely proportional to the recombination fraction. LD and association studies have been carried out in some plant species (reviewed by Gupta et al., 2005). Forage crops contain many accessions with long-established populations derived from a large number of founding parents, in which many rounds of recombination have occurred. Such populations are best suited for candidate gene-based marker analyses (Andersen and Lübberstedt, 2003). Nucleotide variation in genes which are closely linked with the causal mutations that generate variability for key agronomic traits can be used as diagnostics for such variation. Successful correlation of gene haplotype structure and phenotypic variation will provide the basis for a new paradigm in forage crop molecular breeding based on direct selection of superior allele content at targeted genetic loci, allowing highly effective exploitation of germplasm collections for identification of potential parental genotypes.

Recently the availability of abundant markers and the reduction of cost and time of marker analyses through high throughout genotyping system will allow the development of new tools with a more accurate MAS program. For the breeding program of the complex polygenic traits that are crucial for the success of new crop varieties, genomic selection (GS) that predicts the breeding values of lines in a population by analyzing their phenotypes and high-density marker scores appears to be promising breeding schemes (Heffner et al., 2009).

We describe here methods and some preliminarily results for allelic diversity for candidate genes and association studies as well as outlining possibilities for GS as an accurate selection tool.

Use of LD in Plant Genetics

One of the major current and future uses of LD in plants would be to study marker-trait association (without the use of bi-parental mapping population) followed by a MAS scheme. Another important use is the study of genetic diversity in natural populations and germplasm collections and its use in the study of population genetics.

When comparing linkage analysis and LD mapping for QTL detection, it is clear that linkage mapping is more useful for genome-wide scan for QTL, while LD mapping gives more precise location of an individual QTL. One may therefore like to use linkage analysis for preliminary location of QTLs and then use LD for more precise location (Mackay, 2001). Precise location within a very small chromosome region is possible through LD, but not through linkage analysis, since recombination within such a small region is not found in a finite population that is examined (Mackay, 2001).

Limitation of LD and Potential Solutions

There are several limitations in using LD in plant systems as well as the benefits discussed previously. Among them, effects of structured populations, epistasis, gene conversion and ascertainment bias have been discussed by Gupta et al. (2005). The major limitation of LD mapping is that it provides little insight into the mechanistic basis of the LD detected (i.e. LD may or may not be due to linkage), so that, genomic localization and cloning of genes based on LD may not always be successful. This is because a strong LD may sometimes be due to recent occurrence of LD rather than a close physical linkage between the two loci, exhibiting LD. The factors (other than linkage) which lead to an increase in LD, include inbreeding, small population size, drift, genetic isolation between lineages, population subdivision, low recombination rate, population admixture, natural and artificial selection, balancing selection (Rafalski and Morgante, 2004; Gupta et al., 2005). A joint linkage analysis and LD strategy has been devised for genetic mapping, taking advantage of each approach (Wu and Zeng, 2001; Wu et al., 2002).

As the number of known single nucleotide polymorphisms (SNPs) markers in a genome increases, genotyping individuals with all the available SNPs will become a formidable task. The efficiency of SNP haplotype analysis may also be increased by DNA pooling and use of microarrays, which can dramatically reduce the number of genotyping assays (Yang et al., 2003).

Some LD Studies in Forage Crops

Functionally-associated candidate genes have been developed in some forage crops. LD was investigated in the gibberelic acid insensitive gene region in three synthetic varieties of perennial ryegrass chosen for their contrasting number of parents in the initial poly-cross (Auzanneau et al., 2007). Significant LD was observed up to 1.6 Mb in a variety originated from six related parents and not above 174 kb in a variety originating from 336 parents, suggesting that a genome-wide association approach may be possible in synthetic populations with few parents.

SNPs are the most abundant class of genetic marker. Efficient methods for discovery of SNPs and characterization of SNP haplotype structure have been described for perennial ryegrass (Spangenberg et al., 2005; Cogan et al., 2006; Ponting et al., 2007). Multiple SNPs at regular intervals across an amplicon were detected within and between the heterozygous parents and validated in the progeny of the F_1 ($NA_6 \times AU_6$) genetic mapping family. Decay of LD to r^2 values of c. 0.2 typically occurs over 500–3,000 bp, comparable with gene length and with little apparent variation between diverse, ecotypic and varietal population sub-groups (Smith et al., 2009).

A candidate gene approach for associating SNPs with variation in flowering time and water-soluble carbohydrate content (WSC) and other traits has been described for perennial ryegrass (Skøt et al., 2007). Consistent associations between the perennial ryegrass homolog (*LpHD1*) of the rice photoperiod control gene *HD1* and flowering time were identified. One SNP, in the immediate upstream region of the *LpHD1* coding sequence (C-4443-A), was significant in the linear mixed model, and this association has been validated in other populations and a mapping population (Skøt et al., unpublished).

Eleven expressed disease resistance candidate (R) genes including 6 nucleotide binding site and leucine rich repeat (NBS-LRR) like genes and 5 non-NBS-LRR genes were analyzed by allelic diversity (Xing et al., 2008). NBS-LRR like gene fragments showed a high degree of nucleotide diversity. Substantial LD decay was found within 500 bp for most resistance candidate genes. An approach based on in vitro SNP discovery in candidate defence response (DR) genes has been used to develop potential diagnostic genetic markers in perennial ryegrass (Dracatos et al., 2008). SNPs were predicted, validated and mapped for representatives of the pathogenesis-related (PR) protein-encoding and reactive oxygen species (ROS)-generating gene classes.

An EST resource obtained from multiple cDNA libraries constructed from numerous genotypes of a single cultivar has been used for in silico SNP discovery and validation in white clover. A total of 58 from 236 selected sequence clusters (24.5%) were fully validated as containing polymorphic SNPs by genotypic analysis across the parents and progeny of several two-way pseudo-testcross mapping families of white clover (Cogan et al., 2007).

In red clover a consensus linkage map was constructed using six mapping populations originating from eight parental accessions (Isobe et al., 2009). The integrated red clover map was composed of 1,804 loci, including 1,414 microsatellite loci, 181 AFLP loci and 204 RFLP loci, in seven linkage groups. Using the genome-wide allele frequency data of 1,144 red clover individuals and 462 microsatellite loci, a preliminary estimate of the extent of LD has been carried out (Isobe et al., 2009). There was no significant correlation between *D'* and distance between two loci. This result suggests that the extent of LD in red clover is low, highlighting the challenge of obtaining sufficiently dense marker coverage in an out-breeding species.

Genomic Selection

As the cost of genotyping has decreased so the interest in GS methods has increased. It represents an alternative use of MAS compared to the introgression and backcross method, which is unsuited for complex traits controlled by many QTL with small effects. GS is related to genome-wide association mapping in which populations of unknown pedigree are used for associating genotype to phenotype, usually on a marker by marker basis, and very stringent significance criteria. Genomic selection uses all marker locus data across a genome in a joint analysis to capture all locus and haplotype effects. The genetic variance obtained by summing individual marker effects can be used to estimate the breeding value of individual genotypes (the average genotypic value of its progeny) (Heffner et al., 2009). A number of studies indicate that this approach can be superior to phenotypic selection in terms of phenotypic gain per unit cost, even with modest population sizes (50–100) and marker numbers (120–150) (Wong and Bernardo, 2008). However, virtually all evidence is based on simulation studies, so there is an urgent need for proof of principle work with actual crops. The self-incompatibility and high degree of genetic variation within the half-sib breeding populations, combined with the restricted genetic variation within derived synthetic varieties from many perennial ryegrass breeding programs could be potentially useful in a GS approach. The aim of GS is to exploit LD between QTL and markers across the genome in order to estimate the breeding value of a population, and thus enable prediction of phenotypes in the progeny (Meuwissen et al., 2001). Genome-wide markers may also help to identify the best genotypes to use in top crosses for introduction of new genetic variation in ryegrass breeding populations. However, the major stumbling block in populations with little LD is the marker density that is necessary for successfully predicting phenotypes across the genome. The solution may come from continued reduction in genotyping costs, and/or confirming the feasibility of using a low density marker panel. If this can be achieved MAS really could become an integral part of forage breeding programs.

References

Andersen, J.R., Lübberstedt, T. 2003. Functional markers in plants. Trends Plant Sci. 8:554–560.

Auzanneau, J., Huyghe, C., Julier, B., Barre, P. 2007. Linkage disequilibrium in synthetic varieties of perennial ryegrass. Theor. Appl. Genet. 115:837–847.

Cogan, N.O.I., Ponting, R.C., Vecchies, A.C., Drayton, M.C., George, J., Dobrowolski, M.P., Sawbridge, T.I., Spangenberg, G.C., Smith, K.F., Forster, J.W. 2006. Gene-associated single nucleotide polymorphism (SNP) discovery in perennial ryegrass (*Lolium perenne* L.). Mol. Genet. Genom. 276:101–112.

Cogan, N.O.I., Drayton, M.C., Ponting, R.C., Vecchies, A.C., Bannan, N.R., Sawbridge, T.I., Smith, K.F., Spangenberg, G.C., Forster, J.W. 2007. Validation of in silico-predicted genic SNPs in white clover (*Trifolium repens* L.), an outbreeding alloplyploid species. Mol. Genet. Genom. 277:413–425.

Dracatos, P.M., Cogan, N.O.I., Dobrowolski, M.P., Sawbridge, T.I., Spangenberg, G.C., Smith, K.F., Forster, J.W. 2008. Discovery and genetic mapping of single nucleotide polymorphisms in candidate genes for pathogen defence response in perennial ryegrass (*Lolium perenne* L.). Theor. Appl. Genet.117:203–219.

Heffner, E.L., Sorrells, M.E., Jannink, J.-L. 2009. Genomic selection for crop improvement. Crop Sci. 49:1–12.

Isobe, S., Kölliker, R., Hisano, H., Sasamoto, S., Wada, T., Klimenko, I., Okumura, K., Tabata, S. 2009. Construction of a consensus linkage map for red clover (*Trifolium pratense* L.). BMC Plant Biol. 9:57.

Gupta, P.K., Rustgi, S., Kulwal, P.L. 2005. Linkage disequilibrium and association studies in higher plants: Present status and future prospects. Plant Mol. Biol. 57:461–485.

Mackay, T.F.C. 2001. The genetic architecture of quantitative traits. Ann. Rev. Genet. 33:303–339.

Meuwissen, T.H.E., Hayes, B.J., Goddard, M.E. 2001. Prediction of total genetic value using genome-wide dense marker maps. Genetics 157:1819–1829.

Ponting, R.C., Drayton, M.C., Cogan, N.O.I., Dobrowolski, M.P., Spangenberg, G.C., Smith, K.F., Forster, J.W. 2007. SNP discovery, validation, haplotype structure and linkage disequilibrium in full-length herbage nutritive quality genes of perennial ryegrass (*Lolium perenne* L.). Mol. Genet. Genom. 278:585–597.

Rafalski, A., Morgante, M. 2004. Corn and humans: Recombination and linkage disequilibrium in two genomes of similar size. Trends Genet. 20:103–111.

Skøt, L., Humphreys, J., Humphreys, M.O., Thorogood, D., Gallagher, J., Sanderson, R., Armstead, I.P., Thomas, I.D. 2007. Association of candidate genes with flowering time and water-soluble carbohydrate content in *Lolium perenne* L. Genetics 177:535–547.

Smith, K.F., Dobrowolski, M.P., Cogan, N.O.I., Spangenberg, G.C., Forster, J.W. 2009. Utilizing linkage disequilibrium and association mapping to implement candidate gene based markers in perennial ryegrass breeding (pp. 335–340). In: Yamada, T., Spangenberg, G. (eds.), Molecular Breeding of Forage and Turf. Springer Science + Business, New York.

Spangenberg, G.S., Forster, J.W., Edwards, D., John, U., Mouradov, A., Emmerling, M., Batley, J., Felitti, S., Cogan, N.O.I., Smith, K.F., Dobrowolski, M.P. 2005. Future directions in the molecular breeding of forage and turf (pp. 83–97). In: Humphreys, M.O. (ed.), Molecular breeding for the genetic improvement of forage crops and turf. Wageningen Academic Publishers, The Netherlands.

Wong, C.K., Bernardo, R. 2008. Genomewide selection in oil palm: Increasing selection gain per unit time and cost with small populations. Theor. Appl. Genet. 116:815–824.

Wu, R., Zeng, Z.-B. 2001. Joint linkage and linkage disequilibrium mapping in natural populations. Genetics 157:899–909.

Wu, R., Ma, C.-X., Casella, G. 2002. Joint linkage and linkage disequilibrium mapping of quantitative trait loci in natural populations. Genetics 160:779–792.

Xing, Y., Frei, U., Schejbel, B., Asp, T., Lübberstedt, T. 2008. Nucleotide diversity and linkage disequilibrium in 11 expressed resistance candidate genes in *Lolium perenne*. BMC Plant Biol. 7:43.

Yamada, T., Spangenberg, G. (eds.) 2009. Molecular Breeding of Forage and Turf (pp. 1–352). Springer Science + Business Media, New York.

Yang, Y., Zhang, J., Hoh, J., Matsuda, F., Xu, P., Lathrop, M., Ott, J. 2003. Efficiency of single-nucleotide polymorphism haplotype estimation from pooled DNA. Proc. Natl. Acad. Sci. USA 100:7225–7230.

Chapter 57
Polyploidization and Gene Expression in *Medicago sativa*

Stefano Capomaccio, Fabio Veronesi, and Daniele Rosellini

Abstract Polyploidization is a common event in plant evolution and can influence economically important traits. Modifications of gene expression and/or DNA sequence are known to occur as a consequence of polyploidization. To gain insight into the effects of sexual polyploidization on gene expression in alfalfa, we have used two diploid (2x=16) plants of the subspecies *falcata* and *coerulea* that produce 2n eggs and 2n pollen, respectively. From their cross, diploid and tetraploid (4x=32) progenies were obtained. Three 2x and three 4x progeny plants were studied to investigate polyploidization-induced changes in gene transcription using the *Medicago* Genome Array (Affymetrix). Comparisons of gene expression levels were made between the midparent and each diploid and tetraploid progeny plant. Significant expression change as a consequence of polyploidization was observed for 189 genes. Gene ontology analyses of genes deviating from additive expression levels in the progenies permitted to identify functions and processes that are specifically affected by tetraploidization.

Keywords Alfalfa · Microarrays · Polysomic Polyploids · Transcriptome

Introduction

Polyploidization is an increase in genome number and occurs in nature as the consequence of the formation of diploid (2n) gametes (sexual polyploidization), or of somatic genome duplications (somatic polyploidization). Polyploidy is widespread in plants, to the point that, according to Bowers et al. (2003), at least 70% of plant species have experienced polyploidization at some point of their evolution.

D. Rosellini (✉)
Department of Applied Biology, University of Perugia, Borgo XX giugno 74, Perugia 06121, Italy
e-mail: roselli@unipg.it

C. Huyghe (ed.), *Sustainable Use of Genetic Diversity in Forage and Turf Breeding*,
DOI 10.1007/978-90-481-8706-5_57, © Springer Science+Business Media B.V. 2010

Knowledge of the impact of polyploidization on genome expression is scarce. Both genetic and epigenetic mechanisms are involved (Osborn et al., 2003; Chen, 2007; Doyle et al., 2008).

Gene expression changes can be brought about by gene dosage modification (increase of copy number), by the alteration of the interactions among transcription factors, by histone and chromatin state modifications, by cytosine residue methylation of nuclear DNA. All these phenomena may affect important traits such as cell size, growth habit or flowering time (Hegarty and Hiscock, 2008).

The objective of this work was to obtain new knowledge on the phenotypic and transcriptional effects of sexual polyploidization in alfalfa (*Medicago sativa* L.). Cultivated alfalfa is autotetraploid ($2n=4x=32$). We used a diploid ($2n=2x=16$) *M. sativa* subsp. *falcata* genotype, named PG-F9, that produces 55–70% 2n eggs (Barcaccia et al., 1997) and a diploid *M. sativa* subsp. *coerulea* genotype (12-P) that produces 2n pollen (Tavoletti et al., 1991). By crossing PG-F9 (seed parent) with 12-P (pollen parent), diploid and tetraploid progenies were obtained. Tetraploid progenies originate as a consequence of fertilization of a 2n egg by a 2n sperm (bilateral sexual polyploidization, BSP).

Materials and Methods

The *M. sativa* subsp. *falcata* genotype PG-F9, and the *M. sativa* subsp. *coerulea* genotype 12-P were crossed without emasculation using PG-F9 as the female parent.

Progeny plants were reared in flats in the greenhouse. For preliminary screening of ploidy level, a quick test based on chloroplast counts of guard cells was employed (Bingham, 1968). After root tip chromosome counts, three plants were randomly chosen among 2x and 4x progenies displaying variegated flower colour (diagnostic of *falcata* x *sativa* hybrids). Eight rooted cuttings per genotype were reared in pots in the greenhouse with natural light with complete randomization. Fresh and dry matter yield (g per plant) was assessed after clipping the plants at the onset of flowering.

Pollen viability was assessed as the percentage of stained grains after acetocarmine staining. Female fertility was estimated by hand-crossing and selfing each plant without emasculation, using three racemes (replicates). Three unrelated *M. sativa* subsp . *coerulea* and one unrelated cultivated *M. sativa* subsp . *sativa* pollen donors were used for crosses. Self fertility was estimated by tripping florets of 3 racemes per plant. Ovule fertility was estimated by assessing callose deposition in ovules at flower maturity (Rosellini et al., 1998). Seed set was estimated as the number of seeds per floret. Yield and fertility data were subjected to the analysis of variance to estimate the effect of ploidy and the means separated by the lsd test.

Microarray Analyses

Young, fully expanded leaves were harvested from each plant at the vegetative stage, frozen in liquid nitrogen and used for RNA extraction. Plant leaves were lysed in trizol (Invitrogen) and total RNA was further purified With the QIAGEN RNeasy kit. RNA was quantified on the NanoDrop ND-1000 and quality checked with the 2100 Bioanalyzer (Agilent Technologies); cRNA was generated and labelled following the manufacturer's protocol. Twenty μg of cRNA were fragmented and quality checked with the 2100 Bioanalyzer (Agilent Technologies). The biotinylated cRNA was hybridized to the *Medicago* Genome Array (www.affymetrix.com/products_services/arrays/specific/medicago.affx), containing probes of genes and open reading frames from *M. truncatula* (50,900) and *M. sativa* (1,896).

Three arrays were hybridized with cRNA of PG-F9, and 12-P, respectively, and one with each of the three diploid and the three tetraploid progeny plants (12 chips in total). Chips were washed and scanned on the Affymetrix Complete GeneChip® Instrument System.

The data were analyzed by GCOS 1.4 (Affymetrix Inc.). The full data set was normalized by using the Robust Multialignment Algorithm (RMA). The expression values obtained were analyzed by using GeneSpring 7.3 (Agilent Technologies), with a per chip and a per gene normalization. Results were filtered for fold change >1.5. Statistical analysis was performed by ANOVA using as p-value cutoff of 0.05. The following comparisons were made: PG-F9 vs . 12-P; PG-F9P (average of PG-F9 and 12-P) vs. 2x progenies (average of three); PG-F9P vs. 4x progenies (average of three).

The Blast2GO software (Conesa and Gotz, 2008) was used to perform a semi-automatic Gene Ontology annotation and data mining of probe sequence sets of gene whose transcription level was influenced by ploidy. The analysis pipeline is based on that described by Botton et al. (2008).

Results and Discussion

Biomass yield and fertility. Fresh matter yield was slightly but not significantly higher in 4x plants. However, the diploid plants flowered about 10 days later than the tetraploid plants; therefore, the growth period to the time of clipping was 10 days longer for diploid plants. Should the plants have been cut at the same time, the biomass of the 2x would have been significantly lower than that of the 4x plants. The percentage of dry matter was lower in 4x than in 2x plants, and this was phenotypically associated with softer, thicker stems of tetraploids, and probably due to higher cell volume of 4x plants. Overall, the effect of polyploidization on plant biomass at flowering was modest, but precocity of biomass production was increased.

Pollen viability was not significantly different (94.2 and 88.9 % in 2x and 4x plants, respectively). The mean number of ovules per ovary varied between 8.4 and 12.5, and did not differ either between ploidy levels (not shown). On the contrary, ovule sterility, as estimated by callose deposition at anthesis, was significantly lower in 4x with respect to the 2x progenies (Table 57.1). Interestingly, cross fertility of 2x plants in 2x–2x crosses was much lower than that of 4x plants in 4x–4x crosses (0.46 vs. 2.47 seed per floret, Table 57.1); ovule sterility explains only in part this

Table 57.1 Female fertility (seeds per floret) of 2x and 4x plants when hand-crossed without emasculation with unrelated 2x or 4x pollen parents

		Seed set[b]	
Plant	Ovule sterility[a]	2x Pollen parent	4x Pollen parent
PG-F9	21.0	1.09	0.32
12-P	45.3	0.09	0.00
2x Progenies	36.9*A*	0.46*A*	0.20*B*
4x Progenies	15.5*B*	0.01*B*	2.47*A*

[a]Percentage of callosized ovules per pistil based on twenty pistils per plant
[b]Means of four racemes (replications), each with 11–30 florets cross-pollinated without emasculation
Within a pollen parent type (2x or 4x), mean fertility values (bold type) followed by different letters are significantly different at $P<0.01$ according to the lsd test

result, because seed set was much lower than that allowed by the number of non-callosized ovules . The fertility of 2x progenies was intermediate between that of the parents, whereas 4x progenies, on average, greatly exceeded it. This demonstrates a positive effect of chromosome doubling, and no significant heterosis for fertility in this cross combination. All the plants were self-sterile.

Overall, the first-generation tetraploid plants characterized in this study flowered earlier and were more fertile than the parents and the 2x progenies. Faster development can be an advantage both in the wild and in the field to outgrow competitor species. Early flowering associated with fast development can also be an advantage because seed set and maturation has a higher chance to be completed before summer drought stress. This combination of traits may have contributed to the success of tetraploid *M. sativa*, both in the wild and in agriculture.

Microarray analyses. When the average parental gene expression levels were compared with that of the progenies, significantly non-additive gene expression (that is, expression in progeny significantly deviating from the mean expression of the two parents) was found for 528 genes: 189 in 4x progenies only, 178 genes in 2x progenies only, and 161 in both 2x and 4x progenies. Therefore, in this experiment

Table 57.2 Main gene ontology terms over- and under-represented ($P<0.05$) between the groups of genes specifically affected by polyploidization and by hybridization

Description (GO Term)	PAG[a]	HAG[b]
Ion binding (GO:0043167)	9	49
Sulfur metabolic process (GO:0006790)	0	9
Response to radiation (GO:0009314)	6	0
Response to abiotic stimulus (GO:0009628)	6	3
Carbohydrate catabolic process (GO:0016052)	5	2

[a]PAG: Polyploidization-affected genes
[b]HAG: Hybridization-affected genes
Numbers represent the number of genes in each group. Significance was assessed with Fisher exact test

189 genes (0.36% of those tested) were significantly affected, at the transcriptional level, by tetraploidization.

The comparison between the GO annotations obtained for the polyploidization-affected genes group versus the hybridization-affected ones (sum of the other two groups) allowed us to obtain an indication of functions that are under- or over-represented within the two groups (Table 57.2). Further work can now concentrate on the genes that have a higher chance to affect economic traits.

Acknowledgments Funding provided by the Italian Ministry of University and Scientific Research, Project: PRIN 2005, "Effects of polyploidization on gene expression and DNA methylation in alfalfa and potato". The help of Dr. Valentina Passeri with microscope work is gratefully acknowledged.

References

Conesa, A., Götz, S. 2008. Blast2GO: A comprehensive suite for functional analysis in plant genomics. Int. J.Plant Genomics (2008, Article ID 619832, 12 pages).

Barcaccia, G., Tavoletti, S., Falcinelli, M., Veronesi, F. 1997. Environmental influences on the frequency and viability of meiotic and apomeiotic cells of a diploid mutant of alfalfa. Crop Sci. 37:70–76.

Botton, A., Galla G., Conesa A., Bachem C., Ramina A., Barcaccia, G. 2008. Large-scale Gene Ontology analysis of plant transcriptome-derived sequences retrieved by AFLP technology. BMC Genomics 9:347 doi:10.1186/1471-2164-9-347.

Bingham, E.T. 1968. Stomatal chloroplasts in alfalfa at four ploidy levels. Crop Sci. 8:509–510.

Bowers, J.E., Chapman, B.A., Rong, J.K., Paterson, A.H. 2003. Unravelling angiosperm genome evolution by phylogenetic analysis of chromosomal duplication events. Nature 422:433–438.

Chen, Z.J. 2007. Genetic and epigenetic mechanisms for gene expression and phenotypic variation in plant polyploids. Annu.Rev. Plant Biol. 58:377–406.

Doyle, J.J., Flagel, L.E., Paterson, A.H., Rapp, R.A., Soltis, D.E., Soltis, P.S., Wendel, J.F. 2008. Evolutionary genetics of genome merger and doubling in plants. Annu. Rev. Genet. 42: 443–446.

Hegarty, M.J., Hiscock, S.J. 2008. Genomic clues to the evolutionary success of polyploid plants. Curr. Biol. 18:R435–R444.

Osborn, T.C., Pires, J.C., Birchler, J.A., Auger, D.L., Chen, Z.J., Lee, H.-S., Comai, L., Madlung, A., Doerge, R.W., Colot, V., Martienssen, R.A. 2003. Understanding mechanisms of novel gene expression in polyploids. Trends Genet. 19:141–147.

Rosellini, D., Lorenzetti, F., Bingham, E.T. 1998. Quantitative ovule sterility in *Medicago sativa*. Theor. Appl. Genet. 97:1289–1295.

Tavoletti, S., Mariani, A., Veronesi, F. 1991. Phenotypic recurrent selection for 2n pollen and 2n egg production in diploid alfalfa. Euphytica 57:97–102.

Chapter 58
Characterization of Dehydrin Variants Linked to Freezing Tolerance in Alfalfa at the DNA and Post-transcriptional Levels

Yves Castonguay, Wilfried Rémus-Borel, Jean Cloutier, Serge Laberge, Annick Bertrand, and Réal Michaud

Abstract Dehydrins are highly hydrophilic proteins that are thought to play key adaptive roles with regard to tolerance to freezing-induced cell desiccation. We recently identified a DNA polymorphism between alfalfa populations recurrently selected for superior freezing tolerance that was associated with size variations of the C-terminal coding region of a dehydrin homolog (msaCIG). These size variants were caused by the presence of small and large indels.We identified a fragment that was preferentially amplified using pooled DNA from genotypes with the dehydrin polymorphism initially uncovered on Southern blots. In the current study, the amplification of the C-terminal region using cDNA templates from cold-acclimated plants confirmed the cold-induced accumulation of transcripts of the expected molecular sizes. Western blot hybridization with antibodies raised against dehydrins revealed variations in polypeptide profiles that closely matched the allelic pattern uncovered with DNA amplifications.

Keywords Alfalfa · Allelic variation · Dehydrin · DNA polymorphism · Freezing tolerance · Genetic markers · Abiotic stress

Introduction

Breeding alfalfa (*Medicago sativa* L.) cultivars with superior tolerance to freezing (TF) could be accelerated by the identification of molecular markers associated to that trait. Robust markers that can be applied to a wide array of genetic backgrounds require the identification of functional sequences causally related to the trait of interest. Candidate genes potentially associated with freezing tolerance have been tested using restriction fragment length polymorphisms for differences

Y. Castonguay (✉)
Soils and Crops Research and Development Centre, Agriculture and Agri-Food Canada, Québec, QC G1V 2J3, Canada
e-mail: yves.castonguay@agr.gc.ca

C. Huyghe (ed.), *Sustainable Use of Genetic Diversity in Forage and Turf Breeding*, DOI 10.1007/978-90-481-8706-5_58, © Springer Science+Business Media B.V. 2010

in variants frequency among alfalfa populations selectively improved for their freezing tolerance (Castonguay et al., 2005). Polymorphic bands that intensified with the number of selection cycles were identified for a number of candidate genes including *msa*CIG (Rémus-Borel et al., 2009), a Y_2K_4 dehydrin homolog (Close, 1997) previously isolated from cold-acclimated alfalfa cv. Apica. A number of genetic studies have linked dehydrins to superior adaptation to low temperature stress (Marian et al., 2004; Patton et al., 2007). Our objectives were to study the inheritance of these dehydrin variants in relation to freezing tolerance improvement and to assess their expression at the transcriptional and post transcriptional levels.

Material and Methods

Plant Material and Growing Conditions

Genotypes of alfalfa, cultivar Apica (ATF0) and populations ATF2, ATF3 or ATF5 derived from this cultivar after successive cycles of recurrent selection for superior freezing tolerance (Castonguay et al., 2009) were grown and cold-acclimated under either environmentally-controlled or natural conditions in an unheated greenhouse as described in Castonguay and Nadeau (1998).

PCR Amplification

Genomic DNA and RNA were extracted using CTAB procedures. Primers were designed to amplify a specific 380 bp segment of the C-terminal region of *msa*CIG (Fig. 58.1a) using the Oligo Explorer software, version 1.1.0 (T. Kuulasma, University of Kuopio, Finland). The sequences for the forward and the reverse primers were: 5'gacaagatcaaggagaaga3' and 5'gtgttgctcatcatgtcc3' respectively. PCR reactions were performed with 5 prime *Taq* polymerase 5 U/μl (Inter Medico, Markham, ON, Canada) using either 50 ng of genomic DNA or 1 μl of cDNA reaction products synthesized from 5 μg of total RNA according to the manufacturer recommendations (Invitrogen, Burlington, ON, Canada). The conditions for PCR were as follows: an initial denaturating step at 94°C for 3 min followed by 35 cycles at 94°C for 1 min, 50°C for 1 min, 72°C for 1 min and a final extension of 7 min at 72°C. The number of cycles was reduced to 20 with the dehydrin cDNA templates. All the reactions were performed on an Eppendorf Mastercycler ep system (Eppendorf Canada, Mississauga, ON, Canada). Amplicons were separated on 2% (W/V) agarose gels in 1x Tris Borate EDTA running buffer pH 8.0 (70 V for 3 h). Ethidium bromide (0.5 μg/ml) was added to the gel before electrophoresis. Gels images were captured on a gel BioDoc-IT System (UVP, Upland, CA).

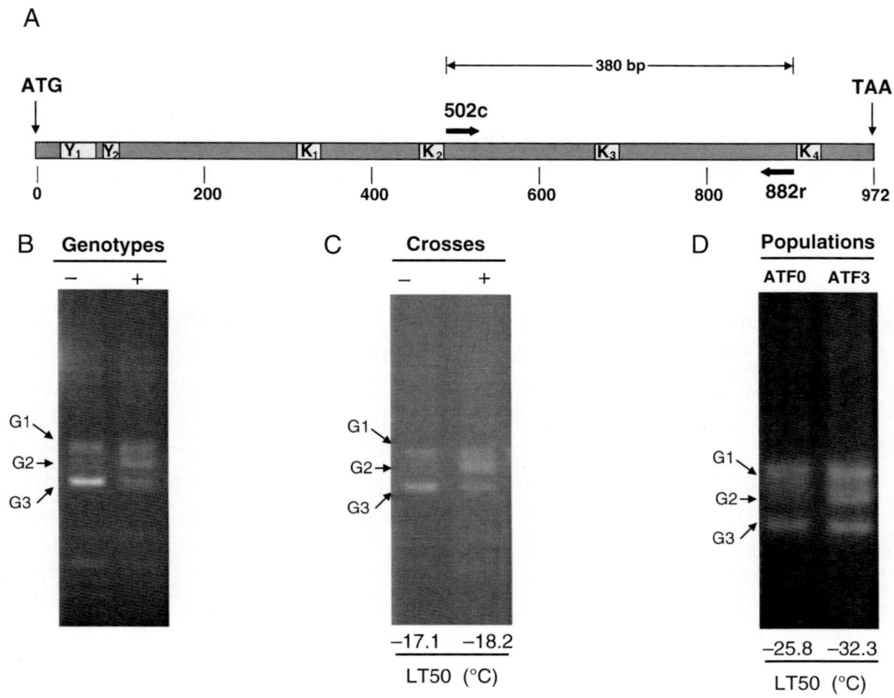

Fig. 58.1 (**a**) Structure of the *msa*CIG open reading frame with position of primers. Conserved Y- and K-segments are indicated. b, c, d: Amplification of a 380 bp-targeted sequence using DNA from: pools (**b**) and crosses (**c**) of genotypes within ATF0 with (+) or without (–) a dehydrin previously found to be polymorphic between TF populations and from populations ATF0 and ATF3 (**d**). Freezing tolerance of crosses (**c**) was assessed under controlled conditions as described in Castonguay et al. (1998) whereas LT_{50} values for populations (**d**) are from Castonguay et al. (2009)

Western Blotting

Proteins were extracted in a 50 mM HEPES buffer at pH 7.0 and separated by SDS-PAGE prior to immunoblotting. Buffer-soluble protein concentrations were determined with a NanoDrop™ 1000 spectrophotometer (Thermo Fischer, Ottawa, ON, Canada). Proteins were transferred electrophoretically onto Immobilon-P™ polyvinylidene fluoride (PVDF) membranes (Sigma-Aldrich., Oakville, ON, Canada) and blocked with TBST buffer (10 mM Tris-HCl, pH 7.5, 150 mM NaCl, 0.05% Tween-20) containing 5% skim milk for 60 min at room temperature. Membranes were incubated with a 1:1000 dilution of rabbit anti-dehydrin polyclonal antibody (PLA-100, Stressgen Bioreagents Corp., Victoria, BC, Canada) in TBST for 2 h. After primary antibody incubation, membranes were immersed in TBST containing the secondary goat anti-rabbit IgG polyclonal antibody (dilution 1:2000) conjugated to alkaline phosphatase (Bio-Rad Laboratories, Mississauga,

ON, Canada). Secondary antibodies were detected using 5-bromo-4-chromo-3-indolyl phosphate and nitro blue tetrazolium substrate solution.

Results and Discussion

Inheritance of a Polymorphic Dehydrin

A cDNA clone (*msa*CIG) with sequence features typical of dehydrins had been previously hybridized to *Dra*I-digested DNA from populations recurrently selected within the cultivar Apica. A polymorphic fragment that markedly intensified in response to recurrent selection for freezing tolerance was detected on Southern blots (Castonguay et al., 2008). Based on the modular nature of dehydrins, a search for intragenic variation associated to the RFLP polymorphism was pursued through the amplification of a 380 bp sequence in the C-terminal region of msaCIG. Three major fragments of $\approx$380 (G1), 330 (G2) and 290 (G3) base pairs were preferentially amplified. DNA sequences analysis showed that size variations were caused by the presence of large (14–23 a.a.) and small (2–7 a.a.) indels in the C-terminal coding region of *msa*CIG (Rémus-Borel et al., 2009). A fragment of intermediate size (G2) was more strongly amplified in pools (Fig. 58.1b) and crosses (Fig. 58.1c) of genotypes possessing the polymorphic dehydrin (+) as well as in population ATF3 recurrently selected for superior freezing tolerance (Fig. 58.1d). The presence of the G2 fragment (Fig. 58.1c and d) was associated with superior levels of freezing tolerance.

Post-transciptional Detection of Dehydrin Variants

Amplification of the 380 bp-targeted sequence from the C-terminal region of *msa*CIG using cDNA templates confirmed the transcription of the three dehydrin sequences that were amplified using genomic DNA as the template. Comparison between non acclimated and cold-acclimated cDNA samples showns the cold inducibility of the dehydrin variants. Fragment G2 was expressed with cDNA from ATF0, ATF2 and ATF3 (Fig. 58.2a and b). Further analyses will be needed to determine whether differential expression of the G2 variant occurs between ATF populations. Western blot analysis with an anti-dehydrin antibody confirmed the cold-induced accumulation of the encoded polypeptides (Fig. 58.2c). It also revealed a profile that closely matched the allelic pattern uncovered with DNA amplifications. Higher polypeptide complexity in the G2 region of the more freezing tolerant ATF5 than in ATF0 is noteworthy (Fig. 58.2c). Experiments are currently underway to confirm the potential of the polymorphic dehydrin in marker-assisted selection for the improvement of freezing tolerance in alfalfa.

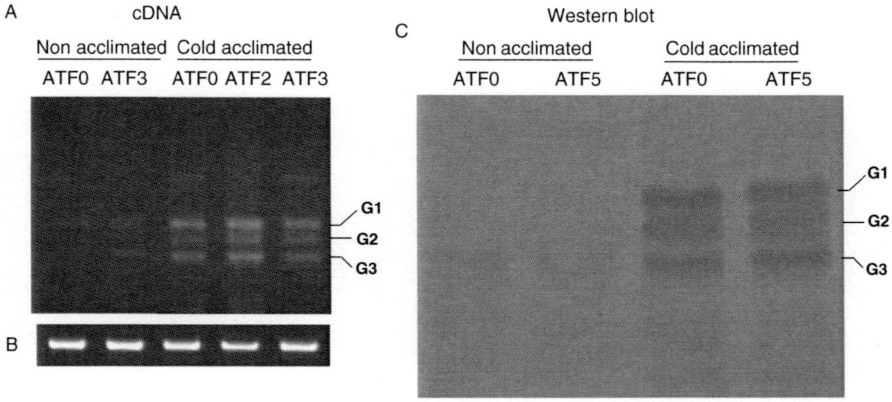

Fig. 58.2 (**a**) Amplification of the 380 bp-targeted sequence in C-terminal region of the *msa*CIG gene using cDNA from non acclimated and cold-acclimated plants of populations ATF0, ATF2 and ATF3; (**b**) Amplifications with a cinnamoyl-CoA reductase gene from a *Medicago sativa* EST collection (Serge Laberge, unpublished), as a constitutive control; (**c**) Western blot from non acclimated and cold-acclimated plants of ATF0 and ATF5 with an anti-dehydrin antibody. Total RNA and proteins were extracted from pooled samples of $\approx$ 30 crowns

Acknowledgements We would like to thank Mrs Josée Michaud and Josée Bourassa for their excellent technical assistance in the realization of this project.

References

Castonguay, Y., Nadeau, P. 1998. Enzymatic control of soluble carbohydrate accumulation in cold-acclimated crowns of alfalfa. Crop Sci. 38:1183–1189.

Castonguay, Y., Cloutier, J., Laberge, S., Bertrand, A., Michaud, R. 2005. A bulk segregant approach to identify genetic polymorphisms associated with cold tolerance in lucerne (pp. 88–102). In: Chen, T. H. H., Uemura, M., Fujikawa, S. (eds), Cold hardiness in plants: Molecular genetics, cell biology and physiology. CABI Publishing, Cambridge, MA, USA.

Castonguay, Y., Rémus-Borel, W., Cloutier, J., Laberge, S., Bertrand, A., Michaud, R. 1998. Analysis of dehydrin sequences from alfalfa popualtions recurrently selected for freezing tolerance. Proc. of the 41st North American Alfalfa Improvement Conference and the 20th Trifolium Conference, Dallas, TX, USA. Available at [http://www.naaic.org/Meetings/National/2008meeting/proceedings/proceedings2008.htm] Updated May 2008.

Castonguay, Y., Michaud, R., Nadeau, P., Bertrand, A. 2009. An indoor screening method for improvement of freezing tolerance in alfalfa. Crop Sci. 49:809–818.

Close, T.J 1997. Dehydrins: A commonality in the response of plants to dehydration and low temperature. Physiol. Plant. 100:291–296.

Marian, C.O., Krebs, S.L., Arora, R. 2004. Dehydrin variability among rhododendron species: A 25-kDa dehydrin is conserved and associated with cold acclimation across diverse species. New Phytol. 161:773–780.

Patton, A.J., Cunningham, S.M., Volenec, J.J., Reicher. Z.J. 2007. Differences in freeze tolerance of zoysiagrasses: I. Role of proteins. Crop Sci. 47:2162–2169.

Rémus-Borel, W., Castonguay, Y., Cloutier, J., Michaud, R., Bertrand, A., Desgagnés, R., Laberge, S. 2009. Dehydrin variants associated to superior freezing tolerance in alfalfa (*Medicago sativa* L). Theor. Appl. Genet. DOI 10.1007/s 00/22-009-1243-7(online first)

Chapter 59
The Complete Chloroplast Genome Sequence of Perennial Ryegrass (*Lolium perenne* L.) Reveals Useful Polymorphisms Among European Ecotypes

Kerstin Diekmann, Trevor R. Hodkinson, Kenneth H. Wolfe, Rob van den Bekerom, Phillip J. Dix, and Susanne Barth

Abstract To date, more than 130 complete chloroplast genome sequences of plants have been published. Thirteen of these sequences belong to Poaceae species. But the complete chloroplast genome sequence of the important forage species, perennial ryegrass, was hitherto unknown. We sequenced, assembled and annotated the entire chloroplast genome of *Lolium perenne* 'Cashel', and searched it for single nucleotide polymorphisms (SNPs) and simple sequence repeats (cpSSRs). The chloroplast genome sequence of *L. perenne* is 135,282 bp long and encodes 128 genes. Forty SNPs were found within the sequenced sample of 'Cashel'. Thirty mononucleotide cpSSRs of more than ten base pairs length were detected and we designed primers to amplify and sequence twelve of the most variable of these regions. Length variation and SNP variation of these markers was assessed on a set of 15 Irish and 15 European ecotypes. The results revealed a total of 21 haplotypes across all ecotypes and these were suitable to detect genetic diversity among and within *L. perenne* accessions.

Keywords Chloroplast genome · cpSSR markers · Diversity · *Lolium perenne* · Perennial ryegrass

Introduction

Chloroplast genomes have a circular structure that is divided into four regions: the large single copy region (LSC), the small single copy region (SSC) and two copies of an inverted repeat (IR) that separate the single copy regions from each other. Its size can vary from 35 kb to 217 kb, but in most angiosperm species a much smaller range of 115–165 kb is found. The gene content can vary from 63 to 209 genes, but in most angiosperm species it contains only 110–130 genes which are generally arranged in the same order (Jansen et al., 2005). Thus the

K. Diekmann (✉)
Teagasc Crops Research Centre, Oak Park, Carlow, Ireland; University of Dublin, Botany Building, School of Natural Sciences, Trinity College Dublin, Dublin, Ireland
e-mail: kerstin.diekmann@teagasc.ie

C. Huyghe (ed.), *Sustainable Use of Genetic Diversity in Forage and Turf Breeding*, DOI 10.1007/978-90-481-8706-5_59,

features of chloroplast genomes are highly conserved and they are suitable tools for genetic resource characterisation and phylogenetic, population genetic, and phylogeographic studies. Due to their maternal inheritance in most species they have some advantages for genetic engineering, because gene transfer via pollen flow does not generally occur. Chloroplast genomes are also suitable tools in plant breeding schemes because they are able to define cytoplasmic breeding pools, to track parentage in interspecific hybrids and to monitor seed mediated gene flow. The number of published chloroplast genome sequences has increased rapidly in recent years and to date more than 130 are publicly available (http://www.ncbi.nlm.nih.gov/genomes/ORGANELLES/plastids_tax.html). Within these 130 sequences 13 are from Poaceae species, including the three forage/turf species *Agrostis stolonifera*, *Festuca arundinacea* and *Lolium perenne*. The chloroplast genome of Poaceae varies only

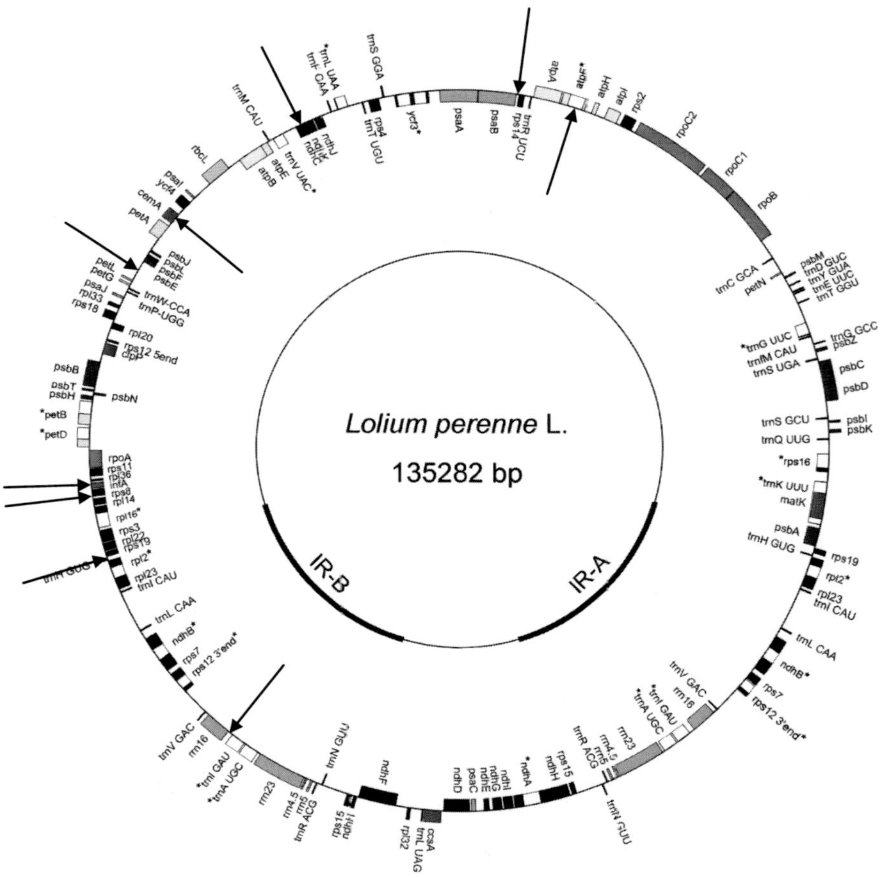

modified after Diekmann et al. 2009

Fig. 59.1 Circular structure of the *Lolium perenne* chloroplast genome. Locations of the newly designed chloroplast microsatellite markers are indicated by arrows. Genes written on the outside are transcribed clockwise, genes on the inside counter-clockwise. IR=Inverted repeat region modified after Diekmann et al. (2009)

from ~134 to 141 kb. The chloroplast genome of *L. perenne* (Fig. 59.1) which was sequenced by our research group measures 135,282 bp and thus is of average size within the Poaceae (Diekmann et al., 2009). Differences in the genome size of *L. perenne* to other Poaceae species are mainly due to length variation in intergenic spacer regions. However, a surprisingly high amount of variation was also found in the length of coding regions between the different Poaceae species (Diekmann et al., 2009). Furthermore, an unexpectedly high amount of variation was also detected at intraspecific level in form of single nucleotide and insertion/deletion polymorphisms within a single cultivar (Diekmann et al., 2009). The aims of this paper are to assess (1) the amount and location of simple sequence repeats (cpSSRs) within the chloroplast genome of *L. perenne* and (2) the potential of newly developed cpSSR markers for detecting inter- and intra-population diversity within a selection of *L. perenne* ecotypes.

Material and Methods

Lolium perenne cpSSRs were detected by using the microsatellite finder tool find_microsat_Win32 (N. Salamin unpublished) searching for mononucleotide repeats with a length of at least 10 bp. Those were likely to show more variation than shorter repeats. Primers were designed around cpSSRs using the program PrimerExpress (version 2.0, Applied Biosystems, Foster City, California, USA). To ensure a broad application range, primers were designed only in regions that were conserved among the Poaceae chloroplast genome. Thus, chloroplast genome sequences of the following species (Genbank accession number) were used in addition to *L. perenne* (AM777385): *Agrostis stolonifera* (EF115543), *Hordeum vulgare* (EF115541), *Oryza nivara* (AP006728), *Oryza sativa 'Indica'* (AY522329), *Oryza sativa 'Japonica'* (X15901), *Saccharum officinarum* (AP006714), *Sorghum bicolor* (EF115542), *Triticum aestivum* (AB042240), *Zea mays* (X86563).

Primers were tested for their use to detect variation within *L. perenne* accessions using 15 Irish and 15 other European *L. perenne* ecotypes of which total genomic DNA was previously extracted by McGrath et al. (2006). On average, six individuals per accession were analysed. PCR reactions for the different primer sets were carried out using a polymerase with proof-reading ability (Phusion High-Fidelity DNA Polymerase, New England Biolabs Inc.) to ensure the correct amplification of the microsatellite regions. Sequencing of the PCR products was sourced to a commercial company (GATC Biotech, Germany). Sequences were aligned and searched visually for variation using Mega 3.1 (Kumar et al., 2004).

Results and Discussion

30 cpSSRs of more than 10 bp lengths were detected in the complete cpDNA genome of *L. perenne* – 5 in coding regions and 25 in non-coding regions. cpSSRs were mainly based on the nucleotides A and T, only three repeats on the nucleotide

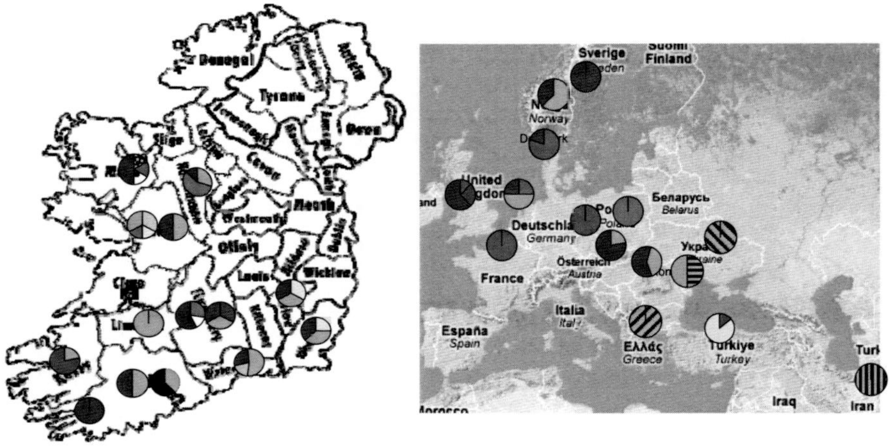

Fig. 59.2 Frequencies of the different haplotypes observed in the Irish and European *Lolium perenne* ecotypes

C. Provan et al. (2004) and McGrath et al. (2006) previously reported cpSSRs for *L. perenne*. However, only a few of these markers were located in regions with mononucleotide repeats of more than 10 bp in length or showed a high amount of variation as observed by Diekmann et al. (2009). We designed nine cpSSR primer pairs around twelve microsatellite regions, amplifying four microsatellites found within genes and eight within non-coding regions (Fig. 59.1). Sequencing with these primer sets detected 16 haplotypes within the 15 Irish ecotype accessions (90 individual plants analysed) and five additional haplotypes within the other European ecotypes (Fig. 59.2). Up to four different haplotypes, within a single accession of in average six analysed individuals, were found. Although there was no length

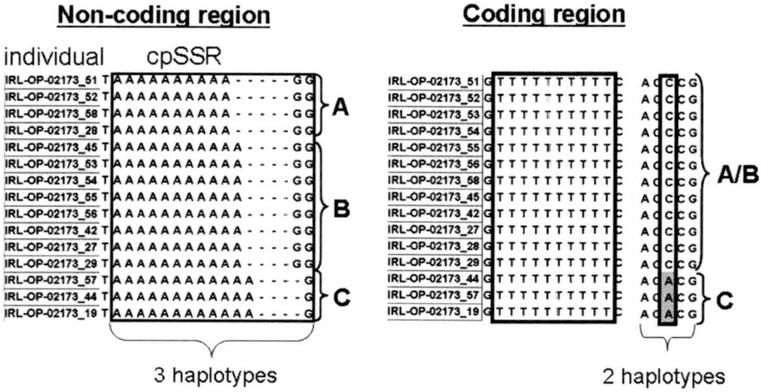

Fig. 59.3 Observed variation for individuals of *Lolium perenne* ecotype IRL-OP-02173 in a non-coding and a coding region

variation in microsatellites of coding regions of the genome, single nucleotide polymorphisms were observed, which added more haplotypes (Fig. 59.3). However, the tested sample set was small. We predict that the application of these newly designed markers to a larger set of plants including more individuals per accession will resolve many more haplotypes and thus better reveal the diversity within Irish and European *L. perenne* populations.

Acknowledgements K.H.W. is supported by Science Foundation Ireland. K.D. and R.vdB. are financed under the Teagasc Walsh Fellowship Scheme. The project funding was obtained from the Teagasc 'Vision' programme.

References

Diekmann, K., Hodkinson, T.R., Wolfe, K.H., van den Bekerom, R., Dix, P.J., Barth, S. 2009. Complete chloroplast genome sequence of a major allogamous forage species, perennial ryegrass (*Lolium perenne* L.). DNA Res. 16(3):165–176.

Kumar, S., Tamura, K., Nei, M. 2004. MEGA3: Integrated software for molecular evolutionary genetics analysis and sequence alignment. Briefings in bioinfo.. 5:150–163.

McGrath, S., Hodkinson, T. R., Salamin, N., Barth, S. 2006. Development and testing of novel chloroplast microsatellite markers for *Lolium perenne* and other grasses (Poaceae) from de novo sequencing and in silico sequences. Mol. Ecol. Notes 6(2):449–452(4).

Provan, J., Biss, P.M., McMeel, D., Mathews, S. 2004. Universal primers for the amplification of chloroplast microsatellites in grasses (Poaceae). Mol. Ecol. Notes 4:262–264.

Chapter 60
Exploring the Potential for Translational Genomics Approaches in Forage Legumes: Regions of Highly Conserved Microsynteny Between White Clover and *Medicago truncatula* Revealed by BAC Sequencing

Melanie Febrer, Michael T. Abberton, Glyn Jenkins, and Dan Milbourne

Abstract The model legume species *Medicago truncatula* is a potentially useful tool for gene discovery in white clover using translational genomics strategies. A prerequisite to the practical implementation of this approach is a good understanding of the extent of conservation of gene order between the species. Previous studies have demonstrated conservation at the macrosyntenic level, but no published information exists on the extent of conserved microsynteny between these species. In a previously published study, we reported the construction of a BAC library of white clover, the end sequencing of approximately 700 clover BACs, and the comparison of these BAC-end sequences to the *M. truncatula* genome. We found that 14 paired BAC-ends were shown to have the equivalent pairs of *M. truncatula* sequence on the same *M. truncatula* BAC clone or contig sequence within a span of 25–200 Kb, suggesting they represent orthologous regions in the two species. In this follow-up analysis, we have chosen five of these BACs, sequenced them to approximately six-fold coverage, and compared the resulting assembled contigs to their putatively equivalent regions of *M. truncatula*. Highly conserved gene content and almost complete conservation of gene order and orientation for all five sequences were found, suggesting that translational genomics approaches for gene discovery using *Medicago* could be successful.

Introduction

Comparative genomics studies between well characterized and resourced model species and key agricultural species continue to be an important discovery route for the genes underlying important traits in the agricultural species. White clover is closely related to the model legume species *Medicago truncatula*, for which the

D. Milbourne (✉)
Teagasc, Crops Research Centre, Oak Park, Carlow, Ireland
e-mail: dan.milbourne@teagasc.ie

C. Huyghe (ed.), *Sustainable Use of Genetic Diversity in Forage and Turf Breeding*,
DOI 10.1007/978-90-481-8706-5_60,

"genespace" sequence is available. Genetic mapping studies have provided evidence that gene content and order is reasonably well conserved between white clover and *M. truncatula*. (George et al., 2006; Zhang et al., 2007) While these studies have begun to address the macrosyntenic relationships between the genome of white clover and the model species such as *M. truncatula*, less work to date has been performed on the microsyntenic scale.

The objective of this work was to gain a preliminary insight into the extent of microsynteny between the two species, and to investigate the utility of a large-scale BES strategy to "tile" a significant proportion of the genome of white clover on to that of *Medicago*. The latter would constitute a useful translational genomics platform for gene isolation in white clover in the absence of significant amounts of publicly available sequence information in this species. To that end, we previously constructed a BAC library of white clover and sequenced both ends of 700 BACs from that library (Febrer et al., 2007). From amongst these BACs we identified those which had a significant BLASTn hit to *Medicago* genome sequence at both ends, and the subset of these BACs for which both BAC-ends had good BLASTn matches in *Medicago* genome sequence at a distance consistent with the span of a BAC (Febrer et al., 2007). Of 204 clover BACs which had a *Medicago* genome sequence match on both ends, only 14 had matches which were separated by a compatible distance in *Medicago*, and, consistent with other studies of this nature, we proposed that these represent regions of conserved microsynteny between white clover and *Medicago*. In this study we extend this analysis by sequencing five of these BACs to approximately sixfold coverage, and comparing the resulting sequence contigs to their proposed *M. truncatula* orthologs.

Materials and Methods

From amongst the 14 potential comparative-tile BACs, five (27B12, 28G20, 27I09, 28F22 and 27K12) were picked for further analysis on the basis of the criterion that at least one of their end-sequences was apparently "genic" in nature, by virtue of a significant match to either a protein or EST sequence using the appropriate BLAST algorithm at a cutoff value of 1e-10.

All five BACs were subjected to shotgun sequencing to approximately six-fold coverage using standard Sanger-sequencing chemistry and assembled using the SeqMan II module of the Lasergene software package (DNASTAR). In this preliminary study, within each BAC assembly, all of the contigs below 5 Kb were eliminated from further analysis. The remaining contigs for each BAC and the associated *M. truncatula* sequence(s) were subjected to a round of *ab initio* gene-prediction with the FGENESH gene structure prediction programme, using the *Medicago truncatula*-trained prediction matrix (www.softberry.com). In order to obtain database support for as many of the predicted gene as possible, their predicted amino-acid sequences were compared the *Arabidopsis* protein database using

BLASTp (E-value cut off 1e-10). To identify repetitive elements, the *Medicago truncatula* and white clover BAC sequences were analysed against the TIGR *Fabaceae* repeats database using BLASTn (E value e-06 and % identity >50%).

Initial sequence comparisons were carried out using the MULAN alignment tool (http://mulan.dcode.org) to generate similarity dotplots of putative orthologous pairs of *Medicago* and clover BACs. Subsequent sequence comparisons were carried out using the Artemis Comparitive Tool (ACT) (www.sanger.ac.uk/Software/ACT/) to generate similarity maps on which the gene annotation data were graphically overlaid manually. To calculate nucleotide identity, predicted mRNA sequences for putative orthologous pairs of clover and *Medicago* genes were aligned using the WATER alignment algorithm in the EMBOSS pairwise alignment tool (http://www.ebi.ac.uk/Tools/emboss/align/).

From the 5 BACs, simple sequence repeats were identified using the Tandem Repeat Finder software application. From amongst the SSRs discovered in the BAC sequences, SSR primer pairs were designed to amplify 2 SSR loci per BAC. The SSRs were analysed on a white clover mapping population (Febrer at al. unpublished) in order to obtain a genetic map location for all of the clover BACs.

Results and Discussion

After assembly and elimination of contigs under 5 kb, two BAC clones were composed of single contigs (BAC 27B12 and 28G20) with sequence lengths of 65,105 bp and 66,130 bp respectively, two BAC clones contained two contigs (BAC 27I09 and 28F22) with total sequence lengths of 81,115 bp and 76,375 bp respectively and the remaining BAC clone 27K12 was composed of five contigs with a total sequence length of 106,241 bp.

A preliminary round of sequence-based analysis showed that all five clover BACs exhibited extensive similarities to their proposed *Medicago* orthologs. For 27I09 and 28F22, a mixture of the position of the BAC vector, read pair-data and clover:*Medicago* alignments allowed the ordering and orientation of the contigs relative to each other and these sequences were concatenated and treated as single "pseudo-sequences" for further analysis. In 27K12, three of five contigs could be ordered reliably, and the remaining sequences were arbitrarily placed at the end of the "pseudo-sequence". The BAC sequences described above were used as input for gene prediction in FGENESH. The total number of predicted genes per BAC ranged from 13 to 26 in clover and from 16 to 39 in the proposed *Medicago* orthologs, with a total of 93 genes predicted in clover sequences and 144 genes predicted in *Medicago*. The larger number of genes predicted for *Medicago* reflected the fact that the *Medicago* sequences being examined are longer. Overall average gene density, average gene size and average numbers of introns/exons per gene for the white clover BACs were very similar to those of the *Medicago* BAC sequences and the summary values for these characteristics in the Mt1.0 sequence build that was used for comparisons in this study (not shown).

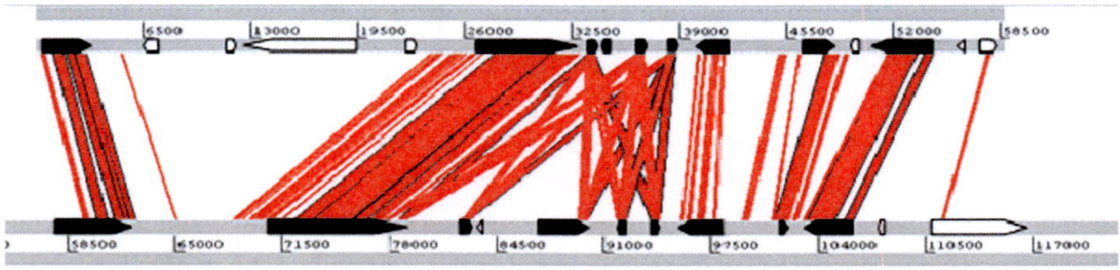

Fig. 60.1 An ACT plot comparing clover BAC 27B12 (*top*) to *Medicago truncatula* BAC AC146852 (*bottom*). Genes with database support are represented as solid arrows, those without database support are white. Arrows indicate transcriptional direction

In order to visualise the relationship between nucleotide similarity and gene content/order, the positions of predicted genes were overlayed on to the ACT plots generated between putatively orthologous pairs of white clover and *Medicago* BACs (an example is given for Clover BAC 27B12 in Fig. 60.1).

Unsurprisingly, regions exhibiting high levels of sequence-based similarity generally correspond to predicted genes shared between the two species, and, on the basis of the BLASTp analysis, pairs of predicted genes exhibiting nucleotide similarity exhibited similarity to the same *Arabidopsis* protein sequences. Average nucleotide identity between pairs of orthologous genes was 84.6%. Little or no similarity was observed in intergenic regions.

Within the window of comparison defined on the ACT plots by the first and last region of nucleotide similarity between the two sets of BACs, a total of 88 and 96 predicted genes were present in white clover and *M. truncatula*, respectively, with 37 (42%) of the predicted genes in white clover exhibiting similarity to genes in *Medicago*, and 45 (49%) of predicted genes in *Medicago* exhibiting similarity with genes in clover. However, there is a strong bias towards conservation of genes with database support (good BLASTp matches to *Arabidopsis*) between the two sets of sequences, suggesting that predicted genes without database support may represent potentially poor gene calls or retroelements. Amongst the 5 clover BACs, in the window of comparison described above, there were 43 supported gene models, of which 33 (77%) exhibited similarity to genes in *Medicago*, and in the equivalent *Medicago* sequences there were 49 supported gene models, 39 (80%) of which exhibit similarity to genes in clover. Thus, excluding potentially poor gene models, approximately 80% of genes are conserved over the five pairs of sequences. In addition, there was complete conservation of gene order between the orthologs in both species, and all had the same transcriptional orientation.

Eight of the 10 clover BAC-derived SSRs described above were successfully mapped on a clover reference mapping population. At least one marker per BAC was successfully mapped. Genetic mapping supported the single locus nature of all of the BAC sequences. All of the clover BACs were mapped to the clover homeolinkage group that is the proposed homeolog of the *Medicago* chromosome from which their equivalent *Medicago* BAC was derived, supporting the idea that the pairs of sequences are truly orthologous.

Our results suggest that, for at least some areas of the white clover genome, extensive conserved microsynteny with *Medicago truncatula* exists, indicating that translational genomics approaches for gene discovery in white clover using *M. truncatula* as a reference species is possible.

References

Febrer, M., Cheung, F., Town, C.D., Cannon, S.B., Young, N.D., Abberton, M.T., Jenkins, G., Milbourne, D. 2007. Construction, characterisation and preliminary BAC-end sequencing of a bacterial artificial chromosome library of white clover (*Trifolium repens* L.) Genome 50(4):412–21.

George, J., Cogan, N.O.I., Smith, K.F., Spangenberg, G.C., Forster, J.W. 2006. Genetic map integration and comparative genome organisation of white clover (*Trifolium repens* L.) with model legumes (p. 214). In: "Plant & animal genome XIV", San Diego, CA.

Zhang, Y., Sledge, M.K., Bouton, J.H. 2007. Genome mapping of white clover (*Trifolium repens* L.) and comparative analysis within the Trifolieae usingcross-species SSR markers. Theor. Appl. Genet. 114:1367–1378.

Chapter 61
Plant Transcription Factors as Novel Molecular Markers for Legumes

Yuanhong Han, Dong-Man Khu, Ivone Torres-Jerez, Michael Udvardi, and Maria J. Monteros

Abstract Legumes represent an important component of the world's crop production. Transcription factors are global regulators of gene networks that play a vital role in many cellular processes and represent excellent targets for developing molecular markers. The goals of this project are to develop a comprehensive resource of plant transcription factors and utilize them to facilitate the transfer of information across multiple model and crop legumes. Primers producing PCR amplicons in alfalfa (*Medicago sativa* L.) were used to further evaluate amplification in a panel consisting of model (*M. truncatula, Lotus japonicus* L.) and eight crop legumes that included parents of existing mapping populations. Amplification, size polymorphism, and sequence variation were evaluated using capillary sequencers. From the total number of primers producing amplicons, 90%, 78%, 38% and 34% of them produced single amplicons in *M. truncatula*, *M. sativa* L., *Glycine max* L., and *L. japonicus*, respectively. In general, the likelihood of successful amplification decreased with increased phylogenetic distance among species. SNP variation both among and within legume species was identified. These transcription-factor based molecular markers can be used to further understand and enhance complex traits of agronomic importance.

Keywords Legumes · Transcription factors · Molecular markers · Comparative mapping

Introduction

Advances in molecular biology and genomics in the model legume species *Medicago truncatula* and *Lotus japonicus* merit further investment in translational genomics to transfer knowledge from models to crop legumes and enhance plant

M.J. Monteros (✉)
Forage Improvement Division, The Samuel Roberts Noble Foundation, 2510 Sam Noble Parkway, Ardmore, OK 73401, USA
e-mail: mjmonteros@noble.org

C. Huyghe (ed.), *Sustainable Use of Genetic Diversity in Forage and Turf Breeding*,
DOI 10.1007/978-90-481-8706-5_61, © Springer Science+Business Media B.V. 2010

breeding programs. Transcription factors (TF) are key components in regulating important plant processes and have an immense potential to optimize these processes for agriculture. Transcription factors are involved in abiotic stress responses including drought (Zhang et al., 2005), freezing, salt (Dai et al., 2007), and aluminum tolerance (Iuchi et al., 2007), in plant defense, detoxification and stress responses (Raffaele et al., 2008), in the development and differentiation of root nodules (Schauser et al., 1999), and in flowering time (Cai et al., 2007). Over-expression of TF sequences has led to increases in freezing, drought, salt, and soil toxicity stress tolerance. The sequence conservation of TF sequences within and across multiple kingdoms and their genome-wide distribution in eukaryotes (Riechmann et al., 2000), facilitates the use of TF primers as anchor markers to transfer information from models to crop species. Comparative genome analyses can reveal genetic conservation among the genomes of related species (synteny) and greatly facilitate gene discovery, which can be used subsequently in either direct genetic engineering or marker-assisted breeding, leading to potential major innovations in crop improvement (Dangl et al., 2008).

Materials and Methods

Plant Materials

A panel of 34 individual genotypes from ten legume species (*M. truncatula, L. japonicus* L., *M. sativa* L., *Glycine max* L., *Pisum sativum* L., *Phaseolus vulgaris* L., *Vigna radiata* L., *Trifolium repens* L., *T. pratense* L., and *Lupinus albus* L.) (Table 61.1) were used for genomic DNA extraction using the Plant DNeasy kit (Qiagen, Valencia, CA). Most entries included in the panel are parents of existing mapping populations.

Table 61.1 Amplification and size polymorphism of 168 TF gene primers in ten legume species

Species	No. of genotypes	Primer pairs with amplification	Polymorphic primer pairs (size only)	Primer pairs with single amplicon	Polymorphic primer pairs with single amplicon (size only)
Medicago truncatula	8	142	45	130	35
Medicago sativa	8	142	47	113	30
Trifolium repens	2	98	17	59	6
Trifolium pratense	2	82	9	64	5
Pisum sativum	2	89	13	70	6
Lotus japonicus	2	88	6	49	1

Table 61.1 (continued)

Species	No. of genotypes	Primer pairs with amplification	Polymorphic primer pairs (size only)	Primer pairs with single amplicon	Polymorphic primer pairs with single amplicon (size only)
Glycine max	4	85	38	55	24
Phaseolus vulgaris	2	82	17	50	7
Vigna radiata	2	79	49	51	22
Lupinus albus	2	86	13	58	5

Primers and PCR Reactions

PCR amplicons from 1,084 transcription-factor based markers (Kakar et al., 2008) obtained using a pooled DNA sample of four alfalfa mapping population parents were separated and visualized using GelGreen-stained (Biotium, Inc., Hayward, CA) agarose gels. PCR reactions were set up as previously described (Zhang et al., 2007) and a total of 770 primer pairs produced successful PCR amplification in alfalfa. One hundred and sixty eight of these 770 primer pairs were used to set up PCR reactions in the legume panel. PCR products and sequencing reactions were analyzed using the ABI PRISM 3730 Genetic Analyzer (Applied Biosystems, Foster City, CA).

Data Analysis

PCR amplicons were visualized and analyzed with GeneMapper 3.7 software (Applied Biosystems, CA) to determine successful amplification and size differences among and within legume species. Sequencer^{TM} 4.8 (Gene Codes Corporation, Ann Arbor, MI) was used to align the sequences and display SNPs. Phylogenetic trees were constructed by parsimony options in PAUP* (Swofford, 2003) using 100 bootstrap replicate searches.

Results and Discussion

Cross-species Amplification

Among the 168 primers pairs evaluated in the legume panel so far, 145 (86%) primer pairs amplified in at least one species other than alfalfa (Table 61.1). Species from Trifolieae and Viceae produced single amplicons with at least 41% of the primer pairs, while 34–40% of the primer pairs produced single amplicons in the species from other tribes. Five primer pairs produced single amplicons across all ten species; thirty two primer pairs produced single amplicons across at least seven species; eighty primer pairs produced single amplicons in at least 5 species and 121 primer pairs produced single amplicons across at least three species. *M. truncatula* and alfalfa share the highest number (110) of single amplicon primer pairs, while *T. repens* and *P. vulgaris* share the least number (21) of primer pairs. In general, at least half of the primer pairs amplified in each legume species evaluated, indicating a high level of conservation of these TF gene sequences across the legume species in the panel.

Polymorphism and Orthology

Many primer pairs revealed amplicon size polymorphism among the genotypes within a species (Table 61.1). Sequencing the amplicons allowed us to verify amplification of the target sequence and detect SNPs both within and among species. As an example, all sequences amplified in the 10 legume species using the primer pair MTTF063 were identical or nearly identical to the sequence used for primer design, and revealed 16 SNP among all species (Fig. 61.1). SNPs within species were identified in *M. truncatula, M. sativa, P. sativum, L. albus* and *P. vulgaris*. Generally, less sequence variation was observed with decreasing phylogenetic distance. The phylogenetic gene trees generated from these sequences using the parsimony method were compatible with the taxonomic tree, supporting the orthologous origin of these sequences (not shown).

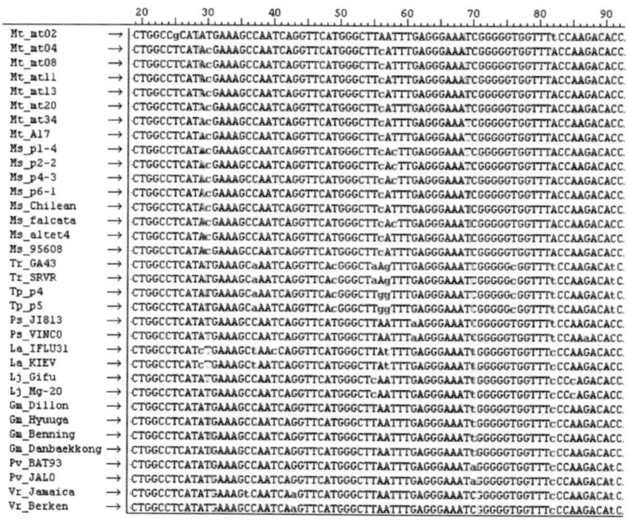

Fig. 61.1 Sequences and SNPs of the PCR amplicon obtained with primer pair MTTF063 in ten legume species (34 genotypes). *Mt*: *M. truncatula*; *Ms*: alfalfa; *Tr:* white clover; *Tp*: red clover; *Ps*: garden pea; *La*: white lupin; *Lj*: *Lotus japonicus*; *Gm*: soybean; *Pv*: common bean; *Vr*: mungbean

A large number of the TF gene primer pairs tested here produced single amplicons across more than three legume species indicating the feasibility of using them as anchor markers to transfer information from models to crop species. The utilization of molecular markers developed from TF genes, which are regulators of gene networks, offers a unique opportunity to improve complex traits of agronomic importance, such as tolerance to abiotic and biotic stress, in multiple legume species in a way that has not been previously feasible using conventional breeding methods. Although markers which amplify across all ten species are rare, the alignment of the markers can be achieved by utilizing *M. truncatula* as the common species; this can facilitate the use of information from linkage and QTL mapping studies across legume species, including those with limited genomic resources.

References

Cai, X., Ballif, J., Endo, S., Davis, E., Liang, M., Chen, D., DeWald, D., Kreps, J., Zhu, T., Wu, Y. 2007. A putative CCAAT-binding transcription factor is a regulator of flowering timing in Arabidopsis. Plant Physiol. 145:98–105.

Dai, X., Xu, Y., Ma, Q., Xu, W., Wang, T., Xue, Y., Chong, K. 2007. Overexpression of an R1R2R3 MYB gene, OsMYB3R-2, increases tolerance to freezing, drought, and salt stress in transgenic Arabidopsis. Plant Physiol. 143:1739–1751.

Dangl, J., Banta, L., Boerma, R., Carrington, J.C., Chory, J., Kay, S., Lewis, S., Mitchell-olds, T., Sinha, N., Snyder, M., Strauss, S., Ward, E. 2008. Achievements of the national plant genome initiative and new horizons in plant biology. The National Academies Press, Washington, D.C., USA.

Iuchi, S., Koyama, H., Iuchi, A., Kobayashi, Y., Kitabayashi, S., Kobayashi, Y., Ikka, T., Hirayama, T., Shinozaki, K., Kobayashi, M. 2007. Zinc finger protein STOP1 is critical for proton tolerance in Arabidopsis and coregulates a key gene in aluminum tolerance. Proc. Natl. Acad. Sci. U. S. A. 104:9900–9905.

Kakar, K., Wandrey, M., Czechowski, T., Gaertner, T., Scheible, W.R., Stitt, M., Torres-Jerez, I., Xiao, Y., Redman, J.C., Wu, H.C., Cheung, F., Town, C.D., Udvardi, M.K. 2008. A community resource for high-throughput quantitative RT-PCR analysis of transcription factor gene expression in *Medicago truncatula*. Plant Methods 4:18.

Raffaele, S., Vailleau, F., Leger, A., Joubes, J., Miersch, O., Huard, C., Blee, E., Mongrand, S., Domergue, F., Roby, D. 2008. A MYB transcription factor regulates very-long-chain fatty acid biosynthesis for activation of the hypersensitive cell death response in Arabidopsis. Plant Cell 20:752–767.

Riechmann, J.L., Heard, J., Martin, G., Reuber, L., Jiang C.Z., Keddie, J., Adam, L., Pineda, O., Ratcliffe, O.J., Samaha, R.R., Creelman, R., Pilgrim, M., Broun, P., Zhang, J.Z., Ghandehari, D., Sherman, B. K., Yu, L.G. 2000. Arabidopsis transcription factors: Genome-wide comparative analysis among eukaryotes. Science 290:2105–2110.

Schauser, L., Roussis, A., Stiller J., Stougaard, J. 1999. A plant regulator controlling development of symbiotic root nodules. Nature 402:191–195.

Swofford, D.L. 2003. PAUP∗, phylogenetic analysis using parsimony (∗and other methods). Version 4. Sinauer Associates, Sunderland, MA.

Zhang, J.Y., Broeckling, C.D., Blancaflor, E.B., Sledge, M.K., Sumner, L.W., Wang, Z.Y. 2005. Overexpression of WXP1, a putative *Medicago truncatula* AP2 domain-containing transcription factor gene, increases cuticular wax accumulation and enhances drought tolerance in transgenic alfalfa (*Medicago sativa*). Plant J. 42:689–707.

Zhang, Y., Sledge, M.K., Bouton, J.H. 2007. Genome mapping of white clover (*Trifolium repens* L.) and comparative analysis within the Trifolieae using cross-species SSR markers. Theor. Appl. Genet. 114:1367–1378.

Chapter 62
From a Model to a Crop Species: Constans is Involved in Aerial Morphogenesis of Lucerne

Bernadette Julier, Doris Herrmann, Philippe Barre, and Jean-Baptiste Pierre

Abstract The development of genomics tools and knowledge in model species can be used for genetic analyses in crop species. QTL detection in the model legume *Medicago truncatula* was carried out for flowering date and stem height, two traits related to aerial morphogenesis. A strong QTL was detected, and after a fine mapping step and bioanalysis of BAC sequences, six candidate genes were listed. One of these genes, Constans, was proved to be differentially expressed among two parental lines differing for flowering date. Constans was sequenced in a population of lucerne (*Medicago sativa*) genotypes that were phenotyped for flowering date and stem height. The low rate of polymorphism of Constans in lucerne shows that this gene is not neutral. Sequence polymorphism was associated to phenotypic variation. Constans is thus involved in flowering date and stem height genetic determinism.

Keywords SNP · *Medicago sativa* · *Medicago truncatula* · QTL · Association genetics

Introduction

Aerial morphogenesis is a key component of biomass production and quality of forage crops. It is also related to competition among plants in the canopies, leading to possible genetic evolution of the populations. The identification of genes or genomic regions (QTL) involved in such traits would offer many prospects for breeding programs and also to understand the mechanisms of genetic evolution of the populations.

B. Julier (✉)
INRA, UR 4, BP 6, 86600 Lusignan, France
e-mail: Bernadette.Julier@lusignan.inra.fr

C. Huyghe (ed.), *Sustainable Use of Genetic Diversity in Forage and Turf Breeding*,
DOI 10.1007/978-90-481-8706-5_62,

Lucerne (*Medicago sativa* L.) genetics is complex because of allogamy and autotetraploidy. *Medicago truncatula*, a related diploid, annual and self-fertilizing species was chosen by the scientific community as a model for genomic studies. Many tools are being developed in this species, such as genomic sequences, EST and markers. SSR markers developed in *M. truncatula* have been successfully used to create a genetic map of lucerne (Julier et al., 2003) and to analyse genetic diversity in lucerne cultivars (Flajoulot et al., 2005). It is expected that genomic resources developed in the model species will accelerate genetic analysis in related crop species. We have hypothesised that a gene involved in trait variation in *M. truncatula* could also explain a part of the variation for the same trait in *M. sativa*.

The objective of this study was to detect QTLs for aerial morphogenesis in *M. truncatula*, to identify candidate genes and to test the variation explained by one candidate gene in lucerne morphogenesis. Association genetics was used to relate sequence polymorphism to phenotypic variation.

Material and Methods

QTL Detection in M. truncatula

Three mapping populations of *M. truncatula*, each of about 200 RILs, were studied over six years in greenhouse: LR1 (DZA315.26 × DZA45.6, 196 RILs in F_5 generation), LR4 (Jemalong6 × DZA315.16, 199 RILs in F_7 generation) and LR5 (Jemalong6 × F83005.5, 173 RILs in F_7 generation). Framework maps made with SSR markers were available for all populations. Flowering date was scored and transformed into a sum of degree-days (°C.D) above 0°C from sowing. The length of the first two emerging primary branches was measured at the end of the experiment. QTL mapping was performed using QTLcartographer with the composite interval mapping (CIM) procedure. A multi-population QTL analysis was carried out with Biomercator software in order to compile the different genetic maps and QTLs.

Among the RILs of LR4 population, a F_6 line that was heterozygous in the region of a strong QTL on chromosome 7 and homozygous elsewhere was identified (line 105). It was self-pollinated and produced about 2000 F_7 seeds. This generation is subsequently called a pseudo-F_2 generation. Seeds were sown in greenhouse and phenotyped for flowering date.

Genomic DNA was extracted from each plant. Two markers (MTIC040 and MTIC714) flanking the confidence interval (7.54 cM) of the QTL detected in F_6 generation were genotyped on the 1,640 plants of the pseudo-F_2 generation to identify recombinant plants.

Ten SSR markers and two polymorphic gene sequences were used to increase the density of the map in the QTL region. QTL analysis by simple interval mapping was performed using the package R-QTL with R software v2.4.1 (Broman et al., 2003). A Bayesian support interval was built to determine the position of the QTL.

Identification of a Candidate Gene in M. truncatula *and Expression Study*

Using sequencing information from the *M. truncatula* website (www.medicago.org) and other bioinformatics resources, candidate genes were selected into the support interval of the QTL defined in the pseudo-F_2 population. The predicted genes were compared to the gene list obtained from literature on flowering date in model species (Putterill et al., 2004; Hecht et al., 2005; Komeda, 2004). For six candidate genes, an expression study was performed on the parents of the LR4 population.

Association Genetics in Lucerne

This gene was used as a candidate for association genetics in lucerne. Primers were designed to amplify two regions of about 500 bp each. Direct sequencing of PCR products was performed in a population of 400 lucerne genotypes originating from 10 cultivars. In comparison, sequences of a neutral gene was obtained on the same genotypes. No genetic structure was evidenced with SSR markers in this population. Only the SNPs with an allelic frequency higher than 0.1 were considered. The genotypes were evaluated for flowering date and plant height in two locations during four years. An ANOVA was performed for each trait with each SNP as a factor.

Results and Discussion

QTL Detection in M. truncatula

QTLs for flowering date and stem length were previously identified on chromosome 7 for all populations and years (Julier et al., 2007; Pierre et al., 2008). They explained between 11 and 60% of the variation for flowering date. In addition, other QTLs were detected on chromosomes 1, 2, 4, 5, 8, depending on population, trait and year. A meta-analysis of the QTLs on chromosome 7 indicated than all QTLs belonged to the same confidence interval.

In the fine mapping strategy (analysis of a pseudo-F_2 population of 1640 plants), the position of the QTL was more precisely established and a confidence interval of 2.4 cM was calculated.

Identification of a Candidate Gene in M. truncatula *and Expression Study*

Using sequencing information from the *M. truncatula* website (www.medicago.org) and other bioinformatics resources, a list of 44 BACs carrying 573 genes in the QTL confidence interval was established. Among them, six genes (Constans, three

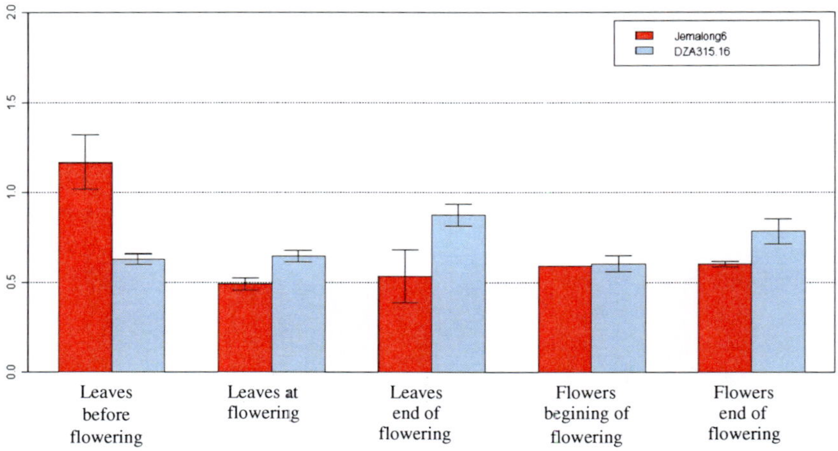

Fig. 62.1 Expression level of Constans in organs sampled at different stages in two lines of *M. truncatula*, Jemalong6 (*early flowering*) and DZA315.16 (*late flowering*)

copies of FT, PKS and FD) were described in the literature to be involved in flowering date. For Constans and FTb, the genomic sequences of the two parents were different. For FTb, PKS and FD, no difference in expression was observed between the two parental lines with contrasting flowering date; for FTa and FTc, no expression was detected (not shown). For Constans, the expression level was higher in the early flowering than in the late line, in leaves sampled before flowering (Fig. 62.1). Constans thus seems to be a good candidate gene to explain differences in flowering date between these two lines.

Association Genetics in Lucerne

Constans was chosen as a candidate gene for association genetics in lucerne. Primers were designed to amplify two regions of about 500 bp each. Direct sequencing of PCR products was performed in a population of 400 lucerne genotypes. Within-genotype allelic variation was evidenced and the allelic doses were scored. Due to indels, reading of the chromatograms from direct sequencing was labour intensive. The genotypes were evaluated for flowering date and plant height in two locations during four years. Only eight SNPs with a frequency higher than 0.1 were identified (1 SNP every 125 pb), which was clearly less than in a neutral gene (1 SNP every 30 pb). Significant association of SNPs with flowering date and plant length recorded in some conditions were detected (Table 62.1). Constans is thus also involved in plant morphogenesis in lucerne.

In conclusion, we have shown that a strategy of QTL detection, fine mapping and gene expression analyses lead to the identification of Constans as candidate gene controlling flowering date and stem height in *M. truncatula*. This gene was clearly

Table 62.1 Percentage of variation for flowering date and stem height explained by SNPs in Constans

	Flowering date		Stem height				
SNP	Lusignan 2004 Cut 2	Lusignan 2005 Cut 2	Connantre 2004 Cut 1	Lusignan 2007 Cut 1	Connantre 2006 Cut 2	Lusignan 2004 Cut 2	Lusignan 2005 Cut 2
b71	2.4*	–	–	–	–	3.0**	2.1*
b84	–	–	–	–	–	–	–
b271	–	3.9**	–	–	–	2.7*	3.8**
b308	–	–	–	–	2.6*	4.0**	–
e73	4.2**	–	–	–	–	3.1*	–
e205	–	4.1**	2.8*	–	–	3.8**	2.8*
e332	–	–	–	–	–	–	–
e446	–	–	–	2.9*	–	–	–

non neutral in perennial lucerne and explained a part of variation of flowering date and stem height. The use of a model species to identify genes involved in plant architecture of crop species is valuable.

Acknowledgements This work was supported by Contrat de Branches from Ministry of Agriculture and Fisheries (2005–2008). D. Herrmann received a grant from the Plant Breeding Division of INRA. J.B. Pierre received a PhD grant from Région Poitou-Charentes.

References

Broman, K.W., Wu, H., Sen, S., Churchill, G.A. 2003. R/qtl: QTL mapping in experimental crosses. Bioinfo. 19:889–890.

Flajoulot, S., Ronfort, J., Baudouin, P., Barre, P., Huguet, T., Huyghe, C., Julier, B. 2005. Genetic diversity among alfalfa (*Medicago sativa*) cultivars coming from a single breeding program, using SSR markers. Theor. Appl. Genet. 111:1420–1429.

Hecht, V., Foucher, F., Ferrandiz, C., Macknight, R., Navarro, C., Morin, J., Vardy, M.E., Ellis, N., Beltran, J.P., Rameau, C., Weller, J.L. 2005. Conservation of *Arabidopsis* flowering genes in model legumes. Plant Physiol. 137:1420–1434.

Julier, B., Flajoulot, S., Barre, P., Cardinet, G., Santoni, S., Huguet, T., Huyghe, C. 2003. Construction of two genetic linkage maps in cultivated tetraploid alfalfa (*Medicago sativa*) using microsatellite and AFLP markers. BMC Plant Biol. 3:9.

Julier, B., Huguet, T., Chardon, F., Ayadi, R., Pierre, J.B., Prosperi, J.M., Barre, P., Huyghe, C. 2007. Identification of quantitative trait loci influencing aerial morphogenesis in the model legume *Medicago truncatula*. Theor. Appl. Genet. 114:1391–1406.

Komeda, Y. 2004. Genetic regulation of time to flower in *Arabidopsis thaliana*. Annu. Rev. Plant Biol. 55:521–535.

Pierre, J.B., Huguet, T., Barre, P., Huyghe, C., Julier, B. 2008. Detection of QTLs for flowering date in three mapping populations of the model legume species *Medicago truncatula*. Theor. Appl. Genet. 117:609–620.

Putterill, J., Laurie, R., Macknight, R. 2004. It's time to flower: the genetic control of flowering time. Bioessays 26:363–373.

Chapter 63
QTL for Water Use Efficiency in Alfalfa

Bernadette Julier, Karine Bernard, Chrystel Gibelin, Thierry Huguet, and François Lelièvre

Abstract Alfalfa is the most important forage crop cultivated in semi-arid areas, both in rainfed and irrigated conditions. Water available for crops is a limited resource and improvement of water use efficiency (WUE) is an important goal for plant breeding. The objective of this study was to detect quantitative traits loci (QTL) for WUE in a mapping population of alfalfa. A F_1 mapping population was obtained and a genetic map based on 85 SSR markers was built. The F_1 plants and the parents were transplanted in soil columns, and WUE was measured during six regrowth cycles under well-watered conditions. QTL detection was carried out by single and multiple factor analysis of variance. The two parents significantly differed for WUE as expected. The F_1 population showed quantitative variation for this trait. In single factor ANOVA, nine markers/alleles had a significant effect on WUE variation. In multiple factor ANOVA, six markers/alleles had a significant effect and explained 31.0% of the variation. These QTL will be useful to better understand adaptation to water stress conditions in lucerne and to breed improved varieties.

Keywords Medicago sativa · QTL · Water stress · Genetic variation

Introduction

Alfalfa is the most important forage crop cultivated in semi-arid areas, both in rainfed and irrigated conditions. Water available for crops is a limited resource. Plants have first to acquire water and then use it to accumulate biomass. Under irrigated conditions, the improvement of water use efficiency (WUE) is important. WUE is the efficiency of a plant or a crop to transform transpired water

B. Julier (✉)
INRA, UR 4, BP 6, 86600 Lusignan, France
e-mail: Bernadette.Julier@lusignan.inra.fr

C. Huyghe (ed.), *Sustainable Use of Genetic Diversity in Forage and Turf Breeding*,
DOI 10.1007/978-90-481-8706-5_63, © Springer Science+Business Media B.V. 2010

into biomass and yield. The objective of the study was to evaluate genetic variation for WUE and detect QTL contributing to this trait in a mapping population of alfalfa.

Materials and Methods

A F_1 mapping population was obtained by crossing two plants. The female plant, named Magali-A (or MF) was taken in the French Provence type cultivar Magali, and is assumed to be drought-susceptible. The male plant, named Gabès-2355 (or GT) originated from a Gabès population of Tunisia, known to be drought and salt resistant (Thami-Alami et al., pers. com.). The two plants were manually crossed and 450 seeds were obtained.

SSR markers, mostly originating from *Medicago truncatula* ESTs, were chosen among those already described in alfalfa to cover the eight linkage groups (Julier et al., 2003, Eujayl et al., 2004). One hundred and sixty six markers were tested for amplification and polymorphism among the parents. The polymorphic markers were selected with an objective to reach from 10 to 15 markers per chromosome. The F_1 population was genotyped and the map was calculated with TetraploidMap software (Hackett and Luo, 2003).

A total of 230 seeds were germinated and 224 F_1 plants were obtained. Cloning via cuttings was performed and three to six rooted cuttings per plant were obtained. Each of them was transplanted in a soil column (height 2 m, diameter 8 cm). After full establishment (1 year, 4 cuts), WUE was measured during six regrowth cycles under well-watered conditions. Aerial dry matter (DMA, g) was measured at each cut. Each column was conducted as a lysimeter: they were weighted at each cut, and water supply and drainage between two cuts were also weighted. Evapotranspiration (ET, g) was calculated from the water balance for each plant and each growth cycle. WUE of each plant was calculated as $(\mathrm{DMA/ET})^{*}10^3$. The average value of WUE over the six cycles was calculated.

QTL detection was carried out by analysis of variance. In a first step, a single factor ANOVA with each allele was performed. The significant alleles were then submitted to a multiple factor ANOVA, using the option SS2 of proc GLM of SAS.

Results and Discussion

Eighty five markers were selected for mapping, which generated 280 alleles. The map comprised 8 linkage groups for each parent (not shown).

The two parents significantly differed for WUE under irrigated conditions (Fig. 63.1). As expected, Gabes2355, adapted to drought, had a higher WUE than Magali-A.

The F_1 population showed a quantitative variation for this trait (Fig. 63.2). Most F_1 genotypes had a WUE intermediate between those of the parents. A few genotypes had a WUE lower than that of Magali or higher than that of Gabes, indicating possible transgressions.

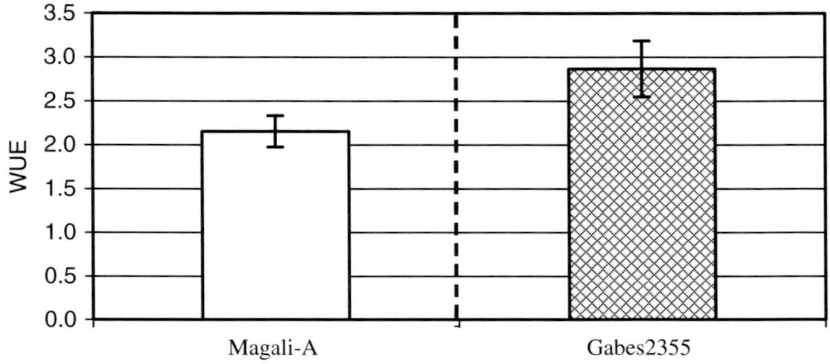

Fig. 63.1 Water use efficiency (WUE, g DM/100 g water) of the two parents of the mapping population of alfalfa, average over six growth cycles

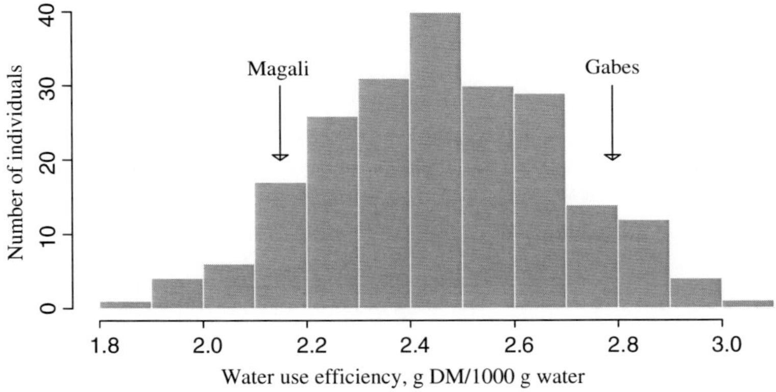

Fig. 63.2 Distribution of the F_1 population for WUE

In single factor ANOVA, nine markers/alleles had a significant effect on WUE variation. In multiple factor ANOVA, six markers/alleles had a significant effect and explained 31.0% of the variation. They were located on chromosomes 2, 3 7 and 8 (Table 63.1). Four alleles having a positive effect were carried by both Gabes2355 and Magali-A. The other two alleles were specific of Gabes2355, one had a negative

Table 63.1 Markers involved in variation for WUE in a mapping population of alfalfa

Marker	Chromosome	Parent	Effect	Mean square
B21M04A	8	M/G	0.976	0.610
B21M04B	8	G	−0.086	0.093
MTIC220B	2	M/G	0.192	0.235
MTIC263C	7	M/G	0.243	0.646
MTIC338C	3	M/G	0.453	0.136
MTIC469A	3	G	0.137	0.182

effect and the other one a positive effect. It seems that alleles generating a high WUE were present in both parents. New genotypes carrying positive alleles from both parents could be created, that could have an improved WUE.

This QTL detection experiment will be completed by the analysis of other traits, such as WUE in dry conditions and drought tolerance in field conditions. The QTLs will be useful to better understand adaptation to water stress conditions in lucerne and to breed improved varieties.

Acknowledgements This study was supported by the European Union, program INCO "PERMED" (Improvement of native PERennial forage plants for sustainibility of MEDiterranean farming systems), 2004–2008. We thank Joël Jousse and Sandrine Bouchier for plant management.

References

Eujayl, I., Sledge, M.K., Wang, L., May, G.D., Chekhowskiy, K., Zwonitzer, J.C.Z., Mian, M.A.R. 2004. *Medicago truncatula* EST-SSRs reveal cross-species genetic markers for *Medicago* spp. Theor. Appl. Genet. 108:414–422.

Hackett, C.A., Luo, Z.W. 2003. TetraploidMap: Construction of a linkage map in autotetraploid species. J. Hered. 94:358–359.

Julier, B., Flajoulot, S., Barre, P., Cardinet, G., Santoni, S., Huguet, T., Huyghe, C. 2003. Construction of two genetic linkage maps in cultivated tetraploid alfalfa (*Medicago sativa*) using microsatellite and AFLP markers. BMC Plant Biol. 3:9.

Chapter 64
QTL Mapping of Aluminum Tolerance in Tetraploid Alfalfa

Dong-Man Khu, Rafael Reyno, E. Charles Brummer, Joseph H. Bouton, Yuanhong Han, and Maria J. Monteros

Abstract Aluminum (Al) toxicity in acid soils is one of the factors limiting crop production. Alfalfa (*Medicago sativa* L.) is one of the most important forage legumes worldwide and is susceptible to Al toxicity. Al tolerance in alfalfa was identified in a diploid *Medicago sativa* subs. *caerulea* accession (Sledge et al., 2002) and has been successfully integrated at the tetraploid level in the Al-tolerant genotype Altet-4. The goals of this study are to identify and confirm quantitative trait loci (QTL) for Al-tolerance in tetraploid alfalfa. Two populations of at least 190 individuals each were developed from crosses between Altet-4 (Al-tolerant) and the Al-susceptible genotypes 95-608 derived from CUF-101, and NECS141, a semi-dormant breeding line developed in Iowa. The parental lines and the progeny from the mapping populations were screened using a callus bioassay and a whole plant assay. Genetic linkage maps constructed using EST-SSR markers (Sledge et al., 2005) were developed and used to identify QTL associated with Al tolerance. Comparison of QTL identified using different screening methods will be discussed. The long-term goal of this research is to use molecular markers associated with the Al-tolerance trait to accelerate the development of alfalfa cultivars with improved productivity in acidic and Al-toxic soils.

Keywords Aluminum · Alfalfa · *Medicago sativa* · QTL mapping

Introduction

Aluminum (Al) is an important factor limiting crop productivity in acid soils. At low pH, Al is soluble in the soil and is accessible to the plant which leads to inhibition of root growth and plant productivity (Kochian, 1995; Rechcigl et al.,

---

M.J. Monteros (✉)
Forage Improvement Division, The Samuel Roberts Noble Foundation, 2510 Sam Noble Parkway, Ardmore, OK 73401, USA
e-mail: mjmonteros@noble.org

C. Huyghe (ed.), *Sustainable Use of Genetic Diversity in Forage and Turf Breeding*,
DOI 10.1007/978-90-481-8706-5_64,

1988). Surface lime application is commonly used to prevent growth losses associated with Al toxicity, but these amendments are often not economical or practical due to high energy inputs and the increase in establishment costs (Haby et al., 2002; McLay et al., 1994; Summers, 1998). Therefore, a viable alternative to lime application is the development of Al tolerant crops (Foy, 1988). Alfalfa is one of the most important forage legume species in agricultural production systems. Alfalfa requires little or no nitrogen fertilizer because of its ability to carry out symbiotic nitrogen fixation. Its biomass can be harvested multiple times during the growing season (Barnes, 1993). Due to the production of more protein per hectare than grain or oilseed crops, alfalfa is highly desirable for hay production and pasture for livestock. Alfalfa is set to become an integral component of a sustainable bio-fuel industry (Lamb et al., 2007), especially if factors limiting its productivity are overcome. Alfalfa is very sensitive to soil acidity and Al toxicity; therefore, its biomass yields and ability to persist are compromised due to inhibited root growth in soils with low pH. Limited variation for Al tolerance exists within tetraploid *Medicago sativa* germplasm (Bouton, 1996; Bouton and Sumner, 1983; Dall'Agnol et al., 1996; Hartel and Bouton, 1989), and no commercially available Al-tolerant alfalfa cultivar has been developed. A single Al-tolerant clone, Al-4, was identified in the diploid *M. sativa* subsp. *caerulea* (PI464724-25) germplasm and used to develop a diploid F_2 mapping population (Bouton, 1996; Sledge et al., 2002). Efforts to introgress Al tolerance from the diploid PI464724-25 at the tetraploid level in alfalfa produced the Al-tolerant tetraploid alfalfa Altet-4, which was obtained from 2X × 4X crosses between Al-4 and CUF101-derived clones. Here we utilize three laboratory and greenhouse screening protocols to assess Al-tolerance in two tetraploid alfalfa mapping populations segregating for Al-tolerance and to identify quantitative trait loci (QTL) for acid and Al tolerance. We also developed a molecular linkage map of tetraploid alfalfa using simple sequence repeat (SSR) markers and candidate gene-based markers.

Materials and Methods

Plant Materials

The diploid Al-tolerant alfalfa Al-4 (Narasimhamoorthy et al., 2007) was crossed with a single genotype from tetraploid alfalfa CUF 101 (Lehman et al., 1983). The resulting progeny (Altet-1 through Altet-4) and the CUF101-derived genotypes 95-608 and 95-653, were tested for Al-tolerance using a callus growth bioassay in Blaydes media as previously described (Parrott and Bouton, 1990). The Al-tolerant genotype Altet-4 was crossed with 95-608 and the NECS-141 breeding line. At least 190 individuals from each cross were randomly selected and each group of individuals was designated as population 608Altet4 and NECS141Altet4.

Al-tolerance Evaluations

Individual seeds from the 608Altet4 F_1 mapping population were aseptically germinated by soaking in 70% EtOH for 5 min and rinsed 3X with double distilled sterile water for 5 min. All genotypes were clonally propagated using axillary meristem and terminal meristem subcultured in MS media (Murashige and Skoog, 1962) containing 2 mg/l of Indole-3-butyric acid (IBA) to induce root formation (Ligaba et al., 2006). The three evaluation methods include: (1) callus growth bioassay described by Parrot and Bouton (1990), (2) whole plant assay using a $CaCl_2$ culture media adapted from Ma et al. (1997) containing 1% Gelrite and 0 or 50 μM $AlCl_3$ at a pH of 4.0, and (3) whole plant soil-based assays (Bouton, 1996). For the callus bioassay, leaves and petioles were used to induce callus formation in Blaydes callus induction media and then transferred to either Al- or Al+ Blaydes media (400 μM of $AlCl_3$, pH = 4) (Parrot and Bouton, 1990). The experimental design for the phenotypic evaluations in culture media was a randomized complete block design with three replications.

Molecular Markers, Linkage Map and QTL Mapping

Genomic DNA from two mapping populations was extracted using DNeasy Plant Mini Kit (Qiagen, Valencia, USA). A total of 1024 EST-SSR primer pairs distributed throughout the alfalfa linkage groups (Sledge et al., 2005) were used to evaluate polymorphism between Altet-4, 95-608, and NECS-141 using PCR reactions as previously described (Narasimhamoorthy et al., 2007). Primer pairs that amplified polymorphic fragments were visualized and scored using GeneMapper software. Polymorphic markers were scored based on segregation in the population to achieve maximum resolution on the parental linkage map as described by Hackett et al. (2001). Development of the parental linkage maps and QTL interval mapping were performed using TetraploidMap software (Hackett and Luo, 2003).

Results and Discussion

The relative growth rate of Altet-4 using the callus bio-assay was consistently higher than any other genotype evaluated, including the other Altet genotypes (Fig. 64.1). The CUF101-derived genotype 95-608 had the lowest relative growth rate among the genotypes tested and was consistently the most Al-susceptible germplasm evaluated. Therefore, Altet-4 was used as the Al-tolerant common parent to develop the populations 608Altet4 (non-dormant background) and NECS141Altet4 (semi-dormant background). The phenotypic evaluations for Al tolerance in the 608Altet4 population exhibited a continuous and normal distribution consistent with polygenic inheritance (Fig. 64.2). The relative growth rate values (+Al/–Al) were 0.7 and 1.2 for the 95-608 and Altet-4 parents, respectively. The relative growth rates of the

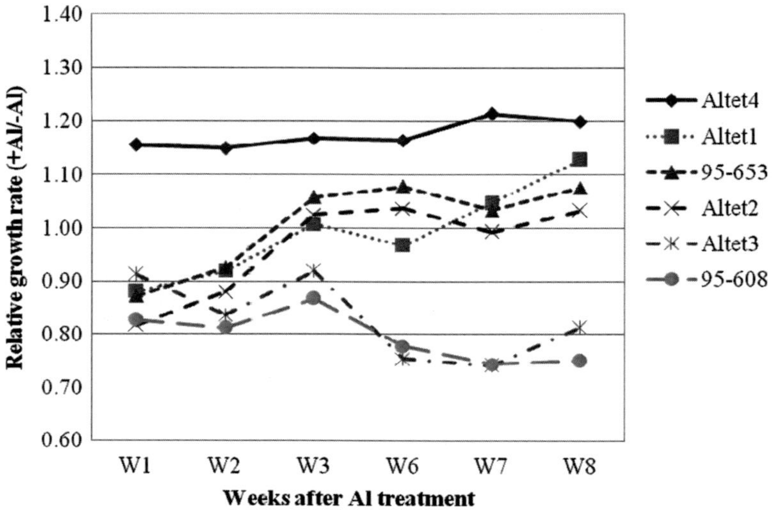

Fig. 64.1 Callus relative growth rate (+Al/–Al) from diverse parental genotypes in +Al and –Al modified Blaydes callus induction media

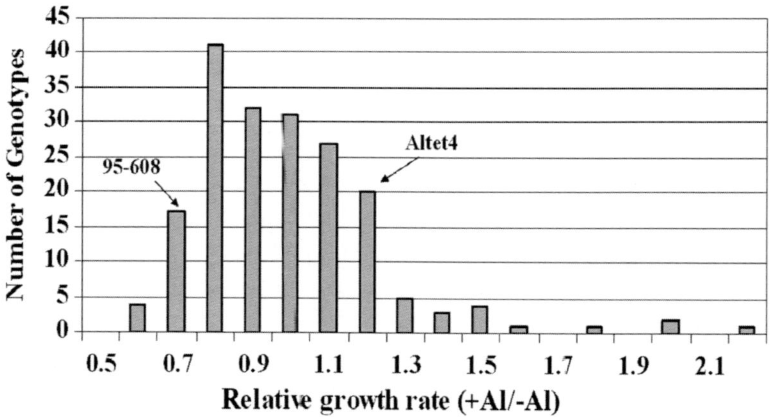

Fig. 64.2 Frequency distribution of Al tolerance in the 608Alt4 mapping population from the callus growth bioassay

progeny indicate transgressive segregation for Al tolerance in this population and confirm the ability of the assay to detect quantitative differences in Al tolerance. Intriguingly, these findings suggest that both parents may be contributing positive alleles for Al tolerance.

Phenotypic evaluations using the whole plant assay and soil-based assays for the two mapping population will be forthcoming. We will compare the correspondence between and reliability of the three phenotypic evaluation methods to screen for Al-tolerance and help determine which methods are best suited for high throughput screening of large numbers of individuals in populations with relevance to soil.

A complete SSR map and identification of QTL associated with Al tolerance will also be updated. These molecular breeding tools will be used to efficiently integrate sources of Al tolerance in alfalfa cultivars.

Acknowledgements This research has been funded by The Samuel Roberts Noble Foundation, Inc.

References

Barnes, D.K. 1993. Alfalfa, Traditional Crop Breeding Practices: An Historical Review to Serve as a Baseline for Assessing the Role of Modern Biotechnology (pp. 135–146). OECD, Paris.

Bouton, J.H. 1996. Screening the alfalfa core collection for acid soil tolerance. Crop Sci. 36:198–200.

Bouton, J.H., Sumner, M.E. 1983. Alfalfa, *Medicago sativa* L., in highly weathered, acid soils. V. Field performance of alfalfa selected for acid tolerance. Plant Soil 74:431–436.

Dall'Agnol, M., Bouton, J.H., Parrott, W.A. 1996. Screening methods to develop alfalfa germplasms tolerant of acid, aluminum toxic soils. Crop Sci. 36:64–70.

Foy, C. 1988. Plant adaptation to acid, aluminum toxic soils. Comm. Soil Sci. Plant Anal. 19:958–987.

Haby, V.A., Rouquette, F.M., Leonard, A.T. 2002. Requirements for successful alfalfa establishment on acid soils. Res. Center Tech. Rep. 2002:31–32.

Hackett, C.A., Luo, Z.W.. 2003. TetraploidMap: construction of linkage map in autotetraploid species. J. Heredity 94:358–359.

Hackett, C.A., Bradshow, J.E., McNichol, J.W. 2001. Interval mapping of quantitative trait loci in autotetraploid species. Genetics 159:1819–1832.

Hartel, P.G., Bouton, J.H. 1989. Rhizobium-meliloti inoculation of alfalfa selected for tolerance to acid aluminum-rich soils. Plant and Soil 116:283–285.

Kochian, L.V. 1995. Cellular mechanisms of aluminum toxicity and resistance in plants. Annu. Rev. Plant Physiol. Plant Mol. Biol. 46:237–260.

Lamb, J.F.S., Jung, H.-J.G., Sheaffer, C.C., Samac, D.A. 2007. Alfalfa leaf protein and stem cell wall polysaccharide yields under Hay and Biomass management systems. Crop Sci. 47:1407–1415.

Lehman, W.F., Nielson, M.W., Marble, V.L., Stanford, E.H. 1983. Registration of CUF 101 alfalfa. Crop Sci. 23:398.

Ligaba, A., Katsuhara, M., Ryan, P.R., Shibasake, M., Matsumoto, H. 2006. The BnALMT1 and BnALMT2 genes from rape encode aluminum-activated malate transporters that enhance the aluminum resistance of plant cells. Plant Physiol. 142:1294–1303.

Ma, J.F., Zheng, S.J., Li, X.F., Takeda, K., Matsumoto, H. 1997. A rapid hydroponic screening for aluminum tolerance in barley. Plant and Soil 191:133–137.

McLay, C.D.A., Ritchie, G.S.P., Porter, W.M., Cruse, A. 1994. Amelioration of subsurface acidity in sandy soils in low rainfall regions 2. Changes to soil solution composition following the surface application of gypsum and lime. Aust. J. Soil Res. 32:847–865.

Murashige, T., Skoog, F. 1962. A revised medium for rapid growth and bioassays with tobacco tissue cultures. Physiol. Plant 15:473–497.

Narasimhamoorthy, B., Bouton, J.H., Olsen, K.M., Sledge, M.K. 2007. Quantitative trait loci and candidate gene mapping of aluminum tolerance in diploid alfalfa. Theor. Appl. Genet. 114:901–913.

Parrott, W.A., Bouton, J.H. 1990. Aluminum tolerance in alfalfa as expressed in tissue culture. Crop Sci. 30:387–389.

Rechcigl, J.E., Reneau, J.R.B., Zelazny, L.W. 1988. Soil solution Al as a measure of Al toxicity to alfalfa in acid soils. Comm. Soil Sci. Plant Anal. 9:989–1001.

Sledge, M., Ray, I., Jiang, G. 2005. An expressed sequence tag SSR map of tetraploid alfalfa (*Medicago sativa* L.). Theor. Appl. Genet. 111:980–992.

Sledge, M.K., Bouton, J.H., Dall'Agnoll, M., Parrott, W.A., Kochert, G. 2002. Identification and confirmation of aluminum tolerance QTL in diploid *Medicago sativa* subsp. *coerulea*. Crop Sci. 42:1121–1128.

Summers, C.G. 1998. Integrated pest management in forage alfalfa. Integrated Pest Manag. Rev. 3:127–154.

Chapter 65
DArTFest – A Platform for High-Throughput Genome Profiling Within the *Festuca – Lolium* Complex

David Kopecký, Jan Bartoš, Adam J. Lukaszewski, James H. Baird, Vladimír Černoch, Roland Kölliker, Simen Rød Sandve, Odd Arne Rognli, Helene Blois, Vanessa Caig, Jaroslav Doležel, and Andrzej Kilian

Abstract With the aim to facilitate high-throughput genome profiling and genetic and physical mapping within the *Festuca-Lolium* complex, we have developed a Diversity Arrays Technology (DArT) array for five important species: *Festuca pratensis*, *Festuca arundinacea*, *Festuca glaucescens*, *Lolium perenne* and *Lolium multiflorum*. The DArTFest array contains 7,680 probes derived from methyl-filtered genomic representations. Of 3,884 polymorphic DArT markers identified in the first marker discovery experiment, over 1,000 markers detected a positive allele in each species.

We assigned DArT markers to individual chromosome regions of *F. pratensis* using a series of single chromosome substitution and recombinant lines of *F. pratensis* in *L. multiflorum*. Moreover, we enriched the existing genetic map of *F. pratensis* by over 200 DArT markers and existing genetic map of *L. multiflorum* by over 500 DArT markers. The resources developed in this project will facilitate development of genetic maps in *Festuca* and *Lolium*, the analysis of genomic constitution in *Festuca* × *Lolium* hybrids, as well as marker-assisted selection for multiple traits.

Keywords DArT · Fescue · Hybrids · Introgression · Mapping · Ryegrass

Introduction

Grasses are among the most important and widely cultivated plants on Earth, with a total area of grassland estimated to be twice that of cropland. Among the cultivated grasses, ryegrasses (*Lolium* spp.) and fescues (*Festuca* spp.) predominate, especially in temperate climate conditions (Jauhar, 1993). Decades of breeding resulted in superior ryegrass and fescue cultivars outperforming their wild progenitors. However, there is a risk that some desirable alleles of progenitors were lost during

D. Kopecký (✉)
Laboratory of Molecular Cytogenetics and Cytometry, Institute of Experimental Botany, Olomouc CZ-77200, Czech Republic
e-mail: kopecky@ueb.cas.cz

C. Huyghe (ed.), *Sustainable Use of Genetic Diversity in Forage and Turf Breeding*, DOI 10.1007/978-90-481-8706-5_65,

the breeding process. The risk of erosion of species' gene pools calls for characterization of the natural genetic variability and its conservation in gene banks. The banks with well described accessions will allow for more effective selection of materials for breeding improved cultivars. In order to characterize the existing genetic diversity in detail and to provide sufficient information on the genetic make-up of a particular accession, thousands of genomic loci need to be quickly interrogated. Thus, high-throughput and cost-effective genotyping platforms are needed.

In addition to the analysis of genetic diversity, a high-throughput genotyping system is also required to speed up the development of genetic linkage maps. Until now, several genetic maps were constructed for *Lolium* spp. (e.g. Hayward et al., 1998). Within the fescue species, genetic maps were generated for the two agronomically important species – *F. arundinacea* and *F. pratensis* (Saha et al., 2005; Alm et al., 2003). However, the number of markers is too low for routine identification of markers tightly linked to genes underlying traits of interest. Finally, high-throughput genotyping platforms are needed to speed up the development of new cultivars with desirable attributes using marker assisted selection.

The need for high-throughput genotyping led to the development of various DNA arrays and chips (reviewed in Gupta et al., 2008). Although they are based on different principles, all of them can be used to screen thousands or hundreds of thousands of genomic loci in a single pass. Diversity Array Technology (DArT) is a microarray hybridization based technique that permits simultaneous screening of thousands polymorphic loci without any prior sequence information (Jaccoud et al., 2001). DArT is high-throughput, low-cost, quick and reproducible.

In this study we demonstrate the utility of the approach for the estimation of intra- and interspecific genetic diversity and genetic and physical mapping.

Material and Methods

Plant Material

For the development of the DArT array and analysis of intra- and interspecific diversity, 40 accessions each of *L. perenne* L., *L. multiflorum* Lam., *F. pratensis* Huds. and *F. arundinacea* Schreb. were used, plus all seven available accessions of *F. glaucescens* Boiss. The choice of accessions (ecotypes, cultivars and parents of mapping populations) aimed at the discovery of the maximum genetic variability. In order to map DArT markers to individual chromosomes and chromosome bins of *F. pratensis*, we used single chromosome monosomic substitutions and recombinant lines with various lengths of *Festuca* chromatin present in tetraploid *L. multiflorum*, as described by Kopecký et al. (2008). Mapping population of *F. pratensis* with 138 genotypes was used for genetic mapping (Alm et al., 2003).

Development of the DArTFest Array

DarTFest array was developed as described in Kopecky et al. (2009).

Analysis of Genetic Diversity

The DArTsoft-generated 0–1 scores were used as input for the RESTDIST and NEIGHBOR programs of the PHYLIP 3.6 software package to construct a dendrogram based on the Unweighted Pair Group Method with Algorithmic Mean (UPGMA) and Felsenstein's modification of the Nei/Li restriction fragment distance (Felsenstein, 2004).

Genetic and Physical Mapping of DArT Markers in F. pratensis

To anchor markers to individual chromosomes of *F. pratensis*, DNA isolated from chromosome substitution and recombinant lines were hybridized to the DArT array. A marker present in *F. pratensis* and absent in *L. multiflorum* was assigned to a chromosome if it was present in at least one substitution line for a particular chromosome. The same approach was used to anchor markers allocated on bins of *F. pratensis* using the recombinant lines described above. DNA from a mapping population of *F. pratensis* and *L. multiflorum* already used for the development of genetic maps (Alm et al., 2003; Studer et al., 2006) was hybridized to the array.

Results and Discussion

Development of DArT Array

We developed a DArT array containing 7,680 probes derived from methyl-filtered (through the use of *Pst*I restriction enzyme) genomic representations. In the first marker discovery experiment performed with 40 genotypes from each of the species *L. perenne*, *L. multiflorum*, *F. pratensis* and *F. arundinacea*, and seven genotypes of *F. glaucescens*, we identified 3,884 polymorphic markers with standard DArTsoft settings.

Analysis of Genetic Diversity Using DArT Markers

Of the 3,884 polymorphic markers detected, 2,629 markers gave unequivocal scores in the five species tested. Using these markers, we compiled a dendrogram including all 167 tested accessions of fescue and ryegrass. This differentiated two major groups, representing the fescue and ryegrass genera (Fig. 65.1). Both ryegrass species analyzed (*L. perenne* and *L. multiflorum*) were closely related, but divergent enough to form separate groups. Fescue species formed two major groups. The first one included *F. pratensis* forming a tight group in the dendrogram. The second group included two subgroups, one representing *F. arundinacea* accessions and the second *F. glaucescens* accessions. One accession of *F. glaucescens* Fg-07 clustered with the subgroup of *F. arundinacea*. Another inconsistent accession *F. pratensis* Fp-40 was located outside of all other species in the dendrogram, probably due to

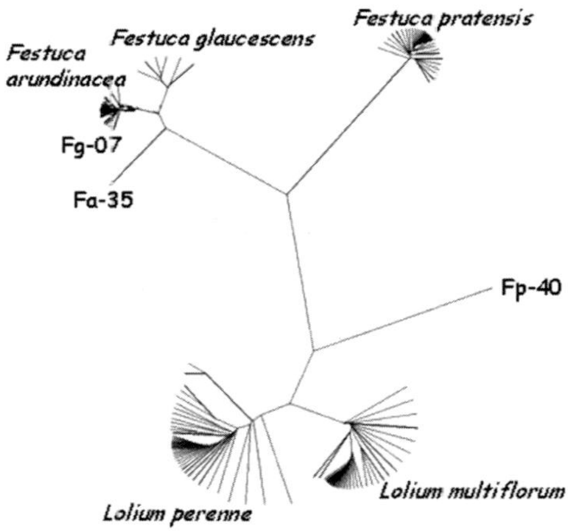

Fig. 65.1 UPGMA dendrogram (shown as radial cladogram) based on hybridization of 80 *Lolium* and 87 *Festuca* genotypes to 2,629 DArT markers and Felsenstein's modified Nei/Li restriction fragment distance. Two major groups representing the fescues and ryegrasses are clearly differentiated. Both ryegrass species display higher genetic diversity than fescue species. Note that the accession of *F. arundinacea* Fa-35 (Moroccan ecotype 599,533) was found distant of the major group. Similarly, one accession of *F. glaucescens* (Fg-07) clustered with the subgroup of *F. arundinacea*. Another inconsistent accession *F. pratensis* Fp-40 (cultivar Norild) was located separately outside of all other species

contamination of its DNA. Both accessions (Fg-07 and Fp-40) were excluded from further analyses.

Genus- and Species-Specificity of DArT Markers

Reliable discrimination of DNA markers from both grass genera tested here, and possibly also from individual species, would greatly expand the utility of the DArT array. As the DArTFest array contains markers derived from both genera and all five species tested, our subsequent analysis focused on identifying genus- and species-specific markers. Of 3,884 polymorphic DArT markers identified, over 1,000 markers detected a positive ("1") allele in each species. However, a large proportion of markers present in all five species tested reduced the numbers of species-specific markers (Table 65.1).

Physical and Genetic Mapping of DArT Markers in F. pratensis and L. multiflorum

We used a complete set of *Festuca-Lolium* single chromosome substitution lines of *F. pratensis* into *L. multiflorum* to assign DArT markers to individual chromosomes of *F. pratensis*. In total, 160 DArT markers were anchored using this approach with

Table 65.1 Number of species-specific DArT markers identified on a DArT array containing 7,680 probes of 40 accessions each of *L. perenne* L., *L. multiflorum* Lam., *F. pratensis* Huds., *F. arundinacea* Schreb. and seven accessions of *F. glaucescens* Boiss

Species	Scored markers[a]	Positive markers[b]	Polymorphic markers[c]	Species-specific markers[d]
Lolium perenne	2,638	1,725 (821–1127)	1,407	52
Lolium multiflorum	3,883	2,761 (1507–1852)	2,148	82
Festuca pratensis	3,884	2,257 (1619–1821)	1,078	123
Festuca glaucescens	2,630	1,346 (1059–1101)	387	9
Festuca arundinacea	2,638	1,572 (1000–1351)	512	34

[a]Note that some markers were not scored in all species, [b]Range of positive markers for individual accessions is in brackets, [c]Markers polymorphic among the accessions, [d]Identified after scoring 2,629 markers

between six and 34 DArT markers anchored to a particular chromosome. This represents 56% of all markers present in *F. pratensis* but absent in *L. multiflorum*. Up to date, we have dissected chromosomes 3 and 6 of *F. pratensis* into 7 and 9 bins, respectively, with between one and nine DArT markers anchored to each bin. The existing genetic map of *F. pratensis*, which includes AFLPs, RFLPs, isozyme and other markers (Alm et al., 2003) has been enriched for 204 DArT markers and its total linkage length increased by 170 cM to 775 cM. In case of *L. multiflorum*, existing genetic map was enriched for 531 DArT markers.

Acknowledgements This work was supported by the Ministry of Agriculture of the Czech Republic (grant award NAZV QH71267) and the Czech Science Foundation (grant award 521/07/P479).

References

Alm, V., Fang, C., Busso, C.S., Devos, K., Vollan, K. et al. 2003. A linkage map of meadow fescue (*Festuca pratensis* Huds.) and comparative mapping with other *Poaceae* species. Theor. Appl. Genet. 108:25–40.

Felsenstein, J. 2004. PHYLIP (Phylogeny Inference Package) version 3.6. University of Washington, Seattle.

Gupta, P.K., Rustgi, S., Mir, R.R. 2008. Array-based high-throughput DNA markers for crop improvement. Heredity 101:5–18.

Hayward, M.D., Forster, J.W., Jones, J.G., Dolstra, O., Evans, C. et al. 1998. Genetic analysis of *Lolium* I. Identification of linkage groups and the establishment of a genetic map. Plant Breed. 117:451–455.

Jaccoud, D., Peng, K., Feinstein, D., Kilian, A. 2001. Diversity Arrays: a solid state technology for sequence information independent genotyping. Nucleic Acids Res. 29(4):e25.

Jauhar, P.P. 1993. Cytogenetics of the *Festuca-Lolium* complex. Relevance to Breeding. Monographs on Theor. Appl. Genet. Vol. 18, Springer-Verlag, Berlin, 255 p.

Kopecký, D., Lukaszewski, A.J., Doležel, J. 2008. Meiotic behavior of individual chromosomes of *Festuca pratensis* in tetraploid *Lolium multiflorum*. Chromosome Res. 16:987–998.

Kopecký , D., Bartoš, J., Lukaszewski, A.J., Baird, J.H., Černoch, V., Kölliker, R., Rognli, O.A., Blois, H., Caig, V., Lübberstedt, T., Studer, B., Doležel, J., Kilian, A. 2009. Development and mapping of DArT markers within the *Festuca – Lolium* complex. BMC Genomics 10:473.

Saha, M.C., Mian, R., Zwonitzer, J.C., Chekhovskiy, K., Hopkins, A.A. 2005. An SSR- and AFLP-based genetic linkage map of tall fescue (*Festuca arundinacea* Schreb.) Theor. Appl. Genet. 110:323–336.

Studer, B., Boller, B., Herrmann, D., Bauer, E., Posselt, U.K., Widmer, F., Koelliker, R. 2006. Genetic mapping reveals a single major QTL for bacterial wilt resistance in Italian ryegrass (*Lolium multiflorum* Lam.). Theor. Appl. Genet. 113:661–671.

Chapter 66
Vegetative Plant Height QTLs in Elite Perennial Ryegrass Material

Laurence Pauly, Sandrine Flajoulot, Philippe Barre, and Jérôme Garon

Abstract Leaf length in perennial ryegrass meadows is a limiting factor for feeding grazing cows. In spite of the phenotypic selection efficiency to improve this trait, synthetic varieties hamper fixing favourable alleles. Our aim was to detect QTLs of vegetative plant height which is highly correlated to leaf length in order to start Molecular Assisted Selection (MAS). Two hundred plants from a cross between two elite plants were used to build a genetic map for each parent with 39 SSR and 47 AFLP markers. The maps consisted of seven linkage groups for both parents, with a length of 408 cM and 548 cM. Plant height was measured in a nursery on spaced plants during spring 2008. Using the Composite Interval Mapping method, we detected two QTLs of plant height in the parent RA958 B on linkage group 2 and 5. The one on LG2 co-localized with a QTL of earliness in vegetative growth after winter explaining 21% of variance. Moreover, two QTLs of plant growth rate were found on linkage groups 4 (RA958 F) and 7 (RA958 B), each of them explaining 11% of variance.

Keywords Marker-assisted selection · Molecular markers · Perennial ryegrass · Quantitative trait loci · Synthetic variety

Introduction

In temperate regions, perennial ryegrass is the most commonly sown species used to feed ruminants. Leaf length is an important trait for increasing yield and intake rate by grazing animals (Barre et al., 2006; Horst et al., 1978; McGilloway et al., 1999; Rhodes, 1969; Reeder et al., 1984,). Moreover, leaf elongation rate has to be high in order to maintain a high yield while continuous grazing is applied. Thus,

L. Pauly (✉)
Jouffray-Drillaud, Station expérimentale La Litière, Saint Sauvant 86 600, France
e-mail: lpauly@lusignan.inra.fr

C. Huyghe (ed.), *Sustainable Use of Genetic Diversity in Forage and Turf Breeding*,
DOI 10.1007/978-90-481-8706-5_66,

breeding for varieties with long leaves and a high leaf elongation rate is a target for forage breeders. Selection for lamina length in nurseries of spaced plants can be an effective means of modifying lamina length in dense canopies (Rhodes and Mee, 1980; Hazard et al., 1996). Due to a high effect of inbreeding in *L. perenne*, commercial varieties are synthetics, what leads to a high level of diversity within varieties (Balfourier et al., 1992). In such varieties, it should be particularly powerful to use molecular markers for breeding in order to fix several regions of the genome involved in the control of key traits while maintaining a wide diversity in the rest of the genome.

In this context, the aim of this study was to detect QTLs (Quantitative Trait Loci) for leaf growth in elite material with the final goal of including Marker Assisted Selection (MAS) in a classical selection scheme.

Material and Methods

Plant Material and Phenotypic Evaluation

The mapping population was derived from a cross between two elite plants: RA958 B and RA958 F. Two hundred individuals of this progeny were planted as spaced plants in a field nursery in May 2007. They were split into 10 micro-plots of 20 individuals each. This cross was included into an experiment with seven other crosses.

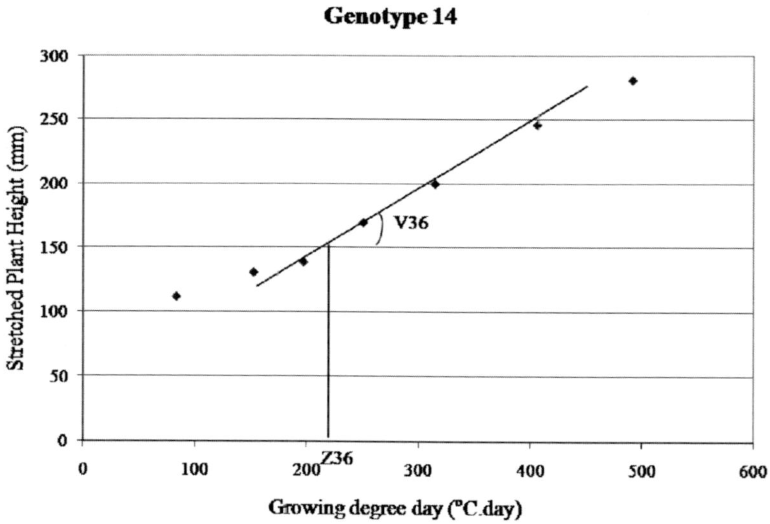

Fig. 66.1 Kinetics of growth during spring 2008 for the genotype 14 issued from the cross between RA958 B and RA958 F

After a first defoliation at the beginning of spring, the stretched height of the plants (SPH) was measured at regular time intervals from March to May 2008 in order to build a kinetic model of plant growth. We calculated two parameters: V36 and Z36 which correspond to the plant growth rate between the third and the sixth measurements, and the beginning of vegetative growth, respectively (Fig. 66.1). Z36 can be comparable to a score of vegetative earliness. We considered that a plant had started its linear growth when it had reached a height of 150 mm (Fig. 66.1).

Molecular Analysis

DNA was extracted using the CIMMYT protocol (http://www.cimmyt.org). Each plant was genotyped using 39 SSR markers and 47 AFLP markers.

Map Construction and QTLs Detection

The genetic maps of RA958 B and RA958 F were built using the two-way pseudo testcross strategy (Grattapaglia and Sederoff, 1994). The Haldane function was used to estimate map distances.

The software Windows QTL Cartographer version 2.5 (http://statgen.ncsu.edu/qtlcart/) was used for QTL detection. A Composite Interval Mapping method was performed using the co-factors found in a stepwise multiple linear regression. The LOD score threshold was estimated using 1,000 bootstraps.

Results and Discussion

Genetic Mapping

For both maps, we found seven linkage groups (LG). The genetic map of RA958 B covered 408 cM with 29 SSR markers and 26 AFLP markers (Fig. 66.2). The map for RA958 F covered 548 cM with 29 SSR markers and 20 AFLP markers (Fig. 66.2).

These maps were smaller than the reference map built by Jones et al. (2002). It may be due to an insufficient number of markers in our study, a difference in the software or in the function used to calculate the genetic distances.

The presence of a high number of skewed markers on LG1 and 2 could come from their proximity to the two self-incompatibility loci (*S* and *Z*) revealing a possible relatedness between the parents (Thorogood et al., 2002).

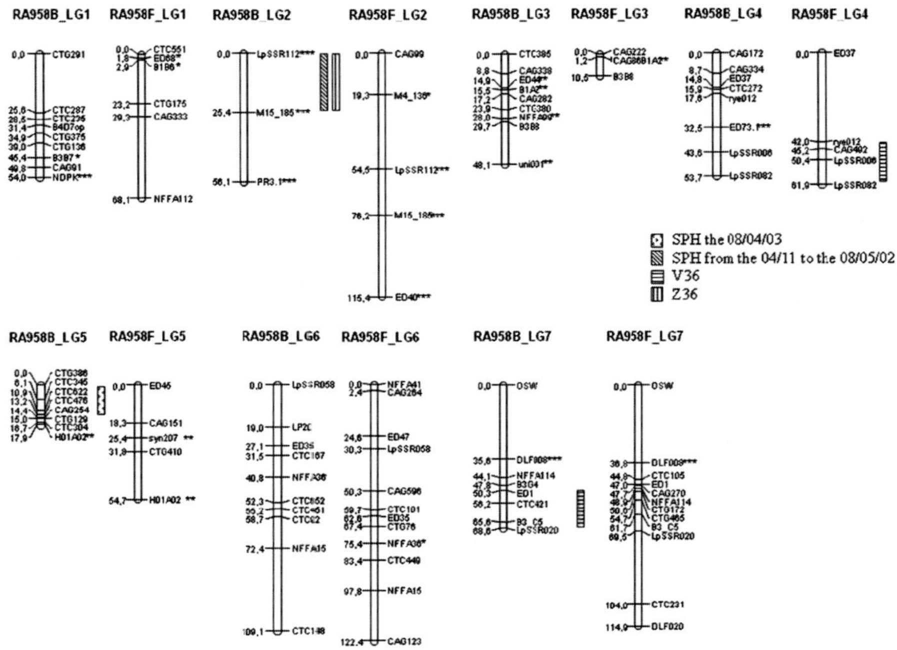

Fig. 66.2 Genetic maps of parents RA958 B and RA958 F and QTLs detected in each case. Markers showing segregation distortion are indicated by: significant distortion at: *, $P < 0.05$; **, $P < 0.01$; and ***, $P < 0.001$. (SPH: Stretched Plant Height; V36 and Z36: see text)

QTL Detection

Four QTLs were identified in this study (Fig. 66.2, Table 66.1).

On LG2, one QTL is expressed at different measurement dates (from the 11th of April to the 2nd of May 2008). This QTL also co-localized with a QTL of vegetative earliness (Z36) which explains 20.9% of the total variance. Two QTLs for growth rate (V36) were detected on LG4 (RA958 F) and LG7 (RA958 B). Each of them explained 10.8% of the variance. None of them co-localized with QTLs of SPH situated on LG2 and 5. Plant height in spring was mainly explained by vegetative earliness (correlation between –0.53 and –0.89) and a common QTL was observed on LG2. Nevertheless, plant height at the end of spring (080424-080507) was also correlated to plant growth rate (0.46–0.63) but no common QTL was detected. Moreover no significant correlation was observed between vegetative earliness and plant growth rate. So, it should be possible to select independently QTLs of vegetative earliness and plant growth rate in order to obtain a stronger effect on plant height.

The relatively low number of QTLs detected could have different origins. The first hypothesis is that the two parents come from elite material and some plant height QTLs should be already fixed. An unsaturated map is the second hypothesis. A third one could be the presence of many QTLs with small effect.

Table 66.1 QTLs detected by composite interval mapping (CIM) for both parents RA958 B and RA958 F

Trait	Parents	LG	Distance (cM)	Marker	LOD Threshold	LOD score	Interval –1 LOD (cM)	Additive effect	R^2 (%)
SPH080403	RA958 B	5	6	CTG386	3.82	5.48	2–13.2	–11.59	12.3
SPH080411	RA958 B	2	12	LpSSR112	4.07	5.14	0–24	25.49	24.7
SPH080418	RA958 B	2	8	LpSSR112	4.09	4.67	0–20	19.85	16.5
SPH080424	RA958 B	2	10	LpSSR112	4.26	5.52	0–24	26.71	21.7
SPH080502	RA958 B	2	8	LpSSR112	4.86	4.88	0–24	22.69	13.4
SPH080507	RA958 B	2	8	LpSSR112	4.16	5.46	0–22	35.17	19.4
V36	RA958 B	7	60.2	CTC421	4.03	4.44	50.3–67.6	0.06	10.8
	RA958 F	4	50.3	LpSSR006	4.26	4.53	42–58.3	0.07	10.8
Z36	RA958 B	2	10	LpSSR112	4.81	5.42	0–22	–50.43	20.9

Conclusion and Perspectives

Finally, we found one major QTL for plant height and vegetative earliness and two other QTLs of growth rate different from the previous one. There is also one QTL for SPH on LG5 of parent RA958B. However, the number of markers used was clearly not sufficient and the map needs to be saturated.

A direct MAS on the mapping population is possible but would lead to inbreeding. Other populations should be studied and QTLs from different parents should be combined.

This method of QTL mapping presents several constraints such as the creation of a mapping population. An alternative could be an association mapping directly applied to breeding plant material.

References

Balfourier, F., Charmet, G., Betin, M., Bourgoin, B. 1992. Les ray-grass. In: Gallais, A., Bannerot, H. (eds.), Amélioration des plantes cultivées: objectifs et critères de sélection (pp. 310–321). INRA Edition, France.

Barre, P., Emile, J.C., Betin, M., Surault, F., Ghesquière,M., Hazard, L. 2006. Morphological characteristics of perennial ryegrass leaves that influence short-term intake in dairy cows. Agron. J 98:978–985.

Grattapaglia, D., Sederoff, R. 1994. Genetic linkage maps of *Eucalyptus grandis* and *Eucalyptus urophylla* using a pseudo-testcross: mapping strategy and RAPD markers. Genetics 137: 1121–1137.

Hazard, L., Ghesquière, M., Barraux, C. 1996. Genetic variability for leaf development in perennial ryegrass populations. Can. J. Plant Sci. 76:113–118.

Horst, G.L., Nelson, C.J., Asay, K.H. 1978. Relationship of leaf elongation to forage yield of tall fescue genotypes. Crop Sci. 18:715–719.

Jones, E.S., Mahoney, N.L., Hayward, M.D., Armstead, I.P., Jones, J., Humphreys, M.O., King, I.P., Kishida, T., Yamada, T., Balfourier, F., Charmet, G., Forster, J.W. 2002. An enhanced molecular marker based genetic map of perennial ryegrass (*Lolium perenne*) reveals comparative relationships with other Poaceae genome. Genome 45:282–295.

McGilloway, D.A.; Cushnahan, A.; Laidlaw, A.S.; Mayne, C.S.; Kilpatrick, D.J. 1999. The relationship between level of sward height reduction in a rotationally grazed sward and short-term intake rates of dairy cows. Grass Forage Science 54: 116–126.

Reeder, L., Sleper, D., Nelson, C. 1984. Response to selection for leaf area expansion rate of tall fescue. Crop Sci. 24:97–100.

Rhodes, I. 1969. The relationship between productivity and some components of canopy structure in ryegrass (*Lolium* spp.). I. Leaf length. J. Agric. Sci. 73:315–319.

Rhodes, I., Mee, S. 1980. Changes in dry matter yield associated with selection for canopy characters in ryegrass. Grass For. Sci. 35:35–39.

Thorogood, D., Kaiser, W.J., Jones, J.G., Armstead, I. 2002. Self-incompatibility in ryegrass12. Genotyping and mapping the *S* and *Z* loci of *Lolium perenne* L. Heredity 88:385–390.

Chapter 67
Genotyping Unknown Genomic Terrain in Complex Plant Genomes

Simen R. Sandve, Heidi Rudi, Guro Dørum, Magnus D. Vigeland, Paul R. Berg, and Odd Arne Rognli

Abstract High-Throughput Genotyping (HTG) has limitations in discriminating between true SNP alleles and paralogous loci. We present a case study and theoretical modelling of paralog evolution that exemplify the inherent difficulties of HTG in plants. Our study underlines the importance of genomic information and careful experimental design.

Keywords High-throughput genotyping · Paralog · MassArray · *Lolium perenne* · LpIRI

Introduction

Duplicated genes, i.e. paralogs, are often the cause of genotyping errors (GE) in HTG (Hyten et al., 2008). Many plant genomes are extremely rich in gene duplicates due to single gene- and whole genome duplications and polyploidy (Lockton and Gaut, 2005; Leitch and Leitch, 2008). Thus HTG of plant genomes are more prone to GE than HTG of other eukaryotes. The aim of this study is to provide an instructive case study for the non-specialist researcher of how unknown paralogs or homoeologs, in the case of polyploids, can interfere with HTG.

SNP Genotyping of Unknown Paralogs

In the quest for novel genes associated with frost tolerance we investigated allele frequencies of candidate genes in experimental populations of the diploid forage grass *Lolium perenne* L. The genotyping was conducted on the Sequenom mass

O.A. Rognli (✉)
Department of Plant and Environmental Sciences, Norwegian University of Life Sciences, N-1432 Ås, Norway
e-mail: odd-arne.rognli@umb.no

C. Huyghe (ed.), *Sustainable Use of Genetic Diversity in Forage and Turf Breeding*,
DOI 10.1007/978-90-481-8706-5_67,

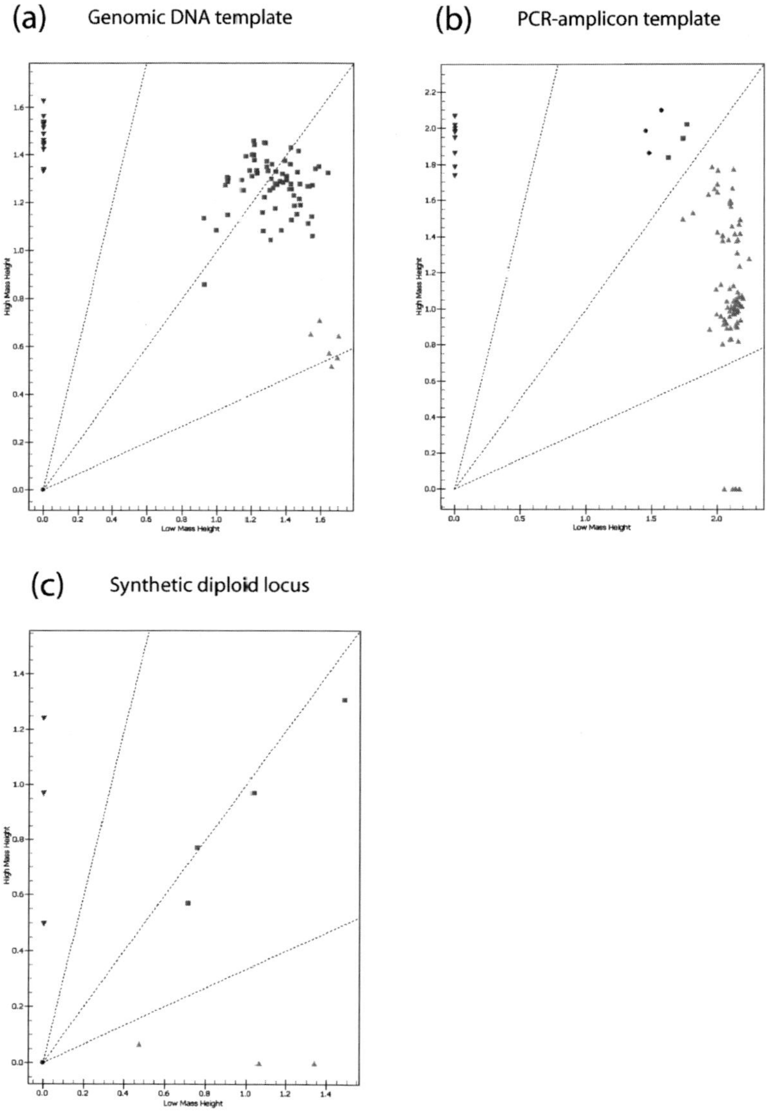

Fig. 67.1 MassArray genotyping scores of SNP-locus 1 of *LpIRI1* based on (**a**) genomic template, (**b**) PCR amplicons, and (**c**) a synthetic diploid locus made up by combining two cloned LpIRI1 fragments. The graphs depicts the ratios between the molecular masses of SNP-alleles as detected by MALDI-TOF MS (matrix-assisted laser desorption/ionization time-of-flight mass spectrometry). Heterozygotes should fall on the 45° line dividing the graph. Homozygotes with two low-molecular mass alleles should locate in the triangle below the lower line (towards the *X*-axis); homozygotes with two high-molecular mass alleles should locate in the left triangle towards the *Y*-axis

spectrometry based MassARRAY (MA) genotyping platform (Bray et al., 2001). Genotyping of 5 SNPs in the gene *LpIRI1*, an intronless ice re-crystallisation inhibition (IRI) gene, produced well spaced genotype clusters (Fig. 67.1a), but two of the SNP loci deviated from Hardy-Weinberg equilibrium (HWE) ($p < 0.0001$). We suspected that the deviations from HWE were caused by paralog-related GE. MassARRAY genotyping data for one of these SNP's are presented in Fig. 67.1a–c.

Two additional *LpIRI*-paralogs, *LpIRI3* and *LpIRI4*, were subsequently identified and sequenced from a BAC-library (Sandve et al., 2008). To avoid simultaneous genotyping of the paralogs we now tried to genotype PCR-amplicons of *LpIRI1*. However, the PCR-amplicon HTG yielded no clear genotype clustering (Fig. 67.1b).

To rule out technical problems, we cloned and sequenced the *LpIRI1* amplicons. Sequencing data revealed a third unknown paralog, almost identical to *LpIRI1*, putatively responsible for the PCR-amplicon genotyping failure (Fig. 67.1b). HTG on synthetic diploid genotypes (SDG) made up by combining pairs of *LpIRI1* clones had 100% accuracy compared to the sequencing data, and produced well clustered genotype classes and (Fig. 67.1c).

Modelling Paralog Divergence

Ultimately, it is the sequence differences of paralogous sites surrounding the SNP loci that affect the chance of paralog related GE. We modelled paralog divergence using a codon model (Yang et al., 2000) to estimate expected divergence levels for paralogs under different evolutionary scenarios (Fig. 67.2).

For example, if two paralogs are both functionally important (i.e. under strong purifying selection; scenario 4) we expect very slow paralog divergence. SNPs in coding regions of such paralogs have high probability of contributing to GEs, even after >50 million years (my) of divergence (Fig. 67.2) (<20% differences). If SNP genotyping assays are designed in intron-sequences (scenario 1), i.e. paralogs evolving neutrally, the expected differences between two paralogs after 50 my have increased two-fold compared to scenario 4.

Conclusions

As we have shown, well clustered genotype groups can mask paralog-related GE. Beyond the standard data quality assessments, the validity of HTG results can only be ascertained by prior expectations of genotype segregation or by re-sequencing. A HTG experiment in a non-sequenced plant genome should use SNPs localized in regions with surrounding intron sequences to maximize paralog divergence. This can be difficult, if not impossible, to achieve in intronless genes as the one described herein. Also, it is important to identify potential unknown paralogs by comparative genomics in related species through bioinformatics approaches.

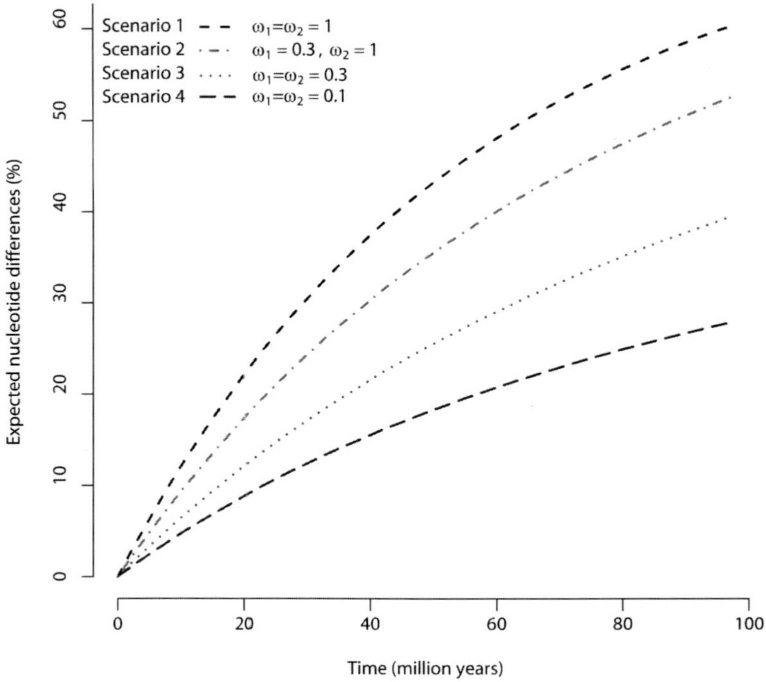

Fig. 67.2 Paralog divergence under different evolutionary scenarios. ω=ratio of non-synonymous to synonymous substitutions pr synonymous site pr year. Scenario 1: neutral evolution for both paralogs; Scenario 2: intermediate strength purifying selection; Scenario 3: one paralog under intermediate purifying selection and the other pseudogene (neutral); Scenario 4: both paralogs evolve under strict purifying selection

Recent research indicates that different HTG technologies have different sensitivities to paralog-related GE. In two SNP genotyping experiments in soybean, estimated paralog-related GE rates of 33% and 10% were found for MassARRAY and GoldenGate, respectively (Hyten et al., 2008). A side by side comparison of the performance of HTG platforms in paralog rich genomes would be valuable and provide the plant researchers guidelines for choosing the HTG technology best suited for "their genome".

References

Bray, M.S., Boerwinkle, E., Doris, P.A. 2001. High-throughput multiplex SNP genotyping with maldi-tof mass spectrometry: practice, problems and promise. Hum. Mutat. 17(4):296–304.

Hyten, D., Song, Q., Choi, I-Y., Yoon, M-S., Specht, J., Matukumalli, L., Nelson, R., Shoemaker, R., Young, N., Cregan, P. 2008. High-throughput genotyping with the goldengate assay in the complex genome of soybean. Theor. Appl. Genet. 116:945–952.

Leitch, A.R., Leitch, I.J. 2008. Genomic plasticity and the diversity of polyploid plants. Science 320:481–483.

Lockton, S., Gaut, B.S. 2005. Plant conserved non-coding sequences and paralogue evolution. Trends Genet. 21:60–65.

Sandve, S., Rudi, H., Asp, T., Rognli, O.A. 2008. Tracking the evolution of a cold stress associated gene family in cold tolerant grasses. BMC Evol. Biol. 8(1):245.

Yang, Z., Nielsen, R., Goldman, N., Pedersen, A.-M.K. 2000. Codon-substitution models for heterogeneous selection pressure at amino acid sites. Genetics 155:431–449.

Chapter 68
Fine Mapping of Quantitative Trait Loci for Biomass Yield in Perennial Ryegrass

Céline Tomaszewski, J.S. (Pat) Heslop-Harrison, Ulrike C.M. Anhalt, and Susanne Barth

Abstract Biomass yield is one of the most important agronomic traits in forage crops, and increasing biomass yield is the most important objective of breeding programmes. In a precursor study, a genetic map of *Lolium perenne* (perennial ryegrass) was constructed based on an inbred-derived F_2 population using AFLP and SSR markers. The aim was to map quantitative trait loci (QTL) for biomass heterosis in the F_2 population. Dry weight biomass QTL were identified on linkage groups (LGs) 2, 3 and 7. This work focuses on the fine mapping of the QTL positions by mapping additional ryegrass SSR markers and rice-sequence derived STS markers. The STS markers will be used to reveal fine synteny for LGs 2, 3 and 7 between ryegrass and rice. The QTL positions were recalculated for dry weight data collected in the field and the greenhouse over two years. In accordance with the preliminary analysis, biomass QTL were localized on LGs 2, 3 and 7 with a LOD score value between 3.6 and 14.1 and an explained variance from 3.8% to 22.5%. The addition of markers permitted a reduction in the length of QTL intervals. The combination of the experimental approach of fine-mapping QTL positions and exploiting synteny relationships to rice will benefit marker assisted selection and the future identification of biomass-related genes.

Keywords *Lolium perenne* · Mapping · Quantitative trait loci · Biomass · Forage grass · Rice

Introduction

Biomass yield is an important criterion for forage grass breeding and selection. This study uses a dual approach to study biomass yield in perennial ryegrass. One approach is a Quantitative Trait Loci (QTL) mapping approach and the second

---

C. Tomaszewski (✉)
Teagasc Crop Research Centre, Oak Park, Carlow, Ireland
e-mail: celine.tomaszewski@teagasc.ie

C. Huyghe (ed.), *Sustainable Use of Genetic Diversity in Forage and Turf Breeding*,
DOI 10.1007/978-90-481-8706-5_68,

approach uses synteny relationships among the Poaceae. Good conservation of gene content and gene order has been shown between *Lolium perenne* and rice despite their variation in chromosome number and genome size (Sim et al., 2005). Genomic knowledge from rice could be transferred to *L. perenne* which has a 20-fold larger genome and is much less genetically characterized.

A genetic map had already been developed for an F_2 *L. perenne* population and biomass QTL were detected on linkage groups (LGs) 2, 3 and 7 (Anhalt et al., 2008, 2009). The aims of this study are (1) to refine this preliminary map by the addition of more co-dominant molecular markers on selected chromosomal regions of LGs 2, 3 and 7 (2) to better define QTL positions with a reduction of the QTL intervals to a few cM, and (3) to identify molecular markers linked to dry weight biomass to facilitate in the long term selection in early stages of plant breeding programs.

Material and Methods

Two inbred lines with different genetic backgrounds have been crossed to produce the F_1 generation which was selfed to generate the F_2 mapping population (Anhalt et al., 2008). Ryegrass specific SSR markers and rice STS markers were mapped on this linkage map based on 360 genotypes. The STS primers were designed by aligning rice sequences with Poaceae sequences published on the rice genome annotation project webpage (http://www.tigr.org/tdb/e2k1/osa1/). According to the synteny between rice and ryegrass, STS primers for ryegrass LG2 and LG3 were designed from sequences of rice chromosomes 4 and 1, respectively. The refined genetic map was calculated with JoinMap 3.0 (Van Ooijen and Voorrips, 2001) using the Kosambi mapping function, and was drawn using MapChart 2.2. Previous reported QTL positions were recalculated for dry weight data collected in the field and the greenhouse over a two years period (Anhalt et al., 2009). The positions were generated in MapQTL 4.0 (Van Ooijen et al., 2002) using Interval Mapping and Multiple QTL Model mapping (MQM).

Results and Discussion

Twenty two ryegrass SSR markers and five STS markers were added to the 37 markers already mapped on the QTL containing regions, resulting in a marker density of 5 cM. Dry weight biomass QTL were identified for the greenhouse and field experiments on LGs 2, 3 and 7. The logarithm of odds (LOD) score values varied from 3.6 to 14.1 for QTL with a percentage of explained variance between 3.8% and 22.5%. The QTL positions were stable across environments and replications. These results are in accordance with previously obtained results (Anhalt et al., 2009). After addition of markers, QTL intervals have been overall reduced to a better resolution of the QTL on LGs 2, 3 and 7. For example on LG 3 (Fig. 68.1) one single major QTL of 24 cM has been broken down in three smaller QTL, but one still remains substantial in size, but with a precision of 10 cM and a LOD score larger than 10.

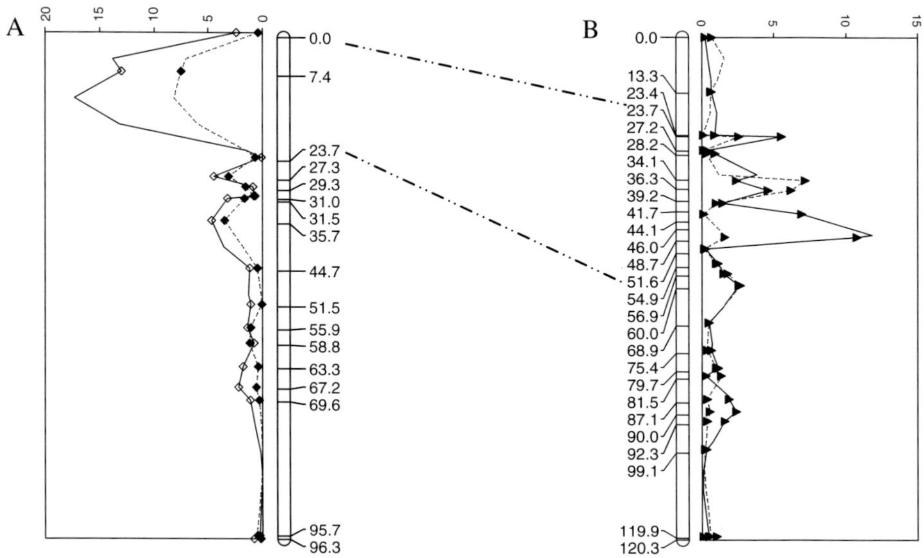

Fig. 68.1 Dry weight biomass QTL (displayed with LOD scores) obtained on LG3 (**a**) (Anhalt et al., 2009) and including additional molecular markers (**b**) with the position of all the markers along the chromosome. QTL were calculated with MQM mapping for the field experiment (—) and for the greenhouse experiment (. . . . . .). The connecting line (–··–) between (**a**) and (**b**) indicates the marker position of the two QTL flanking markers

Overall the fine mapping work has confirmed dry weight biomass QTL positions on three LGs and also the positions of these QTL were mapped with much better precision. We show that rice STS markers which are mapped by polymorphism are useful to study synteny between ryegrass and rice for these three chromosomes. The rice STS markers and corresponding rice BACs will be used to develop a physical map of the biomass QTL regions in ryegrass, and to characterise the QTL and candidate genes. This work will also contribute to reveal how chromosomal rearrangements contribute to the expression of strong phenotypes like heterosis for biomass yield.

Acknowledgements Céline Tomaszewski is financed by a Teagasc Walsh fellowship and thanks Teagasc for support to present this work at this Conference. We acknowledge funding support by the National Development Plan and Teagasc.

References

Anhalt, U.C.M., Heslop-Harrison, P. (J.S.), Byrne, S., Guillard, A., Barth, S. 2008. Segregation distortion in *Lolium*: evidence for genetic effects. Theor. Appl. Genet. 117:297–306.

Anhalt, U.C.M., Heslop-Harisson P. (J.S.), Piepho, H.P., Byrne, S.L., Barth, S. 2009. Quantitative trait loci mapping for biomass yield traits in *Lolium* inbred line derived F_2 population. Euphytica DOI:10.1007/s10681-009-9957-9.

Sim, S., Chang, T., Curley, J., Warnke, S.E., Barker, R.E., Jung, G. 2005. Chromosomal rearrangements differentiating the ryegrass genome from the Triticeae, oat, and rice genomes using common heterologous RFLP probes. Theor. Appl. Genet. 110:1011–1019.

Van Ooijen, J.W., Boer, M.P., Jansen, R.C., Maliepaard, C. 2002. MapQTL 4.0: Software for the Calculation of QTL Positions on Genetic Maps. Plant Research International, Wageningen, The Netherlands.
Van Ooijen, J.W., Vorrips, R.E. 2001. JoinMap® 3.1: Software for the Calculation of Genetic Linkage Maps. Plant Research International, Wageningen, The Netherlands.

Chapter 69
Testing a QTL Index for Marker Selection of Water-Soluble Carbohydrate Content in Perennial Ryegrass

Lesley Turner, Andrew Cairns, Markku Farrell, and Mervyn Humphreys

Abstract Water-soluble carbohydrate provides an easily fermentable energy source in perennial ryegrass fodder. Previously the considerable genetic variation available within the species has been characterised by quantitative trait locus mapping and regions of the genome with significant control over trait expression validated by individual test crosses. However, only limited extents of the large variation available within the source population were recovered. Here a quantitative trait locus marker index has been devised to test the potential of this strategy to exploit the available variation more efficiently. Parents for polycrossing were chosen on the basis of an index weighted for the size of leaf water-soluble carbohydrate effects. Populations of one hundred plants from the crosses were maintained in a glasshouse and the water-soluble carbohydrate content of tiller bases and leaves was measured. The fructan and total water-soluble carbohydrate content of the high populations were always greater than in the low populations. However the differences were not always significant and were generally lower than for the parental populations. This highlights the confounding effects of genome-wide, uncontrolled, segregation for carbohydrate content outside marker-selection regions and the challenges ahead for marker selection of polygenic, complex traits in outbreeding species.

Keywords Forage grass · *Lolium perenne* · Marker selection · QTL index · Water-soluble carbohydrate

Introduction

Grasslands supply around 75% of cattle and sheep feed requirements in the UK (Wilkins and Humphreys, 2003). An important component of the nutritional value of forage is an easily fermentable energy source (Miller et al., 2001). Water-soluble

---

L. Turner (✉)
Crop Genetics, Genomics and Breeding, Institute of Biological, Environmental and Rural Sciences (IBERS), Aberystwyth University, Gogerddan, Aberystwyth SY23 3EB, UK
e-mail: lbt@aber.ac.uk

C. Huyghe (ed.), *Sustainable Use of Genetic Diversity in Forage and Turf Breeding*,
DOI 10.1007/978-90-481-8706-5_69,

carbohydrates (WSC) supply this energy in perennial ryegrass (*Lolium perenne* L.), a grass which is widely-used in temperate regions (Burgon et al., 1997). Increased milk and meat production have been experimentally confirmed when ruminant animals were fed grasses with elevated sugar content (Miller et al., 2001). In addition high-sugar grasses can ameliorate environmental problems by reducing nitrogen losses in urine and manures, and, in the future, grasses with high carbohydrate content may be important for use in the production of bio-energy.

Considerable genetic variation for water-soluble-carbohydrate (WSC) content has been found in perennial ryegrass (Humphreys, 1989; Turner et al., 2002). This has been characterised by quantitative trait locus (QTL) mapping (Turner et al., 2006). Several regions of the genome with basic control over carbohydrate metabolism were identified. To benefit future breeding programmes this increased understanding of the genetic control of sugar metabolism must be applied to the development of markers that can be used for selection. Candidate genes which might explain these QTL have not currently been identified. Therefore, the most appropriate way to proceed at present is with random markers from within the genomic regions of interest. The main QTL regions identified have been validated by individual test crosses (Humphreys and Turner, 2003). However, in general, only limited extents of the large variation available within the source population were exploited by these individual selections. The work reported here describes further experimental crosses carried out to test the potential of a QTL marker index to exploit the available variation. Parental populations were chosen on the basis of an index weighted for the size of the different QTL.

Materials and Methods

Plant Material and Growth Conditions

Parents were identified from within the 188 genotypes of the WSC F2 mapping family on the basis of an index created from marker values for QTL regions. Five QTL for leaf WSC (QTL 1, sucrose on chromosome 2; QTL 2, glucose on chromosome 2; QTL 3, fructan on chromosome 6; QTL 4, fructan and total WSC on chromosome 6 and QTL 5, glucose and fructose on chromosome 6 (Turner et al., 2006)) were allocated factors to represent the size of their effect (QTL 1, $\times 2$; QTL 2,4,5, $\times 1$; QTL 3, $\times 3$) and weighted marker scores calculated on the basis of purely additive effects for the two alleles. No allowance was made for positive and negative dominance effects or epistatic interactions. The selection index (SI) was

$$\mathrm{SI} = \sum \text{mean marker score for QTL1 to QTL5}$$

Heading date varied considerably, so two sets of crosses were carried out; one set with early-heading parents and one set with late-heading parents. Table 69.1 describes the four parental groups. Polycrosses were carried out during summer

Table 69.1 Number of parental genotypes, heading date (days after April 1st), selection index (score), leaf fructan content (mg/g dm) and leaf total WSC content (mg/g dm) for the parental populations. Data are means with standard errors. Carbohydrate data are main effects over season taken from 3 years of data for the WSC F2 mapping family (Turner et al., 2006)

	early-heading		late-heading	
	low	high	low	high
number of parents	8	8	8	8
heading date	40.4 ±0.86	39.5 ±0.50	70.5 ±1.09	56.9 ±1.03
selection index	4.1 ± 0.34	12.2 ± 0.40	3.6 ± 0.63	11.7 ± 0.29
fructan	81.9 ± 10.21	111.2 ± 9.48	83.3 ± 7.34	130.6 ± 8.23
total WSC	201.7 ± 7.07	224.1 ± 5.84	212.3 ± 6.26	256.2 ± 12.76

2005 in pollen isolation houses. Seed was harvested in July, cleaned, and sown in August. One hundred low-selection and one hundred high-selection progeny were maintained in 15 cm pots in a frost-free, unlit glasshouse throughout the year and re-cloned each year from a small sub-set of tillers.

Measurements and Data Analysis

Tiller base and leaf material were sampled for analysis of WSC in spring (late February/early March) and autumn (late October/early November) from the plants in the glasshouse. Carbohydrates were extracted and analysed following the procedures of Turner et al. (2006). All statistical analyses were carried out in GenStat with the menu-driven procedures standard within the programme.

Results and Discussion

Data analysis is on-going but preliminary results are available. Currently there were no significant interactions between selection population and sampling time and therefore main effect population means are presented. Differences between the high and low populations produced by the early-heading crosses were smaller than for the parental populations and few were significant, although the high population contained more leaf fructan and total WSC than the low population (Table 69.2). There were larger differences between the high and low populations from the late-heading crosses (Table 69.3). Tiller base as well as leaf carbohydrate contents showed significant effects, although the differences for leaves were again smaller than for the respective parental populations. The reasons for the better recovery of significant effects in the late-heading crosses are not entirely clear at present, but there was greater divergence between the initial parental populations for these crosses.

Table 69.2 Tiller base and leaf fructan (mg/g dm) and total WSC content (mg/g dm) for the early-heading low and high populations. Data are means over year and season. $n = 100$

tissue	WSC content	low	high	P
tiller base	fructan	185.5	190.3	0.370
	total WSC	257.6	258.7	0.835
leaf	fructan	49.4	57.0	0.064
	total WSC	134.7	142.7	0.034

Table 69.3 Tiller base and leaf fructan (mg/g dm) and total WSC content (mg/g dm) for the late-heading low and high populations. Data are means over year and season. $n = 100$

tissue	WSC content	low	high	P
tiller base	fructan	177.7	200.5	0.002
	total WSC	251.1	271.5	0.004
leaf	fructan	72.7	88.9	0.036
	total WSC	193.9	201.3	0.473

The direction of the selection effect with a QTL index was more consistent than with selection on individual QTL (Humphreys and Turner, 2003). Controlling segregation at a larger number of loci improved the efficiency of selection to some extent. However the recovery of potential variation with this relatively simple marker index was still poor. In all cases there was greater divergence between the parental populations than between the progeny populations. It is expected that this resulted from uncontrolled segregation of numerous loci (with effects on carbohydrate content) located outside the marker-selection regions. This highlights the complexity of the regulation of WSC content in ryegrass. Improvement of such polygenic traits by marker selection poses major challenges. Many genes with small individual effects need to be identified and selected for together. The data presented here suggest that further QTL would have to be identified and (perhaps considerably) more than five QTL regions selected concurrently with a more sophisticated selection index taking into account dominance and epistatic effects (which are known to be widespread for WSC in ryegrass (unpublished data)) for substantial improvement of WSC content. The literature contains various estimates of the number of QTL regions it is feasible to select concurrently (Araus et al., 2008; Bernardo, 2008). What is not in doubt is that the population size required for marker selection increases rapidly as more QTL are added. The position is more straightforward for major-gene traits. The identification of markers for major genes or QTL can inform introgression by standard breeding practices, meaning marker selection is, and will continue to be, relatively straightforward. Unfortunately, however, many traits of agronomic importance such as yield, yield stability, product quality and adaptation to environment are under complex polygenic control. Moreover further complications can arise from epistatic effects and interactions with environment (Babu et al., 2004). In the short term it may be more efficient to continue to breed for traits that are easily measured by phenotypic selection. In the future marker selection will be most useful for traits that are particularly difficult or expensive to phenotype.

Acknowledgements We thank Jenny Ashton for technical assistance with running the HPLC. Kirsten Skot carried out the marker analysis and Ian Armstead did the linkage mapping. This work was funded by BBSRC and Defra.

References

Araus, J.L., Slafer, G.A., Royo, C., Serret, M.D. 2008. Breeding for yield potential and stress adaptation in cereals. Crit. Rev. Plant Sci. 27:377–412.

Babu, R., Nair, S.K., Prasanna, B.M., Gupta, H.S. 2004. Integrating marker-assisted selection in crop breeding – prospects and challenges. Curr. Sci. 87:607–619.

Bernardo, R. 2008. Molecular markers and selection for complex traits in plants: learning from the last 20 years. Crop Sci. 48:1649–1664.

Burgon, A., Bondesen, O.B., Verburgt, W.H., Hall, A.G., Bark, N.S., Robinson, M., Timm, G. 1997. The forage seed trade. In: Fairey, D.T., Hampton, J.G. (eds.), Forage Seed Production. 1. Temperate Species (pp. 271–286). CAB International, Wallingford, UK.

Humphreys, M.O. 1989. Water-soluble carbohydrates in perennial ryegrass breeding. I. Genetic differences among cultivars and hybrid progeny grown as spaced plants. Grass For. Sci. 44: 231–236.

Humphreys, M.O., Turner, L.B. 2003. Nutritive quality QTL and marker assisted selection in ryegrass. Vortr. Pflanzenzuchtg. 59:280–288.

Miller, L.A., Moorby, J.M., Davies, D.R., Humphreys, M.O., Scollan, N.D., Macrae, J.C., Theodorou, M.K. 2001. Increased concentration of water-soluble carbohydrate in perennial ryegrass (*Lolium perenne* L.). Milk production from late-lactation dairy cows. Grass For. Sci. 56:383–394.

Turner, L.B., Humphreys, M.O., Cairns, A.J., Pollock, C.J. 2002. Carbon assimilation and partitioning into non-structural carbohydrate in contrasting varieties of *Lolium perenne*. J. Plant Physiol. 159:257–263.

Turner, L.B., Cairns, A.J., Armstead, I.P., Ashton, J., Skøt, K., Whittaker, D., Humphreys, M.O. 2006. Dissecting the regulation of fructan metabolism in perennial ryegrass (*Lolium perenne*) with quantitative trait locus mapping. New Phytol. 169:45–58.

Wilkins, P.W., Humphreys, M.O. 2003. Progress in breeding forage grasses for temperate agriculture. J. Agric. Sci. 140:129–150.

Chapter 70
Dissecting *Festulolium* Chromosome 3 to Locate Rooting and Drought Resistance Traits

Lesley Turner, Kit Macleod, Chris Watts, Richard Whalley, Andy Binley, Tolis Papadopoulos, Sally O'Donovan, Phil Haygarth, Julie King, and Mike Humphreys

Abstract Breeding for drought resistance/tolerance traits that favour continued growth and yield in forage grasses rather than just survival would be of great economic value. *Festuca* chromosome 3 has proved a rich source of alleles for introgression to enhance the expression of beneficial traits in a *Lolium* background. Three drought resistant introgression lines and a chromosome 3 recombination series were grown in 1 m deep pipes of compost and drought imposed by withholding water. Root traits were shown to be important in drought resistance and the presence of genes on chromosome 3 with some control over relevant traits has been confirmed. Furthermore it is concluded that *Lolium* and *Festuca* should be considered as a single genetic complex of grasses comprising a useful source of variation for breeding.

Keywords Chromosome 3 · Drought · Fescue · Rooting · Ryegrass

Introduction

Over half the land used for agriculture in the UK is grassland for grazing animals and drought poses major challenges. Plant response to water shortage is complex and although it is vital that perennial crops survive and re-grow rapidly when autumn rains set in, it is equally important that yields should not be reduced greatly during mild drought. The major crops of UK grasslands are perennial and Italian ryegrasses (*Lolium perenne* L. (Lp) and *Lolium multiflorum* L. (Lm). Ryegrasses are shallow rooting and less well adapted to drought than many other grasses (Crush et al., 2002; Humphreys et al., 2006). In general fescues have deeper roots and better soil water extraction characteristics and this is believed to contribute to their greater drought resistance (Durand et al., 2007). Genetic variation for rooting and drought

L. Turner (✉)
Crop Genetics, Genomics and Breeding, Institute of Biological, Environmental and Rural Sciences (IBERS), Aberystwyth University, Gogerddan, Aberystwyth SY23 3EB, UK
e-mail: lbt@aber.ac.uk

C. Huyghe (ed.), *Sustainable Use of Genetic Diversity in Forage and Turf Breeding*,
DOI 10.1007/978-90-481-8706-5_70,

tolerance/resistance traits is available within both *Festuca* and *Lolium* (Bonos et al., 2004; Turner et al., 2008; Alm et al., 2009). The benefit of *Festuca* alleles on chromosome 3 (C3) has been demonstrated by introgression mapping (Humphreys and Pašakinskienė, 1996; Humphreys et al., 1997; Humphreys et al., 2005). Here we report the use of a recombination series family and three introgression lines to examine the role of genes on C3 in rooting and plant performance during drought.

Materials and Methods

Plant Material and Growth Conditions

Festulolium chromosome substitution line C3 (King et al., 2002) was further backcrossed to the Lp Liprior parent to produce a C3 recombination series. Genomic *in situ* hybridisation (GISH) was used to identify 19 recombinant line genotypes with *Festuca* segments covering the whole chromosome (Fig. 70.1). The *Festulolium* introgression lines have been described previously: LmFa (with *F. arundinacea*) genotype, prefix P135 (Humphreys and Pašakinskienė, 1996) and LmFg (with *F. glaucescens*) genotypes, prefix P194 (Humphreys et al., 2005). Plants were grown

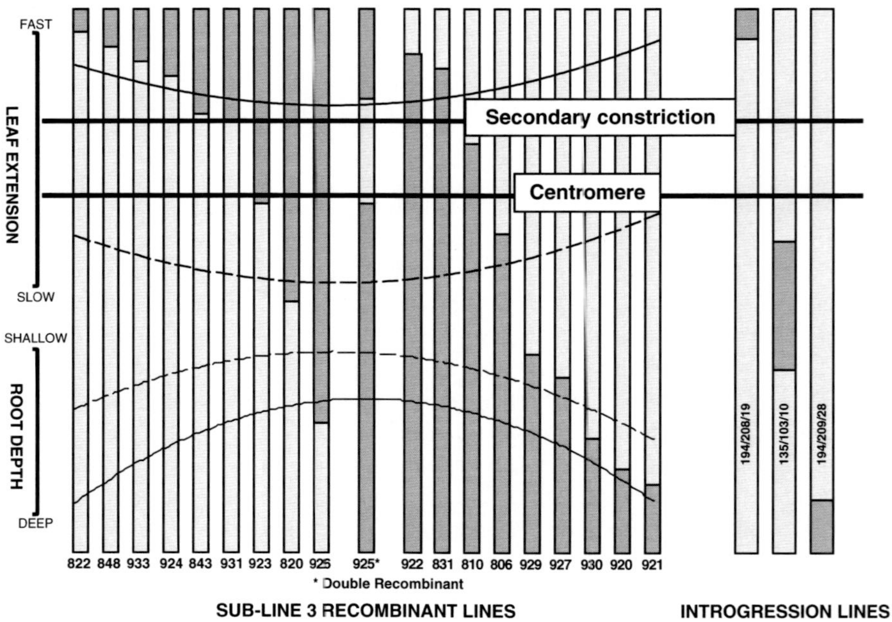

Fig. 70.1 Genetic structure of the sub-line-3 recombinant and introgression lines. *Festuca* DNA (*dark bars*) and *Lolium* DNA (*pale bars*) as identified by genomic *in situ* hybridisation (GISH). Fitted curves for root depth and leaf extension are overlaid. Watered control (*solid line*) and drought treatment (*dashed line*)

in 1 m deep pipes (110 mm wide) lined with polythene tubing and filled with potting compost. Five individual tillers were planted in each pipe and the pipes arranged in a block design with 4 replicates. Drought was imposed by withholding water for 45 days.

Measurements and Data Analysis

Above ground growth was cut back to 5 cm stubble on July 2nd immediately before the drought treatment. Herbage dry mass production after five weeks of drought was measured by cutting back again to 5 cm stubble and analysed with initial plant mass as a covariate. Regrowth at the end of the drought period was the length of new leaf extension growth 7 days after the cut. Below ground growth was examined in mid-April before the drought to assess establishment and again in mid-September following the period of drought treatment and regrowth. The polythene sleeve was slid out of the pipe to enable non-destructive observations of the roots around the sides of the column of potting compost and measurements of root traits. The maximum depth at which roots were visible was recorded. The column of compost was marked into 10 cm sections and root density scored for each section on a scale of 0 (no roots visible) to 4 (extremely dense rooting). Further traits were derived from these measurements. Root system size was the sum of scores for all the sections and root profile was the regression coefficient for a line fitted through the root scores from the lowest section containing roots to the top section. The September-data were analysed with the April-data as covariate. Anova was carried out with the menu-driven options in GenStat.

Results and Discussion

There were significant effects and interactions between genotype and drought treatment for many of the traits examined, so this study has confirmed the wide genetic variation available in fescues and ryegrasses for rooting and its response to water deficit (Crush et al., 2007; Turner et al., 2008). C3 recombinant line 822 had the deepest roots, both under well watered and dry conditions (Table 70.1). Line 931 had the largest root system in the controls, but under drought the value for line 920 was relatively highest at 139% of the control. Leaf extension growth during moderate water stress varied from 7% of the control in line 929 to 85% in line 806. The identification of the best performing genotype from the C3 recombinant lines was rather trait-dependent as no single line performed well for all below and above ground growth traits. Two lines, 922 and 924, did have largely contrasting root characters and tended to rank towards opposite ends of the data sets for root depth and root system size. These have reciprocal ryegrass and fescue segments (Fig. 70.1). However, as there were few clear differences in above ground growth for these particular recombinant lines, they do not help identify the role of roots in shoot growth during drought.

Table 70.1 Below ground growth after drought and above ground growth during drought. Deepest root, (mm); root system size (sum of scores); root system profile (regression coefficient). Dry mass production (g); regrowth (mm). C3 recombinant genotypes are in the same order as Fig. 70.1. Data are means ($n = 4$)

	Deepest root		Root system size		Root system profile		Dry matter week 5 with week 0 as covariate		Leaf extension-regrowth week 6	
Genotype	Control	Drought	Control	Drought	Control	Drought	Control	Drought	Control	Drought
Lp Liprior	811.6	670.5	14.73	17.19	0.316	0.291	0.97	0.41	56.5	38.1
Lp Meltra	368.6	408.1	8.21	5.38	0.522	0.327	0.31	0.28	37.5	3.2
Fp Bf1183	666.3	648.3	13.35	12.35	0.363	0.289	0.91	0.34	91.0	35.0
F1	565.1	422.8	13.58	14.19	0.443	0.299	0.47	0.27	36.5	4.5
Sub-line C3	802.5	647.3	12.3	10.85	0.185	0.157	0.55	0.76	36.8	18.7
822	908.1	878.9	8.77	8.92	0.261	0.294	1.13	1.29	67.5	30.0
848	669.7	661.6	15.45	16.7	0.381	0.221	0.50	0.49	45.0	18.7
933	818.9	595.4	7.47	8.28	0.300	0.382	0.59	0.35	47.5	24.5
924	534.2	247.8	15.49	11.5	0.448	0.608	0.34	0.37	27.5	3.7
843	838.8	687.7	11.64	11.84	0.305	0.286	0.48	0.82	60.8	32.2
931	705.5	553.4	15.94	10.59	0.247	0.309	0.58	0.46	34.0	7.0
923	744.9	625.2	12.24	11.64	0.309	0.301	0.84	0.57	50.1	8.5
820	642.6	521.5	7.37	9.18	0.256	0.289	0.31	0.47	37.6	21.2
926	670.2	541.6	11.34	11.64	0.245	0.246	0.45	0.44	43.5	16.5
925	731.6	677.3	7.31	5.26	0.287	0.326	0.66	0.54	48.5	23.5
922	439.9	539.8	11.07	11.78	0.690	0.203	0.31	0.41	65.5	13.9
831	553.3	488.9	10.68	9.48	0.298	0.227	0.53	0.41	51.7	15.1
810	763	620.4	12.24	10.6	0.285	0.331	0.63	0.44	38.5	8.4
806	776	806	8.13	10.18	0.306	0.272	0.49	0.61	51.5	44.0
929	613.3	649.8	10.12	10.18	0.250	0.260	0.39	0.46	40.5	3.0
927	884.6	766.9	11.03	9.38	0.244	0.253	0.88	0.57	60.0	17.7
930	643.9	690.5	13.59	10.75	0.305	0.175	0.49	0.65	48.5	35.8

Table 70.1 (continued)

Genotype	Deepest root		Root system size		Root system profile		Dry matter week 5 with week 0 as covariate		Leaf extension-regrowth week 6	
	Control	Drought	Control	Drought	Control	Drought	Control	Drought	Control	Drought
920	845.8	541.4	14.41	20.06	0.330	0.348	0.76	0.31	47.5	23.5
921	770.9	654.8	10.98	11.54	0.317	0.332	0.78	0.97	59.0	22.5
P194/208/19	542.7	372	8.31	8.22	0.432	0.101	0.22	0.28	81.4	38.3
P135/103/10	859.6	918.7	10.53	10.38	0.232	0.229	0.52	0.00	71.0	10.7
P194/209/28	858.7	721.4	15.3	17.7	0.275	0.325	0.56	0.00	86.8	9.0
Lm AberEpic	896.6	845.1	6.92	7.18	0.358	0.251	0.00	0.00	126.0	0.0
Fa Dovey	878.7	777.7	13.9	11.14	0.239	0.290	0.00	0.00	71.0	3.0
Fg Bn354	534.3	480.5	5.81	6.72	0.325	0.308	0.69	0.50	26.5	17.4
P drought	0.017		0.669		0.297		0.020		<0.001	
LSD (P < 0.05)	65.3		NS		NS		0.10		4.3	
P genotype	< 0.001		< 0.001		0.001		< 0.001		< 0.001	
LSD (< 0.05)	147.4		2.8		0.090		0.34		19.8	
Pinteraction	0.597		0.169		0.006		0.026		<0.001	
LSD (P < 0.05)	NS		NS		0.193		0.46		28.1	

Relationships between roots and shoots are complex. Correlation analysis was carried out for the C3 recombinant line data to look for associations between traits. Leaf extension during week 6 was correlated with root depth ($r = 0.509$; $P < 0.05$) and root system size ($r = 0.482$; $P < 0.05$). Generally plants with large root systems do produce greater shoot dry mass, but it has proved possible to simultaneously select for small shoots and large roots and break this linkage (Bonos et al., 2004). Drought adds a further layer of complexity. Despite correlations between root and shoot responses in this plant material, which implicate a role of roots and water uptake in shoot growth, the lack of significant correlations between other traits shows the complexity of the suite of responses involved during water deficit.

The patterns of inheritance of beneficial alleles were also complex. Lp Liprior appeared to be the most likely source of alleles conferring large, deep root systems to the C3 recombinant lines in both well-watered and water-deficient conditions. High expression of leaf extension growth during moderate water stress appeared to be derived from Fg in the introgression line P194/208/19, but could originate from either Fp and Lp Liprior in the C3 recombinant series. The presence of heritable and selectable genetic variation for deep roots within both *Festuca* and *Lolium* was expected, but the overlap in the respective ranges of trait values proved to be considerable. However Wilman et al. (1998) have reported that the order of suitability for conditions of severe water deficit in the field was Fa > Lp > Lp × Fp > Fp > Lm × Fp > Lm × Lp > Lm. It may be that *Festuca* and *Lolium* should really be considered as one genetic complex. This does have the advantage that suitable variation for variety improvement may be found within ryegrasses and hence the need for introgression from fescues avoided.

The variable origin of positive and negative alleles for different traits does not prevent physically mapping relevant QTL regions onto C3 with the recombinant line family, except that, particularly for *Lolium* derived positive effects, it is necessary to allow for increased noise (variability) from segregation of the *Lolium* background on C3 as well as on the other chromosomes. The aim was to detect patterns in the phenotype data which matched the genetic structure of the plants and thus to identify regions of C3 with particular roles in trait expression. Data for the recombinants were arranged in order of the position of the *Festuca* segment. Parabolic curves were found to give the best fit to smooth the data, although the curve parameters were not significant in all cases as a result of the predicted high level of noise. However, these curves did identify some interesting patterns (Fig. 70.1). Recombinant lines in the central part of the sequence had shorter roots and lower leaf extension growth under conditions of both water deficit and adequate water supply. These curves suggest the correlation between the two traits may have a genetic basis, due either to pleiotropic effects or to linkage of functional genes. The recombinant lines with shorter roots and lower leaf extension growth are uniquely distinguished by the absence of ryegrass-derived alleles on the central part of the chromosome; in this material it was alleles from ryegrass that conferred deeper roots and higher leaf extension rates. Although the parabolic curves did not indicate any particular interactive response of these genes on C3 to the rate of water supply, it does appear that there are genes controlling traits relevant to drought resistance/tolerance on the region of C3 around the centromere.

Acknowledgements We thank BBSRC for funding this work through the SuperGRaSS project (BB/D011612/1).

References

Alm, V., Busso, C.S., Ergon, A., Rudi, H., Larsen, A., Humphreys, M.W., Rognli, O.A. 2010. QTL analysis and comparative genetic mapping of frost tolerance, winter survival and drought tolerance in meadow fescue (*Festuca pratensis* Huds.). Genetics (In press).

Bonos, S.A., Rush, D., Hignight, K., Meyer, W.A. 2004. Selection for deep root production in tall fescue and perennial ryegrass. Crop Sci. 44:1770–1775.

Crush, J.R., Easton, H.S., Waller, J.E., Hume, D.E., Faville, M.J. 2007. Genotypic variation in root distribution patterns, nitrate interception, and moisture stress response of a perennial ryegrass (*Lolium perenne* L.) mapping population. Grass For. Sci. 62:265–273.

Crush, J.R., Ouyang, L., Eerens, J.P.J., Stewart, A.V. 2002. The growth of roots of perennial, Italian, hybrid and annual ryegrasses through a high-strength root medium. Grass For. Sci. 57:322–328.

Durand, J.L., Bariac, T., Ghesquiere, M., Biron, P., Richard, P., Humphreys, M., Zwierzykowski, Z. 2007. Ranking of the depth of water extraction by individual grass plants, using natural ^{18}O isotope abundance. Env. Exp. Bot. 60:137–144.

Humphreys, M.W., Pašakinskienė, I. 1996. Chromosome painting to locate genes for drought resistance transferred from *Festuca arundinacea* into *Lolium multiflorum*. Heredity 77:530–534.

Humphreys, M.W., Thomas, H-M., Harper, J., Morgan, G., James, A., Ghamari-Zare, A., Thomas, H. 1997. Dissecting drought- and cold-tolerance traits in the *Lolium-Festuca* complex by introgression mapping. New Phytol. 137:55–60.

Humphreys, J., Harper, J.A., Armstead, I.P., Humphreys, M.W. 2005. Introgression-mapping of genes for drought resistance transferred from *Festuca arundinacea* var. *glaucescens* into *Lolium multiflorum*. Theor. Appl. Genet. 110:579–587.

Humphreys, M.W., Yadav, R.S., Cairns, A.J., Turner, L.B., Humphreys, J., Skøt, L. 2006. A changing climate for grassland research. New Phytol. 169:9–26.

King, J., Armstead, I.P., Donnison, I.S., Thomas, H-M., Jones, R.N., Kearsey, M.J., Roberts, L.A., Jones, A., King, I.P. 2002. Physical and genetic mapping in the grasses *Lolium perenne* and *Festuca pratensis*. Genetics 161:315–324.

Turner, L.B., Cairns, A.J., Armstead, I.P., Thomas, H., Humphreys, M.W., Humphreys, M.O. 2008. Does fructan have a functional role in physiological traits? Investigation by QTL mapping. New Phytol. 179:765–775.

Wilman, D., Gao, Y., Leitch, M.H. 1998. Some differences between eight grasses within the *Lolium-Festuca* complex when grown in conditions of severe water shortage. Grass For. Sci. 53:57–65.

Chapter 71
Functional Analysis of Genes Involved in Cell Wall Biosynthesis of the Model Species *Brachypodium distachyon* to Improve Saccharification

Steven Van Hulle, Isabel Roldán-Ruiz, Erik Van Bockstaele, and Hilde Muylle

Abstract Members of the grass family are important as resource for the production of first generation bio-ethanol, which is based on the fermentation of starch and sucrose which is obtained from food crops like corn (US), wheat (Europe) or sugarcane (Brazil). If Europe wants to meet the objective of the new directive on the promotion of the use of energy from renewable sources, it will be necessary to make the transition from first to second generation conversion technologies for the production of bio-ethanol. These conversion techniques are based on the use of recalcitrant lignocellulosic biomass as feedstock. The energy contained in lignocellulosic biomass is largely entrapped in the plant cell wall, which is built up of cellulose, hemicellulose and lignin and can make up to 70% of the total plant biomass. To be able to produce ethanol from these rigid cell walls, the cellulose and hemicellulose need to be degraded first into monosaccharides. For the moment, this degradation constitutes a bottleneck in the process. Especially lignin is a disturbing factor. Therefore, an interesting approach to improve lignocellulosic crops is to reduce their lignin content. In this study, we use *Brachypodium distachyon* as a model to study the effect of up- or down regulation of genes with a key-role in the monolignol biosynthesis pathway on the saccharification efficiency. The general strategy and preliminary results of this study will be discussed.

Keywords Bio-ethanol · *Brachypodium distachyon* · Functional analysis · Lignin

Introduction

The Poaceae is a large plant family comprising over 10,000 species (Kellogg, 2001). Traditionally, grass crops have been used for the production of food and feed, as fodder or as turf. Recently, they are increasingly used in bio-ethanol production.

S. Van Hulle (✉)
Plant Sciences Unit – Growth and Development, Institute for Agricultural and Fisheries Research (ILVO), Caritasstraat 21, Melle B-9090, Belgium
e-mail: Steven.VanHulle@ilvo.vlaanderen.be

C. Huyghe (ed.), *Sustainable Use of Genetic Diversity in Forage and Turf Breeding*,
DOI 10.1007/978-90-481-8706-5_71, © Springer Science+Business Media B.V. 2010

Bio-ethanol is already commercially available in several countries all over the world as a first generation bio-fuel. First generation bio-ethanol is produced by fermentation of sugar-rich food crops such as wheat or sugarcane. However, the cultivation of these 'traditional' crops is energetically demanding, and first-generation transformation processes exploit only a small part of the biomass produced. As a consequence first generation bio-ethanol is not a sustainable alternative for fossil fuels. Second generation bio-ethanol on the other hand is derived from lignocellulosic biomass. Several good candidates for the production lignocellulosic biomass can be found within the grass family, such as *Miscanthus x giganteus* (elephant grass), or *Panicum virgatum* (switchgrass). These are perennial crops producing high yields under low input regimes. Although these crops have a lot of potential, they still can be genetically improved. As second-generation conversion processes exploit mainly the sugars contained in cell walls, significant gains can be achieved through alteration of the composition of cell walls in crops used as feedstock. The three main components of plants cell walls are cellulose (β-1,4 linked glucose), hemicellulose (group of branched polysaccharides) and lignin (a polyphenolic matrix). Cellulose and hemicellulose render the fermentable sugars while lignin is a disturbing factor for sugar release. Lignin disturbs the process by blocking the access of cellulases to cellulose and by irreversibly binding with cellulases (Berlin et al., 2005). Lignin can mainly be found in the secondary cell wall, where it is synthesized by oxidative coupling of p-hydroxycinnamyl alcohols (Boerjan et al., 2003). The three main p-hydroxycinnamyl alcohols are hydroxycinnamyl alcohol, coniferyl alcohol and sinapyl alcohol, which are the products of the monolignols biosynthetic pathway. Lignin modification, through alternation of the monolignols biosynthetic pathway, is a promising strategy to improve the saccharifiability of lignocellulosic biomass.

Research Strategy

According to literature data, 4-coumarate CoA ligase (4CL) and caffeic acid/5-hydroxyconiferaldehyde O-methyltransferase (COMT) are two of the enzymes of the monolignols biosynthetic pathway whose alteration will have the highest influence on the lignin content of cell walls without limiting growth. For example, down regulation of 4CL in poplar resulted in a decrease in lignin content of 45% and an increase in cellulose of 15% and an increase in the relative abundance of arabinan, galactan and rhamnan (20, 46, 23% increase respectively) (Hu et al., 1999). COMT on the other hand has been shown to be disrupted (lacks different conserved domains and has an altered expression) in brown midrib mutant 3 (bm3) of maize which results in a 20–30% reduction in lignin content and an increased digestibility (Guillaumie et al., 2008).

The effect of lignin modification on saccharification will be studied in *B. distachyon*, a model for the energy grasses. *B. distachyon* has gained relevance in recent years as model for Poaceae because of its close phylogenetic relationship with important temperate grasses like wheat and switchgrass. It possesses all the necessary features to be a good model species. It is small, has a short life cycle, has a small genome approximately 355 Mb, (Hong et al., 2008), is self-fertile, has diploid accessions and transformation protocols are published (Christiansens et al., 2005; Vogel et al., 2006; Păcurar et al., 2008; Vogel and Hill, 2008; Vain et al., 2008).

Through bioinformatics, sequences of the genes of interest in *B. distachyon* are traced using the 4x genome sequence of *Brachypodium* (http://blast.brachybase.org) starting from the *Arabidopsis thaliana* sequences of the desired genes.

The down regulation of the genes in *Brachypodium* will be accomplished using RNAi. The transgenic lines will be evaluated using chemical analysis, to determine cell wall composition and a small-scale saccharification assay. With this assay, the effect of the modified lignin on the enzymatic hydrolysis will be analyzed.

First Results

A. thaliana contains four 4CL genes of which three are members of the class I 4CL genes: 4CL1, 4CL2 and 4CL4 (Raes et al., 2003). These three genes are involved in lignin production, therefore the protein sequence of 4CL1 (At1g51680) of *A. thaliana* was used to initiate the search for the *Brachypodium* 4CL sequences with the tblastn program on the BrachyBase BLAST page (http://blast.brachybase.org). This resulted in four hits with a bit score above 250 (super_3.2108, super_5.1175, super_8.27 and super_6.2326). The following hits have less homology. The isolation of these four sequences is in progress.

A. thaliana has only one actual COMT and thirteen COMT-likes (Raes et al., 2003). Using the *A. thaliana* protein sequence of COMT (At5g54160) in the tblastn program resulted in three hits (super_3.1632, super_0.4389, super_9.1215) with a bit score above 250. Those three copies are currently being isolated in *Brachypodium*.

Future Work

For all copies of both enzymes, RNAi constructs will be synthesized with the Gateway® system which will be used for transformation. The resulting transgenic plants will be evaluated and may reveal potential directions for the improvement of grasses for bio-ethanol production.

Acknowledgements This work is done in collaboration with the department Plant Systems Biology of the VIB (Flanders Institute for Biotechnology). The authors wish to thank Katrien Liebaut for outstanding technical assistance.

References

Berlin, A., Gilkes, N., Kurabi, A., Bura, R., Tu, M., Kilburn, D., Saddler, J. 2005. Weak lignin-binding enzymes: a novel approach to improve activity of cellulases for hydrolysis of lignocellulosics. Appl. Biochem. Biotechnol. 121–124:163–170.

Boerjan, W., Ralph, J., Baucher, M. 2003. Lignin biosynthesis. Annu. Rev. Plant Biol. 54:519–546.

Christiansen, P., Andersen, C.H., Didion, T., Folling, M., Nielsen, K.K. 2005. A rapid and efficient transformation protocol for the grass *Brachypodium distachyon*. Plant Cell Rep.23:751–758.

Guillaumie, S., Goffner, D., Barbier, O., Martinant, J.P., Pichon, M., Barrière, Y. 2008. Expression of cell wall related genes in basal and ear internodes of silking brown-midrib-3, caffeic acid O-methyltransferase (COMT) down-regulated, and normal maize plants. BMC Plant Biol. 8: 71–87.

Hong, S.Y., Seo, P.J., Yang, M.S., Xiang, F., Park, C.M. 2008. Exploring valid reference genes for gene expression studies in *Brachypodium distachyon* by real-time PCR. BMC Plant Biol. 8:112–123.

Hu, W.J., Harding, S.A., Lung, J., Popko, J.L., Ralph, J., Stokke, D.D., Tsai, C.J., Chiang, V.L. 1999. Repression of lignin biosynthesis promotes cellulose accumulation and growth in transgenic trees. Nat. Biotechnol. 17:808–812.

Kellogg, E.A. 2001. Evolutionary history of the grasses. Plant Physiol. 125:1198–1205.

Păcurar, D.I., Thordal-Christensen, H., Nielsen, K.K., Lenk, I. 2008. A high-throughput *Agrobacterium*-mediated transformation for the grass model species *Brachypodium distachyon* L. Transgenic Res. 17: 965–975.

Raes, J., Rohde, A., Christensen, J.H., Van de Peer, Y., Boerjan, W. 2003. Genome-wide characterization of the lignification toolbox in arabidopsis. Plant Physiol. 133: 1051–1071.

Vain, P., Worland, B., Thole, V., McKenzie, N., Alves, S.C., Opanowicz, M., Fish, L.J., Bevan, M.W., Snape, J.W. 2008. *Agrobacterium*-mediated transformation of the temperate grass *Brachypodium distachyon* (genotype Bd21) for T-DNA insertional mutagenesis. Plant Biotechnol. J. 6:236–245.

Vogel, J.P., Garvin, D.F., Leong, O.M., Hayden, D.M. 2006. *Agrobacterium*-mediated transformation and inbred line development in het model grass *Brachypodium distachyon*. Plant Cell Tissue Organ Cult. 84:199–211.

Vogel, J., Hill, T. 2008. High-efficiency *Agrobacterium*-mediated transformation of *Brachypodium distachyon* inbred line Bd21-3. Plant Cell Rep. 27:471–478.

Chapter 72
Identification of Genes Induced in *Lolium multiflorum* by Bacterial Wilt Infection

Fabienne Wichmann, Torben Asp, Franco Widmer, and Roland Kölliker

Abstract *Xanthomonas translucens* pv. *graminis*(*Xtg*) causes bacterial wilt in many forage grasses including Italian ryegrass (*Lolium multiflorum* Lam), seriously reducing yield and quality. Breeding for resistance is currently the only practicable means of disease control. Molecular markers closely linked to resistance genes or QTL could complement and support phenotypic selection. We used comparative gene expression analysis of a partially resistant *L. multiflorum* genotype infected and not infected with *Xtg* to identify genes involved in the control of resistance to bacterial wilt. The genes differentially expressed upon infection will serve as the basis for the identification of key genes involved in bacterial wilt resistance and to develop molecular markers for marker assisted breeding. Fluorescently labelled cDNA prepared from plant leaves collected at four different time points after infection was hybridized to a cDNA microarray containing 10,000 unique genes from *L. perenne*. Comparisons and statistical analyses of the gene expression profiles revealed 0, 20, 52 and 124 differentially regulated genes 8, 48, 192 and 288 h after infection compared to non-infected controls and considering a p-value threshold of 0.01. This is the first genome-wide transcriptome analysis of *L. multiflorum* investigating the reaction to *Xtg* infection.

Keywords *Lolium multiflorum* Lam · *Xanthomonas translucens* pv. *graminis* · Microarray · Disease resistance

Introduction

Bacterial wilt caused by the pathogen *Xanthomonas translucens* pv. *graminis* (*Xtg*) is a major disease in fodder crop production and in breeding populations of Italian ryegrass (*Lolium multiflorum* Lam.). Infection occurs through contaminated

F. Wichmann (✉)
Agroscope Reckenholz-Tänikon Research Station ART, Zürich 8046, Switzerland
e-mail: fabienne.wichmann@art.admin.ch

C. Huyghe (ed.), *Sustainable Use of Genetic Diversity in Forage and Turf Breeding*,
DOI 10.1007/978-90-481-8706-5_72,

mowing equipment and after the bacteria enter the vascular tissue, they multiply in the xylem and cause wilting of the leaves and in severe cases death of the plant. Shortly after the discovery of bacterial wilt, breeding for resistant cultivars was commenced and proved to be the only practicable means of disease control (Schmidt, 1989). However, when breeding for resistant cultivars, progress usually stagnates after several cycles of recurrent selection (Michel, 2001) and, so far, complete resistance has never been achieved.

In a previous study localizing loci associated with bacterial wilt resistance, one major QTL on linkage group 4 was identified, explaining 67% of the total phenotypic variance. This indicated one or only a few genes with major effects to be responsible for resistance (Studer et al., 2006). However, the molecular basis leading to partial resistance is still unknown in *L. multiflorum*.

In other crops, effector triggered immunity to *Xanthomonas* spp. based on *R* gene specific recognition of bacterial *Avr* gene products is well studied and some of the resistance genes are cloned (e.g. Wu et al., 2007). In rice, also a member of the *Poaceae*, at least 29 race-specific and wide-spectrum genes and QTL conferring complete or partial resistance to bacterial blight (*X. oryzae* pv. *oryzae*) are known (reviewed by Nino-Liu et al., 2006).

With the availability of microarray-based expression profiling, substantial progress in understanding pathogenesis-related responses upon *Xanthomonas* infection has been achieved (Cernadas et al., 2008). Understanding the genetic mechanisms of resistance and affected pathways in *L. multiflorum* in more detail may support phenotypic selection and provide necessary tools for the development and implementation of marker assisted breeding. Therefore, the aim of this study was to further investigate defense responses in the *L. multiflorum* – *Xtg* interaction by using transcriptome analysis.

Material and Methods

A *L. multiflorum* genotype with high partial resistance to bacterial wilt was clonally propagated and used for artificial inoculation with the *Xtg*29 isolate, a representative isolate of moderate to high pathogenicity (Kölliker et al., 2006). Plant samples were collected from four experimental replicates at four time points (8, 48, 192 and 288 h) after infection or control treatment. After RNA extraction and reverse transcription, the cDNA was fluorescently labelled and hybridized to a cDNA microarray which contains 10,000 unique genes from *L. perenne* and has been designed by *Det Jordbrugsvidenskabelige Fakultet* (DJF, Aarhus University, Denmark). After the slides were scanned, the signals were quantified with the GenePix Pro software 6.0. Statistical analysis was performed with the R software (http://www.r-project.org) and the LIMMA package provided by Bioconductor (http://www.bioconductor.org). In order to generate lists of differentially expressed genes, a linear model was fitted to the data and the genes with a p value < 0.01 were regarded as differentially regulated according to the moderated t-statistics (Lonnstedt and Speed, 2002).

Results and Discussion

Of the 10,000 genes spotted on the *L. perenne* cDNA microarray, approximately 5,700 genes were detectable in *L. multiflorum* (57%). The transcriptome analysis upon bacterial wilt infection of the partially resistant genotype revealed in total 158 differentially expressed genes (p value < 0.01). Twenty up-regulated genes were observed 48 hours post inoculation (hpi) 52 genes were differentially expressed 192 hpi (42 up- and 10 down-regulated) and 124 genes were differentially expressed 288 hpi (76 up- and 48 down-regulated). No genes were differentially expressed 8 hpi. This is in congruence with findings of an earlier gene expression study based on cDNA-AFLP where only few differentially regulated Transcript Derived Fragments were observed in *L. multiflorum* earlier than 12 h after infection with *Xtg* (Rechsteiner et al., 2006).

Of the 158 differentially regulated genes in total, 5 were up-regulated at the three sampling time points 48, 192 and 288 hpi, 2 were up-regulated at 48 and 192 hpi, 6 were up-regulated at 48 and 288 hpi and 20 were up-regulated at 192 and 288 hpi (Fig. 72.1). Some of the up-regulated genes detected at more than one time point after infection were also induced in other plant species challenged by *Xanthomonas* infection (Balaji et al., 2007). For example, studies with rice and *X. oryzae* pv. *oryzae* also showed up-regulation of genes encoding ankyrin, germin and malate dehydrogenase (Kottapalli et al., 2007; Li et al., 2006). Furthermore, a synaptobrevin-like protein was up-regulated in tomato leaves infected with *X. campestris* pv. *vesicatoria* (Balaji et al., 2007). In this study, the most significantly up-regulated gene with the greatest fold change encodes the Low silicon protein 1

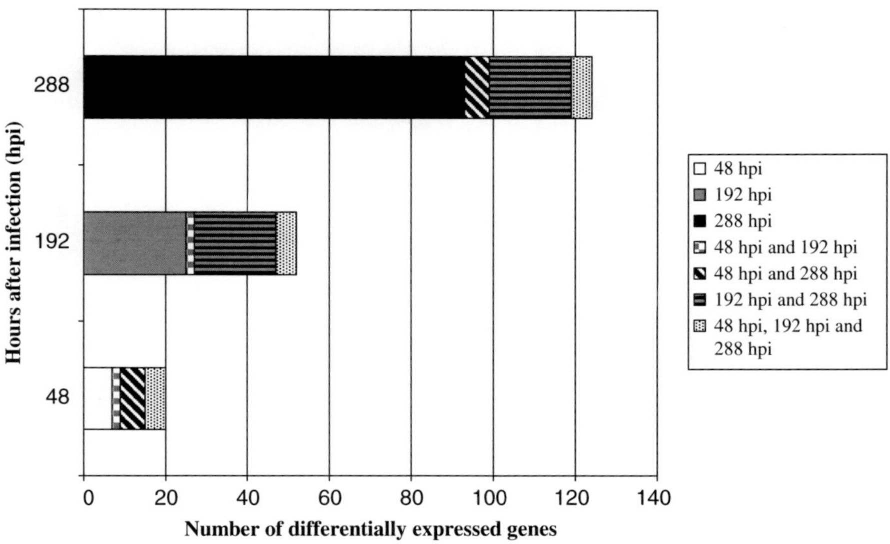

Fig. 72.1 Number of significantly up-regulated genes (p-value < 0.01) detected in *L. multiflorum* after 48, 192 and 288 h of infection by *Xanthomonas translucens* pv. *graminis*

(Lsi1) belonging to a Nod26-like major intrinsic protein sub-family of aquaporins. Lsi1 is known to be responsible for Silicon uptake in rice and barley. This is thought to be important for resistance against biotic and abiotic stresses (reviewed by Ma and Yamaji, 2006). Similar to our results, aquaporins were up-regulated in cotton upon *X. campestris* pv. *malvacearum* infection (Patil et al., 2005). Additionally, another gene significantly up-regulated in the present study showed high sequence similarity to a germin-like protein, known to be involved in basal defense against various pathogens and induced upon abiotic stress (Manosalva et al., 2009).

In general, the number of differentially expressed genes is low compared to other explanatory studies dealing with pathogen infection. This may be partially due to high number of *L. perenne* genes that did not hybridize to *L. multiflorum* cDNA (43%). These genes may not be detectable in *L. multiflorum* or they are only expressed in tissues not investigated. Verification of selected differentially expressed genes by Real-time PCR and Northern blot is currently in progress and has, so far, confirmed the up-regulation of some genes.

Acknowledgements This research has been financed by Swiss National Science Foundation SNF Project (3100A0-112582). We would like to thank L. B. Jensen for technical assistance in the lab of DJF, H.-P. Piepho and A. Schützenmeister for help with the experimental design and Hubert Rehrauer from the Functional Genomics Center Zürich for assistance with statistical analyses.

References

Balaji, V., Gibly, A., Debbie, P., Sessa, G. 2007. Transcriptional analysis of the tomato resistance response triggered by recognition of the *Xanthomonas* type III effector AvrXv3. Funct. Integr. Genomics 7:305–316.

Cernadas, R.A., Camillo, L.R., Benedetti, C.E. 2008. Transcriptional analysis of the sweet orange interaction with the citrus canker pathogens *Xanthomonas axonopodis* pv. *citri* and *Xanthomonas axonopodis* pv. *aurantifolii*. Mol. Plant Pathol. 9:609–631.

Kölliker, R., Kraehenbuehl, R., Boller, B., Widmer, F. 2006. Genetic diversity and pathogenicity of the grass pathogen *Xanthomonas translucens*pv. *graminis*. Syst. Appl. Microbiol. 29:109–119.

Kottapalli, K.R., Satoh, K., Rakwal, R., Shibato, J., Doi, K., Nagata, T., Kikuchi, S. 2007. Combining in silico mapping and arraying: an approach to identifying common candidate genes for submergence tolerance and resistance to bacterial leaf blight in ice. Mol. Cells 24:394–408.

Li, Q., Chen, F., Sun, L.X., Zhang, Z.Q., Yang, Y.N., He, Z.H. 2006. Expression profiling of rice genes in early defense responses to blast and bacterial blight pathogens using cDNA microarray. Physiol. Mol. Plant Pathol. 68:51–60.

Lonnstedt, I., Speed, T. 2002.Replicated microarray data. Statistica Sinica 12:31–46.

Ma, J.F., Yamaji, N. 2006. Silicon uptake and accumulation in higher plants. Trends Plant Sci. 11:392–397.

Manosalva, P.M., Davidson, R.M., Liu, B., Zhu, X.Y., Hulbert, S.H., Leung, H., Leach, J.E. 2009. A Germin-like protein gene family functions as a complex quantitative trait locus conferring broad-spectrum disease resistance in rice. Plant Physiol. 149:286–296.

Michel, V.V. 2001. Interactions between *Xanthomonas campestris* pv. *graminis* strains and meadow fescue and Italian rye grass cultivars. Plant Disease 85:538–542.

Nino-Liu, D.O., Ronald, P.C., Bogdanove, A.J. 2006. *Xanthomonas oryzae* pathovars: model pathogens of a model crop. Mol. Plant Pathol. 7:303–324.

Patil, M.A., Pierce, M.L., Phillips, A.L., Venters, B.J., Essenberg, M. 2005. Identification of genes up-regulated in bacterial-blight-resistant upland cotton in response to inoculation with *Xanthomonas campestris* pv. *malvacearum*. Physiol. Mol. Plant Pathol. 67:319–335.

Rechsteiner, M.P., Widmer, F., Kölliker, R. 2006. Expression profiling of Italian ryegrass (*Lolium multiflorum* Lam.) during infection with the bacterial wilt inducing pathogen *Xanthomonas translucens* pv. *graminis*. Plant Breed. 125:43–51.

Schmidt, D. 1989. Epidemiological aspects of bacterial wilt of fodder grasses. EPPO Bulletin 19:89–95.

Studer, B., Boller, B., Herrmann, D., Bauer, E., Posselt, U.K., Widmer, F., Kölliker, R. 2006. Genetic mapping reveals a single major QTL for bacterial wilt resistance in Italian ryegrass (*Lolium multiflorum* Lam.). Theor. Appl. Genet. 113:661–671.

Wu, X.M., Li, Y.R., Zou, L.F., Chen, G.Y. 2007. Gene-for-gene relationships between rice and diverse avrBs3/pthA avirulence genes in *Xanthomonas oryzae* pv.*oryzae*. Plant Pathol. 56: 26–34.

Chapter 73
Identification of QTLs Associated with Morphological and Agronomic Traits in White Clover (*Trifolium repens* L.)

Yan Zhang, Christy Motes, Mary K. Sledge, Joseph H. Bouton, Yuanhong Han, and Maria J. Monteros

Abstract White clover (*Trifolium repens* L.) is an important cool-season perennial forage legume species used in pastures to improve forage quality. The identification of molecular markers linked to morphological and agronomic traits could facilitate the development of superior white clover cultivars. The objectives of this study were to map quantitative trait loci (QTL) associated with morphological and agronomic traits using a F_1 population from a double pseudo-testcross between two highly heterozygous genotypes. Phenotypic data was collected from multiple field locations and years for morphological traits (leaf length and width, petiole length, stolon diameter and inter-node length), and for growth traits (plant spreading, plant height, and stolon number). Analysis of variance indicated there were significant effects from location, genotype, and genotype $\times$ location for the traits evaluated. Correlation coefficients showed that growth traits in the field were highly correlated with each other. Broad sense heritability estimates for all traits evaluated were less than 25%. The population was genotyped using simple sequence repeat (SSR) markers and multiple QTL model (MQM) analysis was used to identify 37 QTLs on eight linkage groups associated with five morphological traits and four plant growth traits. The consistent location of QTLs for the same traits or highly correlated traits across different locations and years indicates the potential value of utilizing marker-assisted breeding for white clover improvement.

Keywords White clover · *Trifolium repens* · QTL mapping

Introduction

Allopolyploid ($2n = 4x = 32$) white clover (*Trifolium repens* L.) is an important cool-season perennial forage legume species used in pastures to improve forage quality. Microsatellite or simple sequence repeat (SSR) markers are widely used for

---

M.J. Monteros (✉)
Forage Improvement Division, The Samuel Roberts Noble Foundation, 2510 Sam Noble Parkway, Ardmore, OK 73401, USA
e-mail: mjmonteros@noble.org

C. Huyghe (ed.), *Sustainable Use of Genetic Diversity in Forage and Turf Breeding*,
DOI 10.1007/978-90-481-8706-5_73,

plant genome analysis. They are PCR-based, co-dominant markers that are widely distributed throughout eukaryotic genomes (Morgante et al., 2002). A white clover genetic linkage map was developed from a cross between GA02-56 derived from 'Durana' (Bouton et al., 2005) and GA02-15 derived from 'SRVR' (Gibson et al., 1989) in which all eight homoeologous chromosome pairs were detected (Zhang et al., 2007). Many economically important traits are quantitative indicating that they are controlled by multiple genes. Marker-based technologies have already proven powerful for identification and mapping individual genetic components or quantitative trait loci (QTL). The objective of this study was to map QTLs associated with morphological and agronomic traits in white clover using a segregating mapping population.

Materials and Methods

Plant Materials

The GA02-56 genotype derived from 'Durana' is an intermediate-type clover with high stolon density, short plant height with prostrate growth habit, small leaflets, short petioles, an early heading date and a high frequency of cyanogenesis (Bouton et al., 2005). The GA02-15 genotype from SRVR is a ladino-type clover with medium-large leaves, long stolons and upright growth habit and is non-cyanogenic (Gibson et al., 1989). These were used as parents (Fig. 73.1) to develop an F_1 mapping population consisting of 179 individuals.

Fig. 73.1 Parental white clover genotypes used to develop the mapping population

Phenotypic Evaluations

The two parents and the 179 F_1 progenies were evaluated in the greenhouse and in multiple field sites and years using a randomized complete block design with six replications per site in Ardmore, Oklahoma. Phenotypic data collected included:

Table 73.1 Analysis of variance for leaf and stolon morphology from the mapping population based on greenhouse (2004) and field (2005, 2007) data

		Mean squares				
Source of variation	df†	LL‡	LW‡	PL‡	IL‡	SD‡
Environment§	3	60.04***	73.56***	4939.37***	41.32***	34.92***
Rep (env)	13	3.02***	1.54***	246.28***	11.44***	1.22***
Genotype	178	0.30***	0.23***	14.10***	0.76***	0.35***
Genotype × environment	497	0.08***	0.07***	6.52***	0.40***	0.10***
Error		0.06	0.05	3.87	0.26	0.07
R^2		0.73	0.77	0.77	0.56	0.65
CV		13.18	14.98	25.9	33.84	13.62

†Degrees of freedom (df) based on the mean from the data collected in the greenhouse in 2004, Paddock B in 2005, Research Park II and Research Park V in 2007, ‡Leaf length (LL), leaf width (LW), petiole length (PL), internode length (IL) and stolon diameter (SD), §Environment represents the data collected in each location × year combination, *, **, *** Significant at $p < 0.05$, $p < 0.01$, $p < 0.001$, respectively

leaf length (LL), leaf width (LW), petiole length (PL), stolon diameter (SD), internode length (IL), plant spread length (SprL), plant spread width (SprW), plant height (SprH) and number of stolons (Sn). All statistical analyses were conducted using SAS V. 9.1. Analysis of variance (ANOVA) was used to estimate environment, genotype, replication, and genotype × environment effects for the traits evaluated (Table 73.1).

Molecular Markers and QTL Mapping

The segregating population was genotyped with 343 simple sequence repeat (SSR) markers and analyzed using the multiple QTL model (MQM) method. For each trait, the within-year mean value for each genotype was analyzed in MapQTL 5.0 software (Kyazma B.V., Wageningen, The Netherlands) for interval mapping (IM) QTL analysis. Forward selection (Jansen, 2004) was used to select the cofactor for multiple-QTL mapping (MQM) approach also in MapQTL 5.0. The LOD threshold for significance was based on permutation tests (n=1000 permutations) calculated for each individual QTL using statistical significance at p <0.05.

Results and Discussion

We identified transgressive segregation for the traits evaluated and ANOVA showed significant environment, replicate, genotype and genotype × environment effects for leaf and stolon morphology traits (Table 73.1). We have identified 37 QTLs associated with five morphological traits and four plant growth traits (leaf length, leaf width, petiole length, stolon diameter, inter-node length, plant spreading and height, and stolon numbers) on eight linkage groups (not shown). A number of QTLs across multiple years and locations for multiple plant growth traits mapped to linkage group B2 (not shown). The consistent location of QTLs for the same traits or highly correlated traits across different locations and years indicates the potential value of utilizing marker-assisted breeding for white clover improvement. Current efforts include mapping additional, recently developed SSR markers (Zhang et al., 2008) to enhance the marker density in regions of the white clover genome associated with morphologic and agronomic traits.

Acknowledgements We thank Shauna Smith, Dusty Pittman and Jarrod Smith for assistance collecting field data and maintaining field plots. This research was funded by The Samuel Roberts Noble Foundation, Inc.

References

Bouton J.H., Woodfield, D.R., Caradus, J.R., Wood, D.T. 2005. Registration of 'Durana' white clover. Crop Sci. 45:797.

Gibson, P.B., Barnett, O.W., Pederson, G.A., McLaughlin, M.R., Knight, W.E., Miller, J.D., Cope, W.A., Tolin, S.A. 1989. Registration of southern regional virus resistant white clover germplasm. Crop Sci. 29:241–242.

Jansen, R.C. 2004. Quantitative Trait Loci in Inbred Lines. Handbook of Statistical Genetics. Part 3. Animal and Plant Breeding. John Wiley & Sons, New York.

Morgante, M., Hanafey, M., Powell, W. 2002. Microsatellites are preferentially associated with nonrepetitive DNA in plant genomes. Nat. Genet. 30:194–200.

Zhang, Y., Sledge, M., Bouton, J.H. 2007. Genome mapping of white clover (*Trifolium repens* L.) and comparative analysis within the Trifolieae using cross-species SSR markers. Theor. Appl. Genet. 114:1367–1378.

Zhang, Y., He, J., Zhao, P.X., Bouton, J.H., Monteros, M.J. 2008. Genome-wide identification of microsatellites in white clover (*Trifolium repens* L.) using FIASCO and phpSSRMiner. Plant Methods 4:19.

Part VI
Type and Structure of Varieties to Better Exploit Genetic Diversity

Chapter 74
Festulolium Hybrids: Results, Limits and Prospects

Marc Ghesquière, Mike Humphreys, and Zbigniew Zwierzykowski

Abstract *Festulolium* refers to natural or synthetic intergeneric hybrids between obligate outbreeding species of the *Festuca* (fescue) and *Lolium* (ryegrass) genera, species considered frequently as ideal components of agricultural or turf-grass systems. *Festulolium* provide specialist function and novel alternatives to existing grass cultivars that may lack resilience against abiotic or biotic stresses. So far, 23 amphiploid *Festulolium* cultivars have been registered onto national lists as well as 18 cultivars resulting from introgression either into tall fescue or into Italian and perennial ryegrass. Although dispersed throughout the world, *Festulolium* breeding has considerably stimulated research on genetics of the grasses. This has contributed to the development of numerous new technologies among which *Genome In-Situ Hybridization* has played an essential role for precision breeding, i.e. monitoring the transfer of selected traits from fescues into diploid *Lolium* sp. In the future, it would seem very likely that breeding amphiploid *Festulolium* will also benefit from the genomic advances achieved in diploids using introgression. As future *Festulolium* cultivars are expected to be extremely diverse, this will require that regulations for registration rely on a genome-based classification and suitable official tests for better acknowledgement of agronomic advances by end-users.

Keywords Amphiploidy · Introgression · Homoeologous recombination · GISH · Abiotic stress tolerance · *Festulolium*

Introduction

The *Lolium* genus contains eight diploid ($2n = 2x = 14$) species, whereas the *Festuca* genus includes many more species and a polyploid series comprising species from diploid ($2n = 2x = 14$) to dodecaploid ($2n = 12x = 84$). Intermediate

M. Ghesquière (✉)
UR4 (Unité de Recherche Pluridisciplinaire Prairies et Plantes Fourragères), INRA, Centre Poitou-Charentes, Lusignan, F-86600, France
e-mail: Marc.Ghesquiere@lusignan.inra.fr

C. Huyghe (ed.), *Sustainable Use of Genetic Diversity in Forage and Turf Breeding*,
DOI 10.1007/978-90-481-8706-5_74,

forms between the two genera have long been recognized in nature and considered as hybrids (×*Festulolium* spp.) by taxonomists mostly on the basis of the inflorescence shape and their suspected progenitor species' combinations.

Lolium and *Festuca* share many complementary characters. In general, *L. perenne* and *L. multiflorum* are considered the optimal species for relatively intensive grassland agriculture as they provide high yields of nutritious forage. However, they lack resilience against abiotic stresses and it is primarily for this reason that the first artificial *Festuca* sp. × *Lolium* sp. crosses initiated by Jenkin (1933) stimulated breeding programmes with the release of *Festulolium* cultivars in the early 1970s (Lewis et al., 1973).

Until 2004, all hybrid cultivars derived from 6x *F. arundinacea* were registered in tall fescue national lists of the European Union while *Festulolium* cultivars were formally restricted only to the *L. multiflorum*× *F. pratensis* (= ×*Festulolium braunii* (K. Richter) A. Camus) hybrid combination (66/401/EEC & 92/19/EEC directives). In 2004, the definition of *Festulolium* was extended by the European Commissionn (2004/55/UE) *to include all hybrids resulting from the crossing of a species of a genus Festuca with a species of a genus Lolium* regardless of chromosome number and whether hybrids were intentionally backcrossed to one parental species. Consequently, *Festulolium* may at present include any amphiploid or introgression cultivars derived from any *Lolium*× *Festuca* hybrid combination.

Genetics

Although diploid *Lolium* plants when intercrossed with a diploid or polyploid *Festuca* species may produce F1 hybrids, these have low female, and very low or virtually no male fertility, which limits their use in crop improvement programmes. Strategies for restoration of fertility in F1 hybrids comprising entire sets of *Lolium* and *Festuca* chromosomes have involved polyploidy via application of colchicine. Once F1 amphiploid hybrids are recovered, these are sufficiently fertile to derive next generations by polycrossing under controlled conditions and without requiring further embryo rescue.

Festuca species used in *Festulolium* breeding (except *F. pratensis*) are themselves polyploid comprising homoeologous genome sets which inheritance, at least in *F. arundinacea* and in *F. a.* var. *glaucescens*, was shown to be of disomic nature with homologous pairing under genetic control (Jauhar, 1975; Lewis, 1980). However, the chromosome pairing control is hemizygous ineffective thereby allowing homoeologous *Festuca* chromosome to pair in polyploid hybrids as well as opportunities for interspecific recombination with *Lolium*. This explains how *Festulolium* breeding has been able to succesfully exploit the two contrasting interspecific genetic strategies, amphiploidization and introgression. *Festulolium* cultivar development thus far, has largely been through an amphiploidy approach using the greater chromosome pairing affinity of the disomic sets of homologous *Lolium* chromosomes that restrict instances of intergeneric chromosome pairing (Zwierzykowski et al., 2008). In contrast, introgression breeding approach has encouraged interspecific genome recombination through high frequencies of homoeologous chromosome pairing between *Lolium* and *Festuca* genomes (Humphreys, 1989; Humphreys and Ghesquière, 1994). Amongst all crop plants the *Lolium-Festuca* complex offers the greatest opportunity for homoeologous chromosome pairing (Jauhar, 1975; King et al., 1999; Naganowska et al., 2001; Kosmala et al., 2006a; Zwierzykowski et al., 2008) making the introgression breeding approach a promising strategy for future crop improvement initiatives.

Breeding

To date 41 cultivars have been quoted at least once as *Festulolium* in literature, of which 25 are registered in the OECD list of 2009. In many cases, the origin of cultivars is poorly referenced. Some cultivars among the very first *Festulolium* are no longer registered on the list (e.g., 'Prior' or 'Elmet') while *Festulolium* cultivars registered in former Eastern or Central Europe have been recently incorporated in the EU list following German reunification and the extension of the European Union to 27 countries.

Amphiploidisation

Twenty-three *Festulolium* cultivars, all tetraploids ($2n = 4x = 28$), have been bred using an amphiploidy approach (Fig. 74.1). They derive mostly from reciprocal hybrids of *L. multiflorum* × *F. pratensis*. Casler et al. (2001) reported the registration of cv 'Spring Green' derived from the intercrossing of four cultivars involving the two types of amphiploid hybrids: 'Elmet', 'Tandem', 'Kemal' (*L. multiflorum* × *F. pratensis*) and 'Prior' (*L. perenne* × *F. pratensis*).

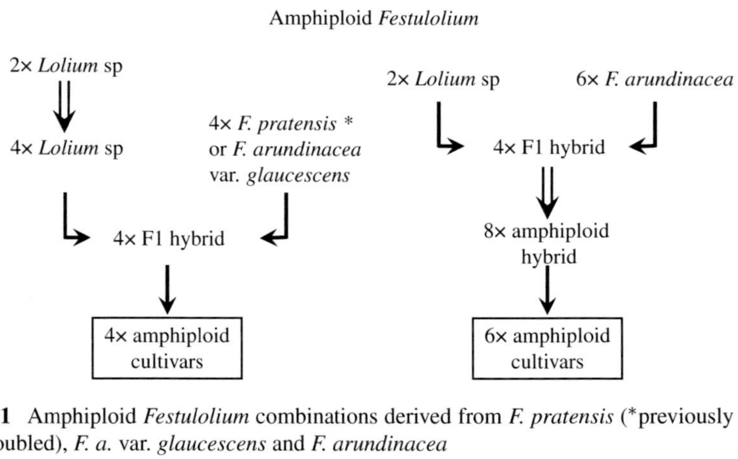

Fig. 74.1 Amphiploid *Festulolium* combinations derived from *F. pratensis* (*previously chromosome doubled), *F. a.* var. *glaucescens* and *F. arundinacea*

Approaches in development of *Festulolium* hybrids between *Lolium* and *F. arundinacea* have targeted the generation of 8x amphiploids by chromosome doubling 4× hybrids of *L. multiflorum*× *F. arundinacea* (Buckner et al., 1961; Zwierzykowski, 1980; Kleijer, 1984).

However, attempts at stabilizing chromosome number over generations failed to prevent rapid chromosome loss down to about $2n = 6x = 42$, at the expense of chromosomes of *Lolium* origin (Kleijer, 1987). The genome composition of these 6× cultivars is not documented but likely, it should not differ much from cultivars such as 'Hykor' and 'Felina' derived after back-crossing into tall fescue so that they should be considered more as introgressive forms than true *Lolium* × *Festuca* amphiploids.

From *F. arundinacea* var. *glaucescens*, only 4x amphiploids with *L. multiflorum* have been released as *Festulolium* cultivars (Ghesquière et al., 1993, 1996) and first registered on the French list in 2007 (e.g. cv 'Lueur'). Tetraploid F1 hybrids from *L. perenne* have also been successfully produced in the past (Zwierzykowski and Zwierzykowska, 1994) but they were of insufficient fertility for further breeding. True amphiploid ($2n = 6x = 42$) *L. multiflorum* × *F. a.* var. *glaucescens* were obtained only after chromosome doubling of the parental species. Although they were phenotypically similar to the 4x F1 hybrids and of potential agronomical value (Ghesquière et al., 1994), their chromosome number was quite unstable over generations similarly to the 8x amphiploids developed from *F. arundinacea*.

Introgression

This breeding approach has been practically investigated at all ploidy level into all parental species, except into *F. a.* var. *glaucescens*, from 3x, 4x or 5x F1 hybrids or advanced amphiploids, and by eventually incorporating an initial androgenesis stage (e.g. Zwierzykowski et al., 1998; Leśniewska et al., 2001; Kopecký et al., 2005), (Table 74.1). Eighteen cultivars have been developed following this approach, including the 6 cultivars 'Evergreen', 'Duo', 'Tandem', 'Barfest' 'Kemal' and 'Matrix' for which no publication reported whether they actually derived from back-cross. All are tetraploid except cv 'Matrix' which is diploid. Those cultivars were classified as introgression on the basis that they phenotypically appeared quite *Lolium*-like.

Kopecký et al. (2006) were unable to detect the presence of any *Festuca* chromatin introgression in cultivars 'Duo', 'Kemal' and 'Matrix' while Yonemaru et al. (2004) found that the rate of natural fluorescence of root tips (a highly *Lolium*-specific phenotypic trait) may range from 34 to 50% within these cultivars. *F. pratensis*-specific SSR markers were also evidenced in some of them by Momotaz et al. (2004).

Tetraploid F1 hybrids between 2x *L. multiflorum* and 6x *F. arundinacea* were extensively used by Fojtik (1994) for introgression into both parent species. Four cultivars were released following introgression into tetraploid *L. multiflorum*. However, again no *Festuca* chromatin was detected by GISH in cultivars 'Lofa' and

Table 74.1 *Festulolium* F1 (or amphiploid) hybrids used for introgression either into di- and tetraploid ryegrass or into tall fescue

Parent species combination	F1 hybrid genome composition[a]	Introgression[b] into:	
Lolium sp. (4x) × *F. pratensis* (2x)	LLC (3x)	*Lolium* sp. (2x)	
Lolium sp. (2x) × *F. pratensis* (4x)	LCC (3x)		*F. pratensis* (2x)
Lolium sp. (4x) × *F. glaucescens* (4x)	LLAB (4x)	*Lolium* sp. (2x or 4x)	
Lolium sp. (2x) × *F. arundinacea* (6x)	LABC (4x)	<u>*Lolium*</u> sp. (4x)	<u>*F. arundinacea*</u> (6x)
Lolium sp. (4x) × *F. arundinacea* (6x)	LLABC (5x)	*Lolium* sp. (2x)	

[a]L, C, A & B are the genomes of *Lolium* sp., *F. pratensis* and *F. a.* var. *glaucescens* (resp.), all A, B & C genomes being included in the 6x tall fescue *F. arundinacea.*
[b]Underlined are introgression ways having resulted into cultivars.

'Bečva', which consistently display a pronounced *Lolium*-like aspect for most traits. In the converse approach, i.e. introgression of *Lolium* into *F. arundinacea,* four cultivars 'Hykor', 'Felina', 'Lesana' and 'Korina' were bred in the Czech Republic. In this case, maintenance of the introgressed *Lolium* genome was found to be much higher with up to 7.5 intact *Lolium* chromosomes (e.g. cv 'Lesana') or up to 10.21 translocated chromosomes (e.g. cv 'Hykor') (Kopecký et al., 2006). Using double probing GISH, it was further confirmed that interspecific recombination between *L. multiflorum* and *F. arundinacea* genomes involved more frequently the *F. pratensis* related subgenome than those of *F. a.* var. *glaucescens* as reported previously (Humphreys and Ghesquière, 1994; Ghesquière et al., 2000).

Varietal groups

The genome classification of almost all current *Festulolium* cultivars by Kopecký et al. (2006) gave an evolutionary insight of *Festulolium* genetics (Fig. 74.2). It clearly appears that no present *Festulolium* cultivar achieves a perfectly balanced genome composition. All 6x *Festulolium* cultivars from *F. arundinacea* have intact *Festuca* chromosomes in excess while 4x *Festulolium* display predominantly translocated chromosomes mixed with intact *Lolium* chromosomes. Versatility of present *Festulolium* cultivars also gives evidence for the role of the *Festuca* species used initially as parent and of the number of generations by which *Festulolium* cultivars are from the primary F1 hybrids. The most genome-balanced *Festulolium* and also the most stable over generations, most closely resembling the F1 hybrid, was the only available amphiploid *Festulolium* derived from *F. a.* var. *glaucescens* (cv 'Lueur') while all amphiploid cultivars from *F. pratensis* indicate a significant and progressive *Festuca* chromosome loss. This was further demonstrated over 6

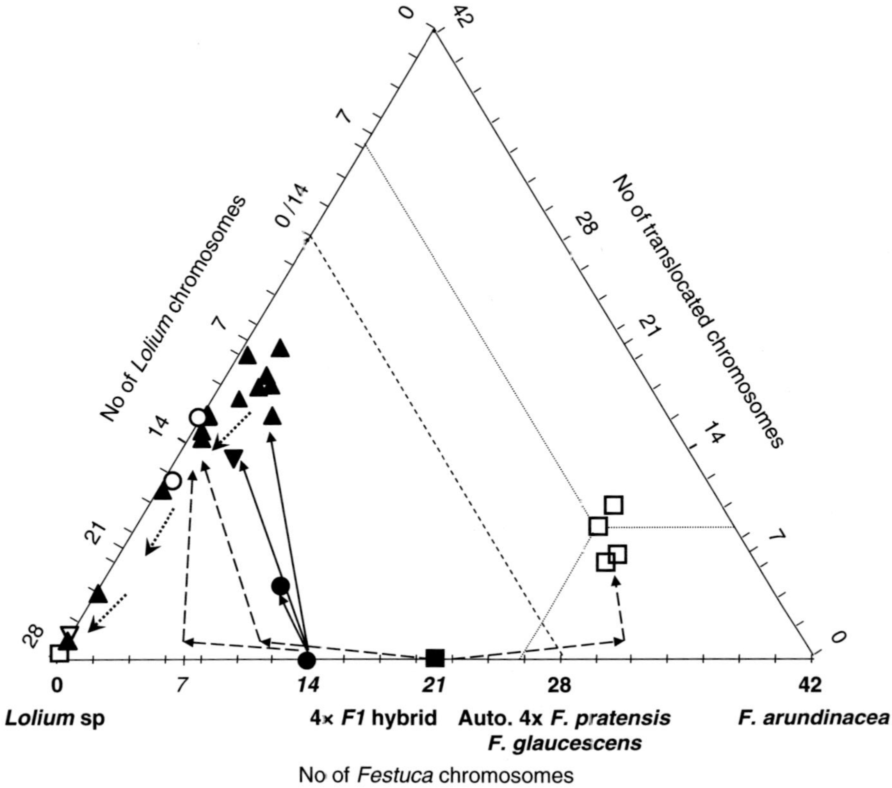

Fig. 74.2 Genome composition and evolution in present *Festulolium* cultivars. Solid symbols are 4x-amphiploid cultivars or F1 hybrids from *L. multiflorum* × *F. arundinacea* (■), *L. multiflorum* × *F. pratensis* (▲) and reciprocals, *L. perenne* × *F. pratensis* (▼) and *L. multiflorum* × *F. a.*var. *glaucescens* (•). Open symbols refer to cultivars resulting from introgression of corresponding amphiploid hybrids (same symbol) into *Lolium* sp. or *F. arundinacea*. Solid and large dotted arrows are the way of genome evolution following amphiploidisation or introgression (resp.). Small dotted arrows indicate the *Lolium*-genome drift observed in advanced generations of 4x-*Festulolium* or from introgression (adapted from Kopecký et al., 2006)

successive generations of breeding a same *F. pratensis* × *L. perenne* population by Zwierzykowski et al. (2006). The loss of *Festuca* chromosomes is even more dramatically enhanced after backcrossing into 4x *Lolium* sp., or adding further generations of breeding and/or of seed multiplication in amphiploid *Festulolium*, so that eventually no more *Festuca* genome is evidenced by GISH as if introgression has been performed straight into diploid ryegrass (e.g. cv 'Matrix').

Genome composition in *Festulolium* is likely associated with variability of fertility within cultivar and selection mechanisms over generations of seed multiplication. The drift to *Lolium* genome could be also linked to the fact that *Festulolium* hybrids having a disomic *Lolium* genome complement are generally more fertile than those having an alternative disomic *Festuca* complement. Current researches in IBERS and IPG/PAS based on the use of populations derived in Poland from dihaploid

F. pratensis/ L. multiflorum genomes produced by androgenesis, demonstrate clear evidence for a meiotic drive that favoured transmission of *Lolium* compared to *Festuca* chromosomes.

From an end-user viewpoint the current practice for all available *Festulolium* cultivars to all share the same definition whilst ranging widely in their genome composition and overall value of utilisation may be confusing. In the future, registration of *Festulolium* cultivars onto national lists could usefully rely on the discrimination between 'Festucoid' and 'Lolioid' types of *Festulolium* as initially suggested by Fojtik (1994). In this way, all 6x-*Festulolium* cultivars which necessarily involve a predominant genome contribution of tall fescue should be classified as 'Festucoid' while all cultivars resulting from introgression into *Lolium* sp. as well as 4x-amphiploids of advanced generation should be classified as 'Lolioid'.

Achievements

The achievements and overall benefit of *Festulolium* are still difficult to identify because of the unusual range of variability for most traits under assessment and of non-standardisation of protocols and regulations in official trials. Likely, *Festulolium* breeding can be illustrated at the best by gains in climatic adaptation and better understanding of abiotic stress tolerance. Generally, 6x-*Festulolium* cultivars as 'Felina' and 'Hykor' are more winter tolerant than amphiploid cultivars derived from *F. pratensis* (e.g., 'Felopa' and 'Perun') (Østrem and Larsen, 2008). The *Festulolium* cv 'Felina' was found also to be the most summer tolerant in Japan among various *Festulolium* derived from *F. pratensis* (Ushiyama et al., 2004). This suggests that *Festulolium* cultivars derived from *F. arundinacea* could be of wider range of climatic adaptation because they accumulate sources of abiotic stress tolerance from both *F. pratensis* and *F. a.*var. *glaucescens* genomes included in tall fescue. However, as shown by Kopecký et al. (2006), *Festulolium* derived from *F. arundinacea* have an overall *Festuca* genome contribution much higher than those derived from *F. pratensis*.

The comparison over contrasted climates carried out in the SAGES project of the 5th PCRDT of the EU enabled a better discrimination of the contribution of each *Festuca* species as progenitor to climatic adaptation of amphiploid cultivars (SAGES, 2004). *Festulolium* involving *F. pratensis* were found to perform better than alternative *Festulolium* hybrids containing *F. a.* var. *glaucescens* under continental climate with cold winter. Conversely, in field trials under south oceanic climate that requires an enhanced summer drought tolerance, the amphiploid cultivar derived from *F. a.* var. *glaucescens* (e.g. 'Lusilium') had better yield and persistency after summer than *Festulolium* from *F. pratensis*. As respects to quality, 6x-*Festulolium* cultivars from *F. arundinacea* demonstrated significant improvements in digestibility and palatability compared to tall fescue (Berg et al., 1979; Buckner et al., 1979). Early assessment of digestibility in the amphiploid population *L. multiflorum* × *F. a.* var. *glaucescens* having led to cv 'Lusilium' showed that it was at an intermediate level between tall fescue and Italian ryegrass with potential for response of selection into either direction using a palatability test with animals (Ghesquière et al., 1996). In conclusion, rather than better performances of single traits relative to their parental species, the benefit of *Festulolium* would be better assessed by achieving an optimum balance in productivity, quality and stress tolerance that is likely to be unreachable by intraspecific conventional breeding.

Fertility and seed productivity are crucial aspects of breeding *Festulolium*. Generally, first generations of selection for fertility components were found significant, likely in relation with natural selection for more regular meiosis (Ghesquière et al., 1993; Zwierzykowski et al., 1993). However, correlation of seed yield between mother-plant and HS-progeny remained quite low with generally no response over generations of selection for seed yield (Chapter 78). Real seed yield assessment of *Festulolium* cultivars through comparative trials in plots is very little documented. Seed yield of early 6x-hybrids derived from *F. arundinacea* remained low and not liable to improvement, although genetic variability was found (Burner et al., 1991). This probably explains why cultivars like 'Kenhy' or 'Johnstone' were not widely marketed following early release in the USA in the 1980s. When actually resulting from backcross into tall fescue, *Festulolium* cultivars, as 'Hykor' and 'Felina', appear to be of more acceptable seed yield. Seed yield potential is especially well restored when introgression is performed into *L. multiflorum*; cultivars as 'Bečva' and 'Lofa' are currently acknowledged to be the most steadily and highly seed productive *Festulolium* in Europe and New-Zealand, yielding about 10 dt/ha on average.

Prospects

Festulolium breeding has considerably stimulated research on genetics of the grasses and has contributed to the development of new technologies. This has totally renewed the breeding introgression approach, from conventional cytogenetics and isozymes (e.g. Humphreys and Ghesquière, 1994) through use of GISH methods and to more and more diversified molecular markers, to AFLPs and STS markers (e.g. Humphreys et al., 2005), to microsatellites (e.g. Momotaz et al., 2004; Lauvergeat et al., 2005; King et al., 2008), to SNP development (King et al., 2008) and recently the employment of DArT markers within the *Lolium-Festuca* complex (Kopecký et al., 2009a, Chapter 65).

Introgression research incorporated the strategy developed by Morgan et al. (1988) and Humphreys (1989), that is, (i) by using a 3x – 4x or 5x *Lolium* × *Festuca* hybrid and its derivatives as male parent over one or two backcross generations into diploid *Lolium* sp., (ii) by phenotypically selecting among BC2 or BC3 progenies those plants having survived under extreme conditions of temperature or drought and, (iii) by screnning them for *Festuca*-specific chromosome fragment and associate markers. All *Festuca* species have been employed thus far primarily as sources of genes for drought resistance and winter hardiness and coincident evidences are being accumulated with joint QTLs mapping within *Festuca* donor species and/or orthology from sequence data in rice.

Introgression of *Festuca* chromosome 3 and chromosome 5 originating from either *F. arundinacea* or *F. a.* var. *glaucescens* were found to correlate with enhanced drought resistance in the EU funded project SAGES (SAGES, 2004; Humphreys et al., 2005), consistently with marker associations found recently on chromosome 3F of *F. pratensis* with QTL (*QDts3F*) for growth under severe drought stress. Chromosome 3F is orthologous to rice chromosome 1 and is known to have QTL associations with osmotic adjustment, dehydration tolerance and numerous rooting characters (Champoux et al., 1995; Lilley et al., 1996; Yadav et al., 1997; Kamoshita et al., 2008). Durand et al. (2007) demonstrated how *L. multiflorum* × *F. a.* var. *glaucescens* hybrids due to their enhanced rooting capabilities, are able to extract water from greater soil depths than the ryegrass parent when plants are exposed to severe drought stress.

Similarly, introgression of winter hardiness traits into *Lolium* sp. was found to involve gene transfers from *Festuca* onto *Lolium* chromosome 3 (Canter, 2000; Grønnerød et al., 2004), chromosome 2 (Kosmala et al., 2006b, 2007) and chromosome 4 (Humphreys et al., 2006), each leading to significant improvements to *Lolium* in freezing tolerance. New evidence of the importance in *Lolium* and *Festuca* species of the adaptive capabilities of Photosystem II (PSII) during cold acclimation in relation to subsequent freezing-tolerance was demonstrated with a role for genes found on chromosome 4 of *F. pratensis* for increased Non Photochemical Quenching (NPQ) expression. This is consistent with localization of the vernalisation *Vrn-1* locus on chromosome 4 in *Lolium* (Jensen et al., 2005) and its alignement with the regulatory genes for increased NPQ expression. C-repeat binding factor (CBF) family genes also map in this region and are known as important regulators of cold-tolerance in *Arabidopsis* (Thomashow et al., 2001). Yamada et al. (2005) concurred that the distal end of chromosome 4 is a primary site for QTL for frost tolerance.

Conclusion

Festulolium provide specialist functions and novel alternatives to existing grass cultivars that may either lack the quality of *Lolium* or their resilience against abiotic or biotic stresses. They may be viewed as possible alternatives to the use of seed mixtures, or for a specialist use. In addition to the benefits for improved climatic stress tolerance, *Festulolium* breeding also opens new prospects to better address environmental issues, by, for example, emphasizing the role of rooting systems (e.g. MacLeod et al., 2007).

The future of amphiploid *Festulolium* depends closely on a complicated balance between improvement of value of utilisation for end-users and seed productivity for marketing. In this respect, to find the right genome balance between *Lolium* and *Festuca* will be a definite challenge for future breeding of polyploid *Festulolium*. Obviously, there is still a gap between the breeding amphiploidy approach and future precision breeding aimed at the transfer and introgression of selected genes into diploid *Lolium* sp. The generation of complete sets of chromosome substitution lines from *Lolium/Festuca* hybrids now exist both in diploid populations (King et al., 2002), and in tetraploids (Kopecký et al., 2008) and will assist greatly in genomic and phenomic screens to identify key alleles and for map-based cloning and marker-assisted gene transfer for plant breeding. It is not unrealistic that breeding amphiploid *Festulolium* will benefit from the genomic advances achieved in diploids if abiotic stress tolerance is genetically controlled by many co-adapted genes, physically and functionally organized at the scale of the chromosomes.

References

Berg, C.C., Hill, R.R. Jr., Buckner, R.C., Barnes, R.F. 1979. Forage production and quality of synthetics derived from *Lolium* × *Festuca* hybrids. Crop Sci. 19:89–93.

Buckner, R.C., Hill, H.D., Burrus, II, P.B. 1961. Some characteristics of perennial and annual ryegrass × tall fescue hybrids and of the amphiploid progenies of annual ryegrass × tall fescue. Crop Sci. 3:75-80.

Buckner, R.C., Bush, L.P., Burrus, P.B., Jr. 1979. Succulence as a selection criterion for improved forage quality in *Lolium-Festuca* hybrids. Crop Sci. 19:93–96.

Burner, D.M., Eizenga, G.C., Buckner, R.C., Burrus, P.B., Jr. 1991. Genetic variability of seed yield and agronomic characters in *Festuca* hybrids and amphiploids. Crop Sci. 31:56–60.

Canter, P.H. 2000. The use of genomic in situ hybridisation (GISH) to locate introgressed chromosome segments from winter-hardy *Festuca* in a cold-sensitive *Lolium* background. Newsletter Genet. Soc., September 2000:29–30.

Casler, M.D., Pitts, P.G., Rose-Fricker, C., Bilkey, P.C., Wipff, J.K. 2001. Registration of "Spring Green" *Festulolium*. Crop Sci. 41:1365–1366.

Champoux, M.C., Wang, G., Sarkarung, S., Mackill, D.J., O'Toole, J.J., et al. 1995. Locating genes associated with root morphology and drought avoidance in rice via linkage to molecular markers. Theor. Appl. Genet. 90:969–981.

Durand, J.L., Bariac, T., Ghesquière, M., Biron, P., Richard, P., Humphreys, M.W., Zwierzykowski, Z. 2007. Ranking of the depth water extraction by individual grass plants using natural ^{18}O isotope abundance. Environ. Exp. Bot. 60:137–144.

Fojtik, A. 1994. Methods of grass improvement used at the plant breeding station Hladké Životice. Genet. Pol. 35A:25–31.

Ghesquière, M., Zwierzykowski, Z., Poisson, C., Jadas-Hécart, J. 1993. Amphitetraploid *Festulolium*: chromosome stability and fertility over intercrossing generations (pp. 451–453). In: Proceedings of the XVIIth International Grassland Congress, Palmerston North, New Zealand, February 1993.

Ghesquière, M., Mi, F., Hazard, L., Poisson, C. 1994. Leaf growth genetic variability among various polyploid ryegrass × fescue hybrids involving *Festuca arundinacea* var. *glaucescens* (pp. 293–294). In: Rognli, O.A., Solberg, E., Schjelderup, I. (eds.), Breeding Fodder Crops for Marginal Conditions, Proceedings of the 18th Eucarpia Fodder Crops Section Meeting, Loen, Norway, 25–28 August 1993. Kluwer Academic Publishers, Dordrecht.

Ghesquière, M., Emile, J-C., Jadas-Hécart, J., Mousset, C., Traineau, R., Poisson, C. 1996. First in vivo assessment of feeding value of *Festulolium* hybrids derived from *Festuca arundinacea* var. *glaucescens* and selection for palatability. Plant Breed. 115:238–244.

Ghesquière, M., Barre, P., Marhadour, S., Kerlan, M.C. 2000. Estimation of introgression rate of a fescue isozymic marker into tetraploid Italian ryegrass at early generations of backcross. Euphytica 114:223–231.

Ghesquière, M., Bourgoin, T. 2009. Seed yield of new *Festulolium* varieties bred from *F. glaucescens*. This volume.

Grønnerød, S., Fjelldheim, S., Grieg, Z., Jørgensen, Ø., Larsen, A., Østrem, L., Humphreys, M.W., Rognli, O.A. 2004. Application of AFLP and GISH techniques for identification of *Festuca* chromosome segments conferring winter hardiness in a *Lolium perenne* × *Festuca pratensis* population. In: Hopkins, A., Wang, Z.Y., Mian, R., Sledge, M., Backer, R.E. (eds.), Molecular Breeding of Forage and Turf, Developments in Plant Breeding 11:81–86.

Humphreys, J., Harper, J.A., Armstead, I.P., Humphreys, M.W. 2005. Introgression-mapping of genes for drought resistance transferred from *Festuca arundinacea* var. *glaucescens* into *Lolium multiflorum*. Theor. Appl. Genet. 110:579–787.

Humphreys, M.W. 1989. The controlled introgression of *Festuca arundinacea* genes into *Lolium multiflorum*. Euphytica 42:105-116.

Humphreys, M.W., Ghesquière, M. 1994. Assessing success in gene transfer between *Lolium multiflorum* and *Festuca arundinacea*. Euphytica 77:283–289.

Humphreys, M.W., Gąsior, D., Leśniewska-Bocianowska, A., Zwierzykowski, Z., Rapacz, M. 2006. Androgenesis as a means of dissecting complex genetic and physiological controls: selecting useful gene combinations for breeding freezing tolerant grasses. Euphytica 158: 337–345.

Jauhar, P.P. 1975. Chromosome relationships between *Lolium* and *Festuca* (Gramineae). Chromosoma (Berl.) 52:103-121.

Jenkin, T.J. 1933. Interspecific and intergeneric hybrids in herbage grasses. Initial crosses. J. Genet. 28:205-264.

Jensen, L.B., Andersen, J., Frei, U., Xing, Y., Taylor, C., Holm, P.B., Lübberstedt, T. 2005. QTL mapping of vernalization response in perennial ryegrass reveals co-segregation with an orthologue of wheat *Vrn1*. Theor. Appl. Genet. 110:527–536.

Kamoshita, A., Babu, RC., Boopathi, N.M., Fukai, S. 2008. Phenotypicand genotypic analysis of drought-resistance traits for development of rice cultivars adapted to rainfed environments. Field Crops Res. 109:1–23.

King, I.P., Morgan, W.G., Harper, J.A., Thomas, H.M. 1999. Introgression mapping in the grasses. II. Meiotic analysis of the *Lolium perenne/Festuca pratensis* triploid hybrid. Heredity 82: 107–112.

King, J., Armstead, I.P., Donnison, I.S., Thomas, H.M., Jones, R.N., Kearsey, M.J., Roberts L.A., Thomas, A., Morgan, W.G., King, I.P. 2002. Physical and genetic mapping in the grasses *Lolium perenne* and *Festuca pratensis*. Genetics 161:315–324.

King, J., Thorogood, D., Edwards, K.J., Armstead, I.P., Roberts, L.A., Skøt, K., Hanley, Z., King, I.P. 2008. Development of a genomic microsatellite library in perennial ryegrass (*Lolium perenne*) and its use in trait mapping. Ann. Bot. 101:845–853.

Kleijer, G. 1984 Cytogenetic studies of crosses between *Lolium multiflorum* Lam. and *Festuca arundinacea* Schreb. I. The parents and the F_1 hybrids. Z. Pflanzenzüchtg. 93:1-22.

Kleijer, G. 1987. Cytogenetic studies of crosses between *Lolium multiflorum* Lam. and *Festuca arundinacea* Schreb. III. The generations C_1, C_2 and C_3. Plant Breed. 99:144-150.

Kopecký, D., Lukaszewski A.J., Gibeault, V. 2005. Reduction of ploidy levels by androgenesis in intergeneric *Lolium-Festuca* hybrids for turf grass breeding. Crop Sci. 41:274–281.

Kopecký, D., Loureiro, J., Zwierzykowski, Z., Ghesquière, M., Doležel, J. 2006. Genome constitution and evolution in *Lolium* × *Festuca* hybrid cultivars (*Festulolium*). Theor. Appl. Genet. 113: 731–742.

Kopecký, D., Lukaszewski, A.J., Doležel, J. 2008. Meiotic behaviour of individual chromosomes of *Festuca pratensis* in tetraploid *Lolium multiflorum*. Chromosome Res. 16:987–998.

Kopecký, D., Kilian, A., Lukaszewski, A.J., Bartos, J., Baird, J.H., Cernoch, V., Blois, H., Caig, V., Doležel, J. 2009a. Development and mapping of DArT markers within the *Festuca-Lolium* complex. In: Proceedings of the XVIIth Intern. Plant and Animal Genome Conference, San Diego, CA, USA, January 10–14, 2009.

Kopecký, D., Bartoš, I., Lukaszewski, A.J., Baird, J.H., Černoch, V., Koelliker, R., Rognli, O.A., Blois, H., Caig, V., Doležel, J., Kilian, A. 2009b. DArTFest – a platform for high-throughput genome profiling within the *Festuca–Lolium* complex. This volume.

Kosmala, A., Zwierzykowska, E., Zwierzykowski, Z. 2006a. Chromosome pairing in triploid intergeneric hybrids of *Festuca pratensis* with *Lolium multiflorum* revealed by GISH. J. Appl. Genet. 47:215–220.

Kosmala, A., Zwierzykowski, Z., Gąsior, D., Rapacz, M., Zwierzykowska, E., Humphreys, M.W. 2006b. GISH/FISH mapping of genes for freezing tolerance transferred from *Festuca pratensis* to *Lolium multiflorum*. Heredity 96:243–251.

Kosmala, A., Zwierzykowski, Z., Zwierzykowska, E., Łuczak, M., Rapacz, M., Gąsior, D., Humphreys, M.W. 2007. Introgression-mapping of the genes for winter hardiness and frost tolerance from *Festuca arundinacea* into *Lolium multiflorum*. J. Heredity 98:311–316.

Lauvergeat, V., Barre, P., Bonnet, M., Ghesquière, M. 2005. Sixty simple sequence repeat markers for use in the *Festuca-Lolium* complex of grasses. Mol. Ecol. Notes 5:401–405.

Leśniewska, A., Ponitka, A., Slusarkiewicz-Jarzina, A., Zwierzykowska, E., Zwierzykowski, Z., James, A.R., Thomas, H., Humphreys, M.W. 2001. Androgenesis from *Festuca pratensis* × *Lolium multiflorum* amphidiploid cultivars in order to select and stabilise rare gene combinations for grass breeding. Heredity 86:167–176.

Lewis, E.J., Tyler, B.F., Chorlton, K.H. 1973. Development of *Lolium-Festuca* hybrids (pp. 34–37). In: Report Welsh Plant Breeding Station for 1972.

Lewis, E.J. 1980. Chromosome pairing in tetraploid hybrids between *Lolium perenne* and *L. multiflorum*. Theor. Appl. Genet. 58:137-143.

Lilley, J.M., Ludlow, M.M., McCouch, S.R., O'Toole, J.C. 1996. Locating QTL for osmotic adjustment and dehydration tolerance in rice. J. Exp. Bot. 47:1427–1436.

Macleod, C.J.A., Binley, A., Hawkins, S.L., Humphreys, M.W., Turner, L.B., Whalley, W.R., Haygarth, P.M. 2007. Genetically modified hydrographs: what can grass genetics do for temperate catchment hydrology? Hydrol. Process. 21:2217–221.

Momotaz, A., Forster, J.W., Yamada, T. 2004. Identification of cultivars and accessions of *Lolium*, *Festuca* and *Festulolium* hybrids through the detection of simple sequence repeat polymorphism. Plant Breed. 123:370–376.

Morgan, W.G., Thomas, H., Lewis, E.J. 1988. Cytogenetic studies of hybrids between *Festuca gigantea* Vill. and *Lolium multiflorum* Lam. Plant Breed. 101:335-343.

Naganowska, B., Zwierzykowski, Z., Zwierzykowska, E. 2001. Meiosis and fertility of reciprocal triploid hybrids of *Lolium multiflorum* with *Festuca pratensis*. J. Appl. Genet. 42:247–255.

Østrem, L., Larsen, A. 2008. Winter survival, yield performance and forage quality of *Festulolium* cvs. for Norwegian farming. In: Proceedings of the 22nd General Meeting of the European Grassland Federation, Uppsala, Sweden, 9–12 June 2008.

SAGES. 2004. Sustainable Grasslands Withstanding Environmental Stresses, 5th PCRDT shared cost project QLK5-CT-2000-00764 for 2001-2003. Key-Action 5.1.1. Sustainable Agriculture, Technological Implementation Plan. European Commission, Brussels, Belgium, 42 p. (http://www.sages-eu.co.uk/)

Thomashow, M.F., Gilmour, S.J., Stockinger, E.J., Jaglo-Ottosen, K.R., Zarka, D.G. 2001. Role of the *Arabidopsis* CBF transcriptional activators in cold acclimation. Physiol. Plant 112:171–175.

Ushiyama, K., Arakawa, A., Komatsu, T. 2004.Breeding and evaluation of *Festulolium* cultivars in warm region of Japan (pp. 69–74). In: Yamada, T., Takamizo, T. (eds.), Development of a Novel Grass with Environmental Stress Tolerance and High Forage Quality Through Intergeneric Hybridization Between *Lolium* and *Festuca*. National Agriculture and Bio-oriented Research Organization, Tsukuba, Japan.

Yadav, R., Courtois, B., Huang, N., McLaren, G. 1997. Mapping genes controlling root morphology and root distribution on a double-haploid population of rice. Theor. Appl. Genet. 94:619–632.

Yamada, T., Forster, J.W., Humphreys, M.W., Takamizo, T. 2005. Genetics and molecular breeding in *Lolium/Festuca* grass species complex. Grassl. Sci. 51:89–106.

Yonemaru, J., Kubota, A., Ueyama, Y. 2004. Individual variation and selection effectiveness on regrowth after summer of the *Festulolium* cultivars in cold climates. Grassl. Sci. 50:415–420.

Zwierzykowski, Z. 1980. Hybrid of *Lolium multiflorum* Lam. ($2n = 14$) × *Festuca arundinacea* Schreb. ($2n = 42$) and its alloploid derivatives. I. Morphology, fertility and chromosome number of F_1 hybrids and C_0 and C_1 derivatives. Genet. Pol. 21:259–273.

Zwierzykowski, Z., Jokś, W., Naganowska, B. 1993. Amphitetraploid hybrids *Festuca pratensis* Huds. × *Lolium multiflorum* Lam. [= ×*Festulolium braunii* (K. Richter) A. Camus)]. Biuletyn IHAR 188:61–69.

Zwierzykowski, Z., Zwierzykowska, E. 1994. Intergeneric hybridization within the *Lolium-Festuca* complex.Genet. Pol. 35A:65–71.

Zwierzykowski, Z., Lukaszewski, A.J., Leśniewska, A., Naganowska, B. 1998. Genomic structure of androgenic progeny of pentaploid hybrids *Festuca arundinacea* × *Lolium multiflorum*. Plant Breed. 117:457–462.

Zwierzykowski, Z., Kosmala, A., Zwierzykowska, E., Jones, N., Jokś, W., Bocianowski, J. 2006. Genome balance in six successive generations of the allotetraploid *Festuca pratensis* × *Lolium perenne*. Theor. Appl. Genet. 113:539–547.

Zwierzykowski, Z., Zwierzykowska, E., Taciak, M., Jones, N., Kosmala, A., Krajewski, P. 2008. Chromosome pairing in allotetraploid hybrids of *Festuca pratensis* × *Lolium perenne* revealed by genomic in situ hybridization (GISH). Chromosome Res. 16:575–585.

Chapter 75
Creation of Heterotic Groups and Hybrid Varieties

Carla Scotti and E. Charles Brummer

Abstract Capturing heterosis for dry matter yield in forage crops needs to consider both the final product of the breeding process (the type of cultivar) and the breeding method used to produce the cultivar. The discussion is focused specifically on alfalfa and on the semi-hybrid variety model allowing partial expression of heterosis with the technology currently in use and no drawbacks to seed production. The search for heterotic groups is a key point to improve the process of semi-hybrid construction. Among-subspecies diversity (i.e., *Medicago sativa* subsp. *sativa* and *falcata*) and selfing within subspecies are reviewed as tools to manage genetic diversity and to test the value of gene and linkat interactions in crosses. Different breeding methods have been proposed to produce semi-hybrid varieties. Experimental results on heterosis expression and yield gain of semi-hybrids vs the corresponding synthetic variety or elite cultivar are discussed in pointing out the basis of heterotic performance and in defining what are the key points in method and selection procedures to effectively exploit heterosis.

Keywords Alfalfa · Heterosis · Semi-hybrid variety · Yield

Introduction

Yield improvement in forage crops has been slower than in other species, being estimated at 0.15–0.30% per year since the middle of the past century in alfalfa (Holland and Bingham, 1994). Besides the difficulties connected to the characteristics of most forage plants – allogamy, polyploidy, hermaphroditic flowers, plant architecture, and meadow conditions – few studies have focused on the variety construction process (Rotili et al., 1999a). Heterosis, the superiority of progeny relative to parents, is

C. Scotti (✉)
CRA-FLC, Viale Piacenza 29, Lodi 26900, Italy
e-mail: carla.scotti@entecra.it

C. Huyghe (ed.), *Sustainable Use of Genetic Diversity in Forage and Turf Breeding*,
DOI 10.1007/978-90-481-8706-5_75, © Springer Science+Business Media B.V. 2010

partially responsible for the yield gains in many crops. Dominant gene action is generally accepted as the main cause of heterosis, so that different dominant alleles carried by the two genotypes or populations being crossed will complement each other at different heterozygous loci (or linkats) affecting the same trait by masking recessive deleterious alleles. A greater number of complementary gene interactions are possible in tetraploids than in diploids (Bingham et al., 1994).

Capturing heterosis for dry matter yield in forage crops needs to take into account both the type of cultivar ultimately produced and the breeding method used to produce the cultivar. The way to use genetic diversity in variety construction process is of particular importance, as the expression of heterosis depends on (i) directional dominance at loci controlling the trait of interest and (ii) difference in allele frequency at those loci in the populations to be crossed (Brummer, 1999). However, genetic diversity per se is not sufficient for heterosis expression (Kidwell et al., 1999; Riday et al., 2003); the divergent populations must also be 'good combiners' and testing for this aptitude has to be included in the variety construction process. The identification of putative heterotic groups giving consistent and repeated heterotic effects when individuals or populations from these groups are crossed would improve the efficiency of the process by targeting specific populations.

Hybrid or Semi-Hybrid?

Two cytoplasmic-genic types of male sterility have been described and used for the experimental production of alfalfa hybrids in the US and Europe (Pedersen and Stucker, 1969; Staszewski, 1979). The genetic and environmental stability of the male-sterility trait tends to be unreliable, insects may avoid the male sterile genotypes, and low seed yield of male steriles is often obtained; these problems have hampering the widespread use of this technology for the production of commercial hybrids. However, since 2001, commercial alfalfa hybrids with at least 75% hybrid seed have been released in US based on a patented male sterile technology (US Patent 6774280).

As an alternative approach to capture at least partial heterosis in commercial cultivars, various methods using population hybrids have been proposed, variously termed 'semi-hybrids' (Brummer, 1999), 'free-hybrids' (Rotili et al., 1996), or 'chance hybrids' (Bingham, pers. comm.). In this variety model, the probability of within-population and among-population crossing is theoretically the same, and therefore these cultivars will express some level of heterosis using seed production technology currently in use. Due to the large number of complementary gene interactions in tetraploids, semi-hybrids formed from a cross of two populations, though displaying only one-half hybrid progeny, can still express high parent heterosis (Brummer, 1999). Besides, competition for survival at different levels (gametes, developing zygotes, plants in dense sward conditions) is likely to act in semi-hybrids in favour of hybrids, thus reducing the actual inbreeding coefficient compared to the theoretical expectation. Positive selection for desirable traits can be practised by the breeder during the different phases of variety development reducing the potential for inbreeding depression (Rotili, 1976). The semi-hybrid variety model can be a valid alternative to either true hybrids or synthetics, provided that a yield advantage is attainable by the semi-hybrid and that seed production costs are not significantly different among methods.

Heterotic Groups

A prerequisite for the expression of heterosis is the difference in allelic frequencies at loci controlling the traits of interest in the populations or genotypes to be crossed. Typically, breeders can explore extant germplasm sources initially to determine if they possess complementary genetic structures to realize heterosis in their progeny. The structure of genetic diversity in the *Medicago sativa* complex has been considered in order to identify putative heterotic groups. An important subdivision of genetic diversity in alfalfa is in relation to subspecies (*falcata* vs. *sativa*). The cultivated alfalfa pool of the Mediterranean basin, Northern Europe, and the US are largely based on subsp. *sativa*, even though *falcata* sequences have been incorporated over time because of recent common ancestry, gene flow, the use of subsp. *falcata* in alfalfa breeding and the expansion of cultivated alfalfa in areas where subsp. *falcata* was native (Muller et al., 2005). Thus, the alfalfa continental cultivated pool can be considered in a broad sense to be *Medicago* × *varia*, i.e., natural hybrids between subsp. *sativa* and *falcata* and/or *sativa* differently introgressed by *falcata*. It is likely that 'pure' subsp. *falcata* and *sativa* could display high levels of genetic diversity and possible heterotic effects towards the alfalfa cultivated pool. In fact, in a diallel crossing among elite *sativa* genotypes and semi-improved *falcata* populations Riday and Brummer (2002a) found a clear heterotic pattern for yield in *sativa* × *falcata* hybrids with an average heterosis value of 18%; none of the *sativa* × *sativa* crosses had significant deviations from the additive expectation. Despite heterosis for forage yield, however, *sativa* × *falcata* crosses performed only intermediately for important agronomic traits, including regrowth rate, growth habit, height, and maturity (Riday and Brummer, 2002b). Besides, the heterotic pattern *sativa* × *falcata* was not generalized over a broad range of wild and semi-improved *falcata* germplasm but was dependent on the agronomic value of the *falcata* genotype (Riday and Brummer, 2005) and on the level of *falcata* introgression into the elite *sativa* genotype (Riday and Brummer, 2002a).

While *falcata* germplasm presently at disposal of breeders is mainly formed by wild or semi-improved populations of poor agronomic value, *sativa* has a long history of cultivation and practical breeding. Non-dormant germplasm is likely the candidate for 'pure' *sativa* germplasm in contrast to the northern distribution and adaptation to cold climates of *falcata*. Different germplasm sources can be found in the non-dormant *sativa* pool: both Peruvian and African germplasms were distinguished from *sativa* and *varia* clusters on the basis of molecular analyses (Segovia-Lerma et al., 2003; Maureira et al., 2004). Furthermore, Peruvian germplasm showed heterotic effects in diallel crossings with the other nine germplasm sources introduced to North America and with cultivar testers (Segovia-Lerma et al., 2004;

Maureira et al., 2004). Non-dormant germplasm can contribute towards interesting bio-agronomic traits (high regrowth rate, extended growing season) that are expressed mainly additively (Rotili et al., 1999b; Segovia-Lerma et al., 2004). However, winter tolerant populations developed from non-dormant cultivars of Southwestern US by phenotypic recurrent selection showed no heterotic pattern when crossed to elite semidormant *sativa* from the Midwestern states, the hybrids between groups of dormancy being intermediate to the parental populations or to the average of hybrids within groups of same dormancy (Şakiroğlu and Brummer, 2007). At the Lodi Institute (Italy), we are using Egyptian alfalfa germplasm from a Sahara oasis (Carelli et al., 2009) as putative 'pure' ssp. *sativa* in diallel crosses with partly inbred, improved families from Italian ecotypes, but the hybrid progeny have not been evaluated yet.

Heterosis in *sativa* × *sativa* hybrids between existing elite populations has seldom been found in the experiments previously reported. A likely reason for this situation is the intermixing and relatedness of the continental breeding pool (Flajoulot et al., 2005). Similar suggestions have been made for US germplasm (Barnes et al., 1977).

The simple crossing of genetically distinct populations of high agronomic value is not sufficient to express consistent heterosis; a breeding method taking into account the improvement of parents and of their combining abilities seems necessary to identify *de novo* heterotic groups. Instead of relying on identification of pre-existing populations that combine well together, reciprocal recurrent selection can be used to create 'heterotic groups' *de novo*. This strategy could be more effective than the simple use of subspecies diversity especially when adapted populations of high agronomic value from the divergent groups are not available. A breeding program using reciprocal recurrent selection can begin with two different elite populations or even a single population that is subdivided in two. Families are formed by crossing between groups for evaluation, but recombination between parental genotypes is allowed only within each population. In this way, the integrity of each individual population remains while selection is carried out for hybrid performance and that hybrid (or semi-hybrid) cultivars can be developed at each generation.

Breeding Methods of Developing Semi-Hybrids

Different breeding schemes have been proposed for the development of semi-hybrids. A simple, generic scheme to produce broad based population semi-hybrids was proposed by Brummer (1999; Fig. 75.1a). It is based on the identification of complementary germplasm pools that are improved separately by recurrent selection within group and on the assessment of the yield of the semi-hybrid obtained by the polycross of the parents (Fig. 75.1). As parent plants are supposed to be heterozygous, single crosses are expected to be superior to double crosses. Although two populations could produce partially heterotic progeny, as described in Brummer (1999), intercrossing multiple populations (i.e., heterotic groups) would decrease

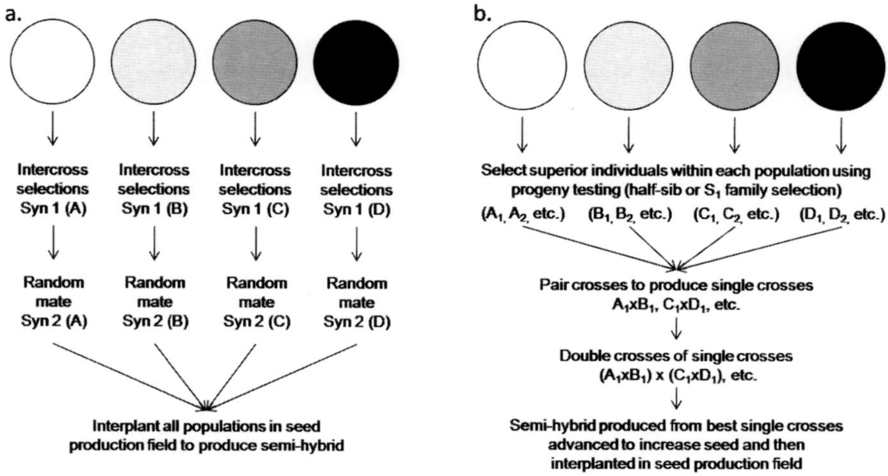

Fig. 75.1 Schemes for the development of semi-hybrid cultivars. (**a**) A simple scheme to create broad-based semi-hybrids whereby populations known to combine well together are intercrossed in the seed production field (after Brummer, 1999). (**b**) A more refined method in which specific genotypes are identified, tested for combining ability, and used to produce a narrow-based semi-hybrid cultivar (after Rotili et al., 1999a)

the amount of non hybrid seed in the semi-hybrid from one half to one fourth in the case of four populations.

A breeding scheme for the production of narrow based free-hybrids was proposed by Rotili in the early 1990s (generically shown in Fig. 75.1b). The specific form of the program in Lodi involves the evaluation in dense conditions simulating swards, of individual genotypes from parental populations of high agronomic value differing about their geographic origin and their morpho-physiological traits. Vigorous plants with yield $> \bar{x} + 2s$ of the population, i.e. 5% of the population, with the desired plant form and stem morphology are selfed and selection for vigour in dense conditions among and within selfed families is applied for two generations. The best individuals in S_2 parental families from the different populations are crossed in a diallel scheme and the single cross hybrids (SH) with the highest general combining ability (GCA) values are multiplied to the Syn3 generation to increase seed (with selection for vigour during the generations of multiplication), and finally producing double cross semi-hybrids (DH) with four unrelated parental genotype constituents (Rotili et al., 1999a).

Selfing is a powerful tool to concentrate in potential genitors genes and linkats favourable to vigour, to decrease negative genetic load, and to homogenize plants for important physiological traits as regrowth rate, earliness, etc. The double cross hybrid variety model allows for the maximum recovery of heterozygosity – and hence, maximum heterosis – because autotetraploids express progressive heterosis if parents are partially inbred. Finally, the crossing of unrelated single cross simple hybrids allows for the maximum expression and evaluation of dominant complementary gene interactions between parents.

Experimental Results on Double Hybrids

Double cross hybrids (DH) were produced from two independent sets of four diverse alfalfa populations with different geographic origins (Fig. 75.2). The corresponding 4-constituent synthetic was obtained with the same plant material and experimental procedures adopted for DHs. All the crosses were made by hand without flower emasculation and selection for vigour was practised in every generation. The DHs, their parental SH (Syn2) populations, and the corresponding 4-constituent synthetics were studied for two years (10 harvests) in the greenhouse at the density of 250 plants m^{-2}. The DHs and the corresponding synthetic are both at the fourth generation far from the initial parental genotypes. The DHs significantly outyielded the best SH (Syn2) parent (high parent heterosis) by 45 and 33% on average in the two diallel crosses. The comparison of DHs and the corresponding synthetics is more interesting as it is an estimation of the role of the variety model on yield, the plant material being exactly the same in DHs and synthetic, and because the synthetic expresses the average heterosis of the six parental SHs. In general, DHs were not significantly different from the synthetic on average but the best DHs outyielded the synthetic by 10–28% (Fig. 75.3). The DHs obtained by unrelated parental SHs performed significantly better than the DHs which SHs had one parental constituent in common and always showed positive and significant SCA effects. The variance

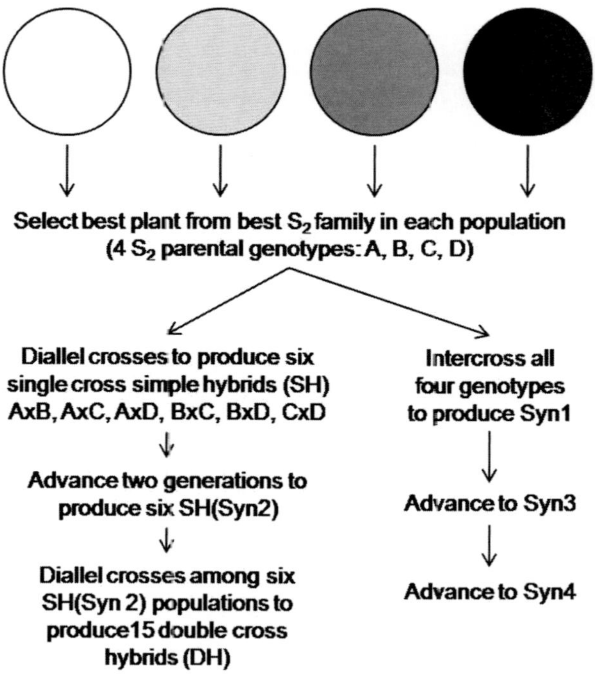

Fig. 75.2 Experimental design to test yield of 4 constituent double cross hybrids versus synthetics. Two separate double cross hybrids were developed

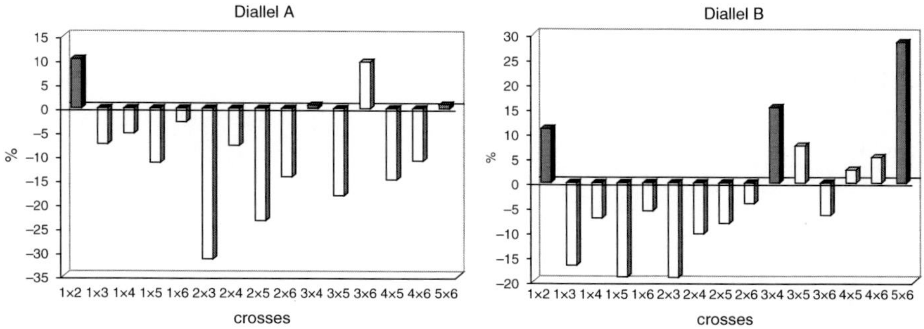

Fig. 75.3 Difference between dry matter yield of double hybrids (DH) and the corresponding synthetic expressed as a percentage of the corresponding synthetic (DHs derived from unrelated single cross hybrids are in grey)

partitioning in diallel crosses producing the DHs indicated that both GCA and SCA sources were highly significant, SCA being about two times larger than GCA. Thus, the best DHs accumulate both the additive effects of the parental SH (Syn2) and the non additive interactions among linkats favourable to yield that are realized in these particular crosses. The maximum yield gain of DH over synthetic was 28%, which seems sufficient to justify field production of semi-hybrid seed for evaluation in field conditions of semi-hybrid performance relative to the corresponding synthetic.

The role of non-additive interactions among linkats on the performance of DHs was confirmed by evaluations of inbreeding depression. Intercrossing DHs to produce F_2 populations showed that, on average, yield declined by 12% in the F_2 compared to the F_1 generation. Similarly, quadruple hybrids (QH) were obtained by crossing F_1or F_2 generations of DHs, and the comparison between QH derived from DH F_2 were 11% lower yielding than QH derived from DH F_1 populations (Table 75.1).

Quadruple hybrids significantly outperformed DHs by 18 and 19% on average at the F_1 and F_2 generations, respectively, further indicating the presence of progressive heterosis recovered when partly inbred parents are hybridized. However, no

Table 75.1 Inbreeding depression for biomass yield in double and quadruple hybrids

	Double Hybrids (DH)	Quadruple Hybrids (QH)
	$F_2 - F_1$	$F_2 - F_1$*
	%	
Average across all DH	−12	−11
DH expressing heterosis	−11	−13
DH not expressing heterosis	−21	−8

*Refers to QH formed by crossing two DH in the F_2 generation – two DH in the F_1 generation

quadruple hybrid outyielded the best DH, confirming the interest of the double cross model in semi-hybrid construction. Li and Brummer (2009) also reported a substantial and generalized inbreeding depression in advancing intra- and intersubspecific crosses among *sativa* and *falcata* to the F_2 generation. The greater depression level found for *sativa* × *falcata* crosses compared to intra subspecific hybrids could be explained by the greater loss of allelic and epistatic interactions in hybrids which parental genomes are more divergent.

Parental genetic diversity in diallel A, estimated by means of 67 SSR markers, showed a significant relationship with yield ($r = 0.59^{*}$), best parent heterosis ($r = 0.70^{**}$) and SCA effects ($r = 0.76^{***}$) among double-cross hybrid progenies. This result reflects the facts that the six SH(Syn2) populations were unambiguously distinguished by molecular analysis, the among population source of variation being estimated as 37% by AMOVA analysis, and that the highest genetic distances was found among SH with no parental genotype in common, as expected (see grey bars in Fig. 75.3). Thus, the genetic distinctiveness and divergence of parents improved for GCA and SCA effects likely result in non-additive interactions that lead to DH performance. Riday et al. (2003) did not find any relationship between genetic diversity, estimated by AFLP and SSR markers, and heterosis for yield in *sativa* × *falcata* crosses, even though subspecies subdivision and morphological distances were correlated with SCA effects of hybrids. This suggests that the construction and identification of parents of high genetic value and genetic distinctiveness is necessary to increase the probability of identifying positive specific combining abilities among parents and putative heterotic groups.

Final Considerations

The maintenance of a series of elite and genetically divergent populations as a 'genetic diversity reservoir' will enhance breeders' ability to recurrently develop semi-hybrid cultivars. These populations should be improved separately in order to maintain and to further explore their diversity and to help identification of genotypes expected to produce heterosis. The improvement of these populations can be achieved by any method of recurrent selection, although progeny testing methods are likely to be of most value for improving quantitative traits like yield. Although half-sib progeny testing or among and within family selection are the most often used (Casler and Brummer, 2008), further consideration should be given to S_1 or S_2 family recurrent selection, selfing being a powerful tool to purge deleterious alleles and to concentrate favourable alleles in the population. Furthermore, selection under sward-like densities is also recommended.

A test for estimating the combining abilities of the improved divergent families could be envisaged; the simple hybrids, evaluated for yield, can be either developed as a commercial variety or used to make double hybrids. The most suitable variety model to express maximum heterosis will depend on the parental inbreeding level. Simple hybrids can be stabilized by means of multiplication and possibly improved for desirable traits (seed yield in particular).

References

Barnes, D.K., Bingham, E.T., Murphy, R.P., Hunt, O.J., Beard, D.F., Skrdla, W.H., Teuber, L.R. 1977. Alfalfa germplasm in the United States: genetic vulnerability, use, improvement, and maintenance. Tech. Bull. 1571. USDA-ARS, U.S. Gov. Print. Office, Washington, DC.

Bingham, E.T., Groose, R.W., Woodfield, D.R., Kidwell, K.K. 1994. Complementary gene interactions in alfalfa are greater in autotetraploids than diploids. Crop Sci. 34:823–829.

Brummer, E.C. 1999. Capturing heterosis in forage crop cultivar development. Crop Sci. 39: 943–954.

Carelli, M., Gnocchi, G., Scotti, C. 2009. Alfalfa germplasm from a Sahara oasis: characterization by means of bio-agronomic traits and SSR markers. Plant Breed. 128:271–277.

Casler, M.D., Brummer, E.C. 2008. Theoretical expected genetic gains for among-and-within-family selection methods in perennial forage crops. Crop Sci. 48:890–902.

Flajoulot, S., Ronfort, J., Baudouin, P., Barre, P., Huguet, T., Huyghe, C., Julier, B. 2005. Genetic diversity among alfalfa (*Medicago sativa*) cultivars coming from a breeding program, using SSR markers. Theor. Appl. Genet. 111:1420–1429.

Holland, J.B., Bingham, E.T. 1994. Genetic improvement for yield and fertility of alfalfa cultivars representing different eras of breeding. Crop Sci. 34:953–957.

Kidwell, K.K., Hartweck, L.M., Yandell, B.S., Crump, P.M., Brummer, J.E., Moutray, J., Osborn, T. C. 1999. Forage yields of alfalfa populations derived from parents selected on the basis of molecular marker diversity. Crop Sci. 39:223–227.

Li, X., Brummer, E.C. 2009. Inbreeding depression for fertility and biomass in advanced generations of inter and intrasubspecific hybrids of tetraploid alfalfa. Crop Sci. 49:13–19.

Maureira, I.J., Ortega, F., Campos, H., Osborn, T.C. 2004. Population structure and combining ability of diverse *Medicago sativa* germplasms. Theor. Appl. Genet. 109:775–782.

Muller, M.H., Poncet, C., Prosperi, J.M., Santoni, S., Ronfort, J. 2005. Domestication history in the *Medicago sativa* species complex: inferences from nuclear sequence polymorphism. Mol. Ecol. 15:1589–1602.

Pedersen, M.W., Stucker, R.E. 1969. Evidence of cytoplasmic male sterility in alfalfa. Crop Sci. 9:767–770.

Riday, H., Brummer, E.C. 2002a. Forage yield heterosis in alfalfa. Crop Sci. 42:716–723.

Riday, H., Brummer, E.C. 2002b. Heterosis of agronomic traits in alfalfa. Crop Sci. 42:1081–1087.

Riday, H., Brummer, E.C. 2005. Heterosis in a broad range of alfalfa germplasm. Crop Sci. 45: 8–17.

Riday, H., Brummer, E.C., Austin Campbell, T., Luth, D., Cazcarro, P.M. 2003. Comparison of genetic and morphological distance with heterosis between *Medicago sativa* subsp. *sativa* and subsp. *falcata*. Euphytica 131:37–45.

Rotili, P. 1976. Performance of diallel crosses and second generation synthetics of alfalfa derived from partly inbred parents. I. Forage yield. Crop Sci. 16:247–251.

Rotili, P., Busbice, T.H., Demarly, Y. 1996. Breeding and variety constitution in alfalfa: present and future. In: Parente, G., Frame, J., Orsi, S. (eds.),Grassland and Land Use Systems, 16th EGF Meeting (pp. 163–180). EGF & ERSA, Gorizia, Italy.

Rotili, P., Gnocchi, G., Scotti, C., Zannone, L. 1999a. Some Aspects of Breeding Methodology in Alfalfa. www.naaic.org/TAG/TAGpapers/rotili/rotili.html.

Rotili, P., Scotti, C., Kertikova, D., Gnocchi, S., Gnocchi, G. 1999b. Performance of diallel crosses of alfalfa derived from partly inbred parents with different levels of genetic diversity. II. Dry matter yield. In: Veronesi, F., Rosellini, D. (eds.), Proc. XIII Eucarpia *Medicago* spp. Group Meeting (pp. 338–342). University of Perugia, Italy.

Şakiroğlu, M., Brummer, E.C. 2007. Little heterosis between alfalfa populations derived from the Midwestern and Southwestern United States. Crop Sci. 47:2364–2371.

Segovia-Lerma, A., Cantrell, R.G., Conway, J.M., Ray, I.M. 2003. AFLP-based assessment of genetic diversity among nine alfalfa germplasms using bulk DNA templates. Genome 46: 51–58.

Segovia-Lerma, A., Murray, L.W., Townsend, M.S., Ray, I.M. 2004. Population-based diallel analyses among nine historically recognized alfalfa germplasms. Theor. Appl. Genet. 109:1568–1575.

Staszewski, S. 1979. Cytoplasmic Male-Sterility and Heterosis in Alfalfa. Final Technical Report, Plant Breeding and Acclimatization Institute, Radzikov, Poland.

Chapter 76
Enhancing the Adaptation to Italian Environments of Egyptian Lucerne Germplasm for Exploitation as a Component of Free-hybrids

Paolo Annicchiarico, Luciano Pecetti, and Sandro Proietti

Abstract Free-hybrids between genetically-distant, well-complementing populations have been proposed for enhancing heterosis in lucerne. Egyptian germplasm is a candidate component of free-hybrids in Italy, owing to its history of cultivation in relatively isolated environments. However, exotic candidate components of free-hybrids need to undergo selection for local adaptation, which, for Egyptian germplasm, mainly concerns improving winter hardiness. Some 74 genotypes belonging to 17 Egyptian landraces and two Egyptian varieties which were developed by preliminary selection across one winter in an open cold greenhouse, and 74 genotypes selected from a locally well-adapted Italian landrace, were evaluated in northern Italy as replicated clonal material for dry-matter yield, disease tolerance and survival after two winters under field conditions. On average, the exotic germplasm displayed distinctly lower disease tolerance and final survival than the Italian germplasm, while showing a non-significant trend towards lower yield. Egyptian variety germplasm was superior to Egyptian landrace material in all respects. Rare, outstanding Egyptian genotypes could be identified, which were used to synthesize a 10-parent synthetic population. However, further improvement of winter hardiness seems necessary to make this germplasm commercially usable for free-hybrids. Schemes for producing two-way or three-way free-hybrids from Egyptian, Italian and another foreign germplasm source are discussed.

Keywords Genetic resources · Heterosis · Semi-hybrids · Winter hardiness · *Medicago sativa*

Introduction

Lucerne may show large heterosis for forage yield mainly because of non-additive, complementary gene interactions between different alleles, which can be quite

---

P. Annicchiarico (✉)
CRA – Centro di Ricerca per le Produzioni Foraggere e Lattiero-Casearie, viale Piacenza 29, Lodi 26900, Italy
e-mail: paolo.annicchiarico@entecra.it

C. Huyghe (ed.), *Sustainable Use of Genetic Diversity in Forage and Turf Breeding*,
DOI 10.1007/978-90-481-8706-5_76, © Springer Science+Business Media B.V. 2010

large in autotetraploids (Bingham et al., 1994). Heterosis may be exploited by hybrid varieties obtained by crossing genetically-distant, relatively narrow-based parent populations through a patented male-sterility technique. These varieties are expected to have at least 75% hybridity (as male-sterile and male-fertile parents are planted at a seed ratio of 4:1), but their selection may be limited by the lack of the patented male-sterile parent. Another avenue for exploiting heterosis is producing free-hybrids (alias semi-hybrids) by free pollination between genetically-distant, well-complementing populations which have previously been subjected to separate selection (Brummer, 1999). Exotic, genetically-distant germplasm is likely to produce heterosis when hybridized with elite locally-adapted germplasm, but it requires selection for local adaptation to avoid that the lower richness in locally-useful adaptive genes that it is expected to provide offsets the advantage of greater heterosis.

A heterotic pattern for yield has been observed in *sativa* × *falcata* free-hybrids, but the hybrid populations have insufficient agronomic value due to unfavourable traits (poor autumn growth, disease susceptibility) conferred by the *falcata* parent germplasm (Riday and Brummer, 2005, 2006). Furthermore, the requirements for variety homogeneity of the EU legislation set a limit to the width of the genetic distance between putative heterotic populations, discouraging the use of *sativa* × *falcata* free-hybrids and suggesting heterotic populations to be derived from within geographically-distant germplasm pools of *M. sativa*. The Egyptian germplasm is a candidate component of free-hybrids in Italy, because of its long history of cultivation in relatively isolated environments such as the desert oases. The comparison among Egyptian landraces or varieties and Italian varieties confirmed the large diversity between these germplasm pools at the morphophysiological and molecular level, while indicating the similar forage yielding ability of these pools in an open cold greenhouse (i.e., under conditions limiting the extent of winter cold stress) (Carelli et al., 2009). Further evaluation and selection of Egyptian germplasm for winter hardiness under field conditions is required to improve its adaptation to northern-Italy conditions before its possible exploitation as a component of free-hybrids.

The objectives of this work are: (i) assessing the variation for forage yield and winter survival under northern-Italy field conditions among and within Egyptian landrace, Egyptian variety and local landrace germplasms which were developed by one cycle of preliminary phenotypic selection; and (ii) discussing problems and opportunities for producing two-way or three-way free-hybrids in Italy from Italian, Egyptian, and other foreign germplasm sources.

Materials and Methods

The Egyptian germplasm included 74 genotypes selected phenotypically for forage yield and winter survival among 2520 genotypes evaluated by Carelli et al. (2009). These selected genotypes belonged to 17 landraces and two varieties (each individual population being represented by three to five genotypes). The Italian germplasm included 74 genotypes obtained by stratified mass selection for dry matter yield applied to 1776 genotypes evaluated over 2 years in Lodi (northern Italy) under field conditions. The Italian genetic base subjected to preliminary selection was provided by landrace '17' in Annicchiarico and Piano (2005) because of its wide adaptation across northern Italy in that study.

The 74 Egyptian and 74 Italian genotypes were evaluated as replicated clonal material. The cloned genotypes were field transplanted in Lodi in late summer 2004 at 12 cm spacing, adopting a randomized complete block design with seven replicates. Observations on the individual plants were performed for: (i) dry matter yield over one harvest in autumn 2004, four harvests in 2005 and one harvest in spring 2006; (ii) winter mortality, assessed each year in early spring; and (iii) tolerance to diseases affecting the aerial part, evaluated in spring and summer 2005 by a visual score ranging from 0 = very susceptible to 3 = highly tolerant. Rust and viruses were the main diseases.

Variation between germplasm pools and between genotypes within pool was assessed by analysis of variance.

Results and Discussion

Comparison of Germplasm Pools

On average, the Egyptian germplasm displayed distinctly lower final survival and disease tolerance than the Italian germplasm, while showing a non-significant trend towards lower total dry matter yield (Table 76.1). Plant counts before and after each winter confirmed that mortality occurred essentially during winters in the Egyptian germplasm.

Because of its importance, winter survival was the main trait which affected the further selection of Egyptian genotypes. Outstanding Egyptian genotypes which responded comparably with the best performing Italian genotypes could be identified with respect to dry matter yield and disease tolerance (Table 76.2).

Within Egyptian germplasm, variety material was superior to landrace material in all respects (Table 76.3). The inference space of this result, however, is limited by the small sample size for the variety germplasm.

Table 76.1 Comparison of Egyptian *vs*. Italian elite germplasm under Lodi field conditions

Item	Egyptian germplasm	Italian germplasm	P level[a]
Number of genotypes	74	74	
Total dry matter yield (g/plant)	16.3 ± 1.6	21.9 ± 4.1	NS
Disease tolerance (score 0–3)	1.0 ± 0.1	1.8 ± 0.1	**
Mortality after two winters (%)	94.6 ± 0.8	30.0 ± 2.4	**

[a]NS, not significant; **, significant at $P < 0.01$

Table 76.2 Genotype range values within Egyptian and Italian elite germplasm under Lodi field conditions

Item	Egyptian germplasm[a]	Italian germplasm[a]
Number of genotypes	74	74
Total dry matter yield (g/plant)	1.7–52.4	0.1–54.2
Disease tolerance (score 0–3)	0.0–2.6	0.0–2.7

[a]Genotype variation within each germplasm pool was significant at $P < 0.001$

Table 76.3 Comparison of Egyptian landrace *vs.* variety elite germplasm under Lodi field conditions

Item	Landrace germplasm	Variety germplasm	P level[a]
Number of populations	17	2	
Number of genotypes	68	6	
Total dry matter yield (g/plant)	14.8 ± 1.6	31.9 ± 2.6	**
Disease tolerance (score 0–3)	0.9 ± 0.1	1.9 ± 0.1	**
Mortality after two winters (%)	97.3 ± 0.8	64.3 ± 6.7	**

[a]NS, not significant; **, significant at $P < 0.01$

Further Selection Work, and Development of Free-Hybrids

The final objective of our research work is to assess the value of a simple and low-cost procedure for developing two-way or three-way free-hybrids derived from narrow-based germplasm pools previously obtained from independent cycles of selection for forage yield and local adaptation. The possible components of free-hybrids for northern Italy would be synthetic varieties selected within: (i) the elite local landrace '17'; (ii) the Egyptian germplasm (featuring lower winter dormancy than the local landrace); and (iii) semi-erect germplasm originated from East European, Canadian and Spanish 'Mielga' populations (featuring higher winter dormancy than the local landrace, and selected mainly for upright growth, relatively deep crown and purple flowers: Pecetti et al., 2006). The two exotic components have winter dormancy levels which can complement well to approach the locally optimal level.

The scheme for producing these free-hybrids is reported in Fig. 76.1. Each two-way hybrid would derive from free pollination between Syn-2 generations of its components (or higher generations multiplied independently, when envisaged on a commercial scale). The three-way hybrid would contemplate a 50% genetic background for the local landrace, which is expected to display greater adaptation to local conditions (Fig. 76.1). The synthetic variety obtained by selection within the local landrace would be the reference germplasm for testing the value of free-hybrids and to verify whether the increase of heterosis provided by hybridization with exotic material could offset the lower richness in locally-useful adaptive genes.

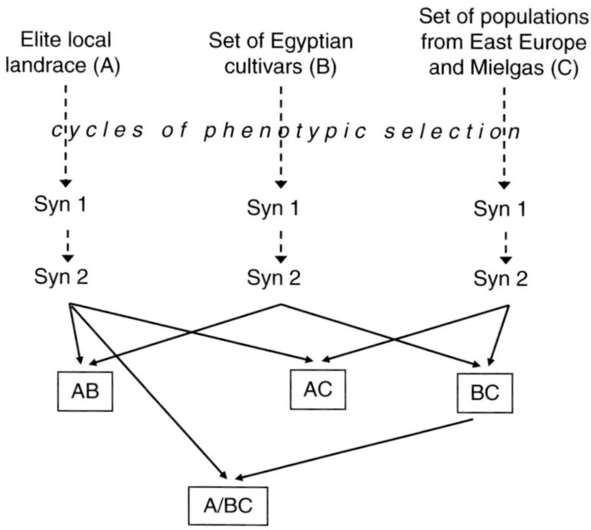

Fig. 76.1 Scheme for producing two-way and three-way free-hybrids in Lodi

Following the described evaluation of cloned genotypes, an experimental synthetic variety including as parents the 10 best-performing genotypes was synthesized for both the Egyptian and the Italian germplasm pools. One more 10-parent synthetic variety was produced by phenotypic selection within East European, Canadian and Mielga germplasm. The preliminary assessment of winter mortality performed in spring 2009 on the Syn-1 generation of the three experimental synthetics suggested that the level of winter hardiness of the Egyptian material is still too low to allow its possible commercial exploitation as a component of free-hybrids (where each component should be multiplied independently before the free-hybrid generation). A drawback emerged also for the East European/Canadian/Mielga selection, i.e., low seed yielding ability, in the preliminary assessment of this trait performed on Syn-1 material in summer 2008. Another phenotypic selection stage will be performed on Syn-2 material within each germplasm pool, to improve forage yield in general, winter hardiness in the Egyptian material, and seed yield in the East European/Canadian/Mielga germplasm.

Acknowledgements The work in Italy was funded by the Italian Ministry of Agricultural, Food and Forestry Policies within the project 'Plant Genetic Resources – FAO Treaty'. We are grateful to Z. Abdelhalim, A. Omar and C. Scotti for providing the Egyptian germplasm.

References

Annicchiarico, P., Piano, E. 2005. Use of artificial environments to reproduce and exploit genotype × location interaction for lucerne in northern Italy. Theor. Appl. Genet. 110:219–227.

Bingham, E.T., Groose R.W., Woodfield D.R., Kidwell K.K. 1994. Complementary gene interactions in alfalfa are greater in autotetraploids than diploids. Crop Sci. 34:823–829.

Brummer, E.C. 1999. Capturing heterosis in forage crop cultivar development. Crop Sci. 39: 943–954.

Carelli, M., Gnocchi, G., Scotti, C. 2009. Alfalfa germplasm from a Sahara oasis: characterization by means of bio-agronomic traits and SSR markers. Plant Breed. 128:271–277.

Pecetti, L., Romani, M., Piano, E. 2006. Persistence of morphophysiologically diverse lucerne under continuous stocking and intensive grazing. Aust. J. Agric. Res. 57:999–1007.

Riday, H., Brummer, E.C. 2005. Heterosis in a broad range of alfalfa germplasm. Crop Sci. 45: 8–17.

Riday, H.; Brummer, E.C. 2006. Persistence and yield stability of intersubspecific alfalfa hybrids. Crop Sci. 46:1058–1063.

Chapter 77
$F1^2$ Performance of Tetraploid Perennial Ryegrass on the Basis of the Composition of a Synthetic Variety

Joost Baert and An Ghesquiere

Abstract Most of the perennial ryegrass cultivars are synthetics composed of 4–20 or more parents. Using component genotypes with a good specific combining ability (SCA) in a synthetic variety may reduce the number of parents and exploit heterosis without inbreeding depression. Determining SCA by carrying out and testing pair crosses is laborious. Is it worthwhile to do all this work?

We carried out a diallel cross with 10 genotypes of tetraploid perennial ryegrass. The F1 seeds were multiplied to $F1^2$ seeds. We determined the dry matter yield of the $F1^2$ populations in a field trial for 2 years. The results showed that the genotypes with the best general combining ability produced the best $F1^2$ progenies. The best 3 $F1^2$ progenies were used to build a synthetic variety that is performing very well in official variety list trials.

This example shows that the use of pair crosses in synthetic variety development of tetraploid perennial ryegrass varieties is valuable but not necessarily superior to other less laborious methods like polycross or topcross where GCA is determined.

Keywords Breeding methods · Combining ability · Perennial ryegrass · Synthetics · Tetraploid

Introduction

Most of the perennial ryegrass cultivars are synthetics composed of 4–20 or more parents. The parents are mostly chosen based on their general combining ability (GCA) in polycrosses or topcrosses. More component genotypes ensure less inbreeding but also reduce the selection intensity and increase the risk of heterogeneity. Synthetics of tetraploid perennial ryegrass based on only 2 components may have a good herbage yield but mostly have a low seed yield (Baert et al., 2007).

J. Baert (✉)
ILVO Plant, Caritasstraat 21, Melle 9090, Belgium
e-mail: joost.baert@ilvo.vlaanderen.be

C. Huyghe (ed.), *Sustainable Use of Genetic Diversity in Forage and Turf Breeding*,
DOI 10.1007/978-90-481-8706-5_77,

Using component genotypes with a good specific combining ability (SCA) in a synthetic variety may reduce the number of parents and exploit heterosis without inbreeding depression. Tamaki et al. (2007) could improve the yield of timothy by exploiting SCA through modified synthetic varieties where syn2 seeds are produced by syn1 seeds of only 2 parental clones out of a polycross with several parental clones. Determining SCA requires making a lot of pair crosses and testing their progenies. Is it worthwhile to do all this work or may we rely on the general combining ability to compose the synthetics of tetraploid perennial ryegrass.

Material and Methods

Ten unrelated genotypes of late-flowering tetraploid perennial ryegrass were used in a diallel crossing in 1999. We carried out the 45 crosses in artificially ventilated isolation cages (one pair of plants per cage) in the greenhouse. The F1 seeds were immediately sown after harvest and planted in the field in Autumn to produce $F1^2$ seeds in 2000. These $F1^2$ seeds were sown in trial in Spring 2001. The dry matter yield was determined at a 5 cuts/year rate in the 2 years after sowing and expressed as relative dry matter yield to the yield of the control varieties Ritz and Roy (mean DM yield: 14147 kg/ha/year). Crown rust resistance was scored in the Autumn of the sowing year (1 = very susceptible; 10 = very resistant). Only 32 progenies were tested because of an insufficient amount of seeds in the F1 or the $F1^2$ generation.

Results and Discussion

Table 77.1 shows the yield performance of the $F1^2$ progenies. All combinations in pairs of three of the ten genotypes (G, H and I) had a high yield. These three had a good specific combining ability.

Based on the DM yield in their $F1^2$ progenies we calculated the potential yield of each of the parents which is an indication of their general combining ability.

Table 77.1 Relative mean dry matter yield of $F1^2$ progenies from diallel crosses among ten tetraploid genotypes A–J (lsd = 7%)

parent	B	C	D	E	F	G	H	I	J
A	–	97	–	97	96	98	98	–	–
B		98	97	96	97	105	99	100	99
C			–	–	92	100	–	95	–
D				–	–	100	100	98	–
E					99	102	97	97	–
F						99	99	103	–
G							**104**	**107**	101
H								**105**	100
I									100

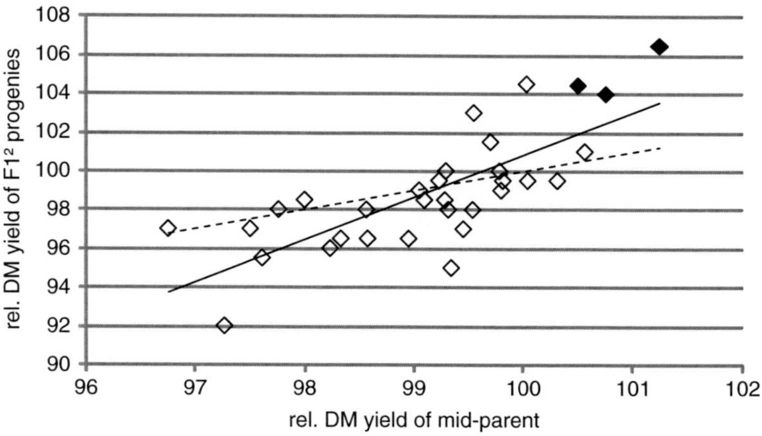

Fig. 77.1 Relationship between the measured relative DM yield of the F1^2 progenies and the calculated relative yield of their mid-parents (*broken line*: $y = x$)

Figure 77.1 shows the relation between observed yield of F1^2 progenies and predicted yield as mid-parent value from GCA estimates.

The correlation coefficient ($r = 0.76$) of this relationship shows that the additive component of the variance of the DM yield of heterozygous tetraploid genotypes is rather high. The 3 genotypes that specifically combine very well are also the genotypes with the highest GCA. These 3 genotypes were used to develop a synthetic variety that entered into official variety list trials in 2006 which gave good results in 2007 and 2008.

The same reasoning was applied for crown rust resistance. Table 77.2 shows the crown rust resistance score of the F1^2 progenies. All combinations in pairs of three of the ten genotypes (A, C and F) were very resistant. These three combinations however had a low yield due to the negative correlation between yield and rust resistance ($r = -0.32$).

Table 77.2 Crown rust resistance score of F1^2 progenies from diallel crosses among ten tetraploid genotypes A–J

parent	B	C	D	E	F	G	H	I	J
A	–	**9,5**	–	7,5	**9,2**	8,4	6,5	–	–
B		7,5	6,6	7,2	7,5	7,0	5,5	7,3	5,3
C			–	–	**9,5**	8,2	–	8,5	–
D				–	–	8,2	5,6	7,0	–
E					8,2	7,4	6,9	8,0	–
F						8,2	6,6	9,0	–
G							6,4	7,7	6,9
H								6,3	4,9
I									7,8

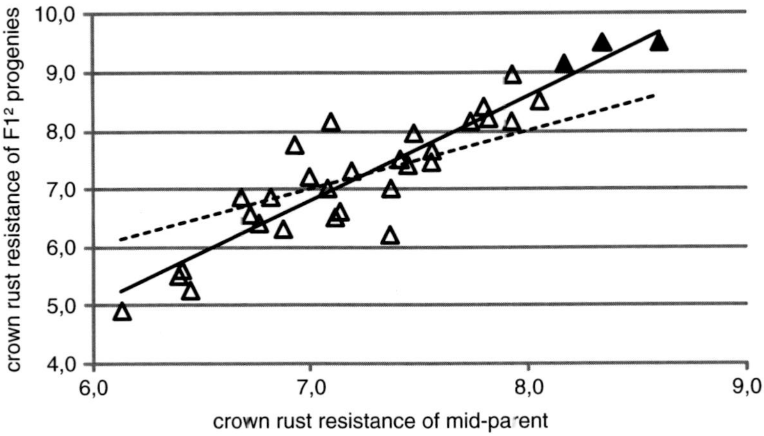

Fig. 77.2 Relationship between the observed crown rust resistance of the $F1^2$ progenies and the calculated crown rust resistance of their mid-parents (*broken line*: $y = x$)

Figure 77.2 shows the relation between the observed crown resistance of $F1^2$ progenies and the predicted resistance as mid-parent value from GCA estimates.

The correlation coefficient ($r = 0.92$) of this relationship shows that the additive component of the variance of the crown rust resistance of heterozygous tetraploid genotypes is even higher than the additive component of the variance of the DM yield. Muylle (2003) reported a high additive effect for crown rust resistance in diploid Italian ryegrass but not in perennial ryegrass. The 3 genotypes that specifically combine very well for crown rust resistance are also the genotypes with the highest GCA for crown rust resistance.

This example shows that the use of pair crosses for synthetic variety development of tetraploid ryegrass is valuable but not necessarily superior to other less laborious methods where GCA is only determined.

References

Baert, J., Ghesquiere, A., Muylle, H. 2008. Comparison between two breeding methods in tetraploid Lolium perenne: polycross versus F2 (pp. 153–155). In: Lübberstedt, T., Studer, B., Graugaard, S. (ed.), Proc. XXVIIth Eucarpia Symposium on Improvement of Fodder Crops and Amenity Grasses. August 19–23, 2007, Copenhagen, Denmark.

Muylle, H. 2003. Genetic analysis of crown rust resistance in ryegrasses (Lolium spp.) using molecular markers. PhD thesis in Applied Biological Sciences, University of Ghent (Belgium).

Tamaki, H., Yoshizawa, A., Fujii, H., Sato, K. 2007. Modified synthetic varieties: a breeding method for forage crops to exploit specific combining ability. Plant Breed. 126:95–100.

Chapter 78
Seed Yield of New *Festulolium* Varieties Bred from *F. Arundinacea* Var. *Glaucescens*

Marc Ghesquière and Thierry Bourgoin

Abstract In 2007 and 2008, three new amphitetraploid ($2n = 4x = 28$) *Festulolium* cultivars were registered on the French National List using for the first time *F. arundinacea*var.*glaucescens* Boiss. as a parent. We report hereafter a wide seed yield assessment of those new cultivars together with back-cross derivatives into *L. multiflorum* or *L. perenne* and previous *Festulolium* cultivars from *F. pratensis*. Seed yield of the new *F. a.*var.*glaucescens*-derived cultivars was found of same magnitude as amphiploid *Festulolium* from *F. pratensis*. However, higher rate of left-over seeds shows that they may be of less intrinsic fertility. We found that total genetic variance was significantly increased between Half-Sibs progenies derived from back-cross into *L. perenne* as compared with amphiploid HS progenies. Higher genetic variance between polycrosses of amphiploid *L. multiflorum* × *F. a.* var. *glaucescens* hybrids also suggests that selection for seed yield could take place effectively without necessarily individualizing progenies. Seed yield tended to increase over generations of seed multiplication especially when populations derived from back-cross into *Lolium sp*. It is concluded that the restoration of full seed productivity in *Festulolium* from *F. a.* var. *glaucescens* could be achieved through new *Lolium*/*Festuca* genome balance with the development of chromosome/marker-assisted selection procedures.

Keywords Interspecific hybrid · Amphiploid · Introgression · Fertility · Chromosome shift · Genome drive

Introduction

Festulolium are hybrids between *Festuca* sp. and *Lolium* sp. If both *L. multiflorum* and *L. perenne* species have been used in crossing so far, present cultivars derive mostly from *F. pratensis*. In this context, *F. arundinacea*

M. Ghesquière (✉)
UR4 (Unité de Recherche Pluridisciplinaire Prairies et Plantes Fourragères), INRA, Centre Poitou-Charentes, Lusignan, F-86600, France
e-mail: Marc.Ghesquiere@lusignan.inra.fr

C. Huyghe (ed.), *Sustainable Use of Genetic Diversity in Forage and Turf Breeding*, DOI 10.1007/978-90-481-8706-5_78, © Springer Science+Business Media B.V. 2010

var.*glaucescens*Boiss.makes a valuable alternative for developing *Festuca* × *Lolium* hybrid cultivars of higher persistency following severe summer drought. This species is tetraploid ($2n = 4x = 28$) and acknowledged as one of the progenitors of the 6*x* cultivated tall fescue, the other one being the diploid meadow fescue *F. pratensis* (Humphreys et al., 1995). When crossed into amphiploid hybrids,*F. a.*var.*glaucescens* confers to *Lolium* sp. better persistency thanks to deep rooting and hence, ability to water uptake deep in the soil when it is dry in summer (Durand et al., 2007). In this aim, a large population of tetraploid ($2n = 4x = 28$) F1 hybrids was produced in 1987 from crosses between *L. multiflorum* and various populations of *F. glaucescens* collected in the French Alps (Ghesquière et al., 1996). Fortunately, fertility in subsequent generations of polycrossing amphiploid hybrids was found to be quite sufficient to undertake a breeding program promoting the combination of high yield, persistency and quality (Ghesquière, 2007). Early meiotic assessment showed that chromosome pairing was mostly formed of bivalents due to high preferential homologous chromosome pairing, conferring high phenotypic stability over generations (Ghesquière et al., 1993). However, disomic-like inheritance in those hybrids may, in counterpart, make fertility partly defective in comparison to amphiploids derived from *F. pratensis* of closer genome relationships with *Lolium* sp. Thus, this paper reports the assessment of seed productivity of those new *Festulolium* cultivars against control cultivars of the pure parent species and some of their *Festulolium* hybrid derivatives.

Material and Methods

A trial comprising 244 genotypes was sown by Oct. 2007 into a 2-replicates unbalanced random design of 3.1 m^2 single plots. The three new *Festulolium* cultivars, *Lueur*, *Lusilium* & *Luxane* were compared to 5 *Festulolium* control cultivars, i.e. derived from *F. arundinacea*or *F. pratensis*, 10 control cultivars of pure species and 12 experimental populations including the very initial *L. multiflorum* × *F. glaucescens* F2 population from which all breeding plant material was produced since 1987. Among the 12 populations, 4 covered up to 3 generations of successive seed multiplication; also, 2 populations were bred following one generation of backcross (BC1) either into 4*x* *L. multiflorum* or *L. perenne*, namely pop. 99.4 and 01.1 (resp.). At last, 212 Half-Sibs progenies (HS) covering all single entries of 15 polycrosses were included in the trial to estimate genetic variance and to infer more effective methodology when breeding for seed yield. The plots were mechanically harvested on the 1st and the 3rd of July 2008 depending on seed maturity. Seed yield were weighted straight from the combine harvester and after fine threshing so that "flat", ergoted seeds, glumes or unfertilized spikelets... were discarded. Re-heading in the plots was scored 0 (nil) to 5 (*Lolium*-like) in August.

Results and Discussion

The highest seed yielding genotype was the control tetraploid Italian ryegrass (*Tonyl*) with 12.3 dt/ha (Table 78.1). All pure fescue control varieties did not yield more than 5.1 dt/ha probably because of late sowing and limiting vernalisation. The two cv *Lofa* and *Paulita* were the best yielding *Festulolium* controls with 11.9 and 10.5 dt/ha (resp.), closely to the very rare seed yield evaluation reports on *Festulolium* (e.g. Fojtik, 1994). On average, populations and varieties from *F. a.*var.*glaucescens* yielded 6.8 dt/ha, not significantly different from the 3 control *Festulolium* cv. *Felopa* (7.6), *Duo* (7.1) and *Lifema* (6.3), all bred from *F. pratensis*. By contrast, the 2 BC1 populations derived from *F. glaucescens* had seed yield (11.0 dt/ha on average) significantly restored up to the level of the best *Festulolium* cultivars *Lofa* and *Paulita*. Re-heading, a typical *Lolium*-like trait was found to correlate with seed yield. BC derivatives thus tended to have better seed yields associated with higher head production in summer than amphiploids (with the exception of

Table 78.1 Seed productivity and quality among contrasted *Festulolium* hybrid combinations and relationships with phenology

Cultivar	Parent combination (ploidy level)	Seed yield after fine threshing (dt/ha)	Rate of discarded seeds (% w/w)	Re-heading in summer (5: high – 0: no)
Tonyl	Lm (4*x*)	12.3^{a}	26^{abc}	5.0^{g}
Lofa	BC1Fa→Lm(4*x*)	11.9^{ab}	31^{abcdef}	4.5^{fg}
99.4	BC1Fg→Lm(4*x*)	11.2^{abc}	26^{abc}	3.0^{cd}
Paulita	Fp × Lm (4*x*)	10.5^{abcd}	33^{abcdef}	3.5^{de}
01.1	BC1Fg→Lp(4*x*)	9.9^{abcde}	20^{a}	3.0^{cd}
Delicial	Lm × Lp (4*x*)	9.0^{bcdef}	29^{abcde}	3.0^{cd}
Fastyl	Lm (2*x*)	8.7^{cdef}	22^{ab}	5.0^{g}
Lueur	Lm × Fg (4*x*)	8.1^{defg}	41^{defg}	2.5^{bc}
Felopa	Fp × Lm (4*x*)	7.6^{defgh}	35^{bcdef}	4.0^{ef}
Aberexel	Lm × Lp (4*x*)	7.3^{efgh}	37^{cdef}	4.5^{fg}
Duo	BC1 Fp→Lp (4*x*)	7.1^{efgh}	30^{abcdef}	3.0^{cd}
Lifema	Lm × Fp (4*x*)	6.3^{fgh}	42^{efg}	3.0^{cd}
Luxane	Lm × Fg (4*x*)	5.7^{gh}	44^{fg}	4.0^{ef}
Lusilium	Lm × Fg (4*x*)	5.5^{gh}	55^{g}	2.0^{b}
Barolex	Fa (6*x*)	5.1^{hi}	22^{ab}	1.0^{a}
Stella	Fp (2*x*)	5.0^{hi}	19^{a}	1.0^{a}
Dulcia	Fa (6*x*)	2.3^{ij}	27^{abcd}	1.0^{a}
F.glaucescens	Fg (4*x*)	2.0^{j}	33^{abcdef}	1.0^{a}
Lunibelle	Fa (ca 10*x*)	1.6^{j}	41^{defg}	1.0^{a}
LSD ($P < 0.05$)		2.9	13.8	0.8

Means sharing a same letter are not significantly different ($P < 0.05$) Lm: *L. multiflorum* – Lp: *L. perenne* – Fa: *F. arundinacea* – Fg: *F. glaucescens* – Fp: *F. pratensis*
BC: back-cross, the arrow indicates the way of introgression.

the introgression cv *Duo* of medium seed yield and the amphiploid cv *Luxane* of relatively high re-heading score).

Regression of seed yield after fine threshing onto crude seed yield gave an estimate of lack of productivity due to interspecific hybridization. Among pure species and *Festulolium* control cultivars, the rate of discarded seeds was 28% while it was 37% among the 36 HS progenies resulting of back-cross into *Lolium* sp. and 45% among the 176 amphiploid *L. multiflorum* × *F. a.*var.*glaucescens* HS progenies.

In 5 instances among 6 (5 amphiploid and one introgression population), seed yield was found to have been increased over generations of seed multiplication, + 20% on average, although without any significant effect of a given generation.

Given an error variance of seed yield of 1.99 $(dt/ha)^2$, the total genetic variance between the 176 HS progenies was estimated to be of 1.35 $(dt/ha)^2$, of which, 42% was found between polycross and 58% between HS within polycross (Fig. 78.1). Within BC1 into *L .perenne* plant material (27 HS progenies), genetic variance was only significant between HS progenies, 2.80 $(dt/ha)^2$, which significantly exceeded the genetic variance between amphiploid HS progenies ($F = 1.72$; $P = 0.023$). This emphasises that interspecific recombination is significantly enhanced when *F. a.*var.*glaucescens* is present only as a haplotype genome in BC derivatives.

Since the initial F2 generation of 6.16 dt/ha of seed yield, no significant genetic improvement was evidenced. This could be due to the fact that only mass selection for highly seed-producer hybrids was primarily applied on the ground of

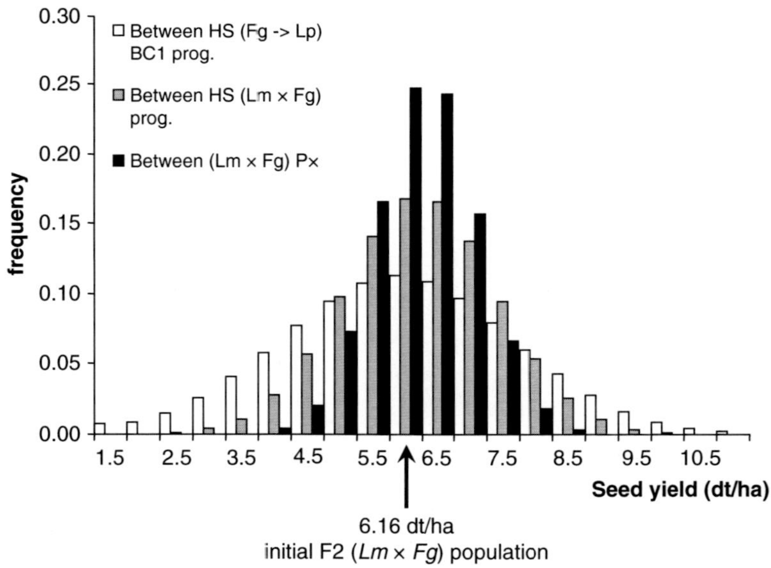

Fig. 78.1 Genetic variance of seed yield between polycrosses and half-sib progenies in *Lolium* sp. × *F. glaucescens* derivatives

high heritability of fertility and seed yield components in the early generations (Ghesquière et al., 1993).

In conclusion, the results show that seed productivity remains of quite low heritability n.s. *per se*. Consequently, selection has to be based definitely upon progeny tests and not upon mass selection. In this aim, progress of seed productivity in *Festulolium*can be expected following selection among progenies, especially among BC1 progenies. As about half of total genetic variance was found between polycrosses of amphiploid HS, it is also suggested to produce more amphiploid polycrosses but without necessarily individualizing them into HS-progenies. By compensating for a higher rate of selection among polycrosses, progress of the same magnitude could be expected in a less expensive and time-consuming way.

Clearly, the *Festuca* species used as parent and the overall interspecific genome balance interact on seed productivity in *Festulolium* cultivars. Furthermore, *Lolium*-genome drive over generations of breeding (e.g. Kopecký et al., 2006; Zwierzykowski et al., 2006) and likely over generations of seed multiplication of no intentional selection, complicate fair comparisons of seed productivity over present *Festulolium* cultivars.

With respect to *Festulolium* derived from *F. glaucescens*, it seems that genome balance would require to be more actively remodelled. One question is to know whether this could be achieved in a reasonable span of time from amphiploid hybrids and suitable methods of selection for seed productivity or this would definitely require deliberate back-cross into *Lolium* sp.

In this case, assisting phenotype selection by any fine monitoring of introgression at the molecular/chromosome level would be crucial to retain all the traits of primary interest from *F. a.*var.*glaucescens* while recovering seed productivity fully compatible with marketing.

References

Durand, J.L., Bariac, T., Ghesquière, M., Biron, P., Richard, P., Humphreys, M., Zwierzykowski, Z. 2007. Ranking of the depth water extraction by individual grass plants using natural ^{18}O isotope abundance. Environ. Exp. Bot. 60:137-144.

Fojtik, A. 1994. Methods of grass improvement used at the Plant Breeding Station Hladké Životice. Genet. Pol. 35A:25-31.

Ghesquière, M., Zwierzykowski, Z., Poisson, C., Jadas-Hécart, J. (1993). Amphitetraploid *Festulolium*: chromosome stability and fertility over intercrossing generations (pp. 451–453). In: Proc. XVIIth Intern. Grassland Congress. Palmerston North, New Zealand.

Ghesquière,M.2007. Note sur la diversité des variétés de *Festulolium*. Fourrages 191:377-379.

Ghesquière, M., Emile, J-C., Jadas-Hécart, J., Mousset, C., Traineau, R., Poisson, C. 1996. First in vivo assessment of feeding value of *Festulolium* hybrids derived from *Festuca arundinacea*var. *glaucescens* and selection for palatability. Plant Breed. 115:238-244.

Humphreys, M.W., Thomas, H.M., Morgan, W.G., Meredith, M.R., Harper, J.A., Thomas, H., Zwierzykowski, Z., Ghesquière, M. 1995. Discriminating the ancestral progenitors of hexaploid *Festuca arundinacea* using genomic *in situ* hybridization. Heredity 75:171-174.

Kopecký, D., Loureiro, J., Zwierzykowski, Z., Ghesquière, M., Doležel, J. 2006. Genome constitution and evolution in *Lolium* × *Festuca* hybrid cultivars (*Festulolium*). Theor. Appl. Genet. 113:731-742.

Zwierzykowski, Z., Kosmala, A., Zwierzykowska, E., Jones, N., Jokś, W., Bocianowski, J. 2006. Genome balance in six successive generations of the allotetraploid *Festuca pratensis* × *Lolium perenne*. Theor. Appl. Genet. 113:539-547.

Chapter 79
Production of Self-Fertile Interspecific Hybrids Between *Lolium temulentum* × *Lolium multiflorum*

Takako Kiyoshi, Akira Arakawa, Kazuhiro Uchiyama, and Tadashi Takamizo

Abstract Three F_1 hybrids plants between *Lolium temulentum* L. (Ba3081) and a genotype of *L. multiflorum* Lam., generated by embryo rescue technique, were proved to be interspecific hybrids (F_1) by analysis with SSR markers for Italian ryegrass (Hirata et al., 2006). Using two of them, we checked the pollen viability of hybrids using 1% acetic carmine staining. Pollen viability of F_1 hybrids was considerably lower than that of parental plants, but normal pollen was observed in F_1 plants. To confirm the potential fertility and self-compatibility of the F_1 plants, those two F_1 plants were self-pollinated in a paper bag in the greenhouse. Self-pollinated F_2 seeds were obtained from the two F_1 hybrids, No.1 and No. 3. We conclude that the two F_1 hybrids are self-fertile.

Keywords Interspecific hybrids · *Lolium multiflorum* Lam. · *Lolium temulentum* L. · Self-fertility

Introduction

Species in genus *Lolium* are divided into two groups; outbreeders and inbreeders. *L. multiflorum* Lam. (Italian ryegrass) and *L. perenne* L. (perennial ryegrass), both used for forage and amenity grasses worldwide, belong to the former group, whereas *L. temulentum* L. (darnel, a cosmopolitan weed of cereal fields), belongs to the latter. Interestingly, *L. temulentum* can hybridize with *L. perenne* and *L. multiflorum*, relatively easily. Thorogood and Hayward (1992) and Yamada (2001) reported that self-compatibility of *L. temulentum* is controlled by a single gene. If the self-compatible gene of *L. temulentum* can be introgressed into *L. multiflorum*, inbred lines could be developed for breeding *L. multiflorum* cultivars. However, there are

T. Kiyoshi (✉)
National Institute of Livestock and Grassland Science, 768 Senbonmatsu, Nasushiobara Tochigi 329-2793, Japan
e-mail: tkiyoshi@affrc.go.jp

C. Huyghe (ed.), *Sustainable Use of Genetic Diversity in Forage and Turf Breeding*,
DOI 10.1007/978-90-481-8706-5_79,

few reports of such self-fertile interspecific F_1 hybrids. In this work, we crossed *L. temulentum* with *L. multiflorum* ($2n = 14$), investigated morphological traits including fertility of the F_1 plants. The results are reported hereafter.

Materials and Methods

Plant material and Crossing

The *L. temulentum* line, Ba3081, from the Institute of Biological, Environmental and Rural Sciences in UK (formerly IGER) was used in this study. The genotype of *L. multiflorum* used was derived from sib F_1 plants of cv. Waseaoba × cv. Axis. Interspecific hybrid plants were produced by crossing hand-emasculated Ba3081 with *L. multiflorum*. Hybrid embryos were rescued on modified B5 or MS medium containing 3% sucrose following the method of Yamada (2001). SSR analysis of the rescued plants was conducted using SSR markers from *L. multiflorum* (Hirata et al., 2006) to confirm hybridity. PCR conditions were the same as the original published ones and the PCR products were separated by electrophoresis on 6% acrylamide gel. We used six SSR primer combinations (LmSSR3-6F, LmSSR7-3E, LmSSR7-5D, LmSSR9-7G, LmSSR10-7A and LmSSR10-8E), after preliminary screening. When the hybrids were sufficiently developed, they were transferred in pots and let to grow in greenhouse.

Fertility of F_1 Plants

Morphological traits of F_1 individuals were examined. We harvested spikes at the flowering stage in greenhouse for checking viability of pollen using 1% aceto-carmine staining. Normally formed and dark stained grains were considered as fertile under microscope observation. To confirm fertility and self-compatibility of F_1 hybrids, we investigated seed set after self-pollination. Spikes of individual plants were bagged before flowering in 10 × 25 cm paper bags and let to mature. *L. multiflorum* cv. Waseaoba was used as control plants instead of parent because the actual parent plants were not available anymore.

Results and Discussion

In total, 40 embryos were rescued and cultured, 17 of them germinated in vitro but only four plants established well. Three of these plants were proved to be interspecific F_1 hybrid according to SSR markers for Italian ryegrass (Fig. 79.1). Thanks to these markers, we can distinguish hybrids from selfed plants.

The F_1 plants grew normally to maturity in greenhouse. Two of them, plant No.1 and 3, were used for pollination survey, plant No.2 was not further investigated. At

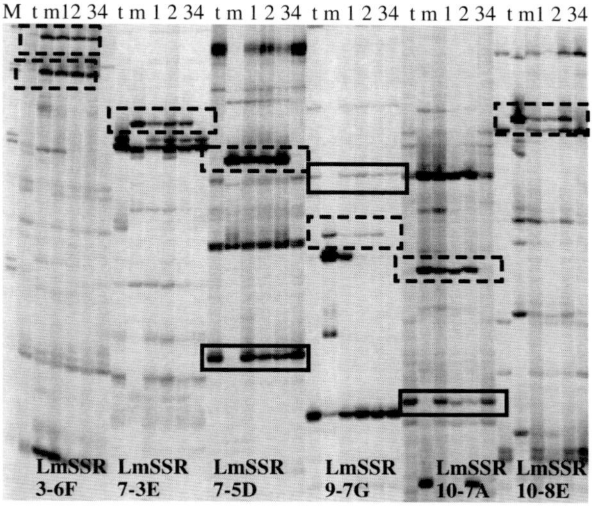

Fig. 79.1 Comparison of SSR marker profiles of *Lolium temulentum, L. multiflorum* and their F_1 hybrids using *L. multiflorum* SSR markers (Hirata et al., 2006). M: size marker; t: *L. temulentum* Ba3081 as female parent; m: *L. multiflorum* as male parent; 1-3: F_1 plants, 4: selfed plant of *L.temulentum*, solid line: *L. temulentum* specific band; broken line: *L. multiflorum* specific band

Fig.79.2 Spikelets of *Lolium temulentum, L. multiflorum* and their F_1 hybrids at flowering stage. (a) *L. temulentum*, Ba3081, (b) *L. multiflorum* cv. Waseaoba, (c) F_1 plant No.1, (d) F_1 plant No.3

flowering stage, anthers of those two F_1 plants were dehiscent (Fig. 79.2). Pollen viability of these F_1 plants estimated by aceto-carmine staining was approximately 40%. It was much lower than that of both parental species but normal pollen grains were observed. Thus, these two F_1 plants appeared to be partially male-sterile. There were many reports about inter-specific F_1 hybrids between *L. temulentum X L. multiflorum or L. perenne* which all reported that F_1 hybrids are male sterile, except for two articles (Jenkin and Thomas, 1938; Jenkin, 1954). Jenkin and Thomas (1938) described the some of F_1 anthers were more or less dehiscent had 20% of good pollen. Jenkin (1954) reported that 55 established F_1 plants between *L. temulentum X L. italicum (= L. multiflorum)* as female parent were female-fertile, and male-sterile except for one individual which have slightly dehiscent anthers. They tried to use the F_1 plant which had dehiscent anthers as pollen parent, but failed completely (Jenkin, 1954). There is no report about F_2 seeds obtained from the F_1 hybrids between *L. temulentum* and *L. multiflorum*. This is first report that we gained seed set of the F_1 hybrids.

After self-pollination in greenhouse, we obtained 340 (17 spikes) and 104 (7 spikes) F_2 seeds from the two F_1 hybrids, plant No.1 and 3, respectively. Yamada (2001) classified BC1 or BC2 plant of *L. temulentum X L. perenne* as (1) self-incompatible (setting no seeds), (2) partially self-compatible (1-5 self-seeds per spike) and (3) fully self-compatible (more than 5 self-seeds per spike), in his experiment. If we follow this criterion, self-seed set per spike from the two F_1 hybrids, No.1 and No.3, 20.0 and 14.9 seeds, respectively, enabled to classify them as fully self-compatible.

Acknowledgements A part of this work was supported by a grant from the Ministry of Agriculture, Forestry and Fisheries of Japan (Genomics for Agricultural Innovation, FBW1103) and by the Genebank Project of the National Institute of Agrobiological Sciences, Japan. We thank Dr. Koya Sugawara, the National Institute of Livestock and Grasslands Science, Japan, for his advices given in the preparation of the manuscript.

References

Hirata, M., Cai, H., Inoue, M., Yuyama, N., Miura, Y., Komatsu, T., Takamizo, T., Fujimori, M. 2006. Development of simple sequence repeat (SSR) markers and construction of an SSR-based linkage map in Italian ryegrass (*Lolium multiflorum* Lam.). Theor. Appl. Genet. 113:270-279.

Jenkin,T.J. 1954. Interspecific and intergenic hybrids in herbage grass.VI. *Lolium italicium* A. Br. intercrossed with other *Lolium* types. J.Genet. 52:282–299.

Jenkin,T.J.,Thomas, P.T. 1938. The breeding affinities and cytology to *Lolium* species. J. Bot.14:10-12.

Thorogood, D., Hayward, M.D. 1992. Self-compatibility in *Lolium temulentum* L.: its genetic control and transfer into *L. perenne* L. and *L. multiflorum* Lam. Heredity 68:71-78.

Yamada T. 2001. Introduction of a self-compatible gene of *Lolium temulentum* L to perennial ryegrass (*Lolium perenne* L.) for the purpose of the production of inbred lines of perennial ryegrass. Euphytica 122:213-217.

Chapter 80
Introgression of Novel Traits into White Clover (*Trifolium repens* L.) from Related *Trifolium* Species

Athole Marshall, Michael Abberton, Matthew Lowe, and Ellen Sizer-Coverdale

Abstract Interspecific hybrids have been developed between white clover and the annual, profuse flowering diploid species *Trifolium nigrescens* Viv.(ball clover) as a strategy to improve the seed yield of white clover. Third generation backcross hybrids have now been developed in different leaf size categories that have the agronomic performance of white clover. Assessment of the seed yield of the hybrids in field experiments conducted over two harvest years showed that the medium and large leaved hybrids produced significantly more inflorescences and had a higher seed yield potential than control varieties of comparable leaf size improving the commercial potential of this material. Introgression of the rhizomatous trait from Caucasian clover (*Trifolium ambiguum* M. Bieb) into white clover has been used to improve persistence and tolerance of moisture stress. Advanced hybrids are now at the stage of development where they will be submitted to official variety trials.

Keywords Interspecific hybrids · Introgression · *Trifolium* · Seed yield · Moisture stress

Introduction

Interspecific hybridization can be used to introduce traits into species where limited variation for that trait exists and in many species introgression from closely related species has been an important route to genetic improvement. Within the *Trifolium* genus, interspecific hybrids have been developed to introduce desirable traits into the agronomically important forage legume white clover (*T. repens* L.). Interspecific hybrids between white clover and the annual profuse flowering species *T. nigrescens* Viv. (ball clover), a self-incompatible diploid

A. Marshall (✉)
Institute of Biological, Environmental and Rural Sciences (IBERS), Aberystwyth University, SY233EB, Aberystwyth, Ceredigion, UK
e-mail: thm@aber.ac.uk

C. Huyghe (ed.), *Sustainable Use of Genetic Diversity in Forage and Turf Breeding*,
DOI 10.1007/978-90-481-8706-5_80,

($2n = 2x = 16$) have been developed to transfer clover cyst nematode resistance from *T. nigrescens* into white clover (Hussain et al., 1997) and to introgress reproductive traits into white clover as a means of improving seed yield (Marshall et al., 2008). Previous studies have shown that introgression of reproductive traits from *T. nigrescens* can be achieved without impairing agronomic performance (Marshall et al., 2002). Analysis of the backcross 3 (BC3; BC2 hybrid crossed with *T. repens*) as spaced plants confirmed it produced significantly more inflorescences than *T. repens*.

Varieties of *T. repens* are characterized by their leaf size which is an important factor determining the livestock sector in which they are most appropriately used. BC3 hybrids have now been developed across the range of leaf size categories (Marshall et al., 2008). This paper describes a field experiment designed to compare the reproductive performance of the BC3 hybrids of different leaf size in comparison with control varieties and the original BC3 hybrid in terms of inflorescence production but also the other components of seed yield.

Materials and Methods

Plant Material and Growing Conditions

Development of the plant material has been described previously (Marshall et al., 2008). Two hundred seeds of the BC_3 hybrids (BC_3, small leaved $BC_3(F_2)_s$,medium leaved $BC_3(F_2)_m$ and large leaved $BC_3(F_2)_l$) and control varieties were sown into John Innes No 2 compost in jiffy pot trays in a glasshouse with an 18 hour photoperiod and day/night temperature of 20°C/15°C. Plants were transferred to an unheated glasshouse when they had three trifoliate leaves and cut to a height of 4 cm prior to being transplanted into 2 m × 1.5 m plots at a density of 20 plants per plot. The experimental design was a fully randomised complete block with four replicates.

Reproductive Characters

Date of appearance of the first reproductive bud and the first inflorescence with a white floret were assessed within permanently sited 0.25 m^2 quadrats within each plot. 35 days after peak flowering, all inflorescences were removed from a cut quadrat in each plot. The total number of inflorescences and those which were ripe were counted. Floret number, seed set, thousand seed weight and seed weight per inflorescence were quantified. Potential seed yield (kg/ha) was calculated: ripe inflorescences/m^2 x seed weight/inflorescence x 10. All data were analysed by analysis of variance (ANOVA) of a randomised block design using Genstat for Windows 6th edition (Baird et al., 2002). Where ANOVA was significant at $P < 0.05$ Duncan Multiple Range test was used for comparison of means.

Results and Discussion

Inflorescence production (ripe and total) of the medium and large leaved BC3 hybrids ($BC_3(F_2)_m$, $BC_3(F_2)_l$) was greater than the control varieties in both harvest years (Table 80.1). However the number of ripe inflorescences produced by the small leaved control variety 'AberDale' was significantly greater than the $BC_3(F_2)$s in 2005 but comparable in 2006.

The seed yield components of the backcross hybrids were generally comparable with the control varieties suggesting that increased numbers of inflorescences were not to the detriment of the other components of seed yield (Table 80.2). Differences between backcross hybrids and control varieties in thousand seed weight were greater in 2006 than in 2005 however in both years thousand seed weight was around 0.6 g which is common for seed of *T. repens*. There was no significant difference in seed set in 2005 and relatively few in 2006 indicating that no loss of fertility was evident. The greatest significant differences were observed in floret number per inflorescence which ranged from 48.7–75.6 in 2005 and from 67.9–87.9 in 2006. Floret number is generally related to leaf size with large leaf types having larger inflorescences with more florets. The control varieties generally followed this trend in 2005 but this was less clear in 2006.

Generally the potential seed yield of the backcross hybrids was greater than the control variety reflecting the increased inflorescence production. However in both years the potential seed yield of the $BC_3(F_2)$s was significantly lower than AberDale (Table 80.2). This confirms that introgression of reproductive traits from the annual species *T. nigrescens* is a successful strategy to increase seed yields. However it is surprising that this was not observed in the small leaved selection. This may be because the small leaved control variety AberDale was itself selected for high

Table 80.1 Total and ripe inflorescences m^{-2} of BC_3 hybrids and control varieties

	Ripe		Total	
Hybrid/variety	2005	2006	2005	2006
$BC_3(F_2)_s$	215c	626b	559b	1150b
AberDale (s)	263a	617b	569b	1343a
$BC_3(F_2)_m$	232bc	784a	549bc	1175b
AberHerald (m)	111e	329d	238e	853d
$BC_3(F_2)_l$	245ab	798a	672a	1293a
Olwen (L)	140d	257e	326d	795d
Menna (m)	148d	511c	322d	1034c
BC3	212c	518c	502c	1150b
s.e.d.	40.7	73.6	83.1	117.0
significance	**	***	***	***

Numbers in each column with letters in common are not significantly different at $p < 0.05$.

+Potential seed yield = ripe inflorescences/m^2 x seed wt./inflorescence

NS, not significant; $^*P = 0.05$; $^{**}P = 0.01$; $^{***}P = 0.001$

Table 80.2 1000 seed weight (g), florets/inflorescence, seed set and potential seed yield ($kgha^{-1}$) of BC3 hybrids and control varieties

	1000 seed wt. (g)		Florets/ inflorescence		Seed set		Potential seed yield ($kgha^{-1}$)	
Hybrid/variety	2005	2006	2005	2006	2005	2006	2005	2006
$BC_3(F_2)_s$	0.58	0.54	61.2	76.4	2.56	2.04	198	524
AberDale (s)	0.60	0.60	48.7	71.9	2.34	2.35	231	626
$BC_3(F_2)_m$	0.61	0.54	71.6	79.5	2.35	1.98	243	663
AberHerald (m)	0.71	0.60	60.0	67.9	2.47	1.84	122	246
$BC_3(F_2)_l$	0.59	0.56	59.9	87.9	2.96	2.38	262	931
Olwen (L)	0.63	0.65	75.6	75.8	2.45	2.79	182	367
Menna (m)	0.66	0.66	50.9	70.8	1.96	2.45	104	564
BC3	0.71	0.62	61.5	72.8	2.35	2.22	218	510
s.e.d.	0.046	0.029	5.58	2.9	0.308	0.236	50.4	128.8
significance	*	***	***	***	NS	*	**	***

Numbers in each column with letters in common are not significantly different at $p < 0.05$. NS, not significant; *$P = 0.05$; **$P = 0.01$; ***$P = 0.001$

seed yield (Marshall, 1995) by selection for enhanced peduncle strength leading to greater survival of the early formed inflorescences giving high seed yields over other small leaved varieties in average and difficult years. That the potential seed yields of the $BC_3(F_2)$s and AberDale were comparable indicates the reproductive potential of this material. A number of these hybrids is now being advanced to official variety trials.

Interspecific hybrids have also been developed between *T. repens* and the rhizomatous species Caucasian clover (*T. ambiguum* M. Bieb) to introgress the rhizomatous trait from Caucasian clover into white clover (*T. repens*) as a means of improving persistence and tolerance of moisture stress (Marshall et al., 2001). Following development of an F1 hybrid several generations of backcrossing with *T. repens* as the recurrent parent have been carried out with the objective of retaining an element of rhizomatous growth. Advanced backcrosses have now been developed which are essentially *T. repens* like in appearance with stolons but with a proportion of their dry matter as rhizomes. A key objective has been to select lines with good fertility therefore seed set has been a major selection criterion. Backcross 3 lines have now been developed and these are currently being tested prior to entry into official variety trials.

Acknowledgements This research was funded by the Department for the Environment, Food and Rural Affairs (Defra).

References

Baird, D.S.S., Harding, S.A., Lane, P.W., Murray, D.A., Payne, R.W., Soutar, D.M. 2002. Genstat for Windows, Introduction (6th ed.). VSN International, Oxford.

Hussain, S.W., Williams, W.M., Mercer, C.F., White, D.W.R. 1997. Transfer of clover cyst nematode resistance from *Trifolium nigrescens* Viv. to *T. repens* L. by interspecific hybridisation. Theor. Appl. Genet. 95:1274–1281.

Marshall, A.H. 1995. Peduncle characteristics, inflorescence survival and reproductive growth of white clover (*Trifolium repens* L.). Grass and Forage Sci. 50:324–330

Marshall, A.H., Rascle, C., Abberton, M.T., Michaelson-Yeates, T.P.T., Rhodes, I. 2001. Introgression as a route to improved drought tolerance in white clover (*Trifolium repens* L.). J. Agron. Crop Sci. 187:11–18.

Marshall, A.H., Williams, T.A., Powell, H.G., Abberton, M.T., Michaelson-Yeates, T.P.T. 2002. Forage yield and persistency of *Trifolium repens* x *T. nigrescens* hybrids when sown with a perennial ryegrass companion. Grass Forage Sci. 57:232–238.

Marshall, A.H., Michaelson-Yeates, T.P.T., Abberton, M.T. 2008. Introgression of reproductive traits from *Trifolium nigrescens* increases the seed yield of white clover (*T. repens*). Plant Breed. 127:597–601.

Chapter 81
Studies on the Expression of Exogenous CBF_4 Gene in Transgenic Wheatgrass

Mi Fugui, Wang Guihua, and Dong Shujun

Abstract Transgenic wheatgrass generated from hybrid wheatgrass (*Agropyron cristatum* × *A.desertorum cv. 'Hycrest-Mengnong'*) were identified by PCR analysis and Southern blot. CBF_4, the exogenous gene, a transcriptional factor which plays an active role in plant during drought adaptation and cold acclimation, was transferred into wheatgrass with phosphinothricin acetyltransferase conferring herbicide resistance as selection gene. Results of Northern blot assay displayed that exogenous gene CBF_4 expressed at transcription level in transgenic plants. Further drought stress detection showed that drought tolerance of transgenic plants was enhanced.

Keywords CBF_4 gene · Northern blot · Drought resistance

Introduction

Wheatgrass which contained about 15 species and originated in Eurasia grassland has a broad ecological adaptation. 5 species and 1 subspecies have been identified in the genus. It is distributed in Northeast and Northwest of China. Wheatgrass has good resistance to drought, cold, diseases and pests. It grows early in spring and turns yellow late in autumn, so it has a long green period. Stem and leaves are soft and have high content of nutrients and good palatability. The breeding of this kind of species has been considered to be of great importance. The livestock industry in developed countries has approved the new varieties obtained by the integrated breeding and selection methods, such as Fairway, Parkway, Ruff, Ephraim, Summit, etc.. Three varieties have been released in China, and the variety 'Hycrest-Mongnong' is now widely used.

Mi Fugui (✉)
College of Ecology and Environmental Science, Inner Mongolia Agricultural University, Hohhot 010018, China
e-mail: mfgui@yahoo.com.cn

C. Huyghe (ed.), *Sustainable Use of Genetic Diversity in Forage and Turf Breeding*,
DOI 10.1007/978-90-481-8706-5_81,

Transgenic technology in the wheatgrass breeding has also been applied with the development of biotechnology. Inner Mongolia Agricultural University conducted its transgenic research attempts, established a regeneration system of tissue culture, successfully transferred the p5cs gene and transcription factor gene CBF4 into the species, and received the transgenic plants (Huo et al., 2004a,b). On the basis of CBF4 transgenic wheatgrass plants, this study using Northern blot detected the expression of transgene in transgenic plants, and measured drought resistance of transgenic plants under drought stress, to provide materials for breeding new varieties suited for use of northwest of China.

Materials and Methods

Plants of wheatgrass cv. 'Hycrest-Mengnong' with CBF4 and bar gene have been tested by PCR and Southern blot with non transgenic plants as controls.

Plant total RNA was abstracted by kit; the RNA electrophoresis was performed on formol denatured gel; RNA was transferred by capillary blotting; hybridized and detected by probe labeled with DIG (Sambrook and Russell, 2002; Wang and Fang, 2002a). Plasmid DNA was amplified by PCR, fragments were evicted by kit as template, labelled with DIG high prime DNA labeling and detection starter Kit by random primer method.

Wheatgrass plants which have been tested by northern blot and the control plants were transplanted into the same plastic pots. Pots were filled with 5 kg cropland soil, and put into greenhouse where temperature was kept between 25 and 28°C day and night. Plants were left to recover growth during 3 weeks. While the plants totally recovered growth, drought stress was applied and the permeability of the plasma membrane using conductivity measurement was measured, Leaf Relative Water Content (RWC = (plant fresh weight – dry weight)/(saturated fresh weight – dry weight) × 100%), free proline content using ethanol extraction and MDA (malonyl dialdehyde) content before and after stress were measured.

Results and Analysis

The results of Northern blot of CBF4 transgenic wheatgrass plants and control plants are shown in Fig 81.1. The hybridization band of CBF4 transgenic plants which tested by PCR and Southern blot hybridized with DIG probe was obvious. It proved that exogenous gene CBF4 expressed at transcription level in transgenic plants.

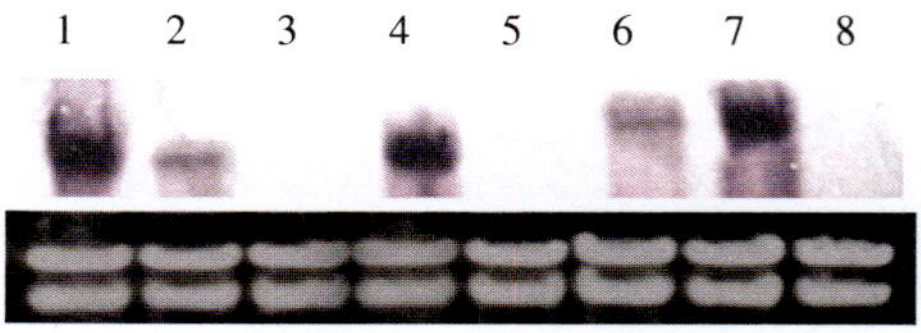

Fig. 81.1 Northern blotting of transgenic plant (CBF4) 1∼7- transgenic plants, 8- negative control

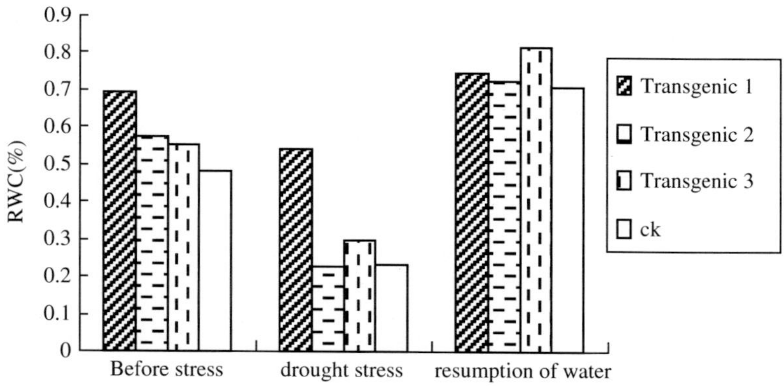

Fig. 81.2 The change of RWC under Drought stress (CBF_4)

Figure 81.2 shows the leaf relative water content of transgenic plants. It can be seen that the transgenic plant leaves contained higher relative water content in the same environment, under drought stress. The RWC decreased but transgenic plants decreased slowly. Leaf RWC could reflect the water status, with the transgenic plants exhibiting smaller RWC changes, stronger plant water-retention capacity and more drought resistance.

The plasma membrane relative permeability increased after drought stress, and in the same environment, the plasma membrane relative permeability of three transgenic plants is less affected than control plants. The membrane permeability of the plasma membrane expresses the extent of damage, with the greater the permeability, the more serious the damages, the worse the drought resistance.

Figure 81.3 showed the free proline content of transgenic plants. The free proline content of different plants was at the same level before stress and after the resumption of water. During the drought stress, free proline content increased, and transgenic plants increased more than the control plants. Because the free proline increases intracellular osmotic potential and hydrophilic ability, free proline content is an index of tolerance to water stress.

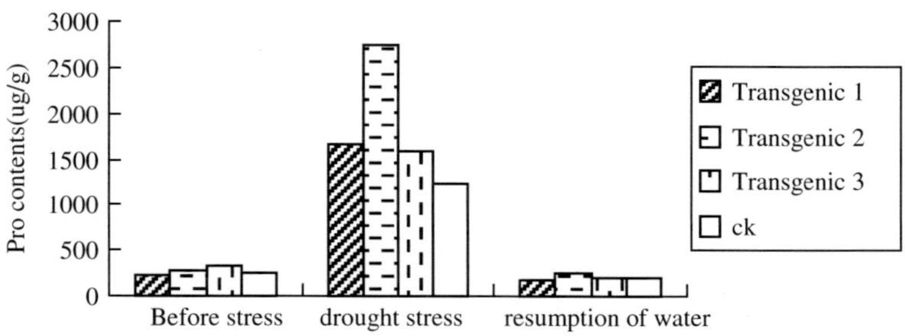

Fig. 81.3 The change of Proline contents under drought stress (CBF4)

The results of MDA content of transgenic plants showed that MDA content increased, with the transgenic plants exhibiting a weaker increase than control plants after stress. Because MDA is a product of ultimate of decomposition in membrane lipid peroxidation when plant get old or in adversity (including drought stress) conditions, content of MDA can reflect the extent of injury of the plant in adversity.

Discussion and Conclusion

When exogenous gene in the plasmid vector through homologous or heterogenous recombination integrated randomly into the chromosomal DNA, it became a practical common problem in genetic engineering techniques to know whether exogenous gene in transgenic offspring has stable expression. The expression of exogenous gene in transgenic plants is subject to many factors, on the basis of the foregone reports, the factors impacted by the integrated approach as follows: (1) the number of transgenic integration loci (copies), (2) gene composing of the inserted loci (position effect), (3) conformation changes of integrated T-DNA (Zeng and Zhan , 2004). Alvarez et al. (2000) and Cervera et al. (2000) found that high-copy T-DNA insertion to the same or different sites often leads to gene silencing. Cervera et al. (2000) found in transgenic citrus that the GUS gene expression level and (UIDA) copies presented a negative correlation. Yang et al. (1998) found in the study of the integration and expression of nuclear shell protein transgenic peanut (*Arachis hypogaea*), the gene high-copy led to silencing. But Dominguez et al. (2000) found in transgenic citrus research that all transgenic lines regardless of the number of GUS gene (UIDA) copies had a similar expression of GUS activity. Hua et al. (2001) found no real link between copy number and expression or silence when they studied the expression and copy number of exogenous gene in gun obtained transgenic rice. Fang et al. (1991) found only some of the anti-viral RNA gene transcript in study of bivalent CMV and TMV tobacco plants. They speculated that this was due to the impact of a reorganization or DNA inserted position.

In the present study, it has been tested by Southern blot analysis that the gene CBF4 integrated into wheatgrass genome in a multi-copy form. Gene composition of the inserted location and structure of integrated T-DNA have not been researched, so the reasons for some of transforming gene inactivation need to be further explored. The Northern hybridization in this study is on the basis of Southern hybridization, so it proved that gene inactivation occurred in the RNA level.

The CBF4 gene used in the paper is another member of CBF family isolated from *Arabidopsis thaliana* reported by Haake et al. (2002), the same as CBF1, CBF2 and CBF3, all of which have a conservative AP2 region, a drought transcription factor and induced by drought stress. The drought and cold resistance increased significantly in transgenic *Arabidopsis* plants which have overexpression of CBF4. In dealing with drought stress or ABA, CBF4 induced expression in the CaMV35S promoter driven, CBF4 excessively express in transgenic plants, and the resistance of transgenic plants to drought and cold tolerance improved (Haake et al., 2000;

Thomashow et al., 2001; Wang and Fang, 2002b; Huo et al., 2005; Huo et al., 2006; Guo et al., 2006). In the present study, after drought stress, leaf relative water content of CBF4 wheatgrass transgenic plants decreased more slowly that control plants; the relative permeability of the plasma membrane was less affected than non transformed plants; the increased extent of free proline content and MDA content was larger in transgenic plants. At present, the plant leaf relative water content, membrane permeability, free proline content and MDA content were considered to be plant water stress indicators because they are related to plant water-retention capacity, extent of plasma membrane damage, ability to regulate the permeability. Taking the above factors, the research can prove that the drought resistance of CBF4 transgenic wheatgrass plants has been enhanced.

The exogenous gene in some transgenic plants expressed at transcriptional level and the drought resistance of expressed plants has increased notably.

References

Alvarez, M.L., Guelman, S, Halford, N.G., Lustig, S., Reggiardo, M., et al. 2000. Silencing of HMW glutenins in transgenic wheat expressing extra HMW subunits. Theor. Appl. Genet. 100: 319–327.

Cervera, M., Pina, J.A., et al. 2000. A broad exploration of a transgenic population of citrus:stability of gene expression and phenotype. Theor. Appl. Genet. 100:670–677.

Dominguez, A., Guerri, J., Cambra, M., Navarro, L., Moreno, P., et al., 2000. Efficient production of transgenic citrus plants expressing the coat protein gene of citrus tristeza virus. Plant Ceall. Rep. 19:427–433.

Fang, R.X., Tian, Y.C., Wang, G.L., Qin, X.F., Yang, M.Z., Li, T.Y., Xu, B.X., Mang, K.Q., Zhang, Z.C., Wu, Q., Zhou, R.H., Wang, F.L., Zhao, G., Han, X.D. 1991. Transgenic tobacco plants resistant to infection of both tobacco mosaic-virus and cucumber mosaic-virus. Chin. Sci. Bull. 36(6):524–526

Guo, X., Wei, Z., Xing, S., Yu,Z. 2006. Progress of CBF Transcription Factors to Improve Plant Tolerance to Abiotic Stresses. Mol. Plant Breed. 4(3):419–424.

Haake, V., Cook, D., Riechmann, J.L., Pineda, O., et al. 2002. Transcription factor CBF4 is a regulator of drought a daptation in *Arabidopsis*. Plant Physiol. 130:639–648.

Hua, Z.-H., Zhu, X.-F., Lin, H.-S., et al. 2001. Studies of the integration and expression of exogenes in transgenic rice obtained via particle Bombardment transformation. Acta Genetica Sinica 2(8):1012–1018.

Huo, X., Mi, F., Yun, J., Wei, J. 2005. Cloning and Functional Analysis of an Induced Promoter of Transcriptional Factor CBF4 from Arabidopsis. Mol. Plant Breed. 3(3):363–368.

Huo, X.W., Wei, J.-H., Xu, C.-B., Mi, F.-G., Yun, J.-F. 2004a, Plant Regeneration and Genetic Transformation in Wheatgrass (*Agropyron cristatum* × *A.desertorum* cv.'Mengnong'). Scientia Agricultura Sinica 37(5):642–647.

Huo, X.-W., Wei, J.-H., Zhang, H., Mi, F.-G., Yun, J.-F. 2004b. Study of Plant Regeneration in Wheatgrass (*Agropyron* Gaertn). Acta Agriculturae Boreali-Sinica 19(1):17–20.

Huo, X.-W., Wei, J.-H., Xu, h.-b., Mi, F.-G., Yun, J.-F. 2006. Plant Regeneration and Genetic Transformation in Wheatgrass (*Agropyron cristatum* × *A.desertorum* cv.'Mengnong') (English). Agricultural sciences in China 5(9):648–654.

Sambrook, J., Russell, D.W. 2002. Molecular cloning 3. Science press, Beijing.

Thomashow, M.F., Gilmour, S.J., Stockinger, E.J., et al. 2001. Role of the *Arabidopsis* CBF transcriptional activators in cold acclimation. Physiol. Plant 112:171–175.

Wang, G.L., Fang H.J., eds. 2002a. Plant Genetic Engineering Principle and Technique (pp. 428–432). Science Press, Beijing, China.

Wang, G.-L., Fang, H.J. 2002b. Plant genetic engineering principle and technology. Science press, Beijing.

Yang, H., Singsit, C., Wang, A., et al. 1998. Transgenic peanut plants containing a nucleocapsid protein gene of tomato spotted wilt virus show divergent levels of gene expression.Plant Cell Rep.17:693–699.

Zeng, F.-S., Zhan, Y.-G. 2004. Integration of Exogenous Genes in Transgenic Plant Genomes: Characteristics and Approaches. Chin. Bull. Bot. 21(5):565–577.

Chapter 82
Heterotic Response from a Diallel Analysis between Alfalfa Cultivars of Different Geographic Origin

Dragan Milić, Slobodan Katić, Aleksandar Mikić, and Đura Karagić

Abstract A semihybrid variety development strategy could capitalize on natural hybrid vigor that exists between alfalfa (*Medicago sativa* L.) germplasms, populations and cultivars. Successful semihybrid model is the identification of improved germplasm with superior agronomic traits as well as good combining ability between designated heterotic groups. Identification of heterotic groups and patterns among breeding populations provides fundamental information to help alfalfa breeders more knowledgeably manipulate heterosis. In this experiment, hybrids among several germplasm sources were evaluated. The significant variation among crosses was attributed primarily to general combining ability (GCA) effects, while specific combining ability effects were also significant. Both mid-parent heterosis (MPH), ranging from 2.6 to 25.4 %, and high-parent heterosis (HPH), ranging from 4.2 to 15%, was detected. The hybrids between French and Spanish, as well as between Iranian and Spanish cultivars, demonstrated the highest cross mean performance in diallel crosses. The results indicate that these crosses should be recognized as a heterotic groups. MPH results suggest that we may have capitalized heterotic response between divergence alfalfa cultivars (different geographic origin) to improve alfalfa forage yield.

Keywords Alfalfa Diallel · GCA · Heterosis · SCA

Introduction

The discovery of heterosis and its utilization for yield increase since the beginning of the 20th century revolutionized and accelerated plant breeding. Although pioneering studies in the 1950s obtained significant hybrid vigor for forage yield,

D. Milić (✉)
Institute of Field and Vegetable Crops, M. Gorkog 30, Novi Sad 21000, Serbia
e-mail: dmilic@ifvcns.ns.ac.rs

C. Huyghe (ed.), *Sustainable Use of Genetic Diversity in Forage and Turf Breeding*,
DOI 10.1007/978-90-481-8706-5_82,

the complex genetic structure (autetraploidy), tetrasomic mode of inheritance and other characteristics prevented the use of the known procedures (for diploid species) for utilization of the heterotic effect in alfalfa breeding. Studies conducted in the USA indicated a possibility of exploiting additive and nonadditive genetic effects including the allelic interactions present in the autotetraploid mode of inheritance for increasing alfalfa yield. In papers published around the end of the 20th century and the beginning of this century, alfalfa breeders reported that significant specific combining abilities existed in hybrids made by crossing divergent genotypes and populations of alfalfa (Riday and Brummer, 2002; Segovia-Lerma et al., 2004). Rotili was the first to understand and propose the effect of heterosis to be partially used for development of free-hybrids by crossing lines obtained from 2–3 generations of selfing (Rotili et al., 1999). The idea on partial utilization of heterosis in alfalfa that emerged in the USA proposed the development of semi-hybrids obtained by crossing genetically divergent germplasms and identifying heterotic groups (Brummer, 1999).

The objective of this study was to conduct a diallel analysis of yield parameters, to determine heterotic effects, and to assess the usefulness of these cultivars as potential heterotic groups for increasing alfalfa yield in Serbia.

Material and Methods

Our research was conducted in Institute of Field and Vegetable Crops Novi Sad, Serbia at experimental field during 2006–2008. A complete diallel including reciprocals was made during 2003 and 2004 between 5 alfalfa cultivars (*Medicago sativa* L. ssp. *sativa*) of different geographic origin, namely NS Banat ZMS II from Serbia, Zuzana from Czech Republic, Ghareh Yon Geh from Iran, RSI 20 from Spain and Pecy from France. These cultivars were tested previously and differed significantly in origin and phenotypic characteristics (Milić, 2007). Cultivars were hand- crossed without emasculation in a diallel mating design. For each pairwise cross, 5 plants were chosen at random from each of the two cultivars (~100 florets per plant) to obtain the F_1 generation. A spaced plant field was established in 2006 which included the five alfalfa cultivars (parents) and their 20 diallel hybrids (F_1) sown in three replications, 20 plants per replication. Heterotic responses were determined by evaluating dry matter yield (g/plant), while all 20 plants per replications were harvested five times in each of 2 years (2007 and 2008). A dialed analysis was used according to Zhang et al. (2005). Midparent heterosis, which describes the performance of crosses relative to the average performance of their parents, and high-parent heterosis (HPH), which describes the performance of crosses relative to the highest-yielding parent, were computed as follows:

$$\mathrm{MPH} = 100^{*}[\mathrm{F1} - \{(\mathrm{P1} + \mathrm{P2})/2\}]/\{(\mathrm{P1} + \mathrm{P2})/2\}$$

$$HPH = 100^*[(F1 - HP)/HP]$$

The absolute MPH and HPH effects were tested using F test, using linear contrast coefficients according to Steel and Torrie (1980).

Results and Discussion

The differences in yield among the parental cultivars were highly significant (Table 82.1).

The yields ranged from 93 g/plant in the cultivar Pecy to 112 g/plant in the cultivar Ghareh Yon Geh. The yields in the cultivars Ghareh Yon Geh and RSI 20 were significantly higher than those of all other cultivars. Five hybrids and eight reciprocal crosses had significantly higher yields than the average of the 5 commercial cultivars used as parents. Yield differences between the hybrids and their reciprocals were significant in some cases, pointing out a possible significance of reciprocal crossing. However, the diallel analysis produced no significant reciprocal effects (Table 82.2). The diallel analysis showed highly significant GCA and SCA effects, revealing that both additive and non-additive effects were significant for determining forage dry matter yield in alfalfa. This suggests that alfalfa yield may be improved by accumulation of favorable genes, but also by using the effect of heterosis. The GCA mean square was significantly higher (three times) than the SCA effects, which reveal a relatively greater importance of additive genes over non-additive ones. Importance of GCA effects was demonstrated by Segovia-Lerma et al., (2004) and Sakiroglu and Brummer (2007). In addition to this, GCA x year interaction was also observed, showing differential reactions of the hybrids to the environmental circumstances of 2007 and 2008 (Table 82.2).

These differences may be a consequence of biological factors acting in the second and the third years of alfalfa stand life.

The GCA effects varied between 4.42 g/plant in the cultivar Pecy to 6.45 g/plant in RSI 20 (Table 82.3)

Two parents, Ghareh Yon Geh and Pecy, had positive GCA effects while the others had negative GCA effects. Differences in GCA effect between genotypes and

Table 82.1 Dry matter yield (g/plant) of parent cultivars (diagonal, underlined) and their diallel hybrids and reciprocals (above and under diagonal) across 2 years (2007–2008)

Cultivar	Banat	Ghareh Yon Geh	Zuzana	Pecy	RSI 20
Banat	<u>109</u>	107	112	100	120**
Ghareh Yon Geh	118**	<u>112*</u>	116**	111	122**
Zuzana	114*	119**	<u>97</u>	99	117**
Pecy	111	119**	106	<u>94</u>	129**
RSI 20	119**	123**	119**	124**	<u>112*</u>

Mean of five checks = 104; LSD (0.05) = 8; (0.01) = 11

*, ** Yields significantly higher than the average of checks at $\alpha = 0.05$ and 0.01 respectively

Table 82.2 Mean squares for dry matter yield (g/plant) from a diallel involving five alfalfa cultivars across 2 year (2007–2008) according Griffing model 1 method 1 (1966)

Source	DF	Across years
GCA	4	1267.53**
SCA	10	523.24**
REC	10	123.45
MAT	4	231.21
NMAT	6	51.61
GCA × Y	4	497.02**
SCA × Y	10	105.06
REC × Y	10	89.22
MAT × Y	4	142.61
NMAT × Y	6	53.62

** Mean squares significant at $\alpha = 0.01$ respectively

Table 82.3 Estimates of diallel effects for general combining ability (GCA) and specific combining ability (SCA) and their respective standard errors (in parentheses) for alfalfa dry matter yield (g/ plant) of the 2 year (2007–2008)

Estimate	Effects	Estimate	Effects	Estimate	Effects
GCA_{11}	−1.17**	SCA_{12}	−2.01	SCA_{15}	3.22
GCA_{22}	2.86	SCA_{13}	4.82	SCA_{25}	6.58
GCA_{33}	−3.72*	SCA_{14}	−1.73	SCA_{35}	11.33
GCA_{44}	−4.42**	SCA_{23}	5.29	SCA_{45}	21.80**
GCA_{55}	6.45**	SCA_{24}	3.57	(6.60)	
	(1.64)	SCA_{34}	−2.68		
		(3.38)			

*, ** Effects are estimates significantly differ from zero at $\alpha = 0.05$ and 0.01 respectively

populations were observed earlier by Segovia-Lerma et al., (2004) and Sakiroglu and Brummer (2007). The best hybrid with significant SCA effects was produced by crossing the parents with the greatest negative and positive GCA effects (Pecy × RSI 20). The hay yield of this hybrid was significantly higher than the average yield of its parental cultivars. Six hybrids had positive SCA effects, 3 had negative effects. Positive MPH (mid-parent heterosis) was found in eight hybrids, negative MPH in two hybrids (Table 82.4). Highly significant MPH was found in two hybrids, Zuzana x RSI 20 and Pecy × RSI 20, while the reciprocals had highly significant MPH in five hybrids (Table 82.4). Disregarding the slight differences, both hybrids and their reciprocals had similar MPH values. A significant HPH (high-parent heterosis) was found in the hybrid Pecy × RSI 20 and its reciprocal cross and their yields were the highest in the trial.

According to the results of Bhandari et al., (2007), parental selection based exclusively on MPH and HPH may sometimes be misleading, because hybrid may express heterosis, but its total yield may not be higher than check cultivars. In that way, the choice of hybrids should be based on their yields relative to check cultivars rather then on their absolute or relative heterosis response. Our results showed that the selection based on MPH and HPH could be successful or at least useful in the

Table 82.4 Percent relative midparent heterosis (MPH) and high-parent heterosis (HPH) for forage dry matter yield in diallel crosses among five alfalfa cultivars

MPH	Banat	Ghareh Yon Geh	Zuzana	Pecy	RSI 20
Banat	–	−2.6	9.1	−0.9	8.7
Ghareh Yon Geh	7.2	–	10.9*	7.7	8.9
Zuzana	11.1*	14.4**	–	3.9	12.1**
Pecy	10.2*	16.0**	11.4**	–	25.4**
RSI 20	7.7	9.4*	13.9**	20.6**	–
HPH	Banat	Ghareh Yon Geh	Zuzana	Pecy	RSI 20
Banat	–	−4.2	3.0	−7.7*	7.0
Ghareh Yon Geh	5.5	–	3.2	−1.2	8.8
Zuzana	5.0	6.4	–	2.2	4.4
Pecy	2.6	6.3	9.7	–	15.0**
RSI 20	6.1	9.4	6.0	10.6*	–

*, ** Heterosis significantly differ from zero at $\alpha = 0.05$ and 0.01 respectively

choice of parental components for alfalfa breeding for high yield. The highest yield, significant SCA effects and MPH and HPH were in the hybrid Pecy × RSI 20 and its reciprocal, i.e., in the cultivars with the smallest and the greatest GCA effects and the highest and the lowest yields, respectively (Table 82.4).

In general, our results show that heterotic effects are significant for alfalfa yields, suggesting that identifying heterotic groups and developing ways to use and maintain them is of value in alfalfa breeding programs.

Conclusions

Our result have confirmed that hybrids superior in yield may be obtained by crossing high-yielding *Medicago sativa* ssp. *sativa* cultivars. GCA and SCA effects are important for explaining yield variations among the hybrids. The parents of the best hybrid had the highest and the lowest GCA effects. The highest yielding hybrid had the highest positive SCA effects, which indicated their importance in determining hybrid productivity. The obtained results showed clearly that alfalfa yield may be improved by exploiting additive and non-additive genetic effects.

Acknowledgements This research has been co-financed by Serbian Ministry of Science and Technological Development within the project BT 20090.

References

Brummer, E. C. 1999. Capturing hetrerosis in forage crop cultivar development. Crop Sci. 39: 943–954.

Bhandari, H.S., Pierce, C.A., Murray, L.W., Ray, I.M. 2007. Combining Abilities and Heterosis for Forage Yield among High-Yielding Accessions of the Alfalfa Core Collection. Crop Sci. 47: 665–671.

Milić, D. 2007. Variability of quantitative traits in genetically divergent alfalfa (*Medicago sativa* L.) genotypes. MSc thesis, University of Novi Sad, Faculty of Agriculture.

Riday, H., Brummer, E.C. 2002. Forage Yeilds Heterosis in Alfalfa. Crop Sci. 42: 716–723.

Rotili, P., Gnocchi, G., Scotti, C., Zannone, L. 1999. Some aspects of breeding methodology in alfalfa. The Alfalfa Genome. http://www.naaic.org/TAG/TAGpapers/rotili/rotilipapers.html

Sakiroglu, M., Brummer, E.C. 2007. Little heterosis between Alfalfa populations derived from the Midwestern and Southwestern United States. Crop Sci. 47: 2364–2371.

Segovia-Lerma, A., Murray, L.W., Townsend, M.S., Ray, I.M. 2004. Population-based diallel analyses among nine historically recognized alfalfa germplasms. Theor. Appl. Genet. 109:1568–1575.

Steel, R.G.D., Torrie, H.J. 1980. Principles and Procedures of Statistics a Biometrical approach. McGraw–Hill Book Company, NY, USA.

Zhang, Y., Kang, M.S., Lamkey, K.R. 2005. DILALLEL-SAS05: A comprehensive program for griffing's and Gardner-Eberhart Analyses. Agron. J. 97: 1097–1106.

Chapter 83
Using Bulk-Hybrids for Breeding Adapted Genotypes of Subterranean Clover

Phillip Nichols, Philip Cocks, and Clive Francis

Abstract Changes were measured over 16 years in a self-regenerating, bulk-hybrid subterranean clover population, comprised of F_2 seed from 253 crosses, sown at Nabawa and Mt Barker, low and high rainfall areas, respectively in south-western Australia. Seed banks were sampled annually and kept in cold storage. Population changes on 26 morphological, agronomic and chemical characters were measured three and 16 years after sowing, in comparison with the ancestral bulk-hybrid population. Changes in population means were observed in 20 characters at one or both sites, with much of this occurring within 3 years. Natural selection at Nabawa favoured early flowering of long duration, thick peduncles, high harvest index and high hardseededness, while at Mt Barker it favoured late flowering of short duration, large leaves and long, thick petioles at flowering, thick stems with long internodes, long, thin peduncles with a high burial angle, large plants at maturity, low hardseededness and high biochanin A and total oestrogenic isoflavone contents. High seed production capacity, with high seed weight and seeds per burr, was important at both sites. The use of bulk-hybrid populations is suggested as a low-input means of breeding and selecting well-adapted subterranean clovers.

Keywords Bulk hybrid populations · Subterranean clover · *Trifolium subterraneum* · Plant breeding · Evolution · Natural selection

Introduction

Subterranean clover (*Trifolium subterraneum* L.) is a self-pollinated, diploid ($2n = 16$) species native to the Mediterranean basin and the Atlantic coast of Western Europe (Katznelson and Morley, 1965). It is well adapted to the

P. Nichols (✉)
Cooperative Research Centre for Future Farm Industries, Crawley, WA 6009, Australia
e-mail: phil.nichols@agric.wa.gov.au

C. Huyghe (ed.), *Sustainable Use of Genetic Diversity in Forage and Turf Breeding*,
DOI 10.1007/978-90-481-8706-5_83,

Mediterranean-type climate of southern Australia, where it is the most widely sown annual pasture legume (Nichols et al., 2007). There are two keys to its widespread use. One is its tolerance of heavy grazing, attributable to its prostrate habit and its ability to burial its burrs and protect against seed predation. The other is a range of cultivars which differ in flowering time, enabling it to be grown in environments with winter growing season lengths ranging from 4 to 9 months. To date 42 cultivars have been released in Australia (Nichols et al., 2007), but further genetic gains can be expected if plant breeders have a greater understanding of the characters important for adaptation to key environments.

Bulk-hybrid populations, in which seed bulked from a set of crosses is repeatedly harvested and re-sown over a number of generations, have been used in barley (*Hordeum vulgare* L.) to indicate the adaptive importance of a range of characters (Allard, 1988) and as a breeding method to improve adaptation (Allard, 1999). No subterranean clover bulk-hybrid studies have been published. Moreover, no bulk-hybrid studies for any species have been reported where the same ancestral population has been sown at more than one site. Subterranean clover is an ideal candidate for such studies, having the essential attributes listed by Allard (1988) of being annual, self-pollinated, diploid, and having broad variation for important agronomic characters. Subterranean clover also has an advantage over crop plants of being a self-regenerating annual that does not require harvesting and re-sowing each generation.

This paper examines changes in 26 morphological, agronomic and oestrogenic isoflavone characters over 16 years in a bulk-hybrid subterranean clover population at two contrasting rainfall sites in south-western Australia. A full analysis is given in Nichols et al. (2009).

Materials and Methods

The ancestral bulk-hybrid population, referred to as the 'Original mixture', consisted of F_2 seed derived from 253 crosses. It was sown at a rate of 25 kg/ha into 0.4 ha plots at Mt Barker (34°38'S, 117°33'E), and Nabawa (28°30'S, 114°47'E), chosen to represent long and short growing season environments, respectively. The long-term mean annual rainfall at Nabawa is 454 mm and at Mt Barker is 743 mm. Plots were allowed to regenerate naturally over a 16-year period. Grazing was conducted by sheep according to local district practices. Further site and management details are given in Nichols et al. (2009). Seed bank samples, consisting of 20 bulked random 20 dm^2 quadrats, were collected each summer after pasture senescence. Threshed seeds were dried, sealed in aluminium foil packets and maintained at 4°C. A sample of the 'Original mixture' was also stored.

Seeds of the 'Original mixture' and from Nabawa and Mt Barker three and 16 years after sowing, were scarified, germinated in the glasshouse and transplanted, 1 m apart, to the field in Perth (31°57'S, 115°50'E) in a randomised block design with four replicates. Total plant numbers consisted of 200 from each

Table 83.1 Mean values for 26 characters measured in bulk-hybrid populations 3 and 16 years after sowing at Mt Barker and Nabawa and in the 'Original mixture' (from Nichols et al., 2009). Significance levels and least significant differences (l.s.d.) are shown for each character. Transformed means and l.s.d. values are presented in parentheses

	Population						
		Mt Barker		Nabawa			
Character	'Original Mixture'	Year 3	Year 16	Year 3	Year 16	Significance level[c]	L.s.d. ($P \leq 0.05$)
Seedling measurements (80 days after sowing)							
Plant diameter (mm)	174	178	184	178	167	n.s.	
Leaf size (cm^2)[a]	3.6(0.64)	3.5(0.65)	3.7(0.66)	3.8(0.67)	4.0(0.69)	(n.s.)	
Specific leaf weight (mg/mm^2)	34.4	32.9	31.8	34.6	33.3	n.s.	
Oestrogenic isoflavone content (% of dry matter)							
Formononetin[b]	0.13 (0.26)	0.23 (0.33)	0.23 (0.33)	0.24 (0.38)	0.19 (0.34)	(n.s.)	
Genistein[a]	0.78 (0.23)	0.84 (0.24)	0.80 (0.23)	0.78 (0.23)	0.67 (0.21)	(n.s.)	
Biochanin A[b]	0.82 (0.82)	1.16 (0.98)	1.56 (1.16)	1.13 (0.99)	1.16 (1.03)	***	(0.107)
Total isoflavones	1.73	2.23	2.60	2.15	2.02	**	0.362
Measurements at first flowering							
Flowering time (days)	110	128	130	87	88	***	3.5
Leaf size (cm^2)[a]	{7.6}[d]	10.0 (0.96)	11.4 (1.03)	4.0 (0.57)	3.9 (0.57)	***	(0.069)
Petiole length (mm)	{119}[d]	136	153	84	84	***	9.7
Petiole diameter (mm)[a]	{1.35}[d]	1.49 (0.39)	1.57 (0.41)	1.13 (0.33)	1.12 (0.33)	***	(0.012)
Internode length (mm)[b]	{46.9}[d]	57.5 (7.50)	62.8 (7.85)	34.2 (5.76)	33.3 (5.70)	***	(0.407)
Stem diameter (mm)	{2.60}[d]	2.66	2.71	2.46	2.47	***	0.048
Peduncle length (mm)	{64.2}[d]	60.8	64.9	50.0	51.8	***	3.96
Peduncle diameter (mm)	{1.23}[d]	1.20	1.21	1.25	1.27	**	0.032
Peduncle angle (degrees)	{42.5}[d]	46.8	49.1	34.5	34.7	***	2.07

Table 83.1 (continued)

Character	Population					Significance level[c]	L.s.d. ($P \leq 0.05$)
		Mt Barker		Nabawa			
	'Original Mixture'	Year 3	Year 16	Year 3	Year 16		
Mature plant and seed characters							
Time to senescence (days)	212	221	222	204	209	***	3.2
Flowering duration (days)	102	92	91	117	122	***	3.6
Length of main stem (cm)	82.2	78.6	82.9	86.0	88.7	n.s.	
Plant weight (g)[b]	187 (12.9)	278 (16.3)	338 (18.0)	129 (10.7)	122 (10.6)	***	(2.37)
Seed yield (g)[a]	14.5 (1.05)	25.5 (1.35)	25.7 (1.33)	11.7 (1.02)	12.8 (1.07)	**	(0.172)
Harvest Index[a]	7.3 (0.88)	8.6 (0.95)	7.1 (0.86)	10.1 (0.99)	10.3 (1.02)	***	(0.070)
Seed weight (mg)	8.6	9.3	9.5	9.6	9.7	*	0.64
Seeds per burr	3.1	3.5	3.7	3.3	3.4	**	0.26
Initial hardseed (% hard at harvest)	64.8	60.0	63.3	57.0	54.2	**	5.63
Hardseededness (% hard after 16 weeks)	28.1	9.1	6.7	39.6	41.3	***	3.29

[a] Log_{10} transformed data in parentheses.
[b] Square root transformed data in parentheses.
[c] Significance levels: * = $P < 0.05$; ** = $P < 0.01$; *** = $P < 0.001$; n.s. = non-significant.
[d] Data from one replicate only. Not included in ANOVA

Nabawa and Mt Barker population and 160 of the 'Original mixture'. The site was hand-weeded and irrigated as required. Twenty six morphological, agronomic and oestrogenic isoflavone characters were measured on each plant (Table 83.1). Analyses of Variance were conducted, following tests for normality and homogeneity of variance. Data was unavailable for three replicates of the 'Original mixture' for petiole, peduncle and stem characters at flowering time; these were excluded from analyses. Further details on characters and methods are given in Nichols et al. (2009).

Results

From one highly variable bulk-hybrid population, two markedly different populations evolved after 16 years. Population changes were observed in 20 characters, with much of this occurring within the first 3 years (Table 83.1). Variability also declined in 11 characters at one or both sites (Nichols et al., 2009). Strong and rapid directional selection occurred for early flowering at Nabawa and late flowering at Mt Barker. Other characters important for adaptation to Nabawa were long flowering duration, thick peduncles, high harvest index and high hardseededness after 16 weeks. At Mt Barker characters important for adaptation included short flowering duration, large leaves and long, thick petioles at flowering, thick stems with long internodes, long, thin peduncles with a high burial angle, large plants at maturity, low hardseededness after 16 weeks and high biochanin A and total isoflavone contents. High seed production capacity, with high seed weight and many seeds per burr, was important at both sites.

Discussion

Evolution at the high rainfall site of Mt Barker towards large plants with large leaves and long petioles at flowering is likely to have been related to higher competitiveness in spring for light than at Nabawa, arising from a longer period of vegetative growth and higher plant densities. Conversely, evolution at Nabawa suggests a trend towards factors leading to greater persistence, particularly early flowering, high harvest index and hardseededness. Early flowering acted to ensure that some seed production occurred before onset of the long summer drought. A higher harvest index and longer flowering duration suggests a much greater proportion of plant resources was devoted to seed production. Hardseededness also conferred an advantage in the more 'risky' environment of Nabawa, whereby the hazards of germination and seed production were spread over several seasons. Selection for increased seed production at both sites was consistent with Allard (1988) in barley bulk hybrids.

Biochanin A has been associated with tolerance to redlegged earth mite (*Halotydeus destructor* (Tucker)) (Wang et al., 1999), the most important pest of subterranean clover. These pests are of greater significance in the cooler climate of

Mt Barker than at Nabawa and may explain why selection for high biochanin A content was stronger in the Mt Barker population.

The bulk-population breeding method (Allard, 1999) is well suited to breeding well-adapted, self-pollinated annual pasture legumes, such as subterranean clover. This study showed that genotypes adapted to test environments can be selected after just three seasons. Nichols et al. (2009) also showed further adaptive fine-tuning with increased generations. The success of the method for selecting adapted genotypes is likely to be dependent on: (i) parents containing genes for desirable characters; (ii) sites being representative of target environments; and (iii) trial management being representative of typical farm practice. The method is cheaper to operate than traditional plant breeding methods, requiring few inputs following the establishment year and can be conducted on small plots. It also allows selection for regional adaptation at sites distant from a main breeding centre. The main disadvantage is that the breeder has limited control over the selection directions. However, it would be possible to impose treatments on populations to select for resistance or tolerance to pests, diseases, or herbicides or to particular farm management practices. Judicious site selection may also allow selection of plants adapted to soil constraints such as salinity, waterlogging, low pH or low nutrient status.

Acknowledgements Drs Bill Collins, John Gladstones and the late Reg Rossiter contributed to the planning and conduct of the bulk-hybrid trial sites. The assistance of Kevin Foster and Peter Skinner in collecting and maintaining seed samples is gratefully acknowledged. Funding for attendance at the Eucarpia conference was provided by the AW Howard Memorial Trust Inc.

References

Allard, R.W. 1999. Principles of plant breeding (2nd ed). John Wiley & Sons Limited, New York

Allard, R.W. 1988. Genetic changes associated with the evolution of adaptedness in cultivated plants and their wild progenitors. J. Hered. 79:225–238.

Katznelson, J., Morley, F.H.W. 1965. A taxonomic revison of sect. *Calycomorphum* of the genus *Trifolium*. I. The geocarpic species. Isr. J. Bot. 14:112–134.

Nichols, P.G.H., Loi, A., Nutt, B.J., Evans, P.M., Craig, A.D., Pengelly, B.C., Dear, B.S., Lloyd, D.L., Revell, C.K., Nair, R.M., Ewing, M.A., Howieson, J.G., Auricht, G.A., Howie, J.H., Sandral, G.A., Carr, S.J., de Koning, C.T., Hackney, B.F., Crocker, G.J., Snowball, R., Hughes, S.J., Hall, E.J., Foster, K.J., Skinner, P.W., Barbetti, M.J., You, M.P. 2007. New annual and short-lived perennial pasture legumes for Australian agriculture – 15 years of revolution. Field Crops Res. 104:10–23.

Nichols, P.G.H., Cocks, P.S., Francis, C.M. 2009. Evolution over 16 years in a bulk-hybrid population of subterranean clover (*Trifolium subterraneum* L.) at two contrasting sites in south-western Australia. Euphytica 169:31–48.

Wang, S.F., Ridsdill-Smith, T.J., Ghisalberti, E.L. 1999. Levels of isoflavonoids as indicators of resistance of subterranean clover to redlegged earth mite *Halotydeus destructor*. J. Chem. Ecol. 25:795–803.

Chapter 84
Fiber Content and Plant Development in *Festulolium*

Liv Østrem and Arild Larsen

Abstract Neutral detergent fiber, NDF, and the indigestible part, iNDF, were observed from market cultivars of *Festulolium* (Hykor, Felopa), Norwegian *Festulolium* candivars, the parent species perennial ryegrass (*Lolium perenne*) (cv. Napoleon, 4*x*), meadow fescue (*Festuca pratensis*) (cv. Fure) and timothy (*Phleum pratense*) (cv. Grindstad). Additionally leaf:stem ratio, developmental stage (msc) and total dry matter yield were observed. Field trials for dry matter yield assessment were established in 2006 at Vågønes (67° 17'N, 14°27'E), Fureneset (61°34'N, 5°21'E), and Bjørke (60°48'N, 11°12'E) and for forage quality assessment at the two first mentioned locations. Genetic variation was identified for NDF and iNDF at early heading and heading stages between the investigated material. Highest content was found in cv. Hykor whereas in candivars only minor content was observed in leaves, however with higher content in stems. Promising candivars (FuRs0463, FuRs0357) from the Norwegian *Festulolium* breeding programme revealed better or equal to cultivars of their parental species as to content of NDF and iNDF. Winter survival which is a prerequisite for approved cultivars in Norway, was at an acceptable level in the candivars, and further breeding should preferably continue at the amphitetraploid level of the combination between perennial ryegrass and meadow fescue.

Keyword *Festulolium* · Indigestible fiber · iNDF · Leaf:stem ratio · NDF · Plant development · Yield

Introduction

In the Nordic countries a new feed evaluation system (NorFor) for ruminants has been developed in which roughage as the main source of energy has increased importance. In this system the indigestible part of neutral detergent fibre (iNDF)

L. Østrem (✉)
Norwegian Institute for Agricultural and Environmental Research, N-6967, Hellevik i Fjaler, Norway
e-mail: liv.ostrem@bioforsk.no

C. Huyghe (ed.), *Sustainable Use of Genetic Diversity in Forage and Turf Breeding*, DOI 10.1007/978-90-481-8706-5_84,

has been found to be the most important factor affecting digestibility of organic matter. For routine assessment of plant material, near-infrared reflectanse spectroscopy (NIRS) gives an acceptable prediction of NDF/iNDF (Nordheim et al., 2007).

The cost benefit of high quality forage grasses is being valued and *Festulolium* cultivars (cvs.) have experienced increased interest as a forage crop in large parts of Norway. High yield and reliable winter survival of cultivars from East European countries have been shown (Østrem and Larsen, 2008), and the regrowth capacity in *Festulolium* cvs. is much appreciated. The Norwegian plant breeding company, Graminor AS (Ltd), has *Festulolium* as one of their breeding priorities. The main objective has been to transfer winter hardiness from meadow fescue (*Festuca pratensis*) into perennial ryegrass (*Lolium perenne*), and forage quality is one of several complementary traits. *Festulolium* is defined as hybrids resulting from the crossing of a species of the genus *Festuca* with a species of the genus *Lolium* (*Festuca spp.* × *Lolium spp.*). In the OECD List of Varieties it is described as 'x *Festulolium*' and the names of the species used in the hybrid should be mentioned.

In the current Norwegian research project we studied NDF and iNDF in DM yield of market cvs. and candivars of *Festulolium* differing in parental origin and ploidy level. Standardised developmental stages (mean stage by count) were estimated as well as leaf:stem ratio in herbage yield during the growing season at two locations.

Material and Methods

Plant material. The cvs. Felopa and Hykor were described by Zwierzykowski (2004). Candivar FuRs0463 (amphiploid *L. perenne* × *F. pratensis*, 4*x*) was selected from synthetic population FuRs9806 after two winters in a single plant field at Bodø. The original population was made of 64 plants from four populations originating from IGER, Aberystwyth, UK. Candivar FuRs0357 (*L. perenne* × *F. pratensis*, 2*x*) originates from a wide genetic pool from several initial hybrids made from either *Festulolium* cv. Prior (LpFp, 4*x*) crossed with perennial ryegrass cv. Rikka (2*x*) or crosses between perennial ryegrass cvs. WIR40697, Einar (4*x*) and meadow fescue cv. Fure (2*x*). The initial hybrids were backcrossed twice onto diploid perennial ryegrass to obtain BC_2 progenies and then followed by two generations of seed propagation.

Field trials for forage quality and plant development assessment were established in 2006 at the two coastal locations; Fureneset, Fjaler, West Norway (61°34'N, 5°21'E), and Vågønes, Bodø, North Norway (67°17'N, 14°27'E), in field trials with three replicates in which *Festulolium* cvs. and candivars were compared to commonly grown cvs. of the parental species, meadow fescue and perennial ryegrass, and to timothy, the most commonly grown forage species in Norway (Table 84.2). In 2007 and 2008, individual cuts according to the developmental stages early heading and heading were performed (Simon and Park, 1981). Samples from plot yield and leaf and stem fractions from all plots were dried at 60°C for 48 h. Samples from location Fureneset in 2007 were analysed by NIRS for NDF and iNDF. At each cut, standardized developmental stage was estimated according to a phenological scale (Moore et al., 1991). The trials were cut three to four times each season.

Field trials for testing of yield potential in different climatic conditions of Norway were established in 2006 at Fureneset, Fjaler, at Vågønes, Bodø and at the inland location Bjørke, Hamar (60°48'N, 11°12'E). Nine market cvs. and seven candivars of *Festulolium* were included and compared to timothy (cvs. Grindstad/Vega), meadow fescue (cvs. Fure/Norild), tall fescue (cv. Retu) and perennial ryegrass (Napoleon).

Results and Discussion

The content of NDF and iNDF. In the first cut, NDF and iNDF in plot DM yield, were significantly higher in 2008 compared with the previous year, and iNDF was more affected than NDF. This was probably due to high temperature and low amount of precipitation the last weeks before harvest. In general, cv. Hykor contained high amounts of NDF and iNDF and was equalled only by timothy cv. Grindstad. Minor differences for NDF and iNDF were seen between the *Festulolium* cv. Felopa and the candivars for leaves, whereas stem content of iNDF in candivars FuRs0463 and FuRs0357 was generally at a lower level (Table 84.1).

Leaf:stem ratio. Hykor differed significantly from the other cultivars by high DM leaf:stem ratio during the growing season, and the *Festulolium* candivars differed significantly from each other in the second and later cuts with diverging leaf:stem ratios.

Phenological developmental stage (msc) is a weighted estimate of tillers in the vegetative, stem elongation and reproductive stage. As a mean of two harvesting systems and two locations, at the first cut, timothy revealed a more advanced stage (2.71) compared with the *Festulolium* cvs. (2.4–1.9) and meadow fescue cv. Fure (2.2). In the second cut, *Festulolium* cvs. were more advanced than timothy cv. Grindstad, and in later cuts the differences were small.

Dry matter yield. Despite increased focus on forage quality, dry matter yield is still the most important character for being listed as an approved cultivar in Norway. Expressed in mean over two years and two locations, significant differences between cultivars occurred. *Festulolium* cv. Hykor and candivar FuRs0463 were the highest yielding, while candivar FuRs0357 yielded at the same level as perennial ryegrass cv. Napoleon. The mentioned cultivars had all significant higher dry matter yield than meadow fescue and timothy. The *Festulolium* cv. Felopa and candivar FuRs0028 yielded significantly lower compared with meadow fescue (Table 84.2). Compared to good performing cultivars of the parental species grown in Norway, the best *Festulolium* candivars were higher yielding.

Testing of candivars. Dry matter yield of *Festulolium* in Norway is highly dependent on winter survival, resulting in different ranking of candivars in different regions of Norway. Cultivars from other European countries were high yielding in Southeast and West Norway, but less productive in North. Wintering ability seems to be a combination of resistance to biotic, physical and physiological factors.

Table 84.1 NDF and iNDF in () of 1st and 2nd cut of plot yield and dm fractions of leaves and stems. Mean of two developmental stages (early heading, heading), two years (2007, 2008), one location (Fureneset). 12 replicates

	Cut 1, NDF (iNDF)						Cut 2, NDF (iNDF)					
Cultivars/ candivars	Plot yield		Leaf		Stem		Plot yield		Leaf		Stem	
Hykor	54.1	(11.5)	51.5	(9.4)	60.4	(13.9)	52.1	(8.8)	55.3	(9.7)	62.8	(11.9)
Felopa	50.6	(7.8)	42.8	(1.7)	59.3	(12.3)	47.7	(5.7)	43.0	(0.9)	56.4	(10.9)
FuRs0463	48.9	(6.6)	45.2	(2.3)	58.2	(9.4)	47.7	(5.5)	45.7	(4.1)	56.5	(9.7)
FuRs0357	48.6	(7.4)	42.7	(2.5)	56.4	(10.5)	48.2	(7.5)	42.9	(3.7)	56.0	(12.2)
Napoleon	49.8	(7.8)	43.3	(2.8)	55.9	(12.2)	49.5	(8.8)	44.6	(3.9)	54.6	(10.4)
Fure	54,8	(7.6)	49.0	(1.3)	65.6	(12.5)	47.1	(3.1)	46.6	(0.3)	54.5	(4.9)
Grindstad	62.3	(10.9)	54.3	(3.9)	71.8	(14.4)	53.8	(6.0)	51.2	(2.9)	67.2	(9.7)
LSD 5%	2.0	(1.3)	2.2	(1.4)	2.7	(2.5)	1.9	(1.4)	1.9	(1.2)	2.4	(2.4)

Table 84.2 Dry matter yield (t hectar-1) and leaf:stem ratio in (). Mean of two locations (Fureneset, Vågønes), two years (2007, 2008) and two developmental stages (early heading, heading). 24 replications

			DMY (t hectar-1) (leaf:stem ratio)								
Cultivars/ candivars	Comb.	Ploidy level	Cut 1		Cut 2		Cut 3		Cut 4		Total DMY
Hykor	*L.m.xF.a.*	6x	4.38	(1.13)	2.54	(2.99)	2.66	(3.64)	1.46	(2.50)	11.03
Felopa	*L.m.xF.p.*	4x	2.63	(0.67)	2.22	(0.87)	2.30	(1.22)	1.36	(1.86)	8.52
FuRs0463	*L.p.xF.p.*	4x	4.37	(0.78)	2.32	(1.50)	2.32	(2.09)	1.67	(1.82)	10.68
FuRs0357	*L.p.xF.p.*	2x	3.99	(0.62)	2.27	(1.09)	2.17	(1.73)	1.46	(1.77)	9.89
Napoleon	*L.perenne*	4x	3.74	(0.52)	2.42	(1.02)	2.27	(1.33)	1.45	(1.59)	9.89
Fure	*F.pratensis*		4.07	(0.64)	2.10	(2.58)	2.19	(2.48)	1.17	(2.08)	9.53
Grindstad	*P.pratense*		4.58	(0.42)	1.73	(1.74)	1.67	(1.78)	1.12	(1.20)	9.09
LSD 5%			0.59	(0.14)	0.48	(0.32)	0.34	(0.36)	0.33	(0.41)	0.99

Physiological factors may be important since some cultivars are still growing in late autumn and thus get low cold hardening. Candivars originated from amphiploid of *L. perenne* × *F. pratensis* showed good winter survival. These cultivars originate from materials which have been multiplied and/or selected in Norway for 3–4 generations. Amphiploid cultivars of *L. multiflorum* × *F. pratensis* origin and androgenic cultivars and families, have mainly showed low winter hardiness.

Conclusion

Genetic variation in leaf and stem fractions for NDF and iNDF was identified and should be combined with the variation found in leaf:stem ratio and standardised developmental stage for selection purposes. Our study has shown that tetraploid candivars of amphiploid origin performs better than diploid candivars developed through introgression mainly because of reduced winter hardiness. This is in accordance with a former study in which different backcross generations were tested. Triploid hybrids resulting from crossing *L. perenne* (4*x*) × *F. pratensis* were used as initial material in backcrosses with *L. perenne* (2*x*). A second backcross increased the winter damage due to loss of winter hardiness genes from the fescues (Østrem et al., 2006). For selection of high quality traits values and simultaneously ensure good winter survival, further breeding should continue at the amphiploid level of the combination between perennial ryegrass and meadow fescue.

Acknowledgements This work has been financed by the Research Council of Norway and the collaborating institutes Norwegian Institute for Agricultural and Environmental Research and Graminor Ltd.

References

Moore, K.J., Moser, L.E., Vogel, K.P., Waller, S.S., Johnson, B.E., Pedersen, J.F. 1991. Describing and quantifying growth stages of perennial forage grasses. Agron. J. 83:1073–1077.

Nordheim, H., Volden, H., Fystro, G., Lunnan, T. 2007. Prediction of in situ degradation characteristics of neutral detergent fibre (aNDF) in temperate grasses and red clover using near-infrared reflectance spectroscopy (NIRS). Anim. Feed Sci. Technol. 139: 92–108

Simon, U., Park, B.H. 1981. A descriptive scheme for stages of development in perennial forage grasses. Proc. Internat. Grassld. Congr. 14:416–418.

Østrem, L., Larsen, A., Pašakinskienė, I. 2007. Prebreeding of *Festulolium* (*Lolium x Festuca* hybrids). In: Rosellini, D., Veronesi, F. (eds.), Breeding for seed production for conventional and organic agriculture, XXVI Eucarpia Fodder Crops and Amenity Grasses Section 47–51. Proceedings of the XXVI EUCARPIA Fodder Crops and Amenity grasses section, Perugia, Italy, 3–7 September 2006.

Østrem, L., Larsen, A. 2008. Winter survival, yield performance and forage quality of *Festulolium* cvs. for Norwegian farming. In: Hopkins, A. et al. (eds.), Biodiversity and Animal Feed. Future Challenges for Grassland Production, Grassland Science in Europe 13:293–295. Proceedings of the 22nd General meeting, EGF, Uppsala, Sweden, 9–12. June 2008.

Zwierzykowski, Z. 2004. Amphiploid and introgression breeding within the *Lolium-Festuca* complex – achievements and perspectives. In: Yamada, T., Takamizo, T. (eds.), Development of a novel grass with environmental stress tolerance and high forage quality through intergeneric hybridization between Lolium and Festuca, (pp.17–29). NARO, Tsukuba.

Chapter 85
Identification of Heterotic Patterns in Perennial Ryegrass

Ulrich K. Posselt

Abstract Perennial ryegrass (*Lolium perenne* L.) is spread all over Europe and ecotypes reflect the large amount of genetic diversity present. Breeders took profit from this diversity in creating new varieties. However, breeders intermated whatever materials they had available and largely ignored the maintenance of divergent materials. With few exceptions, modern varieties as well as ecotypes from Northwestern Europe are built on just one single genepool. To identify heterotic patterns, pre-grouping of 8 populations was done according to their geographic distance (160–1,800 km). The 8 parent populations and their 28 diallel crosses were evaluated for annual dry matter yield (ADMY) for 2 years at two contrasting locations. Mean performance of the parents was 13.6 t/ha ADMY compared to 14.1 t/ha ADMY of the hybrids. The most distant cross yielded highest (15.5 t/ha ADMY), which resulted in a panmictic mid-parent heterosis (PMPH) of almost 13%. These two populations fulfill the heterotic pattern criterion and could be the nucleus of the heterotic groups to be established. A rather high association between GD (geographic distances) and hybrid performance of the diallel crosses was found ($r = 0.64$).

Keywords Diallel crosses · Geographic distance · Heterotic pattern · *Lolium perenne*

Introduction

The importance of heterotic groups and patterns has been discussed in detail by Melchinger and Gumber (1998), as well as the interrelationship between genetic diversity and heterosis (Melchinger, 1999). If heterotic patterns are not yet available,

U.K. Posselt (✉)
State Plant Breeding Institute (720), University of Hohenheim, Stuttgart 70593, Germany
e-mail: posselt@uni-hohenheim.de

C. Huyghe (ed.), *Sustainable Use of Genetic Diversity in Forage and Turf Breeding*,
DOI 10.1007/978-90-481-8706-5_85,

it is suggested to preselect high performing parents according to genetic similarity based on molecular or geographic data, and to produce and test diallel crosses among them. The most promising cross combinations are used to identify heterotic patterns. Adapted populations isolated by time and space are the most promising candidates for heterotic patterns (Melchinger and Gumber, 1998). In a study based on molecular genetic distance (GD) among German ecotypes of perennial ryegrass, distinct gene pools (Northern *vs.* Southern) were identified (Bolaric *et al.*, 2005), although the association between GD and hybrid performance was rather low ($r = 0.3$). The exploitation of heterosis in forage crops has been reviewed by Brummer (1999) and Posselt (2003).

Materials and Methods

Eight perennial ryegrass populations with a similar time of flowering were chosen according to geographic origin (Table 85.1).

Table 85.1 Description of the parent populations

Abbreviation	Name	Type of population	Geographic origin
P1	Aberavon	variety	Wales
P2	Fennema	variety	Austria
P3	RG 5	pre-breeding population	Poland
P4	Weigra	variety	Southern Germany
P5	N9	ecotype	Isle of Poel/Germany
P6	N11	ecotype	NW-Germany
P7	N3	ecotype	NW-Germany
P8	F-10661	pre-breeding population	France

The shortest geographic distance among entries was 160 km (P6/P7) while the largest was 1,800 km (P1/P2). Hybrid seed production of the 28 diallel crosses as well as seed multiplication of the eight parent populations was carried out in seed islands in a field of triticale. The 28 hybrids and their eight parents were sown at two locations (Suebian Alb and Malchow/Poel) in a 6×6 lattice design with three replications. Yield data were collected in two successive years.

Results and Discussion

In the ANOVA the variance component of the entries was highly significant. In the diallel analysis (Griffing, Method 4), GCA-variance was significant only at $P = 0.10$, while the interaction Years $\times$ Locations $\times$ GCA was highly significant. Average annual dry matter yield (ADMY) of the crosses (14.1 t/ha) was only slightly higher than the mean of the parents (13.6 t/ha).

Heterotic Pattern

The highest performing cross with 15.5 t/ha was P1 × P2 (Table 85.2). This combination also showed the highest panmictic midparent heterosis (PMPH) (12.9%). Hybrid parents P1 and P2 originate from Wales and Austria and thus, are the most distant ones. P1 (Aberavon) displayed the largest GCA-effect with 0.72 t/ha ADMY. P1 shows positive heterotic effects in all hybrid combinations and could be defined as the nucleus of a heterotic group.

Table 85.2 Annual dry matter yield (ADMY) in t/ha of 28 population crosses (above diagonal), their 8 parent populations (diagonal, in bold); and panmictic midparent heterosis (PMPH) in % (below diagonal, in italics) of perennial ryegrass averaged over 2 years and 2 locations

P*	1	2	3	4	5	6	7	8	Mean	GCA
1	**13.5**	15.5	14.6	14.4	14.6	14.6	14.2	15.0	14.8	0.72
2	*12.9*	**14.0**	15.0	13.8	14.2	14.7	14.3	14.0	14.6	0.46
3	*10.5*	*11.7*	**12.9**	14.4	13.7	13.2	13.9	14.6	14.2	0.11
4	*4.2*	*–1.8*	*6.2*	**14.1**	13.9	13.4	13.6	13.8	13.9	–0.24
5	*8.5*	*3.5*	*3.2*	*0.6*	**13.5**	13.0	13.7	13.6	13.8	–0.32
6	*11.4*	*9.7*	*2.2*	*0.0*	*–1.2*	**12.8**	13.6	13.6	13.7	–0.43
7	*2.5*	*0.9*	*2.0*	*–3.8*	*–1.5*	*0.4*	**14.3**	13.8	13.9	–0.24
8	*9.9*	*0.7*	*8.9*	*–1.3*	*–0.3*	*2.2*	*–1.4*	**13.8**	14.0	–0.06

P*: 1-'Aberavon', 2-'Fennema', 3-ecotype PL, 4-'Weigra', 5–7-ecotypes D, 8-ecotype F

The second heterotic group, e.g. the opposite pool to P1 would be P2 (Fennema). P3 shows heterotic patterns to both P1 as well as P2, and the question will be whether a third separate group should be established. If a third pool is not desired, then P3 should join the pool of P1 because PMPH is a bit lower in the hybrid with P1 as compared to the hybrid with P2. P8 has a reasonable mean yield and shows heterosis in crosses with P1 and P3, but not with P2. Here the decision is straight forward, and P8 should join the pool with P2. All other parents show negative GCA and should not be considered further.

Broadening the Gene Pools

To assign unknown materials to one of the heterotic groups, two series of testcrosses with the two pools (A and B) as testers are carried out. Populations displaying PMPH with tester A are assigned to pool B and *vice versa*. All materials assigned to a particular pool have to be inter-mated thoroughly to establish the respective base populations A and B.

Using Gene Pools in Practical Breeding

After having identified the heterotic groups the breeder could either start a hybrid breeding program or create a diverse base population for synthetic breeding. In *hybrid breeding* the divergent pools are maintained separately. A series of testcross hybrids (interpool hybrids) is produced and tested for yield performance. In *population breeding* the heterotic groups are intermated to create a base population with large genetic diversity. Several cycles of random mating are needed to reach Hardy-Weinberg equilibrium. From this population, high performing varietal parents are to be selected.

To exploit more heterosis in synthetics, gene pools could be kept separate until final synthesis. This is also preferential if polyploidization is intended to create a new tetraploid variety.

Geographic Distance and Hybrid Performance

A rather high association (Fig. 85.1) between GD (geographic distances) and hybrid performance of the diallel crosses was found ($r = 0.64$).

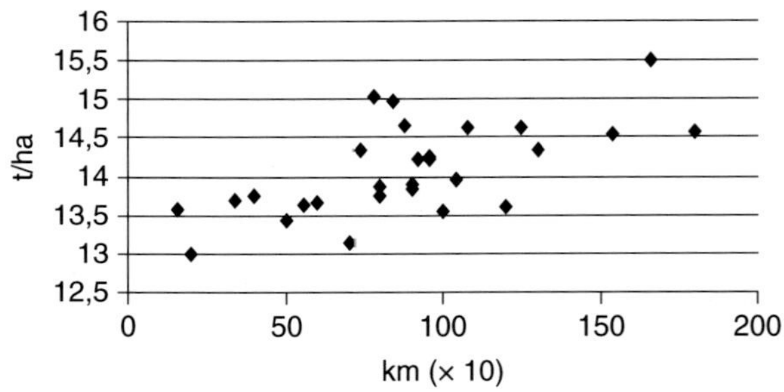

Fig. 85.1 Association between geographic distance (km) of parent populations and the performance (t/ha) of their diallel crosses

References

Bolaric, S., Barth, S., Melchinger, A.E., Posselt, U.K. 2005. Molecular genetic diversity within and among German ecotypes in comparison to European perennial ryegrass cultivars. Plant Breed. 124:257–262.

Brummer, E.C. 1999. Capturing heterosis in forage crop cultivar development. Crop Sci. 39: 943–954.

Melchinger, A.E., Gumber R.K. 1998. Overview of heterosis and heterotic groups in agronomic crops. In: Lamkey, K.R., Staub, J.E. (eds.), Concepts and breeding of heterosis in crop plants. CSSA Special Publ. No. 25, (pp. 29–44). Madison, WI, USA.

Melchinger, A.E. 1999. Genetic Diversity and Heterosis. In: Coors, J.G., Pandey, S. (eds.), The Genetics and Exploitation of Heterosis in Crops. ASA-CSSA-SSSA, (pp. 99–118). Madison, WI, USA.

Posselt, U.K. 2003. Heterosis in grasses. Czech. J. Genet. Plant Breed 39: 48–53.

MIX
Papier aus verantwortungsvollen Quellen
Paper from responsible sources
FSC® C105338

Printed by Books on Demand, Germany